DECISION-MAKING MODELS AND APPLICATIONS IN MANUFACTURING ENVIRONMENTS

AAP Research Notes on Optimization and Decision-Making Theories

DECISION-MAKING MODELS AND APPLICATIONS IN MANUFACTURING ENVIRONMENTS

Edited by

Pushpdant Jain, PhD

Kumar Abhishek, PhD

Prasenjit Chatterjee, PhD

APPLE ACADEMIC PRESS

First edition published 2024

Apple Academic Press Inc.
1265 Goldenrod Circle, NE,
Palm Bay, FL 32905 USA

760 Laurentian Drive, Unit 19,
Burlington, ON L7N 0A4, CANADA

CRC Press
2385 NW Executive Center Drive,
Suite 320, Boca Raton FL 33431

4 Park Square, Milton Park,
Abingdon, Oxon, OX14 4RN UK

Apple Academic Press exclusively co-publishes with CRC Press, an imprint of Taylor & Francis Group, LLC

Library and Archives Canada Cataloguing in Publication

Title: Decision-making models and applications in manufacturing environments / edited by Pushpdant Jain, PhD, Kumar Abhishek, PhD, Prasenjit Chatterjee, PhD.
Names: Jain, Pushpdant, editor. | Kumar, Abhishek, editor. | Chatterjee, Prasenjit, 1982- editor.
Description: First edition. | Series statement: AAP research notes on optimization and decision-making theories | Includes bibliographical references and index.
Identifiers: Canadiana (print) 20230498760 | Canadiana (ebook) 20230498795 | ISBN 9781774913550 (hardcover) | ISBN 9781774913567 (softcover) | ISBN 9781003394723 (ebook)
Subjects: LCSH: Multiple criteria decision making.
Classification: LCC T57.95 .D43 2024 | DDC 658.4/03—dc23

Library of Congress Cataloging-in-Publication Data

Names: Jain, Pushpdant, editor. | Kumar, Abhishek, 1988- editor. | Chatterjee, Prasenjit, 1982- editor.

Title: Decision-making models and applications in manufacturing environments / edited by Pushpdant Jain, PhD, Kumar Abhishek, PhD, Prasenjit Chatterjee, PhD.

Description: First edition. | Palm Bay, FL : Apple Academic Press, [2024] | Series: AAP research notes on optimization and decision-making theories | Includes bibliographical references and index. | Summary: "This new volume applies the multi-criteria decision-making (MCDM) theory to solving problems and challenges in manufacturing environments. In real-time life applications, MCDM has gained vast popularity for its ability to help make decisions in the presence of various similar and conflicting choices. The volume discusses using MCDM computational methods to evaluate and select the most optimal solution or method to real-world manufacturing engineering issues. It details the decision-making process in relation to many manufacturing aspects, including materials selection; identification, assessment, and evaluation of risk; sustainability assessment; selection of green suppliers; and more. The chapter authors demonstrate the application of myriad MCDM techniques in decision-making, including MADM (multiple attribute decision-making), DEA (data envelopment analysis), fuzzy TOPSIS (technique for order preference by similarities to ideal solution), fuzzy-VIKOR (multicriteria optimization and compromise solution); MOORA (multi-objective optimization on the basis of ratio analysis), EWM (entropy weight method), (AHP) analytic hierarchy process, TODIM (TOmada de Decisao Interativa Multicriterio), and others. The volume illustrates these MCDM models in several industries and industrial processes, including for experimental analysis and optimization of drilling of glass fiber reinforced plastic, in the textile industries, for selection of refrigerants for domestic applications, and others. The wide range of topics in Decision-Making Models and Applications in Manufacturing Environments will be valuable for researchers in manufacturing fields who need to use the latest developments and trends in decision-making processes. Professionals as well as faculty and students in manufacturing, industrial, mechanical and production technologies will be able to use, apply, and implement the MCDM models presented in this book"-- Provided by publisher.

Identifiers: LCCN 2023032428 (print) | LCCN 2023032429 (ebook) | ISBN 9781774913550 (hardcover) | ISBN 9781774913567 (paperback) | ISBN 9781003394723 (ebook)

Subjects: LCSH: Multiple criteria decision making. | Production control--Decision making. | Manufacturing industries--Decision making.

Classification: LCC T57.95 .D435 2024 (print) | LCC T57.95 (ebook) | DDC 658.4/03--dc23/eng/20230821

LC record available at https://lccn.loc.gov/2023032428

LC ebook record available at https://lccn.loc.gov/2023032429

ISBN: 978-1-77491-355-0 (hbk)
ISBN: 978-1-77491-356-7 (pbk)
ISBN: 978-1-00339-472-3 (ebk)

Dedications

This book is dedicated to our family members, friends, students, and the entire research fraternity.

—Editors

ABOUT THE BOOK SERIES: AAP RESEARCH NOTES ON OPTIMIZATION AND DECISION-MAKING THEORIES

Most real-world search and optimization problems naturally involve multiple criteria as objectives. Different solutions may produce trade-offs (conflicting scenarios) among different objectives. A solution that is better with respect to one objective may be a compromising one for other objectives. This compels one to choose a solution that is optimal with respect to only one objective. Due to such constraints, multi-objective optimization problems (MOPs) are difficult to solve since the objectives usually conflict with each other. It is usually hard to find an optimal solution that satisfies all objectives from the mathematical point of view. In addition, it is quite common that the criteria of real-world MOPs encompass uncertain information, which becomes quite a challenging task for a decision maker to select the criteria. Also, the complexities involved in designing mathematical models increase. Considering, planning, and appropriate decision-making require the use of analytical methods that examine trade-offs; consider multiple scientific, political, economic, ecological, and social dimensions; and reduce possible conflicts in an optimizing framework. Among all these, real-world multi-criteria decision-making (MCDM) problems related to engineering optimizations are categorically important and are quite often encountered with a wide range of applicability.

MCDM problems are basically a fundamental issue in various fields, including applied mathematics, computer science, engineering, management, and operations research. MCDM models provide a useful way for modeling various real-world problems and are extensively used in many different types of systems, including, but not limited to, communications, mechanics, electronics, manufacturing, business management, logistics, supply chain, energy, urban development, waste management, and so forth.

In the aforementioned cases, modeling of multiple criteria problems often becomes more complex if the associated parameters are uncertain and imprecise in nature. Impreciseness or uncertainty exists within the parameters due to imperfect knowledge of information, measurement uncertainty, sampling uncertainty, mathematical modeling uncertainty, etc. Theories like probability theory, fuzzy set theory, type-2 fuzzy set theory, rough set, grey theory, neutrosophic uncertainty theory available in the existing literature deal with such uncertainties. Nevertheless, the uncertain multi-criteria characteristics in such problems are not explored in depth, and a lot can be achieved in this direction. Hence, different mathematical models of real-life multi-criteria optimization problems can be developed on various uncertain frameworks with special emphasis on sustainability, manufacturing, communications, biomedical, electronics, materials, energy, agriculture, environmental engineering, strategic management, flood risk management, supply chain, waste management, transportations, economics, and industrial engineering problems, to name a few.

Coverage & Approach:

The primary endeavor of this series is to introduce and explore contemporary research developments in a variety of rapidly growing decision-making areas. The editors welcome proposals on the following topics:

- Crisp MCDM models
- Rough set theory in MCDM
- Fuzzy MCDM
- Neutrosophic MCDM models
- Grey set theory
- Mathematical programming in MCDM

- Big data in MCDM
- Soft computing techniques
- Modelling in engineering applications
- Modeling in economic issues
- Waste management
- Agricultural practice
- Material selection
- Renewable energy planning
- Industry 4.0
- Sustainability
- Supply chain management
- Environmental policies
- Manufacturing processes planning
- Transportation and logistics
- Strategic management
- Natural resource management
- Biomedical applications
- Future studies and technology foresight
- MCDM in governance and planning
- MCDM and social issues
- MCDM in flood risk management
- New trends in multi-criteria evaluation
- Multi-criteria analysis in circular economy
- Multi-criteria evaluation for urban and regional planning
- Integrated MCDM approaches for modeling relevant applications and real-life problems

Types of volumes:

This series reports on current trends and advances in optimization and decision-making theories in a wider range of domains for academic and research institutes along with industrial organizations. The series includes the following types of volumes:

- Authored volumes
- Edited volumes
- Conference proceedings
- Short research (thesis-based) books
- Monographs

Features of the volumes include recent trends, model extensions, developments, real-time examples, case studies, and applications. The volumes aim to serve as valuable resources for undergraduate, postgraduate, and doctoral students, as well as for researchers and professionals working in a wider range of areas.

CURRENT & FORTHCOMING BOOKS IN THE SERIES

- **Multi-Criteria Decision-Making Techniques in Waste Management: A Case Study of India**
  Editors: Suchismita Satapathy, Debesh Mishra, and Prasanjit Chatterje
- **Applications of Artificial Intelligence in Business and Finance: Modern Trends**
  Editors: Vikas Garg, Shalini Aggarwal, Pooja Tiwari, and Prasenjit Chatterjee
- **Advanced Computer Science Applications: Recent Trends in AI, Machine Learning, and Network Security**
  Editors: Karan Singh, Latha Banda, and Manisha Manjul
- **Decision-Making Models and Applications in Manufacturing Environments**
  Editors: Pushpdant Jain, Kumar Abhishek, and Prasenjit Chatterjee
- **Precision Agriculture for Sustainability: Use of Smart Sensors, Actuators, and Decision Support Systems**
  Editors: Narendra Khatri, Ajay Kumar Vyas, Celestine Iwendi, and Prasenjit Chatterjee
- **Attacks on Artificial Intelligence: The New Facets of Cyber Ecospace**
  Editors: Kukatlapalli Pradeep Kumar, Boppuru Rudra, Vinay Jha Pillai, and Prasenjit Chatterjee
- **Computational Optimization, Modeling, and Simulation for Engineering Applications**
  Editors: Anupam Shukla, Sourabh Rungta, Mohan Awasthy, and Rakesh L. Himte
- **Advances in Intelligent Systems: Paradigms and Applications**
  Editors: Amit Krishna Dwivedi, Manisha Guduri, Rajanikanth Aluvalu, and Uma Maheshwari V.

ABOUT THE EDITORS

Pushpdant Jain, PhD
Assistant Professor, School of Mechanical Engineering,
VIT Bhopal University, India

Pushpdant Jain, PhD, is currently working as an Assistant Professor in the School of Mechanical Engineering at VIT Bhopal University. He has served in two industries for six years and in various academic institutions for more than three years. Dr. Jain holds a PhD (2020) in Industrial Design from National Institute of Technology, Rourkela (Odisha). He did his Master of Technology in Product Design and Engineering in the year 2012 and B.E. in Mechanical Engineering in the year 2007 from the University Institute of Technology, Barkat Ullah University, Bhopal. He has supervised one MTech student. He has published more than 11 research papers in international peer-reviewed journals. He has actively participated in more than four international conferences and published three book chapters. He is an active member of the European Society of Bio-Mechanics (ESB), life member of the International Association of Engineers (IAENG), Indian Institution of Industrial Engineering (SMIIIE), and Indian Society for Technical Education (ISTE). In addition to this, he was awarded with Research Scholar of the Year in the year 2018 at NIT Rourkela for his PhD Work. His research interest includes new product development, healthcare products, spinal implant, bio-mechanics, and finite element analysis.

Kumar Abhishek, PhD
Assistant Professor, Department of Mechanical and Aerospace Engineering, National Institute of Infrastructure, Technology, Research and Management (IITRAM), Ahmedabad, India

Kumar Abhishek, PhD, is working presently as an Assistant Professor in Mechanical and Aero-Space Engineering Department at IITRAM, Ahmedabad. He has completed both his Doctoral and MTech degree from National Institute of Technology (NIT), Rourkela, India. His area of research mainly focuses on manufacturing and industrial engineering, modeling, and optimization of production processes and composite machining. He has plenty of publications in reputed journal/international conference publications. He organizes a workshop every consecutive year titled: MOOMESA: Multi-Objective Optimization Methods for Engineering and Scientific Applications. He has enlightened various BTech, MTech, and PhD students under his guidance. He also possesses a lifetime membership of International Association of Engineers (IAENG) and Indian Society for Non-destructive Testing (ISNT).

Prasenjit Chatterjee, PhD
Dean (Research and Consultancy), MCKV Institute of Engineering, West Bengal, India

Prasenjit Chatterjee, PhD, is currently the Dean (Research and Consultancy) at MCKV Institute of Engineering, West Bengal, India. He has over 100 research papers in various international journals and peer reviewed conferences. He has authored and edited more than 15 books on intelligent decision-making, supply chain management (SCM), optimization techniques, risk, and sustainability modeling. He has received numerous awards including Best Track Paper Award, Outstanding Reviewer Award, Best Paper Award, Outstanding Researcher Award and University Gold Medal. Dr. Chatterjee is the Editor-in-Chief of the *Journal of Decision Analytics and Intelligent Computing.* He has also been the guest editor of several special issues in different SCIE/Scopus/ESCI (Clarivate Analytics) indexed journals. He is also the Lead Series Editor of *Smart and Intelligent Computing in Engineering,* Chapman and Hall/CRC Press, Founder and Lead Series Editor of *Concise Introductions to AI and Data Science,* Scrivener-Wiley; *AAP Research Notes on Optimization and Decision Making Theories; Frontiers of Mechanical and Industrial Engineering,* Apple Academic Press, co-published with CRC Press, Taylor and Francis Group and *River Publishers Series in Industrial Manufacturing and Systems Engineering.* Dr. Chatterjee is one of the developers of two multiple-criteria decision-making methods called Measurement of Alternatives and Ranking according to COmpromise Solution (MARCOS) and Ranking of Alternatives through Functional mapping of criterion sub-intervals into a Single Interval (RAFSI).

CONTENTS

CONTRIBUTORS

Kumar Abhishek
Department of Mechanical and Aero-Space Engineering, Institute of Infrastructure, Technology, Research, and Management (IITRAM), Ahmedabad, Gujarat, India

Amey Akshay Acharya
School of Mechanical Engineering, VIT University, Vellore, Tamil Nadu, India

Reema Agarwal
Department of Applied Sciences, Shobhit University, Meerut, Uttar Pradesh, India

Ankur Agrawal
School of Business Studies, Sharda University, Greater Noida, Uttar Pradesh, India

Rajeev Agrawal
Department of Mechanical Engineering, Malaviya National Institute of Technology, Jaipur, Rajasthan, India

Deobrat Akhouri
School of Mechanical Engineering, KIIT University, Bhubaneswar, Odisha, India

Alireza Alinezhad
Associate Professor, Department of Industrial Engineering, Qazvin Islamic Azad University, Qazvin, Iran

Bipradas Bairagi
Department of Mechanical Engineering, Haldia Institute of Technology, West Bengal, India

Din Bandhu
Department of Mechanical and Industrial Engineering, Manipal Institute of Technology Bengaluru, Manipal Academy of Higher Education, Manipal, Karnataka, India

Surajit Barad
Department of Mechanical Engineering, National Institute of Technology, Raipur, Chhattisgarh, India

Sumit Bhowmik
Assistant Professor, Department of Mechanical Engineering, National Institute of Technology, Silchar, Assam, India

Ramesh Chand
Industrial and Production Engineering Department, Dr. B. R. Ambedkar National Institute of Technology, Jalandhar, Punjab, India

Suman Chatterjee
Department of Mechanical Engineering, National Institute of Technology, Rourkela, Odisha, India

Nilabhra Ghosh Dastidar
School of Mechanical Engineering, KIIT University, Bhubaneswar, Odisha, India

Hrishikesh Deka
School of Mechanical Engineering, KIIT University, Bhubaneswar, Odisha, India

Yogesh Dhakal
School of Mechanical Engineering, VIT University, Vellore, Tamil Nadu, India

Brajesh K. Dwivedi
Environmental Sciences, Department of Botany, University of Allahabad, Allahabad, Uttar Pradesh, India

Rajiv K Garg
Industrial and Production Engineering Department, Dr. B. R. Ambedkar National Institute of Technology, Jalandhar, Punjab, India

Josy George
School of Mechanical Engineering, VIT Bhopal University, Sehore, Madhya Pradesh, India

Abhey Gupta
Department of Mechanical and Aero-Space Engineering, Institute of Infrastructure, Technology, Research, and Management (IITRAM), Ahmedabad, Gujarat, India

Dileep Kumar Gupta
Department of Mechanical and Aero-Space Engineering, Institute of Infrastructure, Technology, Research, and Management (IITRAM), Ahmedabad, Gujarat, India

Jagadish
Department of Mechanical Engineering, National Institute of Technology, Raipur, Chhattisgarh, India

Jinesh Kumar Jain
Department of Mechanical Engineering, Malaviya National Institute of Technology, Jaipur, Rajasthan, India

Pushpdant Jain
School of Mechanical Engineering, VIT Bhopal University, Sehore, Madhya Pradesh, India

Javad Khalili
MSc, Department of Industrial Engineering, Qazvin Islamic Azad University, Qazvin, Iran

Soni Kumari
Department of Mechanical Engineering, GLA University, Mathura, Uttar Pradesh, India

Nitin Kumar Laxkar
Department of Mechanical and Aero-Space Engineering, Institute of Infrastructure, Technology, Research, and Management (IITRAM), Ahmedabad, Gujarat, India

Banh Tien Long
Hanoi University of Industry, Bac Tu Liem, Hanoi, Vietnam

Siba Sankar Mahapatra
Department of Mechanical Engineering, National Institute of Technology, Rourkela, Odisha, India

Saikat Ranjan Maity
Assistant Professor, Department of Mechanical Engineering, National Institute of Technology, Silchar, Assam, India

Yug Modi
Department of Mechanical and Aero-Space Engineering, Institute of Infrastructure, Technology, Research, and Management (IITRAM), Ahmedabad, Gujarat, India

Gunji Bala Murali
Assistant Professor (Sr.), School of Mechanical Engineering, VIT University, Vellore, Tamil Nadu, India

Vikas Pagey
Department of Mechanical Engineering, Sagar Institute of Research Technology and Science, Bhopal, Madhya Pradesh, India

Dhyanadipta Panda
School of Humanities, KIIT University, Bhubaneswar, Odisha, India

Shubham B. Patil
Department of Mechanical Engineering, National Institute of Technology, Raipur, Chhattisgarh, India

Nguyen Huu Phan
Hanoi University of Industry, Bac Tu Liem, Hanoi, Vietnam

M. Sriram Prasanth
Department of Mechanical Engineering, Saranathan College of Engineering, Tiruchirappalli, Tamil Nadu, India

Jairam Raigar
Industrial and Production Engineering Department, Dr. B. R. Ambedkar National Institute of Technology, Jalandhar, Punjab, India

P. V. Rajesh
Department of Mechanical Engineering, Saranathan College of Engineering, Tiruchirappalli, Tamil Nadu, India

Amitava Ray
Principal Jalpaiguri Government Engineering College, Jalpaiguri, West Bengal, India

Kuldeep K. Saxena
GLA University, Mathura, Uttar Pradesh, India

Minakshi Sehrawat
Research Scholar, Sardar Patel University (SPU), Gujarat, India

Meet Shah
Department of Mechanical and Aero-Space Engineering, Institute of Infrastructure, Technology, Research, and Management (IITRAM), Ahmedabad, Gujarat, India

Bhupendra Sharma
Research Organization of Science and Technology, Ritsumeikan University, Kusatsu, Japan

Vishal S. Sharma
School of Mechanical, Industrial and Aeronautical Engineering, University of Witwatersrand, Johannesburg, South Africa

Abhineet Singh
Department of Mechanical and Aero-Space Engineering, Institute of Infrastructure, Technology, Research, and Management (IITRAM), Ahmedabad, Gujarat, India

Jaivir Singh
Industrial and Production Engineering Department, Dr. B. R. Ambedkar National Institute of Technology, Jalandhar, Punjab, India

Deepak Singhal
School of Mechanical Engineering, KIIT University, Bhubaneswar, Odisha, India

Pankaj Sonia
Department of Mechanical Engineering, Malaviya National Institute of Technology, Jaipur, Rajasthan, India

Ashutosh K. Srivastava
Dr. Shyama Prasad Mukherjee Government PG College, Bhadohi, Uttar Pradesh, India

Shekhar Srivastava
Industrial and Production Engineering Department, Dr. B. R. Ambedkar National Institute of Technology, Jalandhar, Punjab, India

Preetam Subudhi
Department of Mechanical Engineering, National Institute of Technology, Rourkela, Odisha, India

Sandeep Kumar Tembhare
School of Mechanical Engineering, VIT Bhopal University, Sehore, Madhya Pradesh, India

Nguyen Van Thien
Hanoi University of Industry, Bac Tu Liem, Hanoi, Vietnam

Sunil Kumar Tiwari
Department of Mechanical Engineering, Sagar Institute of Research Technology and Science, Bhopal, Madhya Pradesh, India

Sushanta Tripathy
School of Mechanical Engineering, KIIT University, Bhubaneswar, Odisha, India

Rakhi Trivedi
Associate Professor, IILM College of Engineering and Technology, Greater Noida, Uttar Pradesh, India

Vipin Tyagi
Department of Applied Sciences, Shobhit University, Meerut, Uttar Pradesh, India

Jay Vaishnani
Department of Mechanical and Aero-Space Engineering, Institute of Infrastructure, Technology, Research, and Management (IITRAM), Ahmedabad, Gujarat, India

Divya Zindani
Assistant Professor, Department of Mechanical Engineering, Sri Sivasubramaniya Nadar (SSN) College of Engineering, Kalavakkam, Tamil Nadu, India

ABBREVIATIONS

3D	three-dimensional
ABS	acrylonitrile-butadiene-styrene
AC	air-conditioning
AGV	automated guided vehicle
AHP	analytic hierarchy process
AI	artificial intelligence
AM	additive manufacturing
ANN	artificial neural network
ANOVA	analysis of variance
ANP	analytic network planning
ARAS	additive ratio assessment
ASTM	American Society for Testing and Materials
AWNPR	average weighted normalized performance rating
BCC	Banker, Charnes, and Cooper
BJ	binder jetting
BNP	best non-fuzzy performance
BO	best to others
BP-ANN	back-propagation artificial neural network
BRSR	business responsibility and sustainability reporting
BT	build time
BWM	best worst method
CAD	computer-aided design
CBA	cost-benefit analysis
CC_i	closeness coefficient
CCR	Charnes, Cooper, and Rhodes
CFRP	carbon fiber reinforced plastics
CH_4	methane
CNC	computer numerical control
CNN	convolutional NN
COA	center of area
COPRAS	complex proportional assessment
COPRAS-G	complex proportional assessment of alternatives with grey relations
CPS	cyber-physical system

CR	consistency ratio
CRS	constant returns to scale
CSW	common set of weights
D_d	drill diameter
DEA	data envelopment analysis
DED	direct energy deposition
D_f	delamination factor
DfE	design for the environment
DLP	digital light processing
DM	decision-maker
DMLS	direct metal laser sintering
DMU	decision-making unit
DOD	drop on demand
DoE	design of experiment
DT	digital twin
DUTM	digital universal testing machine
DW	dielectric welding
DW	direct write
EA	environment assessment
EBAM	electron beam additive manufacturing
EBM	electron beam melting
ED	environmental design
EDM	electric discharge machining
EER	energy efficiency ratio
EIA	environmental impact assessment
EIS	environment impact statement
EOS	electro-optical systems
EPR	extended product responsibility
ETPs	effluent treatment plants
EWM	entropy weight method
EWR	electrode-wear ratio
FAHP	fuzzy analytical hierarchical process
FDM	fused deposition modeling
FGRC	fuzzy gray relation coefficients
FGT	fuzzy grey theory
FNIA	fuzzy negative-ideal arrangement
F_r	feed rate
FS	flexural strength
FVIKOR	FVIseKriterijumska Optimizacija I Kompromisno Resenje

GFF	generalized feed-forward
GFRP	glass fiber reinforced plastics
GP	goal programming
GRA	gray relation analysis
GRC	grey relational coefficient
GSCM	green supply chain management
GSS	green supplier selection
GTMA	graph theory and matrix approach
HDT	heat deflection temperature
HFCs	hydrofluorocarbons
HIPS	high-impact polystyrene
HIRAC	hazard identification, risk assessment, and control
HTW	hot tool welding
IA	impact assessment
IBA	interpolative Boolean algebra
INS	ideal negative solutions
IoT	internet of things
IPS	ideal positive solution
ISM	interpretive structural modeling
ISO	International Standards Organization
IW	induction welding
JIT	just-in-time
LCA	life cycle assessment
LCVD	laser chemical vapor deposition
LENS	laser engineering net shape
LFF	liquefied filament fabrication
LIFT	laser-induced forward transfer
LMF	lack of management foresight
LP	linear programming
LV	linguistic variable
MADM	multi-attributes decision-making
MAPLE	matrix-assisted pulsed-laser DW
MAUT	multi-attribute theory
MCA	multi-criteria analysis
MCDM	multi-criteria decision making
MDL	modified digital logic
ME	material extrusion
MEKP	methyl ethyl ketone peroxide
MEMS	micro-electromechanical systems

MHD	material handling device
MICMAC	matriced impacts croises-multiplication applique and classification
MJ	material jetting
MJF	multi-jet fusion
MJP	multi-jet printing
MNN	modular neural network
MODM	multiobjective decision-making
MOORA	multiobjective optimization on basis of ratio analysis
MPRI	multi-performance ranking index
MRO	multi-response optimization
MRR	material removal rate
MVGFD	multi-variant general factorial design
MWW	microwave welding
NIS	negative ideal solution
OW	others to worst
PA	polyamide
PBF	powder bed fusion
PC	part cost
PCM	pairwise comparison matrix
PE	processing elements
PEEK	polyetheretherketone
PER	polyester resin
PIS	positive ideal solution
PIV	proximity index value
PLA	polylactic acid
PLC	programmable logic controller
PMC	polymer matrix composite
PMEDM	powder mixed electric discharge machining
PPC	production, planning, and control
PPS	polyphenylene sulfide
PPS	production possibility set
PROMETHEE	preference ranking organization method for enrichment evaluations
PS	polystyrene
PSI	performance selection index
RAC	refrigeration and air conditioning
RBF	radial basis function
RF	radio frequency

RFID	radio frequency identification
RHP	rice husk powder
RIW	resistance implant welding
RLT	recast layer thickness
RM	rapid manufacturing
RNN	recurrent neural network
RP	rapid prototyping
RT	rapid tooling
RTS	returns to scale
S/N	signal-to-noise
SAW	simple additive weighting
SCIW	semi-consistent enlistment welding
SCM	supply chain management
SD	sustainable development
SEA	strategic environmental assessment
SEM	scanning electron microscope
SL	sheet lamination
SLA	stereolithography
SLM	selective laser melting
SLS	selective laser sintering
SMART	simple multi-attribute rating technique
SR	surface roughness
Ss	spindle speed
SSIM	structural self-interaction matrix
STL	standard tessellation language
STM	spring testing machine
T_f	thrust force
TF	transfer functions
TFN	triangular fuzzy number
THFSW	threaded hole friction spot welding
TODIM	TOmada de Decisao Interativa Multicriterio
TOPSIS	technique for order performance by similarity to ideal solution
TPU	thermoplastic polyurethane
T_q	torque
TWR	tool wear rate
UTM	universal testing machine
UTS	ultimate tensile strength
UW	ultrasonic welding

VatPP/VP	vat photopolymerization
VE	value engineering
VH	very high
VIKOR	Vlse Kriterijumska Optimizacija Kompromisno Resenje
VL	very low
VRS	variable returns to scale
WNPR	weighted normalized performance ratings
WPCA	weighted principal component analysis

ACKNOWLEDGMENTS

This book has been compiled to provide state-of-the-art details, recent advancements, methods, and the latest application of MCDM in a manufacturing environment. The editors of this book express their gratitude to all the authors and reviewers who have devoted their time and effort to make this highly comprehensive book possible.

—***Editors***

PREFACE

Decision-making can be considered as an intellectual activity employed in the field of academia and industries to identify and choose from various set of alternatives. It may also be defined as the study of finding and selecting from options based on the preferences of the decision maker. Overall, if an individual's decision is examined, it is mostly concerned with reasoning, with the associated reasons being either rational or irrational based on explicit assumptions. In addition, deriving a logical conclusion is essential to all scientific professions, where experts or professionals apply their expertise to make decisions.

It has been proven that a collective decision is superior to an individual's decision. However, it is also possible that within a group, an individual can exhibit decision biases towards the provided information. Therefore, the solution to a particular problem assessed by a single individual is grouped to enhance the group decision-making process. To solve such types of errors as well as the biases in the decision-making process, it is recommended to include specialized experts in a particular domain to provide justified judgment. Moreover, the decision-making process is recursive and non-linear as various decisions move back and forth due to the criteria and availability of different alternatives. Every decision results from available preferences, information, values, and choices present in the decision environment. Since both alternatives and accessible information are limited due to the constraints of time and effort, it is essential to take decisions within this environment of coercion.

Currently, the major challenge associated with the process of decision-making is uncertainty, and this implies setting a significant goal or target to reduce the amount of uncertainty with regards to all available alternatives. Recently, in the decision-making domain, the idea of robust decision effort is formally integrated with criterion and uncertainty, and due to this, the concept of fuzziness has suddenly become relevant. A major portion of decision-making process involves the analysis of a defined or available set of alternatives specified by some criteria, and this can be a benefit in terms of the associated cost. Further, the problem seeks to categorize these options concerning suitability to the authority of decision-makers (DMs), when all available criteria have been considered parallel.

Furthermore, the next goal is to identify the best option or alternative or to recommend the top priority alternative amongst all present options. Multi-criteria decision making (MCDM) is adopted to solve such problems in the decision-making process. MCDM is defined as the ability to take decisions in the presence of various similar and conflicting choices. When making a decision, there are various and conflicting objectives, such as to reduce cost and to enhance the quality in service. These are the significant parameters that the DMs have to deal with on a daily basis. In fact, in a few cases, the nature of criteria (tangible/non-tangible) calls for uncertainty in the process of decision-making. In real-time application, the process of MCDM faces very critical constraints due to vagueness and imprecision in the available criteria. However, in current real-time life applications MCDM has gained huge popularity.

This book consists of a total of 21 chapters. A brief description of each chapter is presented below.

Chapter 1 advocates data envelopment analysis (DEA) as an MCDM method for an automotive material selection problem for efficient decision making evaluation. A brief description of MCDM and DEA along with classifications of MCDM methods, including Multiple Attribute Decision Making (MADM) and Multiple Objective Decision-Making (MODM) are also presented.

Chapter 2 focuses on various factors associated with risk assessment with respect to manufacturing industries.

Chapter 3 aims to find out the impact of different variables like spindle speed, feed rate, and drill diameter on output attributes, namely thrust force, torque, delamination factor, and circularity of holes at both entrance and exit. To reduce the number of experiments, Taguchi L9 orthogonal array is used and experiments were conducted as per the design of experiment (DoE) approach to obtain values of thrust force and torque using a dynamometer and then the delamination factor and circularity at both entry and exit are deduced.

Chapter 4 deals with the identification of problems and their prioritization in the textile industries across Gujarat, India. The problems and parameters selected in this work are based on the primary data collected from the internet sources as well as from the textile industries in Gujarat through a questionnaire survey with complementary interviews.

Chapter 5 is an endeavor to select the ideal refrigerant from a prepared list of the most common and usable refrigerants using three popular MADM methods namely simple additive weighting (SAW), Grey relation analysis

(GRA), and Vise Kriterijumska Optimizacija I Kompromisno Resenje (VIKOR). All three methods indicated R290 as the best refrigerant to be used for domestic purposes while keeping basic properties and environmental criteria in mind.

In Chapter 6, four different criteria were taken into consideration for vendor selection among ten alternatives. This chapter proposes an MCDM framework to evaluate vendors using an entropy weight method (EWM) for the calculation of the weightage of each criterion.

Chapter 7 is an attempt to use MCDM methods to choose the best possible cutting fluid. Three MCDM techniques, namely analytic hierarchy process (AHP), SAW, and Technique for Order Preference by Similarity to Ideal Solution (TOPSIS), have been implemented to select the ideal cutting fluid from a prepared list of the most common and usable cutting fluids. All three methods suggest LB2000 vegetable-based oil as the best-cutting fluid to be used for machining titanium alloy.

Chapter 8 provides a new systematic and comprehensive framework for sustainability, especially in manufacturing sectors. Guidelines to select and quantify the relevant indicators converts the quantified weighted indicators into dimensionless quantities and allows ranking the alternatives based on the aggregated scores, as presented in this chapter.

In Chapter 9, hybrid PRP composites are fabricated with polyester resin (PER) as matrix material and Alumina (Al_2O_3) as well as rice husk powder (RHP) as reinforcements. Their properties are evaluated by conducting several tests. In the end, the best possible composite specimen is selected from a set of different specimens with varying proportion of reinforcements by additive ratio assessment (ARAS) method.

Chapter 10 deals with the optimization of PMEDM process parameters for MRR, TWR, Ra, and HV using Taguchi method and grey relational analysis for die steel materials.

Chapter 11 proposes various parameters for selection of optimal rapid prototyping (RP) process using multi-variant MCDM-based approaches.

Chapter 12 conducts an investigation into the effect of criteria interaction on TOmada de Decisao Interativa Multicriterio (TODIM) Method.

Chapter 13 highlights the key features for selection of an air-conditioning system via value engineering management tool.

Chapter 14 deals with the comparison of different MCDM methods for the selection of polymer-based additive manufacturing processes.

Chapter 15 focuses on a new approach to alternate the conventional Friction Stir Welding (FSW) process. In the proposed technique, the fixture is mechanically vibrated using DC motors normal to the weld bead to increase the strength of the bead.

Chapter 16 proposes a combined fuzzy AHP with fuzzy grey theory (FGT) for performance evaluation of conveyors under a discrete manufacturing environment.

Chapter 17 aims to compare the performances of artificial neural network (ANN), Taguchi approach, analysis of variance (ANOVA), and GRA methods for process optimization in different polymer welding applications.

Chapter 18 employs Grey COmplex PRoportional ASsessment (COPRAS-G) method combined with fuzzy AHP for performance assessment of material handling devices (MHDs) under a manufacturing environment.

Chapter 19 pinpoints the accomplishment factors of artificial intelligence (AI) with the Internet of Things (IoT) and big data in manufacturing industries and dispenses an exegetical structural model to thrive a map of complicated interconnections.

Chapter 20 concentrates on the relationship of key factors of sustainability supply chain management for Industry 4.0 using a structural self-interaction matrix.

Chapter 21 proposes a TOPSIS-based model to rank green suppliers for an industrial plastic component manufacturing company as well as to suggest how low-rank suppliers can improve their performances by incorporating environmental consciousness in the supply chain process.

CHAPTER 1

MCDM, DEA, AND THEIR RELATIONSHIP IN MATERIALS SELECTION

ALIREZA ALINEZHAD[1] and JAVAD KHALILI[2]

[1]*Associate Professor, Department of Industrial Engineering, Qazvin Islamic Azad University, Qazvin, Iran*

[2]*MSc, Department of Industrial Engineering, Qazvin Islamic Azad University, Qazvin, Iran*

ABSTRACT

Nowadays, materials selection is the most critical activity in the design process. An inappropriate selection can affect the productivity and reputation of a manufacturing system. Therefore, the application of data envelopment analysis (DEA) and multiple criteria decision-making (MCDM) methods can yield different results. MCDM and DEA are the most powerful methods in decision-making and efficiency evaluation. Since the 1980s, there has been much research on the relationship between MCDM and DEA, with the primary purpose being to use the strengths of each one to eliminate or reduce other weaknesses. In this chapter, a brief description of MCDM and DEA is initially presented. Then, given the classification of the MCDM approach into multiple attribute decision-making (MADM) and multiple objective decision-making (MODM), the relationship between MCDM and DEA with the related case studies on the automotive parts manufacturing industry is presented.

Decision-Making Models and Applications in Manufacturing Environments. Pushpdant Jain, Kumar Abhishek, & Prasenjit Chatterjee (Eds.)

1.1 INTRODUCTION

With the advancement of technology, material selections are not always as straightforward as they once were. In the past, some classes of materials (e.g., metals) might have been all an engineer would have needed to consider for a specific application. Nowadays, different classes of materials compete for consideration, in order that a part be as cheap, light, and long-lasting. Material selection includes initial screening, alternatives comparison, and the specification of the ideal solution [1, 2]. It must be discussed from a strategic viewpoint because of its effect on the design. At least 80,000 materials are known in the world [3]. Obviously, due to the presence of almost similar kinds of materials, the material selection problem is one of the most challenging activities for engineers and designers [4, 5], and they take into account various methods for selecting the best material.

Hence, the two methods, multiple criteria decision-making (MCDM) and data envelopment analysis (DEA) methods, are used. MCDM method can be divided into the Multiple Objective Decision-Making (MODM) and Multiple Attribute Decision-Making (MADM) techniques. The MODM used for designing and MADM methods select the ideal alternative [6]. In MCDM, the DM must design the most preferred alternative that is determined implicitly [7]. Using the DEA approach as one of the most powerful performance measurement methods along with the MCDM approach in decision-making and choosing an efficient decision-making unit (DMU) can be considered as one of the alternative evaluation and selection approaches. The DEA could test the function of units against competitors and decide on a better future based on the results [8]. Also, this is an ideal decision-making device for the performance evaluation of units that have similar inputs/outputs [6].

Since the task of both DEA and MCDM, in many cases, is to help the DM in the decision-making process and also concerning the similarities in these two topics, one can use the benefits of each of these two methodologies to overcome some of the disadvantages of other methods. For instance, some of the MADM methods can be used for ranking in the DEA models, or in some cases, in the MADM methods, there is only a ranking. The intensity of each rank is obtained using DEA models.

The chapter is organized as follows; Section 1.2 includes a brief background on MCDM and DEA problems. In Sections 1.3 and 1.4, the literature and the relationship between DEA and MCDM are reviewed, respectively. In Section 1.5, based on the DEA-MCDM methodology, the material selection in the automotive parts manufacturing industry is examined. In the end, conclusions are presented in Section 1.6.

1.2 BACKGROUND

Material selection can be considered as a complex problem. Researchers have emphasized complicated decision problems, and different techniques should be used to achieve a trustworthy decision. A few studies have focused on material selection based on the DEA and MCDM methods. In a previous study, Mousavi-Nasab & Sotoudeh-Anvari [9] used the DEA, COPRAS, and TOPSIS methods for material selection problems.

Many studies have been conducted on material selection using MCDM methods. Thakker et al. [10] presented a material selection strategy using the TOPSIS method, the adapted value engineering (VE) techniques, and the Cambridge material selector-based method. This approach is applied in the optimal material selection of turbine blades. Jiao et al. [11], used the PROMETHEE method for material selection. Then, spearman's correlation coefficient is used to explain the correlation coefficients between results achieved by PROMETHEE and other methods. Jahan [12] suggested a new VIKOR-goal programming (GP) model to select the biomaterial for a pin which is placed in the shaft of the femur. For this purpose, in addition to beneficial and non-beneficial criteria, the target criteria were also considered. Petkovic et al. [13], due to the need for choosing the most appropriate hard coating material, explain the use of MCDM methods, i.e., WASPAS and COPRAS. Marini et al. [14] have proposed the integrated QFD, AHP, and ANP (analytic network planning) approach to select the right material and design concept. This approach leads to minimizing the time and cost. Chaitanya & Srinivas [15] applied the SWING, entropy, SAW, WPM, and AHP methods to the material selection in piston manufacturing. According to the criteria weights for each alternative, the sensitivity analysis was applied to prove the consistency in the ranking of units. Patnaik et al. [16] used the PSI and PROMETHEE methods for choosing the best materials for the design of structural. In this regard, eight numbers of alternative composites and eight criteria, such as impact strength, micro-hardness, flexural strength (FS), and measured density, are suggested for ranking optimal design.

Also, it can be referred to select the right material of the plate in a polymer electrolyte fuel cell by using the ELECTRE IV method as a non-compensatory compromised method [17]; introducing the integrated Simos-ELECTRE III method in the selection of material for a thermal loaded conductor cover sheet [18]; using the UTA method for the material selection in flywheel manufacturing [19]; solving a gear material selection problem by using the ORESTE, OCRA, COPRAS-G, and EXPROM II methods [20]; comparing the Ashby,

VIKOR, and TOPSIS methods for choosing of material in the micro-electro-mechanical systems (MEMS) [21]; presenting a novel approach based on expert judgments and using WASPAS and MOORA methods for selection of hard magnetic materials [22]; developing the PROMETHEE-GAIA method for choosing the best material of turbine strips [23]; and using the GP for minimizing costs, and optimizing recyclable weight in the material selection problem [24].

1.3 LITERATURE REVIEW

1.3.1 MATERIAL SELECTION

Nowadays, a wide variety of materials exists, with each having its own features, applications, advantages, and disadvantages. Material selection is one of the main stages of the design process that has the most important effect on the quality of products [25]. This process will affect procurement costs, assembly costs, recycling costs, and even environmental and economic situations [26]. In selecting the best material, there is not one definite attribute of selection, and the manufacturers should consider lots of attributes. Figure 1.1 shows the overall process of materials selection, which consists of six steps.

Selection approaches have been presented to analyze the data involved in the decision-making so that a principled assessment can be executed [28, 29]. In the stage of material evaluation (Figure 1.1), by using the screening methods, the manageable numbers of alternatives narrow down for subsequent detailed evaluation [30]. Cost per unit property, chart method, questionnaire method, materials in product selection tools, and artificial intelligence (AI) are methods for initial screening of material selection [31]. Then, different ranking methods, such as MCDM and optimization approaches, can be applied to limit the field of materials to some ideal alternatives.

With the advancement of technology, there has never been such a time in which the diversity of materials was so great [32]. In fact, the most important stage of production is the choice of materials, and neglect can cause irreparable damage in all stages of production, sales, and consumption. On the other hand, many factors must be considered in materials selection, which are classified as structural, economic/business, manufacturing, durability, environmental impact, and specialist properties for unique applications [33]. Therefore, designers and engineers pay a lot of attention to the materials database.

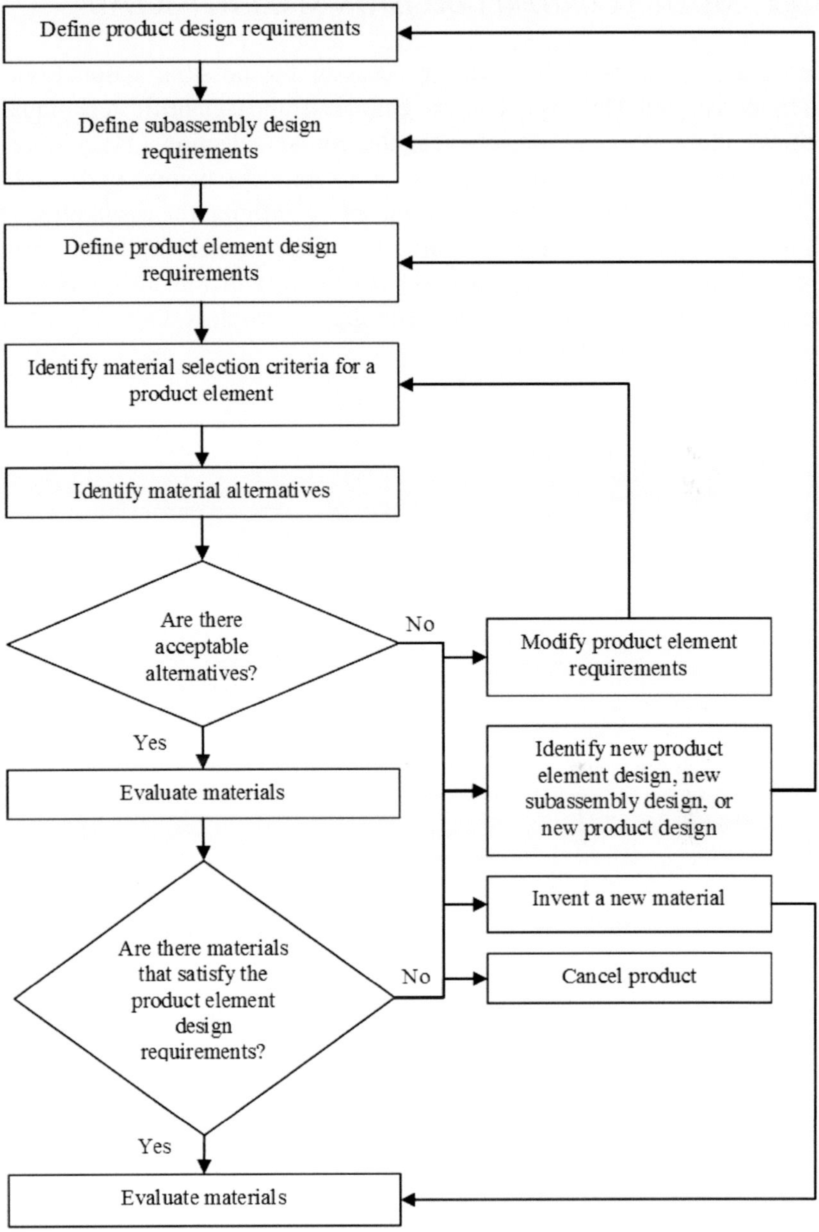

FIGURE 1.1 Material selection process.

Source: Reprinted with permission from Ref. [27]. Copyright 2009. © Elsevier.

1.3.2 MULTIPLE CRITERIA DECISION-MAKING (MCDM)

Decision-making is the selection process of the possible action from the alternatives, and DM uses several empirical and scientific techniques to achieve more than one objective in the course selection [34]. Figure 1.2 shows the decision-making process. This process begins with problem identification. Then, a model is formulated by defining the assumptions that simplify relationships between variables. After model validation, criteria are determined for the alternative assessment, and the model is solved. Successful implementation leads to solving the problem. Otherwise, failure results in a return to the process.

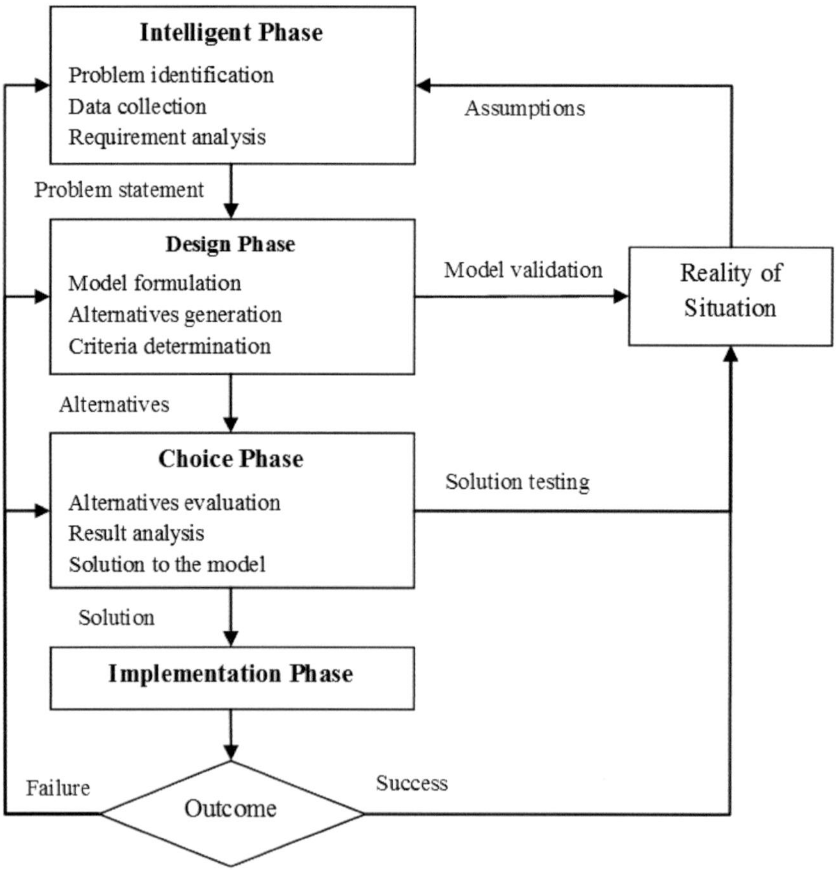

FIGURE 1.2 Decision-making process.

Source: Reprinted with permission from Ref. [35]. Copyright 2007. © Imperial College Press

The use of experiences in one particular field can be useful but cannot be appropriate for decision-making in all fields. In other words, using traditional methods to make decisions is not very accurate and reliable. Nevertheless, the use of a rational method can be efficient in selecting the appropriate candidate in any area.

Experts use scientific and empirical techniques for selecting the ideal alternative among several possible ones. The experiences in one particular area can be useful, but cannot be useful for decision-making in various areas. One of the scientific methods used by managers and experts is MCDM. The MCDM can be classified into two approaches of MADM and MODM. The MODM models use for designing, and MADM methods are used to choose the ideal alternative. The overall process of MCDM is shown in Figure 1.3.

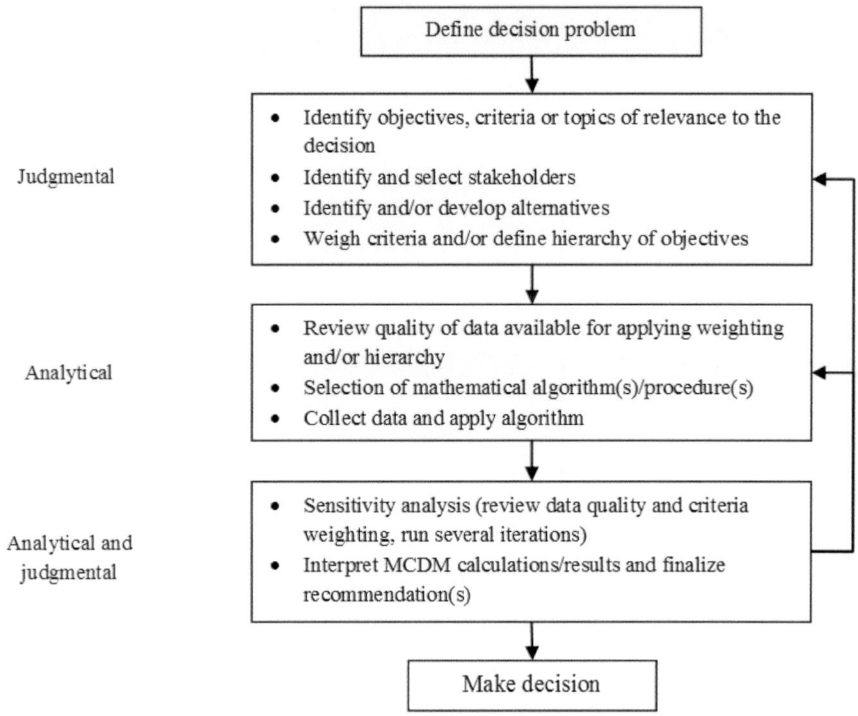

FIGURE 1.3 Overall process of MCDM.

Source: Reprinted with permission from Ref. [36]. Copyright 2013. © Taylor & Francis.

MODM is a continuous type of MCDM. The main features of MODM models are that DMs need to obtain multiple objectives while conflicting with each other [35]. This methodology is used to design and assist DMs in situations with factors such as objectives. The objectives refer to the performance characterization of the alternatives, and decision aids tend to be multiple objective optimizers [37].

Most of the development in MODM has taken place within recent years, and although one of the earliest considerations of MODM can be found by Kuhn & Tucker [38]. The term MODM refers to five steps of problem-solving in Figure 1.4. MODM process begins because of a problem and the need for change. Then, the DM sets objectives and attributes for modeling. The values of attributes and objectives for a specific alternative are achieved from subjective judgments.

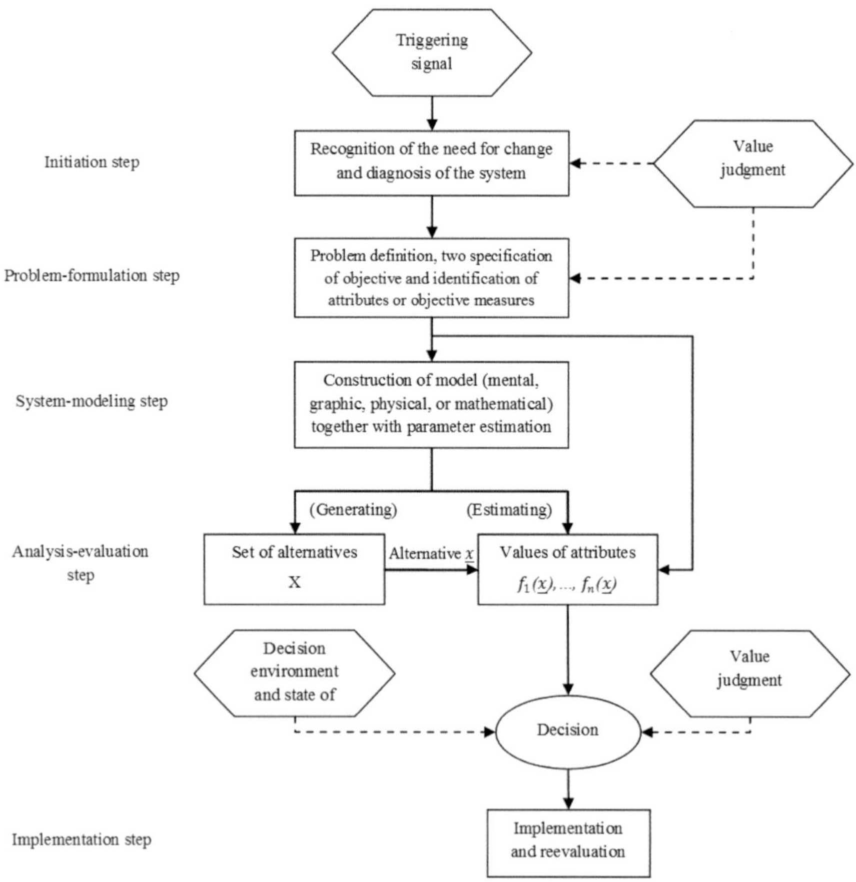

FIGURE 1.4 An illustration of the MODM process.

Source: Reprinted with permission from Ref. [39]. Copyright 2008. © Courier Dover Publications.

In the analysis phase, alternatives are evaluated in terms of a set of principles to rank alternatives. The alternatives with the highest rank are chosen for implementation. If the result is unsatisfactory, this information returns to the problem formulation phase, and it is called a closed-loop process. Also, the process may finish, and it is an open-loop process.

In every decision process, there are valuable elements. Value elements defy all scientific verification and treatment of the value system and prove the assumptions. On the other hand, there are value judgments in three phases (initiation, system modeling, and implementation) of a process. In the first step, for example, determining the need for change and realizing the objectives to be obtained are subjective processes, involving "need" and "wants." It is more important to identify those tasks in the MODM process in which judgment is a separate part [39].

The single-objective decision-making model is shown in Eqn. (1) [40].

$$max(min)\, f(\underline{x}): \mathbb{R}^n \rightarrow \mathbb{R}^m \quad (1)$$

$$s.t.\ \underline{x} \in \mathrm{X}: \mathbb{R}^n \rightarrow \mathbb{R}^m$$

In Eqn. (1), n and m show the numbers of decision variables and functional constraints. $\underline{x} = \begin{pmatrix} x_1 \\ \vdots \\ x_n \end{pmatrix}_{n\times 1}$ represents the space of possible solutions, which equals $\mathrm{X} = \left\{ \underline{x} \in \mathbb{R}^n | g_i(\underline{x}) \begin{bmatrix} \leq \\ \geq \\ = \end{bmatrix} b_i; i = 1,2,\ldots,m \right\}$. Now, if $\underline{x} \in \mathrm{X}$ is a feasible solution and if $\underline{x} \notin \mathrm{X}$ is an infeasible solution. The main focus would be in the discussion of multiple objective models and linear multiple objective functions, which are defined according to Eqn. (2) [40].

$$\underline{F}^t(\underline{x}) = \left(f_1(\underline{x}), f_2(\underline{x}), \cdots, f_k(\underline{x})\right) \quad (2)$$

$$max\ (min)\ \underline{\mathrm{F}}\ (\underline{x}): \mathbb{R}^n \rightarrow \mathbb{R}^k$$

$$s.t.\ \underline{x} \in \mathrm{X}: \mathbb{R}^n \rightarrow \mathbb{R}^k$$

The solution of ($\underline{x}^{*1}, \underline{x}^{*2}, \ldots, \underline{x}^{*k}$) is optimized for all the objective functions; but, such a status, which is due to contradictions in objectives, occurs rarely. If the solution is nonoptimal and feasible, then inefficient and efficient solutions are produced.

MADM methods are selective models used to demarcate the best alternative among available alternatives. In fact, in the MADM methods, choosing the ideal alternative is considered based on the DM's criteria. The decision matrix usually formulates the MADM. The decision matrix is one with the order of $m \times n$, which is shown in Eqn. (3), where A_i, $i = 1,2 \ldots, m$, shows alternatives, C_j, $j = 1,2 \ldots, n$, represents attributes, and r_{ij}, $i = 1,2 \ldots, m$, $j = 1,2 \ldots, n$, shows the value of the ith alternative for j^{th} attribute.

$$\mathbf{D}_{m\times n} = \begin{array}{c} \\ A_1 \\ A_2 \\ \cdot \\ \cdot \\ \cdot \\ A_m \end{array} \begin{array}{c} \begin{array}{cccccc} \mathbf{C}_1 & \mathbf{C}_2 & & & & \mathbf{C}_n \end{array} \\ \begin{pmatrix} r_{11} & r_{12} & \cdot & \cdot & \cdot & r_{1n} \\ r_{21} & r_{22} & \cdot & \cdot & \cdot & r_{2n} \\ \cdot & \cdot & & & & \cdot \\ \cdot & \cdot & & & & \cdot \\ \cdot & \cdot & & & & \cdot \\ r_{m1} & r_{m2} & \cdot & \cdot & \cdot & r_{mn} \end{pmatrix}_{m\times n} \end{array} \tag{3}$$

1.3.3 DATA ENVELOPMENT ANALYSIS (DEA)

Efficiency is one of the most important indicators of the performance evaluation of economic units. So far, various definitions of the term "efficiency" have been made. However, the characteristic feature of all of them points to the fact that it is a business that derives the most from the combination of specific data.

The concept of efficiency means working well in common sense, used most commonly against effectiveness, which signifies proper functioning. However, if a definition is to be provided for efficiency, a precise one should be presented. Accordingly, the ratio of output to input can be specified. If for a particular DMU, the global standard for an input unit is assumed, the output is equal to y^*. If a DMU uses an input unit, y_o produces the output unit, then the absolute efficiency is equal to $\frac{y_0}{y^*}$, because the efficiency of some units is sometimes acceptable far from absolute efficiency, hence relative efficiency is taken into consideration.

If the DMU$_j$, $j = 1,2 \ldots, n$, produce outputs of $y_1, \ldots, y_n$ through in taking $x_1, \ldots, x_n$ inputs, the relative efficiency for the DMU$_j$, represented by θ_j, can be defined in Eqn. (4):

$$\theta_j = \frac{\frac{y_j}{x_j}}{\max\limits_{1 \le k \le n}\left(\frac{y_k}{x_k}\right)}; j = 1,2,\ldots,sn \quad (4)$$

Now, if a DMU such as DMU_j, $j = 1,2 \ldots, n$, uses $x_{1j}, \ldots, x_{mj}$ in inputs to produce $y_{1j}, \ldots, y_{sj}$ outputs and the cost of ith input and the value of rth output are v_i and u_r, respectively. The efficiency of DMU_j is defined as Eqn. (5):

$$Efficiency = \frac{\sum_{r=1}^{s} u_r y_{rj}}{\sum_{i=1}^{m} v_i x_{rj}}; \; j = 1,2,\ldots,n \quad (5)$$

Sometimes the DMU functions in such a way that the price of output and input costs cannot be determined, therefore, DEA models are applied to determine the efficiency. In the last years, DEA has established itself as a popular analytical tool and applied decision instrument. The increasing usage of the DEA is a testament to its reputation in management science and economics in the government institutions and business societies. In 1957, Farrell measured the efficiency of the unit by using a measuring method in engineering topics. The technique that Farrell used to evaluate the efficiency included a single input/output.

Charnes, Cooper, and Rhodes (CCR) applied Farrell's model and provided a model that was able to evaluate the efficiency with multiple inputs/outputs. This template was named "data envelopment analysis (DEA)," and is based on a series of optimization using mathematical programming, which is also called the nonparametric method of measuring efficiency [41]. DEA is a "data-oriented" method for measuring the efficiency of DMUs [42]. In terms of its nature, it is divided into two approaches of input-oriented (minimizing inputs by keeping output levels constant), and output-oriented (minimizing outputs by maintaining input levels). If $x_1, \ldots, x_n$ are DMU inputs and $Q =$ F $(kx_1, kx_2, \ldots, kx_n)$ is the maximum output, and if $Q =$ F $(kx_1, kx_2, \ldots, kx_n) =$ KF$(X_1, X_2, \ldots, x_n)$, the DMU has a constant returns to scale (CRS), and each set of inputs produces the same multiplicative output [43].

If $Q =$ F$(kx_1, kx_2, \ldots, kx_n) \neq$ KF$(X_1, X_2, \ldots, x_n)$ is the DMU with variable returns to scale (VRS), which is a combination of the three modes of CRS, increasing returns to scale (IRS), and decreasing returns to scale (DRS). In the IRS, outputs increase more than tripled with the assumption of tripling the inputs, and in the DRS, outputs decrease more than tripled with the assumption of tripling the inputs [44, 45]. According to concepts of returns to scale (RTS) in Figure 1.5, lines of *oa* and *ab* are referred to as the IRS frontier. The line of *bc* is related to the CRS frontier, and it has a 45° slope. Finally, lines of *cd* and *de* are the DRS frontier.

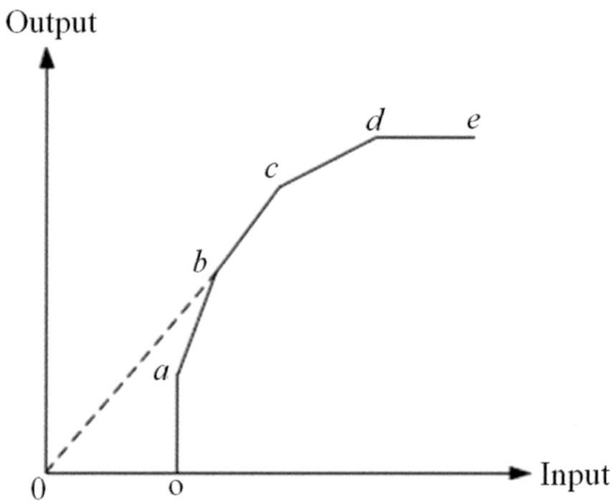

FIGURE 1.5 Concepts of efficiency and RTS.

The input-oriented model of the DEA provided by CCR is defined as Eqn. (6) [46]:

$$max \sum\nolimits_{r=1}^{s} u_r y_{rp} \tag{6}$$

$$s.t. \sum_{i=1}^{m} v_i x_{ip} = 1$$

$$\sum_{r=1}^{s} u_r y_{rj} - \sum_{i=1}^{m} v_i x_{ij} \le 0 \; ; j = 1,2,\ldots,n$$

$$u_r, v_i \ge \varepsilon ; i = 1,2,\ldots,m, r = 1,2,\ldots,s$$

Also, the input-oriented model of Banker, Charnes, and Cooper (BCC) can be mentioned among initial models and are defined as Eqn. (7) [47]:

$$max \sum\nolimits_{r=1}^{s} u_r y_{rp} + u_p \tag{7}$$

$$s.t. \sum_{i=1}^{m} v_i x_{ip} = 1$$

$$\sum_{r=1}^{s} u_r y_{rj} - \sum_{i=1}^{m} v_i x_{ij} + u_p \le 0; j = 1, 2, \ldots, n$$

$$u_r, v_i \ge \varepsilon, u_p \; free; i = 1, 2, \ldots, m, r = 1, 2, \ldots, s$$

1.4 DEA AND MADM

MADM decision matrix in DEA can be presented as Eqn. (8):

$$X_{n\times m} = \begin{pmatrix} x_{11} & x_{21} & \cdots & x_{m1} \\ x_{12} & x_{22} & \cdots & x_{m2} \\ \vdots & \vdots & \vdots & \vdots \\ x_{1n} & x_{2n} & \cdots & x_{mn} \end{pmatrix}_{n\times m} \quad (8)$$

So that $X_{n\times m}$ matrix represents the input vectors in DEA, and each of the rows of the matrix is associated with a DMU.

For example, the first line represents the inputs of the first DM. In Eqn. (9), $Y_{s\times n}$ represents the output vectors in the DEA, and each of the rows of the matrix is associated with a DMU.

$$Y_{n\times s} = \begin{pmatrix} y_{11} & y_{21} & \cdots & y_{s1} \\ y_{12} & y_{22} & \cdots & y_{s2} \\ \vdots & \vdots & \vdots & \vdots \\ y_{1n} & y_{2n} & \cdots & y_{sn} \end{pmatrix}_{n\times s} \quad (9)$$

On the other hand, the first line represents the outputs of the first DMU. The matrix of Eqn. (10) is formed according to Eqns. (8) and (9).

$$D_{n\times(m+s)} = \begin{pmatrix} x_{11} & x_{21} & \cdots & x_{m1} & y_{11} & y_{21} & \cdots & y_{s1} \\ x_{12} & x_{22} & \cdots & x_{m2} & y_{12} & y_{22} & \cdots & y_{s2} \\ \vdots & \vdots & \vdots & \vdots & \vdots & \vdots & \vdots & \vdots \\ x_{1n} & x_{2n} & \cdots & x_{mn} & y_{1n} & y_{2n} & \cdots & y_{sn} \end{pmatrix}_{n\times(m+s)} \quad (10)$$

As it can be clearly seen, in the matrix of Eqn. (10), all references are considered as attributes in MADM and are included in the columns, and all alternatives are included in the matrix rows. Although, it is worth noting that since DEA has lower inputs or less consumed resources, these inputs will be considered as negative attributes in the MADM; since higher the outputs are desirable in DEA, they will be categorized within the group of positive attributes. The other thing is the dependence of inputs and outputs, considering that the appropriate models of MADM should be selected.

Similarly, DEA fields can be used in MADM. As can be seen, the matrix of Eqn. (11) is the decision matrix in the MADM.

$$\begin{array}{c} \qquad\qquad \underline{x}_1 \cdots \underline{x}_n \\ X_{m\times n} = \begin{array}{c} A_1 \\ \vdots \\ A_m \end{array} \begin{pmatrix} x_{11} & \cdots & x_{1n} \\ \vdots & \vdots & \vdots \\ x_{m1} & \cdots & x_{mn} \end{pmatrix}_{m\times n} \end{array} \tag{11}$$

Therefore, the columns represent decision criteria that can be of benefit (y) or cost (x), this matrix can be converted into a DEA model, as shown in Figure 1.6.

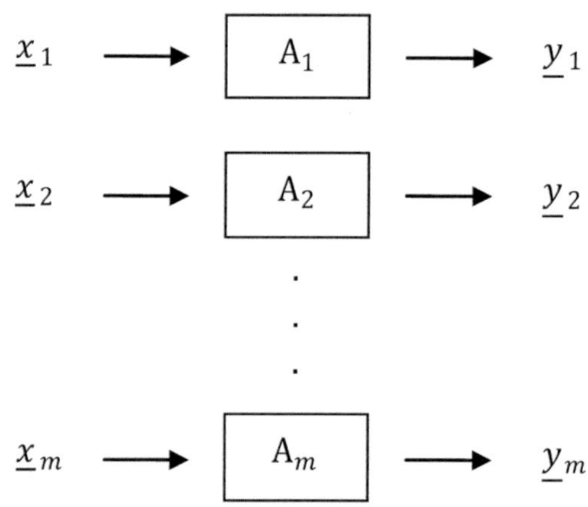

FIGURE 1.6 Representation of a DEA model.

One noteworthy point is the application of this method to issues dependent on each other; on the other hand, if all the attributes are positive or negative, it should be compulsory to reverse some of them to form the DEA model. Both inputs and outputs, of course, and qualitative variables should also be transformed into quantitative variables. The following analyzes the Analytical Hierarchy Process (AHP) and DEA.

1.4.1 DAE AND ANALYSIS HIERARCHY PROCESS (AHP)

Saaty provided the AHP method in 1970 [48]. This method makes it possible to assign a value representing the preference degree for an alternative to other alternatives [49, 50]. In the AHP, after determining the hierarchy, according to experts, the pairwise comparison matrix (PCM) is calculated between the alternatives, and after the normalization of the matrix, the weight of each alternative is determined using the geometric mean [51]. The hierarchy of decision variables is the subject of a pairwise comparison of the AHP method [52]. In this way, when the hierarchy is made, DMs compare alternatives systematically pairwise. DMs usually use their judgments. In AHP, human judgment, and not just necessary information, can be applied in the analysis [53]. One of the problems that have always been raised in the AHP debate is the problems arising from mental judgments on the cost and weighting methods. In this section, a method is presented to resolve this problem.

Initially, the efficiency test is performed in two (pairwise comparison) by the input-oriented multiplier model among all the units. In fact, in this model, the pairwise comparison is running between units 1 and 2, which is in the form of Eqn. (12).

$$max\, z_{1,1} = \frac{\sum_{r=1}^{s} u_r y_{r1}}{\sum_{i=1}^{m} v_i x_{i1}} \tag{12}$$

$$\text{S.t: } \frac{\sum_{r=1}^{s} u_r y_{r1}}{\sum_{i=1}^{m} v_i x_{i1}} \leq 1$$

$$\text{S.t: } \frac{\sum_{r=1}^{s} u_r y_{r2}}{\sum_{i=1}^{m} v_i x_{i2}} \leq 1$$

$$u_r, v_i \geq \varepsilon; i = 1, 2, \ldots m, r = 1, 2, \ldots, s$$

By changing variable $\sum_{i=1}^{m} v_i x_{i1} = 1$ the model of Eqn. (12) will be linear and will be obtained Eqn. (13), with the optimal value shown with $e_{1,1}$.

$$e_{1,1} = max\, z_{1,1} = \sum_{r=1}^{s} u_r y_{r1} \tag{13}$$

$$\text{S.t : } \sum_{i=1}^{m} v_i x_{i1} = 1$$

$$\sum_{r=1}^{s} u_r y_{r1} \leq 1$$

$$\sum_{r=1}^{s} u_r y_{r2} - \sum_{i=1}^{m} v_i x_{i2} \leq 0$$

$$u_r, v_i \geq \varepsilon; i = 1, 2, \ldots, m, r = 1, 2, \ldots, s$$

It is worth noting that if unit 1 is efficient, then Eqn. (14) is obtained.

$$e_{1,1} = \sum_{r=1}^{s} u_r y_{r1} = 1 \tag{14}$$

In this case, the second constraint will only be established in equilibrium, and the resulting slack variable will be zero ($s_2 = 0$). If the first unit is not efficient, then the second unit (unit 2) will be compulsory, and the slack variable corresponding to the third constraint will be zero ($s_3 = 0$). As a result, if $s_2 = 0$, the first unit will be efficient, and the second unit is inefficient ($s_3 > 0$) and if $s_3 = 0$, then the first unit will be nonoperative, and the second unit is working ($s_2 > 0$). Then, the cross-efficiency calculation is followed by Eqn. (15).

$$e_{2,1} = max\, z_{2,1} = \frac{\sum_{r=1}^{s} u_r y_{r2}}{\sum_{i=1}^{m} v_i x_{i2}} \tag{15}$$

$$\text{S.t} : \frac{\sum_{r=1}^{s} u_r y_{r2}}{\sum_{i=1}^{m} v_i x_{i2}} \leq 1$$

$$\frac{\sum_{r=1}^{s} u_r y_{r1}}{\sum_{i=1}^{m} v_i x_{i1}} = e_{1,1}$$

$$u_r, v_i \geq \varepsilon; i = 1, 2, \ldots m, r = 1, 2, \ldots, s$$

Similarly, changing $\sum_{i=1}^{m} v_i x_{i2} = 1$ variable will make the model of Eqn. (15) linear, and Eqn. (16) will be obtained, whose optimal value is shown with $e_{2,1}$.

$$e_{2,1} = max\, z_{2,1} = \sum_{r=1}^{s} u_r y_{r2} \tag{16}$$

$$\text{S.t: } \sum_{i=1}^{m} v_i x_{i2} = 1$$

$$\sum_{r=1}^{s} u_r y_{r2} \leq 1$$

$$\sum_{r=1}^{s} u_r y_{r1} - e_{1,1} \sum_{i=1}^{m} v_i x_{i1} = 0$$

$$u_r, v_i \geq \varepsilon; i = 1, 2, \ldots, m, r = 1, 2, \ldots, s$$

The optimal value of the second unit is equal to $e_{2,1}$. Thus, $e_{1,2}$ and $e_{2,2}$ values are calculated, and finally, the Eqn. (17) is obtained.

$$3_{j,p} = \frac{e_{jj} + e_{jp}}{e_{pp} + e_{pj}}; j = 1, 2, \ldots, n \tag{17}$$

As $3_{j,p}$ represents the element of *the* j^{th} row and p^{th} column of the PCM, it has the concept of the evaluation of the jth unit as compared to the pth unit. In the final step, according to the AHP method, problem-solving will continue.

1.5 DEA AND MODM

MODM model presented is based on the nonlinear DEA model suggested by Charnes et al. [54]. DEA is a decision-making technique based on mathematical programming, which is used for solving decision-making problems that require attention to multiple outputs/inputs to measure the efficiency of DMUs. Despite the attention to multiple outputs in return analysis, DEA uses a clear hypothesis in which each input can act as an alternative to other inputs. This hypothesis does not apply to cases where inputs do not replace each other. Also, considering an input, the problem of overweight on the inputs is resolved [55].

If m inputs consume to produce s outputs in n units (DMUs), then, $x_{ij} \geq 0$ shows the amount consumed of the ith input, and $y_{ij} \geq 0$ shows the amount produced of the rth output by the DMU_j. Also, the production possibility set (PPS) of Eqn. (18) is used the CCR model with CRS characteristics.

$$T_c = \left\{ (\underline{x}, \underline{y}) \middle| \underline{x} \geq \sum_{j=1}^{n} \lambda_j \underline{x_j}, \underline{y} \leq \sum_{j=1}^{n} \lambda_j \underline{y_j}, \lambda_j \geq 0; j = 1, 2, \ldots, n \right\} \tag{18}$$

- **Definition 1:** DMU_j; $j = 1,2,\ldots,n$, is efficient if there does not exist in another $(\underline{x}, \underline{y}) \in T_c$ such that $\underline{x} < \underline{x_j}$ and $\underline{y} < \underline{y_j}$. Moreover, it is

Pareto efficient if there does not exist in another $(\underline{x}, \underline{y}) \in T_c$ such that $\underline{x} < \underline{x_j}$ and $\underline{y} < \underline{y_j}$ and $\exists j : (\underline{x}, \underline{y}) \neq (\underline{x_j}, \underline{y_j})\,; j = 1,2,\ldots,n$.

The efficiency of a DMU is less than or equal to 1 and it is considered as a ratio of a weighted sum of outputs to inputs. Also, the positive weights are selected, and model Eqn. (6) is presented by Charnes et al. [46].

Also, u_r and v_i are the weights of outputs and inputs. The optimal solution is related to the coefficient $(\underline{u}^*, -\underline{v}^*)$ of a hyperplane (it includes the PPS in half-spaces and passes among the points of it). The dual model of Eqn. (6) is used afterward, called the input-oriented CCR envelopment model Eqn. (19), which is typically used to explain and visualize DEA.

$$\begin{aligned} & max\ \theta_p \\ & s.t. \sum_{j=1}^{n} \lambda_j x_{ij} \leq \theta_p x_{ip}\,; i = 1,2,\ldots,m \\ & \sum_{j=1}^{n} \lambda_j y_{rj} \geq y_{rp}\,; r = 1,2,\ldots,s \\ & \lambda_j \geq 0; j = 1,2,\ldots,n \\ & \theta_p\ free \end{aligned} \tag{19}$$

If the objective function values of the optimal solutions are equal to 1, the DMU will be efficient. Also, if in the models all variables take a nonnegative value, slack variables are equal to zero; the DMU will be Pareto efficient.

- **Lemma 1:** If θ_p^* is the solution of the model Eqn. (19), $(\theta_p^* \underline{x_p}, \underline{y_p})$ is the projection point of DMU_p on the efficient frontier.
- **Lemma 2:** If there is $(\underline{u}, \underline{v}) \in R^s \times R^m$ related to the gradient vector of a hyperplane, DMU_p is efficient and $\sum_{r=1}^{s} u_r y_{rp} - \sum_{i=1}^{m} v_i x_{ip} = 0$.

In performance measurement of DMUs, various set of weights is achieved, that it is not acceptable. Kornbluth [56] has noted that the models can be supposed as a multiple objective linear fractional programming problem. According to Eqn. (6), the nonlinear model Eqn. (20) can be written.

$$max \left\{ \frac{\sum_{r=1}^{s} u_r y_{r1}}{\sum_{i=1}^{m} v_i x_{i1}}, \ldots, \frac{\sum_{r=1}^{s} u_r y_{rn}}{\sum_{i=1}^{m} v_i x_{in}} \right\} \tag{20}$$

$$S.t.\ \frac{\sum_{r=1}^{s} u_r y_{rj}}{\sum_{i=1}^{s} v_i x_{ij}} \leq 1;\ j = 1,2,\ldots,n$$

$$u_r, v_i \geq \varepsilon;\ i = 1,2,\ldots,m, r = 1,2,\ldots,s$$

The compromise solution approach is presented by Kao & Hung [57], which was based on Kornbluth's approach. E_j^* shows the optimal objective value of the model Eqn. (6), which is the ideal accessible efficiency for DMU_j. Other weight sets could lead to the efficiency score, and its value is less than or equal to E_j^*. For a common set of weights (CSW), Kao and Hung propose $E^{*t} = (E_1^*, \ldots, E_n^*)$ as the optimal solution, to obtain. On the other hand, the efficiency vector $E(\underline{u}, \underline{v}) = \{E_1(\underline{u}, \underline{v}), \ldots, E_n(\underline{u}, \underline{v})\}$ is obtained from the weights ($\underline{u}$, $\underline{v}$) to be closest to E_j^*. The closeness degree between $E(\underline{u}, \underline{v})$ and E_j^* is estimated from Eqn. (21).

$$D_p\left(E(\underline{u}, \underline{v})\right) = \left[\sum_{j=1}^{n} (E_j^* - E_j(\underline{u}, \underline{v}))^p\right]^{\frac{1}{p}};\ p \geq 1 \tag{21}$$

In Eqn. (21), p is the distance parameter. The efficiency scores $E_j(\underline{u}, \underline{v})$ and the weights ($\underline{u}$, $\underline{v}$) are solved by a compromise solution with parameter p, which is constrained nonlinear programming.

Based on the approach of Kao and Hung to finding CSW in DEA, the CCR multiplier model Eqn. (22) is supposed to look for the efficiency of DMU_p.

$$max \sum_{r=1}^{s} u_r y_{rp} - \sum_{i=1}^{m} v_i \theta_p^* x_{ip} \tag{22}$$

$$s.t. \sum_{r=1}^{s} u_r y_{rj} - \sum_{i=1}^{m} v_i \theta_j^* x_{ij} \leq 0;\ j = 1,2,\ldots,n$$

$$\sum_{r=1}^{s} u_r + \sum_{i=1}^{m} v_i = 1$$

$$u_r, v_i \geq \varepsilon;\ i = 1,2,\ldots,m, r = 1,2,\ldots,s$$

where; $E_j^*, j = 1,2,\ldots,n$, is the optimal solution achieved from the CCR model, when DMU_p is under attention.

➤ **Theorem 1:** The optimal solution $(\underline{u}^*, \underline{v}^*)$ of the model Eqn. (22) is zero and $\frac{\sum_{r=1}^{s} u_r^* y_{rp}}{\sum_{i=1}^{m} v_i^* x_{ip}} = \theta_p^*$.

- **Proof 1:** The PPS produced by $(\theta_j^* \underline{x_j}, \underline{y_j})$, $j = 1,2,\ldots,n,$ is as the original PPS. According to lemmas 1 and 2, $(\theta_p^* \underline{x_p}, \underline{y_p})$ is the input-oriented projection of DMU_p on the efficient frontier, and it is efficient. Also, there exists a hyperplane at $(\theta_p^* \underline{x_p}, \underline{y_p})$ which results in $\frac{\sum_{r=1}^{s} u_r^* y_{rp}}{\sum_{i=1}^{m} v_i^* x_{ip}} = \theta_p^*$.

Based on the model Eqn. (22) and Kornbluth's approach, the idea behind the CSW is formulated, finding a normalized coefficient $(\underline{u}^*, -\underline{v}^*)$of a hyperplane at PPS. So, the model Eqn. (23) is deduced.

$$max\left[\sum_{r=1}^{s} u_r y_{r1} - \sum_{i=1}^{m} v_i \theta_1^* x_{i1}, \ldots, \sum_{r=1}^{s} u_r y_{rn} - \sum_{i=1}^{m} v_i \theta_n^* x_{in}\right] \tag{23}$$

$$s.t. \sum_{r=1}^{s} u_r y_{rj} - \sum_{i=1}^{m} v_i \theta_j^* x_{ij} \leq 0; j = 1,2,\ldots,n$$

$$\sum_{r=1}^{s} u_r + \sum_{i=1}^{m} v_i = 1$$

$$u_r, v_i \geq \varepsilon; i = 1,2,\ldots,m, r = 1,2,\ldots,s$$

1.5.1 COMPROMISE SOLUTION APPROACH

A compromise solution is presented by Zeleny [58]; and Yu [59], which is an approach with various criteria and can help to attain the best decision [60]. This method is the closest solution to the ideal, and an agreement defined by mutual concessions [61]. The compromise solutions have the characteristics of the unique solution and Pareto optimality not benefited by the solutions obtained from the other CSW methods, and this can generate a CSW, which differentiates efficient DMUs and determines the unusual efficiency score [57]. First of all, it can be supposed $f_j^*(\underline{x}), j = 1,\ldots,n,$ is the optimal value of the model Eqn. (24).

$$max\, f_j(\underline{x}); j = 1, 2, \ldots, k \tag{24}$$

$$s.\ t.\ \underline{x} \in X$$

The efficient solution minimizes the deviations from the optimal solution in the model Eqn. (25).

$$min\ D_p = \left\{ \sum_{j=1}^{k} \left[\frac{f_j^*(\underline{x}) - f_j(\underline{x})}{f_j^*(\underline{x})} \right]^p \right\}^{\frac{1}{p}} ;p \geq 1 \tag{25}$$

$$s.t.\ \underline{x} \in X$$

In the model Eqn. (25), p is based upon the utility function of the DM, and for $p = 1$, deviations are weighted equally. By increasing the p, weights are given to the deviations. Finally, the most significant deviation overcomes in $p = \infty$. There exists the values of p,$p = 1,2,\ldots,\infty$, which have unique characteristics.

For solving the model Eqn. (23), the compromise solution uses to generate a vector of deviations closest to the deviations obtained from the model Eqn. (19). Therefore, a vector of zero scores is an optimal solution. Ultimately, Eqn. (26) is obtained.

$$D_p(\underline{u},\underline{v}) = \left[\sum_{j=1}^{n} \left(0 - (\sum_{r=1}^{s} u_r y_{rj} - \sum_{i=1}^{m} v_i \theta_j^* x_{ij}) \right)^p \right]^{\frac{1}{p}} ;p \geq 1 \tag{26}$$

In Eqn. (26), p presents the distance parameter. Then, the model Eqn. (27) is formulated.

$$max\ D_p(\underline{u},\underline{v}) = \left[\sum_{j=1}^{n} \left(0 - (\sum_{r=1}^{s} u_r y_{rj} - \sum_{i=1}^{m} v_i \theta_j^* x_{ij}) \right)^p \right]^{\frac{1}{p}} ;p \geq 1 \tag{27}$$

$$s.t. \sum_{r=1}^{s} u_r y_{rj} - \sum_{i=1}^{m} v_i \theta_j^* x_{ij} \leq 0; j = 1,2,\ldots,n$$

$$\sum_{r=1}^{s} u_r + \sum_{i=1}^{m} v_i = 1$$

$$u_r, v_i \geq \varepsilon; i = 1,2,\ldots,m, r = 1,2,\ldots,s$$

However, the model Eqn. (27) is tantamount to constrained nonlinear programming, for $p = 1$ and $p = \infty$. Hence, for $p = 1$ and $p = \infty$, it is calculated faster than the model of Kao and Hung, which are nonlinear.

The model Eqn. (27) gives a CSW and the efficiency of DMU_j, $j = 1,\ldots,n$ can be achieved by using $\frac{\sum_{r=1}^{s} u_r^* y_{rj}}{\sum_{i=1}^{m} v_i^* x_{ij}}$, $j = 1,\ldots,n$. If $\frac{\sum_{r=1}^{s} u_r^* y_{rp}}{\sum_{i=1}^{m} v_i^* x_{ip}} = 1$, DMU_p is efficient. Hence, by using A= {j: DMU_j is efficient in the model Eqn. (29)}, and the approach of Jahanshahloo et al. [62], DMUs can be ranked.

- Ø **Theorem 2:** If DMUp is efficient by the model Eqn. (27), it is efficient by the input-oriented CCR model.
- Ø **Proof 2:** According to the first inequalities, it can be written $\frac{\sum_{r=1}^{s} u_r^* y_{rp}}{\sum_{i=1}^{m} v_i^* x_{ip}} \leq \theta_p^* \leq 1$. Therefore, if $\frac{\sum_{r=1}^{s} u_r^* y_{rp}}{\sum_{i=1}^{m} v_i^* x_{ip}} = 1$, then DMU$_p$ is efficient, and $\theta_p^* = 1$.
- Ø **Theorem 3:** DMU$_j$, $j = 1,\ldots,n$ will be specified as an efficient DMU in the model Eqn. (25).
- Ø **Proof 3:** $p = 1$ and ∞ are supposed in the model Eqn. (27). If there exists a DMU$_p$, $p \in \{1,\ldots,n\}$ then the inequality is mandatory in the model Eqn. (27). Otherwise, there exists a small value $\varepsilon > 0$ where $(\underline{u}, \underline{v}) = (\underline{u}^* + [\varepsilon, 0, \ldots, 0]^{\mathrm{T}}_{1\times s}, \underline{v}^* + [\varepsilon, 0, \ldots, 0]^{\mathrm{T}}_{1\times m})$ satisfy the constraints in Eqn. (27). Also, the value of $\mathrm{D}_p(\underline{u}, \underline{v})$ is smaller than $\mathrm{D}_p(\underline{u}^*, \underline{v}^*)$, and it is contradicted with the optimal value of $\mathrm{D}_p(\underline{u}^*, \underline{v}^*)$. Thus, there is a DMU$_p$, $p \in \{1,\ldots,n\}$, for which $\sum_{r=1}^{s} u_r^* y_{rp} - \sum_{i=1}^{m} v_i^* \theta_p^* x_{ip} = 0$, such as $(\underline{u}^*, \underline{v}^*)$, is related to the gradient vector of a hyperplane. On the other hand, hyperplane supports the PPS at efficient DMUs, and in the model Eqn. (27), DMUs will be efficient.

1.5.2 MAXIMIN METHOD

Maximin method is used for determining the maximum or minimum value. The highest rank belongs to the minimum of the maximum value, and the minimum of the minimum value obtains the lowest rank [63]. Maximin method applies to linear and nonlinear multiple objective problems and, the goal is to maximize the minimum functions.

In this method, the DM prepares the information during the mathematical formulation. The information can be cardinal or mixed (cardinal and ordinal). In the cardinal information, DM gives some judgment about special objective preference levels. Otherwise, the DM ranks objectives in the order of significance. The Maximin model can be formulated in the form of the model Eqn. (28).

$$\begin{gathered} max\ z \\ s.\ t.\ \underline{x} \in \mathrm{X} \\ f_j = (\underline{x}) - z \geq 0; j = 1,2,\ldots,k \end{gathered} \tag{28}$$

For solving model Eqn. (23), as in Jahanshahloo et al. [62], the Maximin technique use to maximize the objective function. Then, the model Eqn. (29) is determined.

$$\max z \tag{29}$$

$$s.t. \sum_{r=1}^{s} u_r y_{rj} - \sum_{i=1}^{m} v_i \theta_j^* x_{ij} \le 0; j = 1,2,\ldots,n$$

$$\sum_{r=1}^{s} u_r y_{rj} - \sum_{i=1}^{m} v_i \theta_j^* x_{ij} - z \ge 0; j = 1,2,\ldots,n$$

$$\sum_{r=1}^{s} u_r + \sum_{i=1}^{m} v_i = 1$$

$$u_r, v_i \ge \varepsilon, z \text{ free}; i = 1,2,\ldots,m, r = 1,2,\ldots,s$$

Model Eqn. (29) is a problem where sufficient insight can be achieved for obtaining natural results. In the CSW approach, the projection of each DMU is settling on the hyperplane.

This may be happening outside of PPS with negative values for inputs and caused to produce some negative efficiency results. Nevertheless, since model CCR is unit invariance, M time of inputs can be used (for some sufficient large amount M) to solve this problem.

The model Eqn. (29) gives a CSW and the efficiency of DMU_j, $j = 1,\ldots,n$, can be attained by using weights as $\frac{\sum_{r=1}^{s} u_r^* y_{rj}}{\sum_{i=1}^{m} v_i^* x_{ij}}$, $j = 1,\ldots,n$. If $\frac{\sum_{r=1}^{s} u_r^* y_{rp}}{\sum_{i=1}^{m} v_i^* x_{ip}} = 1$, then DMU_p will be an efficient unit. Hence, by using A= {j: DMU_j is efficient in the model Eqn. (29)}, and the approach of Jahanshahloo et al. [62], DMUs can be ranked.

- Ø **Theorem 4:** If DMU_p is efficient by the model Eqn. (29), it is efficient by the input oriented CCR model.
- Ø **Proof 4:** According to the first inequalities, it can be written $\frac{\sum_{r=1}^{s} u_r^* y_{rp}}{\sum_{i=1}^{m} v_i^* x_{ip}} \le \theta_p^* \le 1$. Therefore, if $\frac{\sum_{r=1}^{s} u_r^* y_{rp}}{\sum_{i=1}^{m} v_i^* x_{ip}} = 1$, then $\theta_p^* = 1$, and DMU_p is CCR efficient.
- Ø **Theorem 5:** DMU_j, $j = 1,\ldots,n$, will be specified as an efficient DMU in the model Eqn. (29).

Ø **Proof 5:** If M = m_1 and S = s_1 where vectors of m and s are ones, $[1,\ldots,1]^T$. Then, there is a DMU$_p$, $p \in \{1,\ldots,n\}$, for which the inequality in the model Eqn. (29) is mandatory. If it is not the case, there is $\varepsilon > 0$ where $(\underline{u}, \underline{v}) = (\underline{u} + \varepsilon\underline{M}, \underline{v} - \varepsilon\underline{S})$ satisfy the constraints in Eqn. (29). The value of $\overline{z}$ related to $(\underline{u}, \underline{v})$ and the second constraint are increased, and it is not related to the optimal value. Also, there exists a DMU$_p$, $p \in \{1,\ldots,n\}$, for $\sum_{r=1}^{s} u_r^* y_{rp} - \sum_{i=1}^{m} v_i^* \theta_p^* x_{ip} = 0$. However, $(\theta_p^* \underline{x_p}, \underline{y_p})$ is efficient. Therefore, $(\underline{u}, \underline{v})$ is related to the gradient vector of a hyperplane. Moreover, it should support the PPS at efficient units. Also, these units are demonstrated efficient in the model Eqn. (29).

1.5.3 GOAL PROGRAMMING (GP) METHOD

The goal programming (GP) method was presented by Charnes & Cooper [72] and the goals of each objective should be set that DM wants to achieve [64]. GP model was first developed to attain "constrained regression" for an executive compensation problem, and the aim of the method is the simultaneous satisfaction of goals related to the decision-making problem [65]. The main idea behind the model is the determination of the aspiration levels of the objective function and the minimization of deviations [66].

In this method, collecting information from DMs is combined, and in addition to declaring their ideals of functions (cardinal information), the objective functions must also be ranked (ordinal information). In the ordinal ranking, only the rating of each alternative is given, but in cardinal mode, the scores of alternatives are also mentioned. Also, the search toward the most preferred solution is driven by the DM, and preferences can be expressed at every stage of the solution process. Therefore, this interactive method requires close interaction between the DM and the analyst to converge to the desired efficient solution. The GP model is given by Eqn. (30) [67].

$$min\ D_p = \left\{ \sum_{j=1}^{n} \left[h_j(d_j^-, d_j^+) \right]^p \right\}^{\frac{1}{p}}; p \geq 1 \tag{30}$$

$$S.t.\ g_i(\underline{x}) \leq 0; i = 1,2,\ldots,m$$

$$f_j(\underline{x}) + d_j^+ - d_j^- = b_j; j = 1,2,\ldots,n$$

$$d_j^- \times d_j^+ = 0; j = 1,2,\ldots,n$$

$$d_j^-, d_j^+ \geq 0; j = 1,2,\ldots,n$$

According to model Eqn. (30), $f_j, j = 1, 2, \ldots, n$, and $b_j, j = 1, 2, \ldots, n$, are the goals and objectives, respectively. Also, d_j^- is the under-achievement and d_j^+ is related to the over-achievement of the jth goal, and the p value is associated with the utility function. Firstly, the model Eqn. (31) is considered.

$$max\left\{\frac{\sum_{r=1}^{s} u_r y_{r1}}{\sum_{i=1}^{m} v_i x_{i1}}, \ldots, \frac{\sum_{r=1}^{s} u_r y_{rn}}{\sum_{i=1}^{m} v_i x_{in}}\right\} \tag{31}$$

$$S.t. \frac{\sum_{r=1}^{s} u_r y_{rj}}{\sum_{i=1}^{m} v_i x_{ij}} \leq 1 \quad ; j = 1,2,\ldots,n$$

$$u_r,\ v_i \geq \varepsilon; i = 1,2,\ldots,m, r = 1,2,\ldots,s$$

The objective of solving the model Eqn. (31) is to obtain weights of u_r^* and v_r^* considering DM preferences. The GP model requires the DM to rank the objective functions, and also suppose DM ranks DMUs based on their indexes, to put DMU_1, in the first rank, DMU_2 in the second, ..., DMU_n in the nth one. On the other hand, it is assumed in this method that DMUs are ranked based on their indexes, and $w_j, j = 1, 2, \ldots, n$, are supposed to be the ordinal weights of DMUs by the DM. If all of the goals ($b_j, j = 1, 2, \ldots, n$) be 1, model Eqn. (32) is obtained.

$$min\ \mathrm{D}_p = \left\{\sum_{j=1}^{n}\left[h_j(d_j^-, d_j^+)\right]^p\right\}^{\frac{1}{p}}; p \geq 1 \tag{32}$$

$$s.t. \frac{\sum_{r=1}^{s} u_r y_{rj}}{\sum_{i=1}^{m} v_i x_{ij}} \leq 1 \quad ; j = 1,2,\ldots,n$$

$$\frac{\sum_{r=1}^{s} u_r y_{rj}}{\sum_{i=1}^{m} v_i x_{ij}} + d_j^- - d_j^+ = 1 \quad ; j = 1,2,\ldots,n$$

$$d_j^- . d_j^+ = 0; j = 1,2,\ldots,n$$

$$u_r,\ v_i \geq \varepsilon, d_j^-, d_j^+ \geq 0; i = 1,2,\ldots,m, j = 1,2,\ldots,n, r = 1,2,\ldots,s$$

Model Eqn. (32) is nonlinear programming. Therefore, the model Eqn. (33) is supposed.

$$max \sum_{r=1}^{s} u_r y_{rp} - \theta_p^*(\sum_{i=1}^{m} v_i x_{ip}) \tag{33}$$

$$s.t. \sum_{r=1}^{s} u_r y_{rj} - \theta_j^*(\sum_{i=1}^{m} v_i x_{ij}) \leq 0; j = 1,2,\ldots,n$$

$$\sum_{r=1}^{s} u_r + \sum_{i=1}^{m} v_i = 1$$

$$u_r, v_i \geq \varepsilon; i = 1,2,\ldots,m, r = 1,2,\ldots,s$$

In the model Eqn. (33), $\theta_j^*, j = 1, 2,\ldots,n$, is the optimal solutions attained by the multiplier CCR model for $DMU_j, j = 1, 2,\ldots,n$. Model Eqn. (33) can be transformed into the model Eqn. (34).

$$max \left\{ \sum_{r=1}^{s} u_r y_{r1} - \theta_1^*(\sum_{i=1}^{m} v_i x_{i1}),\ldots, \sum_{r=1}^{s} u_r y_{rn} - \theta_n^*(\sum_{i=1}^{m} v_i x_{in}) \right\} \tag{34}$$

$$s.t. \sum_{r=1}^{s} u_r y_{rp} - \theta_j^*(\sum_{i=1}^{m} v_i x_{ij}) \leq 0; j = 1,2,\ldots,n$$

$$\sum_{r=1}^{s} u_r + \sum_{i=1}^{m} v_i = 1$$

$$u_r, v_i \geq \varepsilon; i = 1,2,\ldots,m, r = 1,2,\ldots,s$$

Using the model Eqn. (34) in place of the model Eqn. (31) and utilizing the GP model with zero goals and $p = 1$, model Eqn. (32) changes to the linear model Eqn. (35).

$$min \sum_{j=1}^{n} (d_j^- + d_j^+) \tag{35}$$

$$s.t. \sum_{r=1}^{s} u_r y_{rj} - \sum_{i=1}^{m} v_i \theta_j^* x_{ij} \leq 0; j = 1,2,\ldots,n$$

$$\sum_{r=1}^{s} u_r y_{rj} - \sum_{i=1}^{m} v_i \theta_j^* x_{ij} + d_j^- - d_j^+ = 1; j = 1,2,\ldots,n$$

$$\sum_{r=1}^{s} u_r + \sum_{i=1}^{m} v_i = 1$$

$$d_j^-, d_j^+ \geq 0; j = 1,2,\ldots,n$$

$$u_r, v_i \geq \varepsilon; i = 1,2,\ldots,m, r = 1,2,\ldots,s$$

In Eqn. (35), there is no need for constraints $d_j^-.\ d_j^- = 0, j = 1, 2,\ldots,n$. The first and second constraint groups in the model Eqn. (35) result in $d_j^+ = 0$, and then model Eqn. (36) is defined.

$$min\sum_{j=1}^{n} d_j^- \tag{36}$$

$$s.t. \sum_{r=1}^{s} u_r y_{rj} - \sum_{i=1}^{m} v_i \theta_j^* x_{ij} \le 0; j = 1,2,\ldots,n$$

$$\sum_{r=1}^{s} u_r y_{rj} - \sum_{i=1}^{m} v_i \theta_j^* x_{ij} + d_j^- = 1; j = 1,2,\ldots,n$$

$$\sum_{r=1}^{s} u_r + \sum_{i=1}^{m} v_i = 1$$

$$d_j^- \ge 0\ \varepsilon; j = 1,2,\ldots,n$$

$$u_r,\ v_i \ge \varepsilon; i = 1,2,\ldots,m, r = 1,2,\ldots,s$$

Model Eqn. (36) not only represents u_r^* and v_i^* as a CSW, but also take into account DMUs ranking as much as possible according to DM's preferences.

1.5.4 LEXICOGRAPHIC METHOD

The lexicographic method is provided by Hell & Huang in 1995 [68]. In this method, preferences are applied by ordering the objective functions according to their importance. Hence, the objectives are maximized and proceed based on the order of significance of objectives. This technique needs that the objectives to be ranked.

The subscripts of the objectives show the priorities and components of the objective vector ($\underline{F}(\underline{x})$). For instance, $f_1(\underline{x})$ is the first component of $\underline{F}(\underline{x})$. Therefore, $f_2(\underline{x})$ is the second most critical objectives and the second component. Then, the model Eqn. (37) is solved.

$$max\, f_1(\underline{x}) \tag{37}$$

$$s.t.\ g_i(\underline{x}) \le 0; i = 1,2,\ \ldots,\ m$$

Let $f_2^*(\underline{x})$ be the solution model Eqn. (37). If in the model Eqn. (37), $\underline{x}$ is a unique solution for $f_1^*(\underline{x})$, the preferred solution will be for the entire problem. Thus, the model Eqn. (38) is to be solved:

$$max\, f_2(\underline{x}) \tag{38}$$

$$s.t.\ g_i(\underline{x}) \le 0;\ i = 1,2,\ \ldots,\ m$$
$$f_1(\underline{x}) = f_1^*(\underline{x})$$

Let $f_2^*(\underline{x})$ be the solution Eqn. (38). If this model gives unique x for $f_2^*(\underline{x})$, the preferred solution will be for the entire problem. Otherwise, the sequence is repeated for all k objectives. Hence, the lth model is determined as Eqn. (39).

$$max\, f_l(\underline{x});\ l = 1,2,\ \ldots,\ k \tag{39}$$
$$s.t.\ g_i(\underline{x}) \le 0;\ i = 1,2,\ \ldots,\ m$$
$$f_p(\underline{x}) = f_p^*(\underline{x});\ p = 1,\ \ldots,\ l-1$$

Since the process is interrupted when an optimal solution is attained at the lth problem, this is a preferred solution and objectives ranked less important than $f_l(\underline{x})$; are rejected. By maximizing the first objective, the second objective, and then the third objective will be maximized, subject to keeping the previous objectives within an optimum percentage. The lth model can be formulated in the form of Eqn. (40).

$$max\, f_l(\underline{x});\ l = 1,2,\ \ldots,\ k \tag{40}$$
$$s.t.\ g_i(\underline{x}) \le 0;\ i = 1,2,\ \ldots,\ m$$
$$f_p(\underline{x}) = f_p^*(\underline{x}) - \Delta_p;\ p = 1,\ \ldots,\ l-1$$

Then, to solve the model Eqn. (31), the Lexicography model is in the form of Eqn. (41).

$$lex\ max \left\{ \frac{\sum_{r=1}^{s} u_r y_{r1}}{\sum_{i=1}^{m} v_i x_{i1}}, \ldots, \frac{\sum_{r=1}^{s} u_r y_{rn}}{\sum_{i=1}^{m} v_i x_{in}} \right\} \tag{41}$$

$$S.t.\ \frac{\sum_{r=1}^{s} u_r y_{rj}}{\sum_{i=1}^{m} v_i x_{ij}} \le 1;\ j = 1,2,\ldots,n$$

$$u_r,\ v_i \ge \varepsilon; i = 1,2,\ldots,m, r = 1,2,\ldots,s$$

If DMU_1 is the most important unit, then, model Eqn. (42) will be obtained.

$$max \frac{\sum_{r=1}^{s} u_r y_{r1}}{\sum_{i=1}^{m} v_i x_{i1}} \tag{42}$$

$$s.t.\ \frac{\sum_{r=1}^{s} u_r y_{rj}}{\sum_{i=1}^{m} v_i x_{ij}} \leq 1;\ j = 1,2,\ldots,n$$

$$u_r, v_i \geq \varepsilon;\ i = 1,2,\ \ldots,\ m, r = 1,2,\ \ldots,\ s$$

The model Eqn. (42) is nonlinear and, the linear model can be written as Eqn. (43)

$$max \sum_{r=1}^{s} u_r y_{r1} \tag{43}$$

$$s.t.\ \sum_{i=1}^{m} v_i x_{i1} = 1$$

$$\sum_{r=1}^{s} u_r y_{rj} - \sum_{i=1}^{m} v_i x_{ij} \leq 0;\ j = 1,2,\ldots,n$$

$$u_r, v_i \geq \varepsilon;\ i = 1,2,\ \ldots,\ m, r = 1,2,\ \ldots,\ s$$

If the optimal solutions (u_r^{*1}, v_i^{*1}) are unique, the weights become common and, the model tries to allocate the DMU_1 in the first rank. Nevertheless, if the solution is not unique, and DMU_2 has the second rank, model Eqn. (44) will be determined.

$$max\ \frac{\sum_{r=1}^{s} u_r y_{r2}}{\sum_{i=1}^{m} v_i x_{i2}} \tag{44}$$

$$s.t.\ \frac{\sum_{r=1}^{s} u_r y_{r2}}{\sum_{i=1}^{m} v_i x_{i1}} = e^{*1} - \Delta_1$$

$$\frac{\sum_{r=1}^{s} u_r y_{r2}}{\sum_{i=1}^{m} v_i x_{i1}} \leq 1;\ j = 1,2,\ldots,n$$

$$u_r, v_i \geq \varepsilon;\ i = 1,2,\ \ldots,\ m, r = 1,2,\ \ldots,\ s$$

Moreover, the linear model is in the form of Eqn. (45).

$$max \sum_{r=1}^{s} u_r y_{r2} \tag{45}$$

$$s.t.\ \sum_{i=1}^{m} v_i x_{i2} = 1$$

$$\sum_{r=1}^{s} u_r y_{r1} = (e^{*1} - \Delta_1)\sum_{i=1}^{m} v_i x_{i1}$$

$$\sum_{r=1}^{s} u_r y_{rj} - \sum_{i=1}^{m} v_i x_{ij} \leq 0;\ j = 1,2,\ldots,n$$

$$u_r, v_i \geq \varepsilon;\ i = 1,2,\ \ldots,\ m, r = 1,2,\ \ldots,\ s$$

Where, e^{*1} is the optimal solution for the model Eqn. (43). If u_r^{*2} and v_i^{*2} resulted from the model Eqn. (45) are unique, the weights are common and, the model states that DMU_2 should at least, stand in the second rank. Otherwise, the solution is proceeded to attain the optimal solutions (u_r^{*1}, v_i^{*1}). In the last stage, the model Eqn. (46) is formed.

$$max \frac{\sum_{r=1}^{s} u_r y_{rn}}{\sum_{i=1}^{m} v_i x_{in}} \tag{46}$$

$$s.t. \frac{\sum_{r=1}^{s} u_r y_{rj}}{\sum_{i=1}^{m} v_i x_{ij}} e^{*j} - \Delta;\ j = 1,2,\ldots,n$$

$$\frac{\sum_{r=1}^{s} u_r y_{rn}}{\sum_{i=1}^{m} v_i x_{in}} \leq 1$$

$$u_r, v_i \geq \varepsilon;\ i = 1,2,\ \ldots,\ m, r = 1,2,\ \ldots,\ s$$

According to the DEA approach, Eqn. (46) can be written in the form of Eqn. (47)

$$max \sum_{r=1}^{s} u_r y_{rn} \tag{47}$$

$$s.t. \sum_{i=1}^{m} v_i x_{in} = 1$$

$$\sum_{r=1}^{s} u_r y_{rj} = (e^{*j} - \Delta_j)\sum_{i=1}^{m} v_i x_{ij};\ j = 1,2,\ldots,n$$

$$\sum_{r=1}^{s} u_r y_{rn} - \sum_{i=1}^{m} v_i x_{in} \leq 0$$

$$u_r, v_i \geq \varepsilon;\ i = 1,2,\ \ldots,\ m, r = 1,2,\ \ldots,\ s$$

Thus, the lexicography model proves the u_r^* and v_i^*, and includes the preferences of DM achieved from DMUs.

1.5.5 MINIMAX METHOD

Minimax method is known as a method for minimizing the deviations and is used for determining the minimum of maximum value and ranks it the highest [69]. On the other hand, it uses the partial rankings for optimal inference.

Efficiency scale is a function of performance deviations. Suppose d_j represents deviations of the efficiency of the DMU_j and E_j are deviations from the ideal efficiency of one ($d_j = 1 - E_j$). By minimizing d_j, the deviation from the efficiency for the DMU_j is equal to maximizing its efficiency E_j, Eqn. (48) is defined.

$$min\ d_p \tag{48}$$

$$s.t. \sum_{r=1}^{s} u_r y_{rj} - \sum_{i=1}^{m} v_i x_{ij} \leq 0;\ j = 1, 2, \ldots,\ n, j \neq p$$

$$\sum_{r=1}^{s} u_r y_{rp} - \sum_{i=1}^{m} v_i x_{ip} + d_p = 0$$

$$\sum_{i=1}^{m} v_i x_{ip} = 1$$

$$u_r, v_i \geq \varepsilon,\ d_p \geq 0;\ i = 1,2,\ \ldots,\ m, r = 1,2,\ \ldots,\ s$$

The objective functions of Eqn. (48) are special to a particular DMU. Also, in order to specify the efficiency of all units, it is necessary to write n models, each aimed at minimizing the deviation of efficiency for a specific DMU. Besides, models that take into account the technical efficiency scale are evaluated by the DMU, which gives the most considerable freedom in the selection of weights and declines the diagnostic power of the model. The most effective returns can be summarized as minimizing the maximum deviation of efficiency between all DMUs. In the following, three models for the relationship between MODM and DEA are examined.

The diagnosis between DMUs is obtained by changing the objective function of Eqn. (48), that the decision model Eqn. (49) is obtained.

$$\min \max_{1 \le j \le n} d_j \tag{49}$$

$$s.t. \sum_{r=1}^{s} u_r y_{rj} - \sum_{i=1}^{m} v_i x_{ij} + d_j = 0; j = 1, 2, \ldots, n$$

$$\sum_{i=1}^{m} v_i x_{ip} = 1$$

$$u_r, v_i \ge \varepsilon, d_j \ge 0; i = 1,2, \ldots, m, j = 1,2, \ldots, n, r = 1,2, \ldots, s$$

If it is assumed that M = *max* d_j, then the model Eqn. (50) is obtained.

$$min\ \mathrm{M} \tag{50}$$

$$s.t. \sum_{r=1}^{s} u_r y_{rj} - \sum_{i=1}^{m} v_i x_{ij} + d_j = 0; j = 1, 2, \ldots, n$$

$$\sum_{i=1}^{m} v_i x_{ip} = 1$$

$$\mathrm{M} \ge d_j; j = 1, 2, \ldots, n$$

$$u_r, v_i \ge \varepsilon, d_j \ge 0, \mathrm{M} \ge 0; i = 1, 2, \ldots, m, j = 1,2, \ldots, n, r = 1,2, \ldots, s$$

In model Eqn. (50), M is the maximum deviation of the efficiency, and $\mathrm{M} \ge d_j$ are units added to the model Eqn. (50) to ensure that the Minimax efficiency scale has a high detection capability; because the utility of all DMUs is considered simultaneously. Besides, since the Minimax efficiency scale is a target function that is not special to a single DMU; therefore, it does not require a solution to several models to determine the efficiency of all DMUs. Effectiveness is measurable for all DMUs by a single model. When the model Eqn. (50) was solved, efficiency for DMUs is specified by calculating $1 - d_j$ for j = 1,2, …, *n*.

This stage calculates the relative efficiency of DMUs by using the CSW, which contradicts the models of DEA in which different weights evaluate each DMU.

When the model Eqn. (21) leads to more than one efficient DMU, the model cannot determine the best DMU, in which case the use of the model Eqn. (51) is suggested.

$$min\,\mathrm{M} - k\sum_{j \in E_f} d_j \tag{51}$$

$$s.t. \sum_{r=1}^{s} u_r y_{rj} - \sum_{i=1}^{m} v_i x_{ij} + d_j = 0; j = 1, 2, \ldots, n$$

$$\sum_{i=1}^{m} v_i x_{ip} = 1$$

$$M \geq d_j; j = 1,2, \ldots, n$$

$$u_r, v_i \geq \varepsilon, d_j \geq 0, M \geq 0; i = 1, 2, \ldots, m, j = 1,2, \ldots, n, r = 1,2, \ldots, s$$

Here, E_f is the set of efficient DMUs by the model Eqn. (50) and are called efficient Minimax DMUs. $k \in [0,1]$ is the diagnostic parameter, which must be specified by the DM.

The goal of the efficiency scale in the proposed model Eqn. (51) is to maximize the total amount of deviations from the efficiency of the efficient DMUs.

1.5.6 LP-METRIC METHOD

Global criterion method or LP-metric method is a method that minimizes the relative deviation of targets from their optimal value, depending on the case, the sum of the first, second, and even infinite powers. This method can be minimized the distance between reference points and the feasible objective region [70]. On the other hand, the LP-metric method is solved the multiple objective problems with conflicting objectives [71].

When multiple outputs and only one input is assumed in the performance evaluation, the DEA model typically comes in the form of Eqn. (52) in the multiplier input-oriented model of the CCR.

$$max\ z = \frac{\sum_{r=1}^{s} u_r y_{rp}}{w x_p} \tag{52}$$

$$s.t.\ \frac{\sum_{r=1}^{s} u_r y_{rj}}{w x_j} \leq 1\ ; j = 1, 2, \ldots, n$$

$$u_r, w \geq \varepsilon; r = 1, 2, \ldots, s$$

In the nonlinear model Eqn. (52), w is related to the weight attributed to each input, although it can be turned into a linear model by changing the variable.

First of all, in order to specify the relative efficiency of DMUs in this model, n (number of DMUs) should be formulated and solved. Therefore, DMUs are not evaluated by CSW, which may lead to undesirable results, because management usually wants to analyze units based on CSW.

Second, the DEA proposes that the DMUs that obtain the efficiency equal to one are called efficient unit and remain on the efficient frontier, whereas DMUs that have an efficiency figure of less than one are called inefficient. In summary, DMUs are divided into efficient and inefficient. Also, when all the DMUs receive an efficiency figure of one, Eqn. (52) cannot distinguish between them. Model Eqn. (52) with low detection ability is a suitable model for cases where the DM has to specify the ideal DMU. In addition, for DMs, the latest model offers the flexibility to select the weights as it wishes in a manner that maximizes its efficiency. The permissibility of such flexibility in the selection of weights may make it possible for a DM to be efficient by giving too much weight to the indicator. The benchmark for doing that is an outstanding performance and giving minimal weight to those who have shown inefficiency. Such an overweight allocation is unusual and makes the DEA model poorly understood. To prevent unusual weight distribution and overcome the weak efficiency, the DEA diagnosis corrects the existing technical efficiency by placing it in the model of weight restrictions formulated according to the value judgments, which reduces the degree of objectivity of DEA.

The suggested method applies efficiency measures that are not special to a single DMU. It is noteworthy that the present model differs from the data coverage analysis with a common goal, which is not the efficiency, but the DMUs can be technically efficient in this manner. The new relationship is transformed by applying the proposed efficiency scale to a multiple decision model with improved diagnostic power. Therefore, the optimal solitary values of Eqn. (53) are determined.

$$max\ f_1 = \frac{\sum_{r=1}^{s} u_r y_{r1}}{x_1} \tag{53}$$

$$max\ f_2 = \frac{\sum_{r=1}^{s} u_r y_{r2}}{x_2}$$

$$\vdots$$

$$max\ f_n = \frac{\sum_{r=1}^{s} u_r y_{rn}}{x_n}$$

$$s.t.\ \frac{\sum_{r=1}^{s} u_r y_{rj}}{x_j} \le 1\ ;\ j = 1, 2, \ldots, n$$

$$u_r \geq \varepsilon; r = 1, 2, \ldots, s$$

And, the Eqn. (54) is obtained for f_l^*.

$$min_{1\leq p\leq\infty} \sum_{l=1}^{k} (\frac{f_l^* - f_l}{f_l^*})^p \tag{54}$$

$$s.t. \frac{\sum_{r=1}^{s} u_r y_{rj}}{x_j} \leq 1 ; j = 1,2,\ldots,n$$

$$u_r \geq \varepsilon; r = 1, 2, \ldots, s$$

In the LP-metric method, when $p = \infty$ is in the form of Eqn. (55).

$$min \sum_{l=1}^{k} (\frac{f_l^* - f_l}{f_l^*})^p \tag{55}$$

$$s.t. \frac{\sum_{r=1}^{s} u_r y_{rj}}{x_j} \leq 1 ; j = 1, 2,\ldots,n$$

$$u_r \geq \varepsilon; r = 1, 2, \ldots, s$$

Then,

$$min \{max \sum_{l=1}^{k} (\frac{f_l^* - f_l}{f_l^*})^p\} \tag{56}$$

$$s.t. \frac{\sum_{r=1}^{s} u_r y_{rj}}{x_j} \leq 1 ; j = 1, 2,\ldots,n$$

$$u_r \geq \varepsilon; r = 1, 2, \ldots, s$$

Therefore, by developing the objective function in Eqn. (56), can write Eqn. (57).

$$min \{max (\frac{f_1^* - f_1}{f_1^*}, \frac{f_2^* - f_2}{f_2^*}, \ldots, \frac{f_k^* - f_k}{f_k^*})\} \tag{57}$$

$$s.t. \frac{\sum_{r=1}^{s} u_r y_{rj}}{x_j} \leq 1 ; j = 1, 2,\ldots,n$$

$$u_r \geq \varepsilon; r = 1, 2, \ldots, s$$

If $max\ \{\frac{f_1^* - f_1}{f_1^*}, \frac{f_2^* - f_2}{f_2^*}, \ldots, \frac{f_k^* - f_k}{f_k^*}\} = \text{M}$, then the Eqn. (58) is determined.

$$min\ \text{M} \tag{58}$$

$$s.t.\ \text{M} \geq \frac{f_1^* - f_1}{f_1^*}$$

$$\text{M} \geq \frac{f_2^* - f_2}{f_2^*}$$

$$\vdots$$

$$\text{M} \geq \frac{f_k^* - f_k}{f_k^*}$$

$$\frac{\sum_{r=1}^{s} u_r y_{rj}}{x_j} \leq 1\ ; j = 1, 2, \ldots, n$$

$$u_r \geq \varepsilon;\ r = 1, 2, \ldots, s$$

1.6 RESULTS

In this section, three case studies on the automotive parts manufacturing industry will be provided to exhibit the applicability and performance of the proposed models.

- Case Study 1: Table 1.1 presents the data of gear manufacturing with materials A, B, C, and D. The purpose is to evaluate the materials based on the AHP-DEA method.

TABLE 1.1 Input and Output Values of Material Evaluation for Gear Manufacturing

Material	Inputs		Outputs	
	Cost of Component ($)	Density (kg/m³)	Hardness (HRB)	Fatigue Limit (ksi)
A	50	55	56	10
B	130	60	78	12
C	68	96	9	45
D	45	30	18	35

Initially, four models are formed to solve the problem:

$$max\, z_{AA} = 10u_1 + 56u_2$$

$$50v_1 + 55v_2 = 1$$

$$10u_1 + 56u_2 \le 1$$

$$12u_1 + 78u_2 - 130v_1 - 60v_2 \le 0$$

$$u_1 = 0, u_2 = \frac{0}{0179}, v_1 = \frac{0}{004}, v_2 = \frac{0}{0145}$$

$$e_{AA} = 1$$

And,

$$max\, z_{BB} = 12u_1 + 78u_2$$

$$130v_1 + 60v_2 = 1$$

$$12u_1 + 78u_2 \le 1$$

$$10u_1 + 56u_2 - 50v_1 - 55v_2 \le 0$$

$$u_1 = 0, u_2 = 0.0128, v_1 = 0.0029, v_2 = 0.0104$$

$$e_{BB} = 1$$

And,

$$max\, z_{BA} = 12u_1 + 78u_2$$

$$130v_1 + 60v_2 = 1$$

$$12u_1 + 78u_2 \le 1$$

$$10u_1 + 56u_2 - e_{AA}\left(50v_1 + 55v_2\right) = 0$$

$$u_1 = 0, u_2 = 0.0128, v_1 = 0.0029, v_2 = 0.0104$$

$$e_{BA} = 1$$

And,

$$max\, z_{AB} = 10u_1 + 56u_2$$

$$50v_1 + 55v_2 = 1$$

$$10u_1 + 56u_2 \le 1$$

$$12u_1 + 78u_2 - e_{BB}(130v_1 + 60v_2) \le 0$$

$$u_1 = 0, u_2 = 0.0179, v_1 = 0.004, v_2 = 0.0145$$

$$e_{AB} = 1$$

The following solution is obtained based on Eqn. (17):

$$3_{j,p} = \frac{1+1}{1+1} = 1$$

Also, the following matrix is created:

$$\begin{array}{c} \\ A \\ B \\ C \\ D \end{array} \begin{array}{c} \begin{array}{cccc} A & B & C & D \end{array} \\ \begin{pmatrix} 1 & 1 & 1 & 1 \\ 1 & 1 & 1 & 1 \\ 1 & 1 & 1 & 0.851 \\ 1 & 1 & 1.754 & 1 \end{pmatrix} \end{array}$$

Then, for the ranking using the AHP method, the values of the matrix are normalized:

$$\begin{array}{c} \\ A \\ B \\ C \\ D \end{array} \begin{array}{c} \begin{array}{cccc} A & B & C & D \end{array} \\ \begin{pmatrix} 0.250 & 0.250 & 0.210 & 0.260 \\ 0.250 & 0.250 & 0.210 & 0.260 \\ 0.250 & 0.250 & 0.211 & 0.220 \\ 0.250 & 0.250 & 0.369 & 0.260 \end{pmatrix} \end{array}$$

By calculating the geometric mean, the values for alternatives A, B, C, and D turn out to be 0.242, 0.242, 0.232, and 0.278, respectively. Therefore, the highest rank is assigned to the material D, and the final ranking is:

$$D > A = B > C$$

- **Case Study 2:** Table 1.2 presents the inputs and outputs for the material evaluation in engine block manufacturing. Four inputs and three outputs are considered ($\varepsilon = 0.00001$).

TABLE 1.2 Input and Output Data of the Material Evaluation

Material	Inputs				Outputs		
	Cost of Component ($)	Density (kg/m³)	Thermal Expansion Coefficient	Specific Heat Capacity (J/kg.C°)	Toughness (J/m³)	Tensile Strength (MPa)	Young's Modulus (MPa)
1	51.62	11.23	49.22	33.52	40.49	14.89	31.71
2	85.87	123.98	55.13	108.46	43.51	173.93	6.45
3	66.65	104.18	257.09	13.65	139.74	115.96	2.01
4	27.87	107.60	14.00	146.43	25.47	131.79	7.09
5	51.28	117.51	32.07	84.50	46.20	144.99	5.01
6	36.05	193.32	59.52	8.23	46.88	190.77	32.92
7	25.83	105.80	9.51	227.20	19.40	120.09	4.10

TABLE 1.2 *(Continued)*

Material	Inputs					Outputs	
	Cost of Component ($)	Density (kg/m³)	Thermal Expansion Coefficient	Specific Heat Capacity (J/kg.C°)	Toughness (J/m³)	Tensile Strength (MPa)	Young's Modulus (MPa)
8	123.02	82.44	87.35	98.80	43.33	125.84	40.69
9	61.95	99.77	33.00	86.37	45.43	79.60	12.62
10	80.33	104.65	53.30	79.06	27.28	132.49	42.67
11	205.92	183.49	144.16	59.66	14.09	196.29	16.15
12	82.09	104.94	46.51	127.28	44.87	108.53	9.02
13	202.21	187.74	149.39	93.65	44.97	184.77	2.04
14	67.55	82.83	44.37	60.85	26.04	85.00	23.95
15	72.60	132.73	44.67	173.48	5.55	135.65	24.13
16	84.83	104.28	159.12	171.11	11.53	110.22	39.09
17	71.77	88.16	69.19	123.14	44.83	74.54	6.14

The five models were solved on LINGO 18.0.44. By solving models, Table 1.3 presents the efficiency scores and ranking of materials.

TABLE 1.3 Efficiency Scores and Final Ranking (in Parentheses)

Material	Compromise Solution		Maximin	GP	Lexicography	Minmax
	($p = 1$)	($p = \infty$)				
1	1.0000(1)	1.0000(1)	1.0000(1)	1.0000(1)	0.9406(4)	1.0000(1)
2	1.0000(1)	1.0000(1)	1.0000(1)	1.0000(1)	1.0000(1)	0.6295(8)
3	0.7433(9)	1.0000(1)	1.0000(1)	1.0000(1)	0.9551(2)	0.5751(12)
4	0.8537(6)	1.0000(1)	1.0000(1)	1.0000(1)	0.9407(3)	0.6012(9)
5	0.9442(4)	1.0000(1)	1.0000(1)	1.0000(1)	0.9281(5)	0.5930(10)
6	0.8351(7)	0.9654(6)	0.9654(6)	0.9659(6)	0.7946(9)	0.7126(4)
7	0.7109(12)	0.8743(8)	0.8743(8)	0.8740(8)	0.8677(7)	0.6437(6)
8	0.9619(3)	0.8469(9)	0.8469(9)	0.8469(9)	0.8930(6)	0.5117(16)
9	0.6494(15)	0.6783(13)	0.6783(13)	0.6791(13)	0.6253(15)	0.5377(14)
10	0.8995(5)	0.8779(7)	0.8779(7)	0.8778(7)	0.8472(8)	0.5454(13)
11	0.7456(8)	0.6526(15)	0.6526(15)	0.6526(15)	0.6337(14)	0.7263(3)
12	0.7265(11)	0.7175(11)	0.7175(11)	0.7173(11)	0.7232(10)	0.5897(11)
13	0.6919(13)	0.6227(16)	0.6227(16)	0.6226(16)	0.6145(16)	0.7411(2)
14	0.7428(10)	0.7126(12)	0.7126(12)	0.7123(12)	0.6947(12)	0.4341(17)
15	0.6691(14)	0.7215(10)	0.7215(10)	0.7214(10)	0.6957(11)	0.6778(5)
16	0.5554(17)	0.6696(14)	0.6696(14)	0.6696(14)	0.6798(13)	0.6301(7)
17	0.5667(16)	0.5925(17)	0.5925(17)	0.5923(17)	0.6116(17)	0.5309(15)

- **Case Study 3:** Table 1.4 involves the results of materials evaluation in ball bearing manufacturing ($\varepsilon = 0.00001$).

TABLE 1.4 Data of the Materials Evaluation in Ball Bearing Manufacturing

Material	Input	Outputs			
	Cost ($/ton)	Corrosion Rate (mmpy)	Toughness (J/m³)	Specific Wear Rate (mm³/Nm)	Young's Modulus (GPa)
1	10,000	0.995	85	1.70	3
2	75,000	0.993	45	2.50	3.60
3	56,250	0.875	18	5	2.20
4	28,125	0.409	16	1.70	1.50
5	46,875	0.818	20	5	1.10
6	78,125	0.664	60	2.50	1.35
7	87,500	0.880	90	2	1.40
8	56,250	0.633	10	8	2.50
9	56,250	0.653	25	4	2.50
10	87,500	0.747	100	2	2.50
11	68,750	0.880	100	4	1.50
12	43,750	0.633	70	5	3

The LP-metric model was coded and solved in LINGO software. The results can be seen in Table 1.5, and materials 5 and 12 will be considered as efficient materials.

TABLE 1.5 Efficiency Scores and Ranking (in Parentheses)

Material	LP-Metric	
	($p = 1$)	($p = \infty$)
1	0.6534(9)	0.6178(9)
2	0.7534(6)	0.8102(5)
3	0.8831(4)	0.9517(3)
4	0.8626(5)	0.9419(4)
5	1.0000(1)	1.0000(1)
6	0.5639(12)	0.5029(12)
7	0.6836(8)	0.5772(10)
8	0.6315(10)	0.7581(7)
9	0.6873(7)	0.7627(6)
10	0.6173(11)	0.5413(11)
11	0.8900(3)	0.7452(8)
12	1.0000(1)	1.0000(1)

1.7 CONCLUSION

In the past few decades, humans have sought to produce various tools and equipment. Since more than 80,000 materials with different structures and features have been identified, material selection has been important. In this regard, engineers and designers were looking for the most appropriate materials. In this chapter, the material selection approach examines the relationship between MCDM and DEA as powerful tools for decision-making and performance evaluation. As mentioned, the main purpose of the combined use of these two methods is to use the strengths of each one to eliminate or reduce other weaknesses. Thus, considering the division of MCDM into two approaches of MODM and MADM, two sections examine the relationship between MODM-DEA and MADM-DEA.

The present study used different methods, including a compromise solution approach, Maximin, GP, Lexicographic, Minimax, LP-metric, and AHP. The problems are treated with case studies on the automotive parts manufacturing industry. These indicated that the proposed models had a high degree of accuracy and interpretation in material selection, which can be of great help to managers by helping them to select the best material. As a result, product failure, cost, and time are minimized, and products are cheap, light, and long-lasting.

KEYWORDS

- **data envelopment analysis**
- **decision-making**
- **decision-making unit**
- **material selection**
- **multiple attribute decision-making**
- **multiple criteria decision-making**
- **multiple objective decision-making**
- **performance evaluation**

REFERENCES

1. Ashby, M. F., (2017). *Materials Selection in Mechanical Design*. Oxford press.
2. Babanli, M. B., Prima, F., Vermaut, P., Demchenko, L. D., Titenko, A. N., Huseynov, S. S., Hajiyev, R. J., & Huseynov, V. M., (2018). Material selection methods: A review. In: *International Conference on Theory and Applications of Fuzzy Systems and Soft Computing* (pp. 929–936). Springer, Cham.
3. Chatterjee, P., Athawale, V. M., & Chakraborty, S., (2011). Materials selection using complex proportional assessment and evaluation of mixed data methods. *Materials & Design, 32*(2), 851–860.
4. Jahan, A., & Edwards, K. L., (2015). A state-of-the-art survey on the influence of normalization techniques in ranking: Improving the materials selection process in engineering design. *Materials & Design, 65*, 335–342.
5. Petkovic, D., Madic, M., Radovanovic, M., & Gecevska, V., (2017). Application of the performance selection index method for solving machining MCDM problems. *Facta Universitatis, Series: Mechanical Engineering, 15*(1), 97–106.
6. Alinezhad, A., (2017). A hybrid method of DEA and MODM in grey environment. *Journal of Industrial Strategic Management, 2*(1), 63–72.
7. Fang, L., & Li, H., (2015). Multi-criteria decision analysis for efficient location-allocation problem combining DEA and goal programming. *RAIRO-Operations Research, 49*(4), 753–772.
8. Jafarian-Moghaddam, A. R., & Ghoseiri, K., (2011). Fuzzy dynamic multi-objective data envelopment analysis model. *Expert Systems with Applications, 38*(1), 850–855.
9. Mousavi-Nasab, S. H., & Sotoudeh-Anvari, A., (2017). A comprehensive MCDM-based approach using TOPSIS, COPRAS and DEA as an auxiliary tool for material selection problems. *Materials & Design, 121*, 237–253.
10. Thakker, A., Jarvis, J., Buggy, M., & Sahed, A., (2008). A novel approach to materials selection strategy case study: Wave energy extraction impulse turbine blade. *Materials & Design, 29*(10), 1973–1980.
11. Jiao, Q., Lan, Y., Guan, Z., & Li, Z., (2011). A new material selection approach using PROMETHEE method. In: *Proceedings of 2011 International Conference on Electronic & Mechanical Engineering and Information Technology* (Vol. 6, pp. 2950–2954). IEEE.
12. Jahan, A., (2012). Material selection in biomedical applications: Comparing the comprehensive VIKOR and goal programming models. *International Journal of Materials and Structural Integrity, 6*(2–4), 230–240.
13. Petkovic, D., Madic, M., Radovanovic, M., & Jankovic, P., (2015). Application of recently developed MCDM methods for materials selection. In: *Applied Mechanics and Materials* (Vol. 809, pp. 1468–1473). Trans Tech Publications Ltd.
14. Marini, C. D., Fatchurrohman, N., Azhari, A., & Suraya, S., (2016). Product development using QFD, MCDM and the combination of these two methods. In: *IOP Conference Series: Materials Science and Engineering* (Vol. 114, No. 1, pp. 1–8). IOP Publishing.
15. Chaitanya, K. L., & Srinivas, K., (2019). Sensitive analysis on selection of piston material using MADM techniques. *Strojnícky Casopis-Journal of Mechanical Engineering, 69*(4), 45–56.
16. Patnaik, P. K., Mishra, S. K., & Ashish, A. T., (2020). Ranking of fiber reinforced composite materials using PSI and PROMETHEE method. In: *2020 International Conference on Computer Science, Engineering and Applications (ICCSEA)* (pp. 1–5). IEEE.

17. Shanian, A., & Savadogo, O., (2006). A non-compensatory compromised solution for material selection of bipolar plates for polymer electrolyte membrane fuel cell (PEMFC) using ELECTRE iv. *Electrochimica Acta, 51*(25), 5307–5315.
18. Milani, A. S., Shanian, A., & Abeyaratne, R. C., (2007). A group decision making approach in multi-criteria material selection. In: *Proceedings of the 18th Conference on Proceedings of the 18th IASTED International Conference: Modeling and Simulation* (pp. 550–555). ACTA Press.
19. Athawale, V. M., Kumar, R., & Chakraborty, S., (2011). Decision making for material selection using the UTA method. *The International Journal of Advanced Manufacturing Technology, 57*(1–4), 11–22.
20. Chatterjee, P., & Chakraborty, S., (2012). Material selection using preferential ranking methods. *Materials & Design, 35*, 384–393.
21. Yazdani, M., & Payam, A. F., (2015). A comparative study on material selection of micro-electromechanical systems electrostatic actuators using Ashby, VIKOR and TOPSIS. *Materials & Design, 65*, 328–334.
22. Yazdani, M., Zavadskas, E. K., Ignatius, J., & Abad, M. D., (2016). Sensitivity analysis in MADM methods: Application of material selection. *Engineering Economics, 27*(4), 382–391.
23. Zindani, D., & Kumar, K., (2018). Material selection for turbine seal strips using PROMETHEE-GAIA method. *Materials Today, 5*(9), 17533–17539.
24. Kinoshita, Y., Yamada, T., Gupta, S. M., Ishigaki, A., & Inoue, M., (2020). Decision support model of environmentally friendly and economical material strategy for life cycle cost and recyclable weight. *International Journal of Production Economics, 224*, 1–50.
25. Zarandi, M. H. F., Mansour, S., Hosseinijou, S. A., & Avazbeigi, M., (2011). A material selection methodology and expert system for sustainable product design. *The International Journal of Advanced Manufacturing Technology, 57*(9–12), 885–903.
26. Biron, M., (2013). *Thermosets and Composites: Material Selection, Applications, Manufacturing and Cost Analysis*. Elsevier.
27. Pfeifer, M., (2009). *Materials Enabled Designs: The Materials Engineering Perspective to Product Design and Manufacturing*. Butterworth-Heinemann.
28. Farag, M. M., (1997). *Materials Selection for Engineering Design.* Prentice Hall.
29. Kutz, M., (2002). *Handbook of Materials Selection*. John Wiley & Sons.
30. Farag, M. M., (2014). Quantitative methods of materials selection. In: *Mechanical Engineers' Handbook* (pp. 1–22). John Wiley & Sons.
31. Jahan, A., Ismail, M. Y., Sapuan, S. M., & Mustapha, F., (2010). Material screening and choosing methods–a review. *Materials & Design, 31*(2), 696–705.
32. Ramalhete, P. S., Senos, A. M. R., & Aguiar, C., (2010). Digital tools for material selection in product design. In: *Materials & Design* (Vol. 31, No. 5, pp. 2275–2287).
33. Mouritz, A. P., (2012). *Introduction to Aerospace Materials*. Elsevier.
34. Hwang, C. L., & Masud, A. S. M., (2012). *Multiple Objective Decision Making—Methods and Applications: A State-of-the-Art Survey* (Vol. 164). Springer Science & Business Media.
35. Lu, J., & Ruan, D., (2007). *Multi-Objective Group Decision Making: Methods, Software and Applications with Fuzzy Set Techniques* (Vol. 6). Imperial College Press.
36. Mosadeghi, R., Warnken, J., Tomlinson, R., & Mirfenderesk, H., (2013). Uncertainty analysis in the application of multi-criteria decision-making methods in Australian strategic environmental decisions. *Journal of Environmental Planning and Management, 56*(8), 1097–1124.

37. Insua, D. R., (1990). *Sensitivity Analysis in Multi-objective Decision Making*. Springer, Berlin, Heidelberg.
38. Kuhn, H. W., & Tucker, A. W., (2014). Nonlinear programming. In: *Traces and Emergence of Nonlinear Programming* (pp. 247–258). Birkhauser, Basel.
39. Chankong, V., & Haimes, Y. Y., (2008). *Multiobjective Decision Making: Theory and Methodology*. Courier Dover Publications.
40. Alinezhad, A., & Khalili, J., (2019). *New Methods and Applications in Multiple Attribute Decision Making (MADM)*. Springer, Cham.
41. Ray, S. C., (2004). *Data Envelopment Analysis: Theory and Techniques for Economics and Operations Research*. Cambridge university press.
42. Cooper, W. W., Seiford, L. M., & Zhu, J., (2011). Data envelopment analysis: History, models, and interpretations. In: *Handbook on Data Envelopment Analysis* (pp. 1–39). Springer, Boston, MA.
43. Banker, R. D., & Thrall, R. M., (1992). Estimation of returns to scale using data envelopment analysis. *European Journal of Operational Research, 62*(1), 74–84.
44. Ji, Y. B., & Lee, C., (2010). Data envelopment analysis. *The Stata Journal, 10*(2), 267–280.
45. Coelli, T. J., Rao, D. S. P., O'Donnell, C. J., & Battese, G. E., (2005). *An Introduction to Efficiency and Productivity Analysis*. Springer Science & Business Media.
46. Charnes, A., Cooper, W. W., & Rhodes, E., (1978). Measuring the efficiency of decision-making units. *European Journal of Operational Research, 2*(6), 429–444.
47. Banker, R. D., Charnes, A., & Cooper, W. W., (1984). Some models for estimating technical and scale inefficiencies in data envelopment analysis. *Management Science, 30*(9), 1078–1092.
48. Saaty, T. L., (2008). Relative measurement and its generalization in decision-making why pair wise comparisons are central in mathematics for the measurement of intangible factors the analytic hierarchy/network process. *RACSAM-Revista de la Real Academia de Ciencias Exactas, Fisicas y Naturales. Serie A. Matematicas, 102*(2), 251–318.
49. Chai, J., & Ngai, E. W., (2020). Decision-making techniques in supplier selection: Recent accomplishments and what lies ahead. *Expert Systems with Applications, 140*, 1–16.
50. Leal, J. E., (2020). AHP-express: A simplified version of the analytical hierarchy process method. *Methods X, 7*, 1–23.
51. Unel, F. B., & Yalpir, S., (2019). Valuations of building plots using the AHP method. *International Journal of Strategic Property Management, 23*(3), 197–212.
52. Alinezad, A., Seif, A., & Esfandiari, N., (2013). Supplier evaluation and selection with QFD and FAHP in a pharmaceutical company. *The International Journal of Advanced Manufacturing Technology, 68*(1–4), 355–364.
53. Amini, A., & Alinezhad, A., (2016). Ranking 7Ms effective factors in Iranian production systems using fuzzy AHP. *Journal of Modern Processes in Manufacturing and Production, 5*(3), 69–82.
54. Charnes, A., Cooper, W. W., Lewin, A. Y., & Seiford, L. M., (2013). *Data Envelopment Analysis: Theory, Methodology, and Applications*. Springer Science & Business Media.
55. Tofallis, C., (1997). Input efficiency profiling: An application to airlines. *Computers & Operations Research, 24*(3), 253–258.
56. Kornbluth, J. S. H., (1991). Analyzing policy effectiveness using cone restricted data envelopment analysis. *Journal of the Operational Research Society, 42*(12), 1097–1104.
57. Kao, C., & Hung, H. T., (2005). Data envelopment analysis with common weights: The compromise solution approach. *Journal of the Operational Research Society, 56*(10), 1196–1203.

58. Zeleny, M., (1973). *Compromise Programming, Multiple Criteria Decision-Making* (pp. 263–301). University of South Carolina Press, Columbia.
59. Yu, P. L., (1973). A class of solutions for group decision problems. *Management Science, 19*(8), 936–946.
60. Yazdani, M., Zarate, P., Zavadskas, E. K., & Turskis, Z., (2019). A combined compromise solution (CoCoSo) method for multi-criteria decision-making problems. *Management Decision, 57*(9), 2501–2519.
61. Opricovic, S., & Tzeng, G. H., (2004). Compromise solution by MCDM methods: A comparative analysis of VIKOR and TOPSIS. *European Journal of Operational Research, 156*(2), 445–455.
62. Jahanshahloo, G. R., Memariani, A., Lotfi, F. H., & Rezai, H. Z., (2005). A note on some of DEA models and finding efficiency and complete ranking using a common set of weights. *Applied Mathematics and Computation, 166*(2), 265–281.
63. Chakraborty, T., (2018). A MCDM-NBO approach for selection of installation location for wave energy power plants. In: *Application of Geographical Information Systems and Soft Computation Techniques in Water and Water-Based Renewable Energy Problems* (pp. 121–140). Springer, Singapore.
64. Schniederjans, M., (2012). *Goal Programming: Methodology and Applications: Methodology and Applications*. Springer Science & Business Media.
65. Romero, C., (2014). *Handbook of Critical Issues in Goal Programming*. Elsevier.
66. Aouni, B., Colapinto, C., & La Torre, D., (2014). Financial portfolio management through the goal programming model: Current state-of-the-art. *European Journal of Operational Research, 234*(2), 536–545.
67. Makui, A., Alinezhad, A., Kiani, M. R., & Zohrehbandian, M., (2008). A goal programming method for finding common weights in DEA with an improved discriminating power for efficiency. *Journal of Industrial and Systems Engineering, 1*(4), 293–303.
68. Francis, M. C., & Jacob, D., (2020). The lexicographic method for the threshold cover problem. In: *Conference on Algorithms and Discrete Applied Mathematics* (pp. 41–52). Springer, Cham.
69. Alinezhad, A., & Amini, M., (2014). Practical common weight maximin approach for technology selection. *Journal of Fundamental and Applied Sciences, 6*(1), 31–47.
70. Behnamian, J., Ghomi, S. F., & Zandieh, M., (2014). Realistic variant of just-in-time flowshop scheduling: Integration of Lp-metric method in PSO-like algorithm. *The International Journal of Advanced Manufacturing Technology, 75*(9–12), 1787–1797.
71. Asadi-Gangraj, E., & Nayeri, S., (2018). A hybrid approach based on LP metric method and genetic algorithm for the vehicle-routing problem with time windows, driver-specific times, and vehicles-specific capacities. *International Journal of Operations Research and Information Systems (IJORIS), 9*(4), 51–67.
72. Charnes, A., & Cooper, W. W., (1961). *Management Models and Industrial Applications of Linear Programming*. John Wiley, New York.

CHAPTER 2

IDENTIFICATION, ASSESSMENT, AND EVALUATION OF RISK IN VARIOUS MANUFACTURING INDUSTRIES

RAKHI TRIVEDI[1], MINAKSHI SEHRAWAT[2], BHUPENDRA SHARMA[3], and KULDEEP K. SAXENA[4]

[1]*Associate Professor, IILM College of Engineering and Technology, Greater Noida, Uttar Pradesh, India*

[2]*Research Scholar, Sardar Patel University (SPU), Gujarat, India*

[3]*Research Organization of Science and Technology, Ritsumeikan University, Kusatsu, Japan*

[4]*GLA University, Mathura, Uttar Pradesh, India*

ABSTRACT

Major activities in the manufacturing industries involve the handling of risks of multiple forms, which can potentially harm people, the environment, and businesses. Industries put energy and investment into identifying and understanding the various types of risks that arise in varying settings. No doubt that there shall be proper management of risk for obligatory reasons, for moral and legal reasons, and to maintain a company's reputation along with the 'licensee to operate.' It involves ensuring that potential threats are identified well and dealt with in an optimistic way by prioritizing actions for the decrease and management of hazards while optimizing the use of all the assets. A significant focal point of danger for the executives is on counteraction of episodes by the presentation of fitting layers of preventive control

Decision-Making Models and Applications in Manufacturing Environments. Pushpdant Jain, Kumar Abhishek, & Prasenjit Chatterjee (Eds.)

measures (obstructions). An auxiliary spotlight has been on the arrangement of different imperatives to limit ramifications for an occasion to happen. This chapter focuses on identifying factors associated with risk, then assessing and analyzing them with respect to various industries.

2.1 INTRODUCTION

The growing number of workforce challenges, process dynamics, and disruptive technology-associated challenges led to rapid changes in the shop floor level's operational realities. Further, according to Abele & Reinhart [1], in manufacturing units, globalization, resource shortage, product life cycle rearrangement, and increased customer expectations have caused companies to re-look at their production planning and control strategies [2]. Technology has made them realize that the routine nature of operations would not sustain for longer. Meanwhile, with the right facts, figures, and calculations – decision-making can be smoothened, and hopes can also be pinned for necessary action in the foreseeable future [3]. If not done, various risks, including control measures. Hazard identification, risk assessment, and control measures (HIRAC), etc. [4] can only add to the risk. Borrowing the lesson learned from previous models on risk identification and assessment, this is the right time to find answers to the questions like – doing a SWOT (strengths, weaknesses, opportunities, and threats) analysis [5]. What are the overall risks involved? What is the current position of technological up-gradation in the company and making realistic assumptions on – how to deal with the potential threats on the level of – shop floor, inventory management, hazardous handling capacity, etc.? However, meeting market demands, exploring new possibilities, sustainable waste management [6] exercises, understanding the full ecosystem, and applying risk management practices and strategies to your key business partners, customers, and suppliers [7].

Developing a plan and bringing an immediate treasury solution is the need of the hour for overall business risk and its continuity plans. The forte of 'risk assessment and control exercise is to:

- Ensuring the viability and operational capacity of your business.
- Timely assessing the risk factors and securing your position.
- Revisit and reduce the varied threats involved in your business for competitive advantage.

- Restructuring the investment plans – whether to continue or stop or pause any capital investment; also, for how long.
- Planning differently for supplier-specific and customer-specific payment performances and identifying the companies who have taken a shift from their traditional payment practices.
- Knowing the pivot alternative business options and also calculating the viability plus risks involved.
- Considering the availability of the extended options of the solutions when the regular streams failed to work.

Thinking out of the box and looking for experts to show you the right path. However, the contemporary scenario requires more flexibility and innovative up-streaming and down-streaming of your operations [8]. To put simply, risk identification of production systems and potential evaluation schemes needs to be an integral part of your overall business risk assessment and action planning for the future [9]. This study was created to add to an increasing body of information that can help reduce risk. Evaluations, risk control, and, finally, position manufacturing firms to achieve greater success [10]. 'The 'Manufacturers are well conscious of these issues, according to the results [11]. The current and potential threats that their businesses face, and have the chance to fine-tune their tactics in order to answer what's up:

- What is the current state of risk?
- What specific risks should manufacturers be aware of?
- Is risk ownership in accordance with the organization's requirements?
- Is it possible to foresee tomorrow's top risks using today's risk management techniques?

In addition to considering these key questions, the study also contemplates the environmental factors manufacturers face, and how those factors influence how they react to take a chance on customer tastes are evolving, for example, and new products are being introduced [12]. Technology's products and software will make a big difference in a short amount of time. Current goods, manufacturing processes, or even whole manufacturing facilities business models are no longer relevant [13]. As a result, executives are quickening their rate of innovation and execution. Risk practitioners face major obstacles as the speed of technological advancements accelerates [14]. The use of computational methods and predictive modeling capabilities makes it possible. Manufacturers have to gain a deeper understanding and guidance from huge quantities of data [15].

2.2 PRODUCTION SYSTEM RISK MANAGEMENT

Operational risk management is concerned with risks that arise directly from a company's processes [16]. The operational risk control in production companies includes the production planning time horizon and the manufacturing process, and product risks are all under control [17, 18].

2.2.1 DESCRIPTION OF THE TERM RISK

First and foremost, the definition of risk must be defined, as it is the foundation for understanding risk management [19]. The word risk can be described in a number of ways. However, for the purposes of this paper, Steinmetz's description is deemed appropriate: Risks are described as "*all negative deviations in objectives that (...) have their source or impact in the environment.*"

2.2.2 PROCESS

Risk management is a continuous management task that can be defined as a loop of control that repeats itself [20]. Risk recognition, risk evaluation, risk treatment, and risk control are the four sub-processes that make up this process [21]. Risk management is not as popular in technological frameworks as it is in project management or other areas of business. Financing or insurance? For safety-critical systems, on the other hand (e.g., nuclear power plants and chemical plants), sophisticated systems (in industry, for example) are available. For procedures like these critical stages [22]. The two most important methods are risk detection and evaluation.

This is also true in the manufacturing area, which is why sections 2.3 and 2.4 go through these two steps in depth [23]. Following risk analysis and risk evaluation, it is determined the threats should be addressed and which should not, as long as the potential harm does not outweigh the effort required to implement countermeasures. Risk monitoring is the final step in the risk management process [24].

Risk monitoring, which has a controlling, control, and knowledge feature, is the final step in the risk management process [25]. The effectiveness of prescribed measures for early detection of hazards, the availability of necessary data, and the observance of defined quantitative or qualitative thresholds or limits are all discussed in this paper [26, 27].

2.2.3 *IDENTIFICATION OF RISKS*

An initial risk assessment is required to begin the risk management process. An evaluation of the current risks is required [28]. The foundation for the solution is understanding of the initial situation. Improvement of the risk situation in a cost-effective manner [29]. Many business decisions, including those in the manufacturing field, are put in jeopardy because their full extent can only be revealed later. This is particularly true for high-risk processes that are interdependent on other (sub)processes [30]. If threats are detected early on, they can be mitigated by taking appropriate steps and being constructive (rather than reactive). Risk management is now feasible (which is referred to as disturbance management) [31]. As a consequence, current and potential individual risks can be defined in a systematic and comprehensive manner known as risk identification, which involves several steps [32]. To begin, the subject of consideration (e.g., product, (sub)process) must be defined. Following that, the available data must be gathered. The most difficult aspect of the research process is collecting knowledge. Employees' implicit knowledge must be supplemented by their recorded, explicit knowledge. In addition to known threats, previously unknown risks may be detected, for example, using innovative methods. After that, a thorough analysis will begin [33].

2.3 CONCEPTUAL ELABORATION

Danger appraisals contemplate help recognize how development organizations survey and respond to risks today and how they plan for the impending time. A leader from the different inside survey and peril the executives' sheets addressed inquiries about their critical danger assessment practice, the danger of prominent business and IT is looked by them. Additionally, they oversaw crossing point of peril the board with arranged dangers [34]. The examination results showed that numerous inward review and danger authorities face a critical need to build up their abilities. These components require a more scientific, deft, and indicative way to deal with danger to successfully demonstrate the unpredictability and speed of critical dangers and business disturbances [35].

Advancement should zero in on better utilization of innovation, changes in the recurrence of danger appraisal cycles, and danger the board rehearses at all association levels [36]. So, hazard appraisal and danger the board procedures should progress at or quicker than the fundamental business on the off chance that they meet your business objectives later on [37].

The current danger executive's techniques will, in general, zero in on dangers (as opposed to circumstances) and mechanical danger (instead of the hierarchical attributes of the issue) [38]. A framework displaying technique was utilized to build up a reasonable model for prophylactic danger to the executives in these circumstances. Therefore, they have restricted materialness to the administration of vital dangers at the authoritative or area level [39]. This report depicts how hazard the executives can be stretched out to chances other than specialized ones, incorporate open doors, close by dangers, and adjust to vital issues [40, 41]. The subsequent model is approximate in nature; however a beginning stage towards a hypothesis for the vital danger the board is. It additionally has suggestions for experts, exhibited by applying the philosophy to a contextual investigation. The article contributes severally [42]. To start with, instead of simply zeroing in on dangers, distinguish the particular chances and components by which they could be caught. The report makes another commitment by building up a hypothetical model for the crossing point of danger the board with procedure definition, in 'essential danger the executives.' A third commitment is standing out. The report explicitly addresses the dangers looked at by an assembling association [43]. The fourth commitment is the arrangement of a hypothesis as well as a coordinated arrangement of techniques for use by specialists [44]. These techniques are not prescriptive but rather offer recommendations that can be adjusted to different circumstances. A recommended project plan for execution is additionally given. Something else, vital assembling the executives is fairly impromptu and inadequately coordinated with hazard the board, and the model gives a methodology that could assist associations with exploring the disturbance in the worldwide assembling industry [45]. While the procedure has been delineated by applying it to key dangers in assembling associations that influence the drawn-out monetary practicality of the association hazards identified with hierarchical methodology are currently viewed as fundamental and gotten progressively significant later on. At the point when got some information about organized future IT and business dangers, development, and digital protection hazards beat out everyone else, individually (Figure 2.1) [46]

Addressing planned risks requires the manufacturer to measure whether hazard measurements are agreed out in a way that profits the manufacturer on the whole [47]. That measurement should raise the question of:

- Whether a risk assessment should be performed more often to identify rising risk.
- Is risk examined on an ongoing basis or only in regular, official presentation?

- And what methods may be needed beyond the conventional discussion and survey methods?

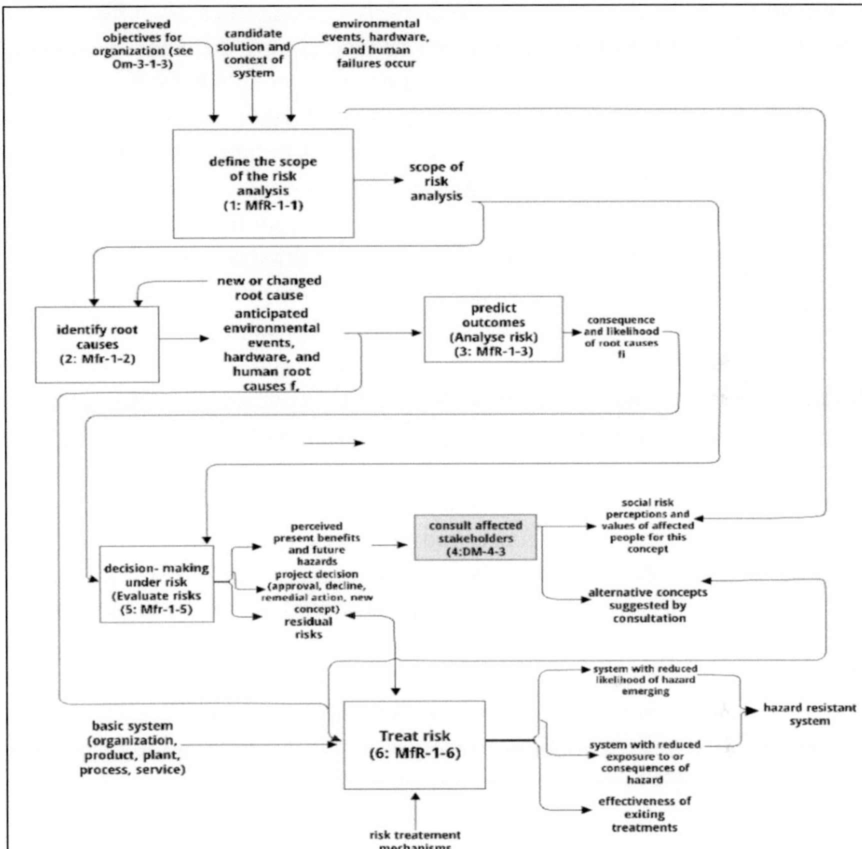

Note:
(1) This set to activities involves anticipating the risk before they arise. This is conventional 'risk management' (RM), as described for example in standard AS/NZ 4360 (2004). It is a structured approach that involves first identifying the potential root causes of a failure. Then the risk is determined. In this context 'risk' has a specific meaning: it is deemed to have likelihood and consequence componets, and these are independent of each other. In other contexts 'risk' can have different meaning including likelihood or hazard event. so some caution is simply the product of the two. For qualitative variables the risk have to be shown graphically. Either way, the more serious risks may be identified and treated. The current model subsumes the activities in the risk mangement standard AS/NZ 4360 which are 'establish the context', identify risk', 'analyse risk' for consequence and likelihood, 'evaluate risk', treat risk' communicate and consult' and monitor and review',

FIGURE 2.1 Risk management process and activities.

Critical strategic concern impact on the speed and innovation, increasing force on manufacturers to achieve predictable profits on investment [48]. Product innovation can quickly make existing products obsolete. The way products are developed, produced, and marketed is the ability to innovate to provide the innovator with tremendous value while facing substantial losses without competition.

Mechanical advancement makes the assembling plan of action more proficient consistently and can offer it. It additionally features the essential significance of R&D; picking the correct degree of venture and viably estimating the profits on those speculations can significantly affect future conditions [49]. Moreover, the essential danger, among different advantages, mechanical improvement empowers organizations to all the more adequately deal with the extension of worldwide stockpile chains and change creation intends to meet changing economic situations. More prominent dependence on innovation additionally implies that specialized dangers can transform into key dangers for producers. Inner review can assume a basic job in giving an autonomous evaluation of cycles and controls identified with development and R&D choices. Measures and measurements are utilized to decide the viability of speculations and screen progress, time, and financial plan (Figure 2.2) [50].

As per the National Science Foundation, makers (barring drug organizations) spent more than $159 million on R&D in 2012, a figure that speaks to 54% of all R&D in the United States [51]. It was about 2.8% for non-family-creating organizations, contrasted with about 3.6% for producers' income. To endure and flourish in a particularly changing danger scene, an organization's danger evaluation approach and practices should line up with those changes. Overall, the assembling business is an innovator in innovative work (R&D) and development in each area in the United States. Besides, roughly 82% of this spending was reflected in the noteworthy degree of fortifying of the US producing industry made through innovative work. Inside inspectors ought to consider making development-related activities in yearly review intends to carry more astounding an incentive to the association by zeroing in on huge dangers to the cycles in question [52].

2.4 EVALUATION OF RISK FACTORS

2.4.1 DRIVERS OF COMPETITIVENESS

They understand that what can differentiate a company competitively, today and in the future, is vitally essential for risk management [53]. Product innovation, driven by changes in customer preferences, technology, or other factors, means that incremental improvements to an existing product may not be enough to address changes in demand. Maintaining a competitive advantage can be costly, and achieving an acceptable ROI can be challenging [54]. For example, 3D printers promise to help companies revolutionize prototyping and perhaps even how production occurs. However, seizing this opportunity will require substantial investments in R&D, where effective governance and sound risk assessment practices are necessary to achieve an acceptable ROI.

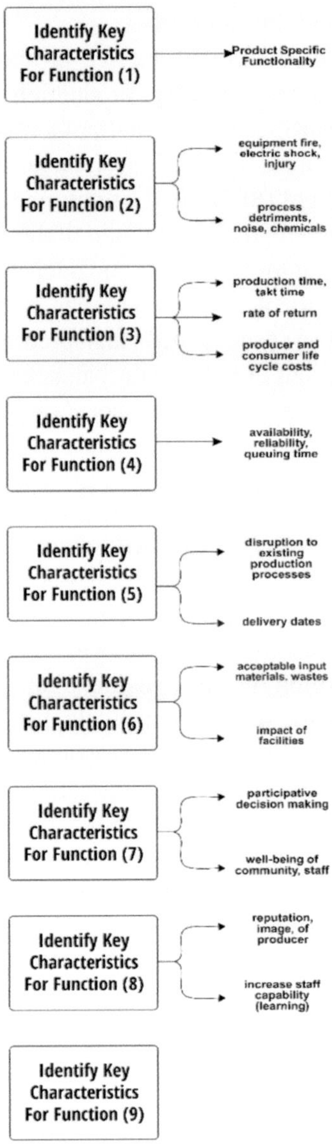

Notes

(1) These key characteristics will vary for different projects, and only a sample are shown here for a generic project, they are based on the key elements of SAA/SNZ HB436 (2004).

(2) The key characteristics form the basis for subsequent evaluation of the risk, i.e. they become dimensions in the risk management decisions, thus they appear in the downstream risk matrix (seeMFR-1-3-6-2).

(3) Risk management is often a political process, where various stakeholders seek to obtain power to influence the outcomes to their own well-being- one of the ways they do this is to define the key characteristics for the decision or at least to influence the definition. they may also seek to redefine those key characteristics deeper into the process if they are dissatisfied with the emerging outcomes. they will seek to have other or modified dimensions included in the decision. To reduce later dispute, it is advantageous to define the key characteristics fully, early, and inclusive of all stakeholders. Doing so, may prevent the downstream process from loosing direction due to late arguments about the key characteristics that will be included.

FIGURE 2.2 Key characteristics of risk evaluation.

2.4.2 COMPLEXITIES OF SUPPLY CHAIN MANUFACTURERS

Supply chains are very complex and are continually exposed to various risks emanating from within and outside their value chains. There are macroeconomic risks around geopolitical pressures, regulatory requirements, environmental/social responsibilities, and challenges in emerging markets. There are also extended value chain risks related to-external service providers, operational risks related to development planning, sourcing, production, and distribution. The supply chain must also consider operational risks associated with financial investments, human resources, and IT. These factors add to the complexity of supply chain management (SCM) and, if not managed carefully, can negatively affect sales and brand reputation. In light of the risks of operating a complex supply chain, manufacturers should consider incorporating resilience into the supply chain. Resilient supply chains can address critical vulnerabilities proactively, with a more specific focus than predicting and preparing for each type of risk. A resilient supply chain balances risk and costs to prevent or quickly recover from many simultaneous, dynamic risk-related disruptions [55].

2.4.3 THE GROWING MANUFACTURING SKILLS GAP

Studies reveal that the talent problem is growing: Over the next decade, nearly three and a ½ million manufacturing jobs will likely need to be filled, and the skills gap is expected to leave 2 million of those positions unfilled. CEOs and manufacturing executives around the world identify talent-driven innovation as a driver of competitiveness [56]. It is for this reason that the implications of such a talent shortage are significant and may affect the growth and profitability of manufacturers. For example, 82% of the executives surveyed indicate that they believe the skills gap will affect their ability to meet customer demand, and 78% think it will affect their ability to implement new technologies and increase productivity. Additionally, executives indicate that the skills gap affects the ability to provide adequate customer service (69%), the ability to innovate and develop new products (62%), and the ability to expand internationally (48%). Around 80% of manufacturing executives reported they are willing to pay more than market rates in areas of the workforce that are reeling from a talent crisis (Figure 2.3) [57].

Still, six out of 10 positions remain unfilled due to a talent shortage. It indicates that there are not enough workers in the manufacturing industry to fill these positions. Additionally, executives reported that it takes an average of 94 days to hire employees in engineer/researcher/scientist and 70 days to employ skilled production workers. Against these numbers, it is not surprising why manufacturers report that the most significant business impact of the talent shortage is their ability to meet customer demand. To address the skills gap, manufacturers not only have to find workers with the necessary skills to meet the advanced manufacturing requirements of today and tomorrow, but they must also develop and engage their existing workforce. Creating a supply of workers with manufacturing skills (engineering, skilled trades, and production) will be critical to the future competitiveness of manufacturing companies and the industry. An essential component in addressing the talent crisis is designing strategies that optimize acquisition and development (Figure 2.4) [58].

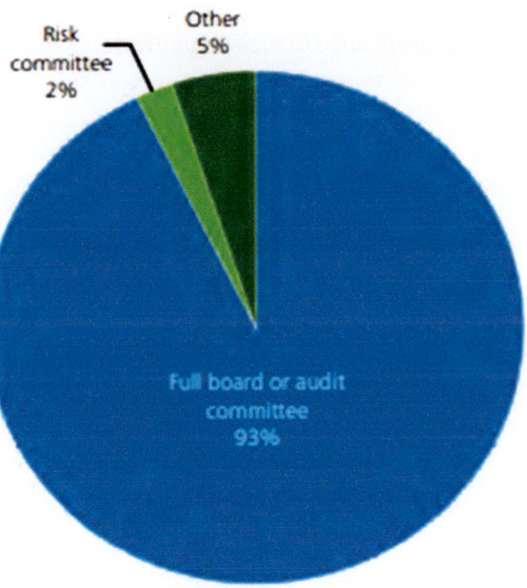

FIGURE 2.3 Percent of respondents indicating where risk oversight of company's risk management activities.

2.5 EFFECTS AND DISCUSSION

2.5.1 EFFECT OF COMPETITIVENESS

Understanding what can separate an organization seriously—today and later on—is basically significant for the risk of the executives. Item development, prodded by changing client inclinations, innovation, or different variables, implies that gradual enhancements to a current item may not be adequate to address changes sought after [59]. Keeping up the upper hand can be expensive, and acknowledging adequate ROI may end up being testing. For instance, 3D printers guarantee to assist organizations with reforming how prototyping and maybe even how creation happens. Be that as it may, catching this open door will require generous interests in research and development, where effective organization and fitting risk assessment practices will also be called to recognize commendable ROI. An ongoing Deloitte Review article—"Cracking the Genetic Code of High-Performing Manufacturers,"—inspects the apparent significance of current and future seriousness abilities, and how those capacities separate high-performing makers from the rest. The discoveries were inferred by heads rating both their organization's present seriousness in every ability comparative with its nearest worldwide opponents and every capacity's significance to their organization's intensity later on [60].

FIGURE 2.4 The percentage chart of respondents indicating the person bears the primary responsibility of ERM in the organization.

Alongside the capacities plotted on the chart, those wherein unrivaled specialists stand isolated from the pack and in which they presumably will continue driving are "particularly favorable position" limits. In the present significance scale, superior workers are fundamentally in a way that is better than their partners today on game-evolving abilities, and alongside the upcoming significance extent, superior workers place significantly more influence on distinct advantage capacities than different organizations in the examination. Outstandingly, top dangers recognized based on reactions to the few review studies line up with superior worker game-evolving abilities. For instance, the development and ability of the executives are in the midst of the game-changing capacities that were additionally positioned as need business chances by respondents to this assessment, showing driving practice in these regions might separate an organization. In addition, hazard the board and information examination—regions where interior review and danger chiefs are making huge speculations—were additionally delegated game-changing capacities that set elite producers apart (Figure 2.5) [61].

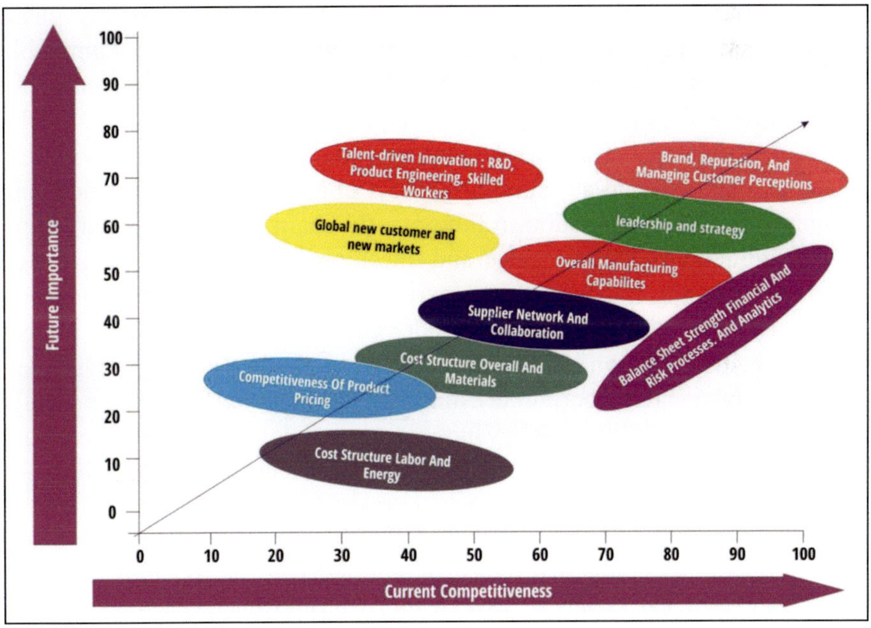

FIGURE 2.5 Significant features – cluster of potential – of future and current potential of competitiveness.

2.5.2 *EFFECT OF COMPLEXITIES OF SUPPLY CHAIN*

Complexities of flexibly chain Manufacturers' benevolent chains are remarkably perplexing and consistently introduced to a combination of risks, emanating from within and outside of their value chains. The developing private sectors are facing difficulties in natural/social issues, administrative necessities, and macroeconomic dangers around international weights [62].

There are, in like manner, loosened-up worth chain perils related to pariah expert associations, and operational threats related to progress masterminding, sourcing, creation, and scattering. Also, the adaptable chain should consider valuable risks related to financial theories, HR, and IT. All of these factors increase the complexities of effortlessly chain the board and, if not supervised warily, can achieve conceivable threatening impacts on arrangements and brand reputation. Considering the risks of working an incredible effortlessly chain, makers should think about how to consolidate strength into the adaptable chain. Intense deftly chains can address essential shortcomings proactively, with a more spotlight on strategy than trying to foresee and anticipate every peril type. An extreme smoothly chain changes peril and costs to prevent or get over quickly from innumerable dynamic and simultaneous threat-related aggravation. This is commonly accomplished through having deceivability and straightforwardness in the graceful chain, adaptability in sourcing, cooperation within as well as remote of the association, and an unmoving direct climate. An inward review can assume a significant function in the graceful chain measures by surveying related guidelines across locales and checking measures on a worldwide premise, assessing import, fare measures, and evaluating outsider dangers. An inside review can likewise consider the danger of the board system strategy, instruments, and innovation utilized by the business [63]. Different territories to consider incorporate estimation strategies for checking provider execution, accessibility, conveyance of materials, and danger-detecting insightful abilities set up by the business to screen hazard introductions inside the graceful chain [64].

2.5.3 EFFECT OF GAP OF SKILLS IN MANUFACTURING

Approximately 4 million vacancies in manufacturing companies are to be filled, and nearly 2 million of these vacancies remain unfilled due to the major problem of the skill gap. These statistics reveal the Skill Gap issue by the Manufacturing Institute and Deloitte Consulting LLP 2015. This talent issue is growing and effects as a major issue over the next decade. The current scenario says that talent and innovation are the top determinant of spirit identified by CEOs and manufacturing executives of the whole world. It is because the talent shortage in the world is visible and directly affecting manufacturing growth and indirectly on the economy. The skill gap will create a problem in meeting the demand of the customers admitted by almost 85% of the manufacturing executives. Approximately 80% of them believe that it will create problems in applying new technologies and efficiency of productivity. Furthermore, chiefs demonstrate the aptitude hole impacts the capacity to give powerful client care (70%), the capacity to improve and grow new items (63%), and the capacity to extend globally (49%). A lot of assembling chiefs (Almost 80%) announced they are glad to pay more than the market rates in workforce domains reeling under capacity crisis [65].

Still, 60% of positions stay vacant because of ability deficiency. This doubtlessly shows there are not a sufficient number of workers gathering to fill these positions. Additionally, bosses declared it takes an ordinary of 94 days to choose delegates in the trained professional/investigator/scientist fields and a typical of 70 days to select talented creation workers. Defying these numbers, it doesn't stun anybody why producers report the principal business impact of the capacity need is their ability to satisfy customer needs. To address the capacities entire, creators not simply need to find workers with the basic aptitudes expected to meet today and the impending advanced delivering requirements, yet they should moreover make and attract their present workforces. Making a smooth of workers with gathering capacities; planning, talented trades, and creation, will be fundamental to the future power of collecting associations, similarly as the business generally. A critical portion of keeping an eye on the capacity crisis is arranging strategies that upgradeability making sure about, progression, and association; and, with seven out of ten reviewed bosses reporting an absence of workers with acceptable development and PC and specific getting ready aptitudes, it is sensible this is a crushing concern [66]. An internal audit can accept an essential part in reviewing the (HR) and capacity estimates set up expected to address predicted capacity inadequacy and aptitudes openings perils. An open entryway exists to accept a vital part in distinctive deficiencies and assessing an affiliation's ability to perceive resources fit for enabling the relationship to meet its objectives. This may incorporate an appraisal of regions, for example, enrolling and maintenance programs, HR IT frameworks, and arrangement of information examination capacities to screen patterns [67].

2.6 CONCLUSION

Increased consumer quality expectations, shorter product life cycles, and a greater variety of variants are among the challenges that manufacturing companies in Germany are increasingly facing. As a result of these advances, dependable production is more critical than ever.

The aim of this study was to create a standardized approach to risk detection and analysis that took into account and utilized the unique characteristics and capabilities of CPPS. The possibilities that arise from continuous digitization of output have been factored into the procedure's growth. The risk factors for improving the risk situation in the production were to be determined, examined, and evaluated based on this principle. The method took into account the estimation of interconnected risks, which are able to spread during the manufacturing process. Current methods do not properly account for this impact.

Finally, it appears that the proposed approach could be used by CPPS to investigate the effects of emerging opportunities, not only for risk identification, risk analysis, and evaluation, but also for risk management and monitoring in the future.

KEYWORDS

- **hazard identification, risk assessment, and control**
- **hierarchical methodology**
- **risk analysis**
- **risk management**
- **strengths, weaknesses, opportunities, and threats**
- **sustainable waste management**

REFERENCES

1. Abele, M., et al., (2020). Simulation-based evaluation of an energy-oriented production planning system. In: *Procedia CIRP* (Vol. 88, pp. 246–251). doi: 10.1016/j.procir.2020.05.044.
2. Satyro, W. C., et al., (2021). Sustainable industries: Production planning and control as an ally to implement strategy. *J. Clean. Prod., 281*, 124781. doi: 10.1016/j.jclepro.2020.124781.
3. Han, J., Liu, Y., Luo, L., & Mao, M., (2020). Integrated production planning and scheduling under uncertainty: A fuzzy bi-level decision-making approach. *Knowledge-Based Syst., 201, 202*, 106056. doi: 10.1016/j.knosys.2020.106056.

4. Mouchtouri, V. A., Goutziana, G., Kremastinou, J., & Hadjichristodoulou, C., (2010). Legionella species colonization in cooling towers: Risk factors and assessment of control measures. *Am. J. Infect. Control, 38*(1), 50–55. doi: 10.1016/j.ajic.2009.04.285.
5. Bonfante, M. C., Raspini, J. P., Fernandes, I. B., Fernandes, S., Campos, L. M. S., & Alarcon, O. E., (2021). Achieving sustainable development goals in rare earth magnets production: A review on state of the art and SWOT analysis. *Renew. Sustain. Energy Rev., 137*, 110616. doi: 10.1016/j.rser.2020.110616.
6. Silvia, P. J., et al., (2021). Measuring everyday creativity: A rasch model analysis of the biographical inventory of creative behaviors (BICB) scale. *Think. Ski. Creat., 39*, 100797. doi: 10.1016/j.tsc.2021.100797.
7. Silva, P. C., & Iturra, H. C., (2021). A conceptual proposal and operational definitions of the cognitive processes of complex thinking. *Think. Ski. Creat.*, 39. doi: 10.1016/j.tsc.2021.100794.
8. Tillema, M., Bouwmeester, S., Verkoeijen, P., & Heijltjes, A., (2021). Psychometric properties of the short-form CART: Investigating its dimensionality through a Mokken scale analysis. *Think. Ski. Creat., 39*. doi: 10.1016/j.tsc.2021.100793.
9. Freitag, G. M., & Martensen, M., (2021). Facilitate design through improv: The qualified eclectic. *Think. Ski. Creat., 40*. doi: 10.1016/j.tsc.2020.100785.
10. Parviainen, T., Goerlandt, F., Helle, I., Haapasaari, P., & Kuikka, S., (2021). Implementing Bayesian networks for ISO 31000:2018-based maritime oil spill risk management: State-of-art, implementation benefits and challenges, and future research directions. *J. Environ. Manage., 278,* 111520. doi: 10.1016/j.jenvman.2020.111520.
11. Aven, T., & Renn, O., (2009). On risk defined as an event where the outcome is uncertain. *J. Risk Res., 12*(1), 1–11. doi: 10.1080/13669870802488883.
12. Goerlandt, F., Khakzad, N., & Reniers, G., (2017). Validity and validation of safety-related quantitative risk analysis: A review. *Saf. Sci., 99*, 127–139. doi: 10.1016/j.ssci. 2016.08.023.
13. Aven, T., & Heide, B., (2009). Reliability and validity of risk analysis. *Reliab. Eng. Syst. Saf., 94*(11), 1862–1868. doi: 10.1016/j.ress.2009.06.003.
14. Goerlandt, F., & Reniers, G., (2018). Prediction in a risk analysis context: Implications for selecting a risk perspective in practical applications. *Safety Science, 101*, 344–351. Elsevier B.V. doi: 10.1016/j.ssci.2017.09.007.
15. Wang, X., Li, D., O'brien, C., & Li, Y., (2010). A production planning model to reduce risk and improve operations management. *Int. J. Prod. Econ., 124*(2), 463–474. doi: 10.1016/j.ijpe.2009.12.009.
16. Lan, Y., Zhao, R., & Tang, W., (2011). Minimum risk criterion for uncertain production planning problems. *Comput. Ind. Eng., 61*(3), 591–599. doi: 10.1016/j.cie.2011.04.014.
17. Chaharsooghi, S. K., & Heydari, J., (2010). LT variance or LT mean reduction in supply chain management: Which one has a higher impact on SC performance? *Int. J. Prod. Econ., 124*(2), 475–481. doi: 10.1016/j.ijpe.2009.12.010.
18. Bodenstein, C., Schryen, G., & Neumann, D., (2012). Energy-aware workload management models for operation cost reduction in data centers. *Eur. J. Oper. Res., 222*(1), 157–167. doi: 10.1016/j.ejor.2012.04.005.
19. Lagodimos, A. G., & Mihiotis, A. N., (2010). Efficient overtime planning in packing shops with lines of identical manning. *Int. J. Prod. Econ., 124*(2), 453–462. doi: 10.1016/j.ijpe.2009.12.008.
20. Xu, X., Guo, W. G., & Rodgers, M. D., (2020). A real-time decision support framework to mitigate degradation in perishable supply chains. *Comput. Ind. Eng., 150*. doi: 10.1016/j.cie.2020.106905.

21. *A Production Planning Model to Reduce Risk and Improve Operations Management*. ScienceDirect. https://www.sciencedirect.com/science/article/abs/pii/S0925527309004563 (accessed on 17 February 2022).
22. Saak, A. E., (2016). Traceability and reputation in supply chains. *Int. J. Prod. Econ., 177*, 149–162. doi: 10.1016/j.ijpe.2016.04.008.
23. Hall, D. C., & Johnson-Hall, T. D., (2021). Recall effectiveness, strategy, and task complexity in the U.S. meat and poultry industry. *Int. J. Prod. Econ., 234*. doi: 10.1016/j.ijpe.2021.108048.
24. Maity, M., Tolooie, A., Sinha, A. K., & Tiwari, M. K., (2021). Stochastic batch dispersion model to optimize traceability and enhance transparency using blockchain. *Comput. Ind. Eng., 154*. doi: 10.1016/j.cie.2021.107134.
25. *Sustainability Risk Management in a Smart Logistics Ecological Chain: An Evaluation Framework Based on Social Network Analysis*. ScienceDirect. https://www.sciencedirect.com/science/article/abs/pii/S0959652620342347 (accessed on 17 February 2022).
26. Liu, W., Wei, W., Yan, X., Dong, D., & Chen, Z., (2020). Sustainability risk management in a smart logistics ecological chain: An evaluation framework based on social network analysis. *J. Clean. Prod., 276*, 124189. doi: 10.1016/j.jclepro.2020.124189.
27. Wide, P., (2020). Improving decisions support for operational disruption management in freight transport. *Res. Transp. Bus. Manag., 37*. doi: 10.1016/j.rtbm.2020.100540.
28. Aboutorab, H., Hussain, O. K., Saberi, M., Hussain, F. K., & Chang, E., (2021). A survey on the suitability of risk identification techniques in the current networked environment. *Journal of Network and Computer Applications, 178*, 102984. Academic Press. doi: 10.1016/j.jnca.2021.102984.
29. Freudenstein, F., Croft, R. J., Wiedemann, P. M., Verrender, A., Böhmert, C., & Loughran, S. P., (2020). Framing effects in risk communication messages – hazard identification vs. risk assessment. *Environ. Res., 190*, 109934. doi: 10.1016/j.envres.2020.109934.
30. Sarabia, M. M., et al., (2020). The challenges of impact evaluation: Attempting to measure the effectiveness of community-based disaster risk management. *Int. J. Disaster Risk Reduct., 49*, 101732. doi: 10.1016/j.ijdrr.2020.101732.
31. Aitsi-Selmi, A., Egawa, S., Sasaki, H., Wannous, C., & Murray, V., (2015). The Sendai framework for disaster risk reduction: Renewing the global commitment to people's resilience, health, and well-being. *Int. J. Disaster Risk Sci., 6*(2), 164–176. doi: 10.1007/s13753-015-0050-9.
32. Maskrey, A., (2011). Revisiting community-based disaster risk management. *Environ. Hazards, 10*(1), 42–52. doi: 10.3763/ehaz.2011.0005.
33. Xu, Y., Yeh, C. H., Yang, S., & Gupta, B., (2020). Risk-based performance evaluation of improvement strategies for sustainable e-waste management. *Resour. Conserv. Recycl., 155*, 104664. doi: 10.1016/j.resconrec.2019.104664.
34. Sanyal, M. K., & Biswas, S. B., (2014). Employee motivation from performance appraisal implications: Test of a theory in the software industry in West Bengal (India). *Procedia Econ. Financ., 11*, 182–196. doi: 10.1016/s2212-5671(14)00187-7.
35. Ibrahim, H., & Almarshed, S. O., (2014). Sporting event as a corporate social responsibility strategy. *Procedia Econ. Financ., 11*, 3–14. doi: 10.1016/s2212-5671(14)00170-1.
36. Suthar, B. K., Chakravarthi, T. L., & Pradhan, S., (2014). Impacts of job analysis on organizational performance: An inquiry on Indian public sector enterprises. *Procedia Econ. Financ., 11*, 166–181. doi: 10.1016/s2212-5671(14)00186-5.

37. Nagendra, A., (2014). Paradigm shift in HR practices on employee life cycle due to influence of social media. *Procedia Econ. Financ., 11*, 197–207. doi: 10.1016/s2212-5671 (14)00188-9.
38. Ashokkumar, S., (2014). Thoughts on business ethics and corporate social responsibility from Vedic literature. *Procedia Econ. Financ., 11*, 15–22. doi: 10.1016/s2212-5671 (14)00171-3.
39. Chauhan, S., & Amit, (2014). A relational study of firm's characteristics and CSR expenditure. *Procedia Econ. Financ., 11*, 23–32. doi: 10.1016/s2212-5671(14)00172-5.
40. Goh, K. Y., Heng, C. S., & Lin, Z., (2013). Social media brand community and consumer behavior: Quantifying the relative impact of user – and marketer-generated content. *Inf. Syst. Res., 24*(1), 88–107. doi: 10.1287/isre.1120.0469.
41. Williams, S. P., Nitschke, C. S., & Hardy, C. A., (2021). Analytical framework for social media risk analysis in organizations. *Procedia Comput. Sci., 181*, 302–309. doi: 10.1016/j.procs.2021.01.150.
42. Walsham, G., (2006). Doing interpretive research. *Eur. J. Inf. Syst., 15*(3), 320–330. doi: 10.1057/palgrave.ejis.3000589.
43. Demek, K. C., Raschke, R. L., Janvrin, D. J., & Dilla, W. N., (2018). Do organizations use a formalized risk management process to address social media risk? *Int. J. Account. Inf. Syst., 28*, 31–44. doi: 10.1016/j.accinf.2017.12.004.
44. Bertot, J. C., Jaeger, P. T., & Hansen, D., (2012). The impact of policies on government social media usage: Issues, challenges, and recommendations. *Gov. Inf. Q., 29*(1), 30–40. doi: 10.1016/j.giq.2011.04.004.
45. Bruno, G., (2021). Data-awareness in B2B process modeling. *Procedia Comput. Sci., 181*, 15–22. doi: 10.1016/j.procs.2021.01.094.
46. *Complying with Best Practice Risk Management Committee Guidance and Performance.* ScienceDirect. https://www.sciencedirect.com/science/article/abs/pii/S1815566920300394 (accessed on 17 February 2022).
47. Alauddin, M., Islam, K. M. A., Khan, F., Imtiaz, S., Ahmed, S., & Amyotte, P., (2020). How can process safety and a risk management approach guide pandemic risk management? *J. Loss Prev. Process Ind., 68*, 104310. doi: 10.1016/j.jlp.2020.104310.
48. Jia, J., & Bradbury, M. E., (2020). Complying with best practice risk management committee guidance and performance. *J. Contemp. Account. Econ., 16*(3), 100225. doi: 10.1016/j.jcae.2020.100225.
49. Lin, S., Xia, H. H., & Ryabova, T., (2020). The effect of analysts' GAAP earnings forecasts on managers' classification shifting. *J. Contemp. Account. Econ., 16*(3). doi: 10.1016/j.jcae.2020.100222.
50. Altuwair, I. A., & Khan, F., (2020). Safety analysis of instantaneous release of compressed natural gas from a cylinder. *J. Loss Prev. Process Ind., 68*, 104284. doi: 10.1016/j.jlp.2020.104284.
51. Iñiguez, L., (2011). The challenges of research and development of small ruminant production in dry areas. *Small Rumin. Res., 98*(1–3), 12–20. doi: 10.1016/j.smallrumres.2011.03.010.
52. *The Challenges of Research and Development of Small Ruminant Production in Dry Areas*. ScienceDirect. https://www.sciencedirect.com/science/article/abs/pii/S092144881100085X (accessed on 17 February 2022).
53. *Hydrogen Productive Chain in Brazil: An Analysis of the Competitiveness' Drivers*. ScienceDirect. https://www.sciencedirect.com/science/article/abs/pii/S0959652618328804 (accessed on 17 February 2022).

54. Soares, L. O., De Moraes, D. R., & Boloy, R. A. M., (2021). Energy-ecological efficiency and TTW emissions of the DFSIE fueled with biofuels. *Bioenergy Res.* doi: 10.1007/s12155-020-10229-1.
55. Kavilal, E. G., Prasanna, V. S., & Harsh, K. K. D., (2017). An integrated fuzzy approach for prioritizing supply chain complexity drivers of an Indian mining equipment manufacturer. *Resour. Policy, 51*, 204–218. doi: 10.1016/j.resourpol.2016.12.008.
56. Gardas, B. B., Mangla, S. K., Raut, R. D., Narkhede, B., & Luthra, S., (2019). Green talent management to unlock sustainability in the oil and gas sector. *J. Clean. Prod., 229*, 850–862. doi: 10.1016/j.jclepro.2019.05.018.
57. Karatop, B., Kubat, C., & Uygun, Ö., (2015). Talent management in manufacturing system using fuzzy logic approach. *Comput. Ind. Eng., 86*, 127–136. doi: 10.1016/j.cie.2014.09.015.
58. Ghobakhloo, M., (2020). Industry 4.0, digitization, and opportunities for sustainability. *Journal of Cleaner Production*, vol. 252. Elsevier Ltd, doi: 10.1016/j.jclepro.2019.119869.
59. *Strategic Decisions, Competition and Cost-Sharing Contract Under Industry 4.0 and Environmental Considerations*. ScienceDirect. https://www.sciencedirect.com/science/article/abs/pii/S0921344920303748 (accessed on 17 February 2022).
60. Galvão, G. D. A., Homrich, A. S., Geissdoerfer, M., Evans, S., Scoleze, F. P. S., & Carvalho, M. M., (2020). Towards a value stream perspective of circular business models. *Resources, Conservation and Recycling, 162*. Elsevier B.V. doi: 10.1016/j.resconrec.2020.105060.
61. *How Do Market Power and Industry Competition Influence the Effect of Corporate Governance on Earnings Management?* ScienceDirect. https://www.sciencedirect.com/science/article/abs/pii/S1062976920300168 (accessed on 17 February 2022).
62. Zengin, Y., Naktiyok, S., Kaygın, E., Kavak, O., & Topçuoğlu, E., (2021). An investigation upon industry 4.0 and society 5.0 within the context of sustainable development goals. *Sustain., 13*(5), 1–16. doi: 10.3390/su13052682.
63. Katerusha, D., (2021). Attitude towards sustainability, study contents and the use of recycled concrete in building construction – case study Germany and Switzerland. *J. Clean. Prod., 289*. doi: 10.1016/j.jclepro.2020.125688.
64. Rauch, E., & Brown, C. A., (2020). Teaching axiomatic design for a long-term sustainable introduction of industry 4.0 in SMEs. In: *Procedia CIRP* (Vol. 96, pp. 169–174). doi: 10.1016/j.procir.2021.01.155.
65. Moldovan, L., (2019). State-of-the-art analysis on the knowledge and skills gaps on the topic of industry 4.0 and the requirements for work-based learning. In: *Procedia Manufacturing* (Vol. 32, pp. 294–301). doi: 10.1016/j.promfg.2019.02.217.
66. Spak, G. T., (2013). Us advanced manufacturing skills gap: Innovative education solutions. *Procedia – Soc. Behav. Sci.* (Vol. 106, pp. 3235–3245). doi: 10.1016/j.sbspro.2013.12.372.
67. Amanda Jacob (2013). ACMA and manufacturing institute partner to address US skills gap. *Reinf. Plast., 57*(6), 10. doi: 10.1016/s0034-3617(13)70174-7. https://www.reinforcedplastics.com/content/news/acma-and-manufacturing-institute-partner-to-address-us-skills-gap (accessed on 12 May 2023).

CHAPTER 3

EXPERIMENTAL ANALYSIS AND OPTIMIZATION OF DRILLING OF GLASS FIBER REINFORCED PLASTICS COMPOSITES USING MADM METHODOLOGY: UTILITY CONCEPT

SUMAN CHATTERJEE[1], PREETAM SUBUDHI[1], DIN BANDHU[2], KUMAR ABHISHEK[3], and SIBA SANKAR MAHAPATRA[1]

[1]*Department of Mechanical Engineering, National Institute of Technology, Rourkela, Odisha, India*

[2]*Department of Mechanical and Industrial Engineering, Manipal Institute of Technology Bengaluru, Manipal Academy of Higher Education, Manipal, Karnataka, India*

[3]*Department of Mechanical and Aero-Space Engineering, Institute of Infrastructure, Technology, Research, and Management (IITRAM), Ahmedabad, Gujarat, India*

ABSTRACT

Composite materials like glass fiber reinforced plastics (GFRP) are immensely utilized in aircraft, automobiles, and several other industries. So, drilling is one of the important manufacturing processes for assembly, and hence care should be taken in order to get damage-free holes. In our research work, we have attempted to find out the impact of a few variables like spindle speed (Ss), feed rate (F_r), and drill diameter (D_d) on the output attributes namely thrust force (T_f), torque (T_q), delamination factor (D_f), and circularity of holes (C_h) at both entrances and exit. To reduce the number of experiments we used

Decision-Making Models and Applications in Manufacturing Environments. Pushpdant Jain, Kumar Abhishek, & Prasenjit Chatterjee (Eds.)

Taguchi L_9 orthogonal array and conducted experiments as per the design of experiment (DoE) approach and obtained thrust force and torque using a dynamometer and then found out D_f and circularity at both entry and exit. To find out optimum parametric setting so as to get minimum D_f and maximum circularity at both entry and exit we used utility concept and then found out the analysis of variance (ANOVA) and main plots for signal-to-noise (S/N) ratios for overall utility value. The optimum results obtained were being confirmed by doing a confirmation test using the optimum parameters. So, for minimum delamination and higher circularity of holes higher spindle speed and low feed rates and D_d must be used. It was also found out that the damage of holes is more at the exit than at the entry. The same experiment can also be performed on carbon fiber reinforced plastics (CFRP) using other cutting parameters namely point angle, material thickness, etc.

3.1 INTRODUCTION

Drilling is a widely used material removal process in which a drill bit is used to cut holes in various materials. The drill bit is a multipoint cutting tool having normally two or more principal cutting edges. Drill bits are made up of high-speed steel, tungsten carbide, cobalt steel alloys, and many more materials. Sometimes drill bits are coated with materials like black oxide, titanium nitride, and diamond, etc., to increase the life and resistance to corrosion of drill bits. The most common parts of a drill bit are shank, flutes, body, chisel edge, cutting edge, etc., common drill bit parameters are nominal diameter, helix angle, point angle (usually 118°), chisel edge angle, etc. Commonly used drill bits are twist drill bits, step drill bit, core drill, etc. Drilling can be done either in metal or in non-metal like wood, plastic, etc. Drilling in metals like aluminum, brass, mild steel, etc., is done using high-speed steel twist drill bits. Material is removed from the metals in the form of chips which is washed away from the flutes of drill bits using a proper coolant [1–3]. Drilling is carried out using a computer numerical control (CNC) machine in which the drill bit is clamped at the holder. For drilling of long holes, peck drilling is generally used which involves pushing the drill partway through the workpiece, for no more than five times the diameter of the drill, and then retracting it back to the surface. Drilling in wood, plastics is easier and quicker as compared to the drilling of metals but the major problem that arises while drilling wood and plastics is that the surfaces around the holes tend to burn at both entry and exit planes. Also, the drill bits tend to chatter or vibrate while drilling wood or plastics; so, backing plates are used to avoid hole damage at the exit [3–5].

Composite materials consist of a combination of two or more than two macro-constituents that differ both in physical form and chemical composition and are insoluble with each other and thus have a distinct interface separating them. One of the macro-constituents is called a matrix, while the other constituent is called reinforcement. So, in polymer matrix composites (PMCs), polymer is the matrix, and fibers are the reinforcement. So, depending on the type of fiber used (usually glass, carbon, and aramid), they have named glass fiber reinforced plastics (GFRP), carbon fiber reinforced plastics (CFRP), etc. Here, the main part of our study is the machining of GFRP composites. GFRP composites materials have been immensely used in the aircraft and automotive industries. These materials are superior in strength, good corrosion-resistant, lightweight construction, and good fatigue strength, and also have low specific gravity as compared to other materials such as metals, wood, etc. As we have already discussed that glass fiber reinforced plastic (GFRP) composites have a wider application in various industrial areas like aerospace, automobile, etc. So, drilling holes becomes an important manufacturing operation in order to make effective assembly of different composite sections. But while drilling of GFRP many problems are generally encountered such as surface delamination at both entry and exit, fiber/resin pull out, surface roughness (SR), and tool wear caused due to abrasive action of fibers with the tool that decrease the quality of holes produced [6–12]. In order to reduce these machining problems, there is a need to develop some methods to select optimum machining control parameters so as to get quality drilled holes in composites. Delamination is a major problem associated with the drilling of fiber-reinforced composite materials as it not only reduces the structural integrity of the material but also results in poor assembly which induces vibrations in parts. Peel-up and pull-out are two different types of delamination observed while drilling composites. In the peel-up delamination mechanism, during the forward movement of the drill towards the workpiece, the layers of the material are pushed towards the cutting faces of the drill bit instead of undergoing cutting action. This type of damage is generally reduced by drilling at low feed rates. The push-down delamination is caused due to the action of indenting by the drill bit chisel edge, acting over the layers of composites. When the thrust force exerted by the drill bit on the composites exceeds the toughness of the layers of composites then, delamination begins to occur. The widely used method to reduce the pushdown delamination is to use a backing plate under the workpiece. Several studies have been done to analyze delamination both at the entry and exit points of the hole so as to obtain good quality holes and proper tolerances.

The literature review was done to find out the research done in this field and also to know the conclusions given by various researchers on various topics related to this field of study and finally their point of view. Here are a few literature reviews which are summarized below.

Byrne & Taguchi [13] experimented to obtain optimal cutting parameters using Taguchi's approach. First, the signal-to-noise (S/N) ratio of each characteristic was found out and then the overall performance was obtained by adding the weighted S/N ratios. By using the utility concept and the Taguchi method for each response, multi-response optimization was found out. Singh et al. [14] worked on the drilling of unidirectional and multi-directional GFRP. Their main aim was to study the effect of thrust force and torque on composites during drilling and then compare the experimentally obtained results with FEA. Mohan et al. [15] worked on the effect of machining control parameters on torque and cutting force during the drilling of glass fiber reinforced composites (GFRP). Their paper described the effect of various machining control parameters during the drilling of composites. Tsao & Hocheng [16] conducted drilling on composites using different types of drill bits like saw drill, twist drill, etc., and then found out delamination factor (D_f) for each case and finally compared all the results with each other. Kilickap et al. [17] conducted drilling on GFRP using various cemented carbide drill bits (grade-K10) and then evaluated SR and delamination for each case. The main objective of their study was to find the effect of cutting velocity, feed rate, and depth of cut on circularity, SR, and delamination while drilling of GFRP composites. Ugo et al. [18] investigated and worked on finding a method to obtain optimal parameters for less damage of holes of composites while drilling. They used both Taguchi design analysis and an optimization method in order to obtain optimal parametric setting. They found out that while drilling at higher spindle speeds and lower feed rates, better results are obtained in the case of composites. Krishnaraj et al. [19] used various combinations of spindle speed and feed rate during the drilling of CFRP laminate and measured the circularity for each experimental run. It was found out that with an increase in spindle speed circularity decreases, while with an increase in feed rates the circularity does not change much. From the ANOVA calculations, it was concluded that spindle speed is the cutting parameter that is significantly affecting circularity, i.e., 86%. Chen [20] worked on the variations of cutting forces during drilling for damaged and undamaged holes and concluded that in order to get an undamaged drilling operation, proper selections of tool geometry and cutting parameters should be made. Capello [21] worked on the influence of feed rate on delamination of a glass fiber reinforced plastic and found that delamination is very less while drilling at low feed rates. Delamination is caused due to an increase of negative back rake angle with an increase of feed rates which decreases the tendency of shearing. Dharan & Won [22] represented a model on taking into account low thrust force and torque in computer numeric control machine tools in order to maximize productivity and generate damage-free holes in composites. Collombet [23] presented a numerical model for the calculation of the thrust force. The numerical model is very good for predicting the influence of plastic deformation of the drill bit and various drill geometry on drilling operation. Sharif & Rahim [24] concluded that drilling is an important manufacturing operation in aerospace industries as nearly 30–50% of the total material removal processes are related to drilling. Durao et al. [25] have investigated the effect of feed rate and different tool geometries on delamination of CFRP while drilling with constant cutting speed and drill diameter (D_d) size. It was observed that minimum delamination was found out while drilling at a lower feed rate and higher spindle speed with a twist drill.

From a number of works done by various researchers, it was found that the defects produced drilling of composites such as delamination, fiber pulls out, etc., are mainly affected by the choice of machining parameters such as D_d, feed, spindle speed, etc. So, the main objective of this chapter is to find out the effects of cutting parameters such as feed, spindle speed, D_d on delamination, and circularity produced while drilling of GFRP composite using Taguchi method design and analysis of ANOVA.

3.2 EXPERIMENTAL METHODS AND MATERIALS

For the experimental investigation, a GFRP composite of the circular cross-section of 50 mm diameter and 150 mm length and 90° orientation was chosen. Then the composites were cut into small pieces of thickness 5 mm using a band saw as shown in Figure 3.1. The GFRP composites have an epoxy matrix and E-glass fibers. Then the drilling tests were conducted on a CNC machine (MTAB-Maxmill) with a dynamometer attached to it (for measurement of torque and thrust force) as shown in Figure 3.2.

The machining control parameters selected for the experimental study are D_d, spindle speed, and feed rate, and each of these parameters was assigned three levels as depicted in Table 3.1.

With these aforementioned cutting parameters, a number of experimental settings can exist, and at all those settings, it will be nearly impossible to conduct all experiments as it will consume lots of time with tiresome computations. Hence, to reduce the number of experimental runs L_9 Taguchi orthogonal array (shown in Table 3.2) was chosen which provides a selective number of experimental runs at selective combinations of cutting parameters. The L_9 Taguchi orthogonal array was generated using Minitab-16.

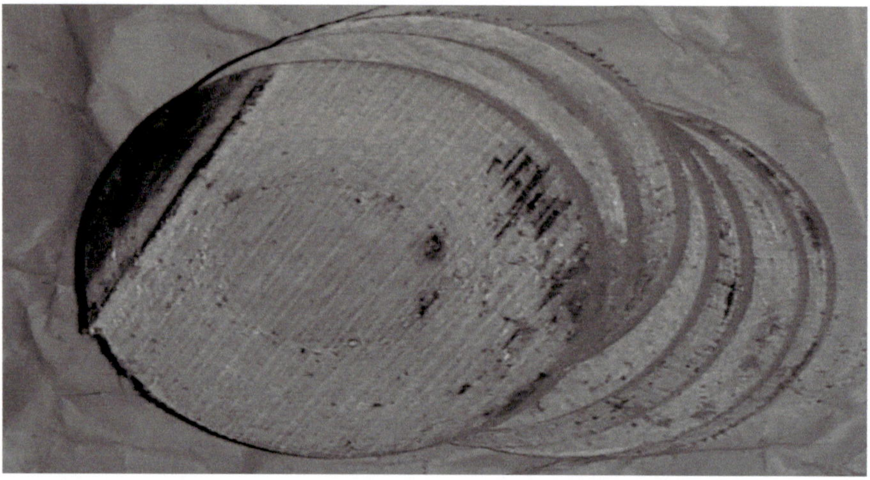

FIGURE 3.1 Workpieces of composites on which drilling operation was carried out.

FIGURE 3.2 Experimental set-up (CNC machine with dynamometer).

TABLE 3.1 Cutting Parameters and Their Levels

Cutting Parameters	Level 1	Level 2	Level 3
Drill diameter (mm)	6	8	10
Spindle speed (rpm)	600	1,200	1,800
Feed rate (mm/min)	100	160	220

TABLE 3.2 Cutting Parameters and Their Levels Using Taguchi OA

SL. No.	Spindle Speed (rpm)	Feed Rate (m/min)	Drill Diameter (mm)
1.	600	100	6
2.	600	160	8
3.	600	220	10
4.	1,200	100	8
5.	1,200	160	10
6.	1,200	220	6
7.	1,800	100	10
8.	1,800	160	6
9.	1,800	220	8

Then using the above set of responses, drilling on GFRP composites was done using a high-speed steel drill bit. A backing plate of Perspex material was used to avoid damage to holes at the exit. The experiment was conducted by keeping the diameter fixed and then changing the spindle speed and feed rates accordingly. Three holes were drilled in each GFRP plate and numbered accordingly at both entry and exit (Figure 3.3).

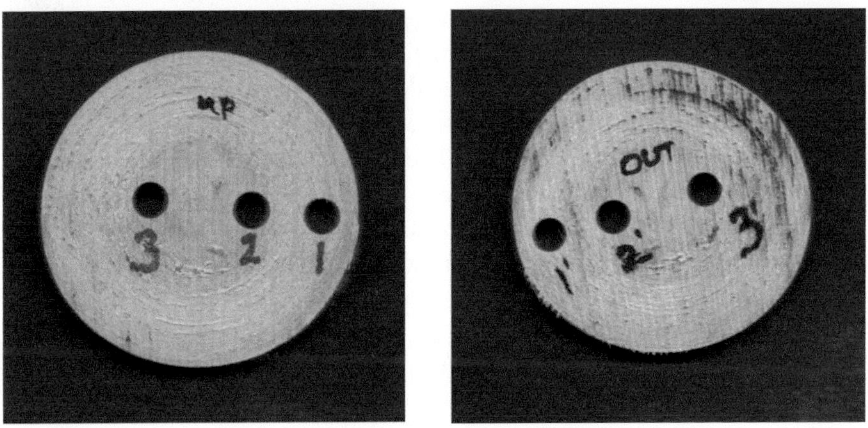

FIGURE 3.3 Drilled holes in each GFRP plate and numbered accordingly.

As is already mentioned earlier that the dynamometer was connected to the CNC machine so, for each response corresponding torque (kN-mm) and thrust force (kN) were measured using the dynamometer. So, the thrust and torque of all the 9 responses are shown in Table 3.3.

TABLE 3.3 Measurement of Thrust and Torque for Each Experimental Run

Spindle Speed (rpm)	Feed Rate (m/min)	Drill Diameter (mm)	Thrust (kN)	Torque (kN-mm)
600	100	6	0.093	0.64
600	160	8	0.195	1.07
600	220	10	0.423	1.92
1,200	100	8	0.15	0.79
1,200	160	10	0.261	0.92
1,200	220	6	0.079	0.2
1,800	100	10	0.205	0.76
1,800	160	6	0.072	0.3
1,800	220	8	0.112	0.87

Delamination is one of the major problems observed while drilling composites. Delamination is observed both at entry and exit planes of holes. Delamination is generally found to be more critical at the hole exit than at the entrance. D_f is found out using the following formula:

$$\text{Delamination factor} = D_{max}/D \tag{1}$$

where; D_{max} is the maximum diameter of the damage hole or delamination zone in mm; and D is the nominal diameter of drill in mm.

So, normally we know the nominal diameter of the drill bit but to measure the maximum diameter of the damaged hole we have to use a microscope. So, using the toolmaker's microscope, D_{max} was found out both at entry and exit of each of the nine holes. The damaged area is obtained through image digitization and processing the picture. To obtain an image with good quality, a series of parameters must be selected properly, such as brightness, intensity, image enhancement, etc. Drawing the circle in the outer periphery touching the foremost damaged point gives D_{max}. Now finally, the D_f was found out using the above formula (Eqn. (1)) at both entry and exit. The data consisting of responses and D_f at both entry and exit is shown in Table 3.4.

Just like delamination, circularity is another problem observed while drilling GFRP. The circularity symbol is used to describe how close an object should be to a true circle. Circularity is independent of any datum feature and is always less than the dimensional tolerance of the diameter of the part.

TABLE 3.4 Calculation of Delamination Factor at Entry and Exit

SL. No.	Spindle Speed (rpm)	Feed Rate (m/min)	Drill Diameter (mm)	Delamination Factor at Entry	Delamination Factor at Exit
1.	600	100	6	1.00333	1.006833
2.	600	160	8	1.01412	1.0151
3.	600	220	10	1.0037	1.0041
4.	1,200	100	8	1.007375	1.0085
5.	1,200	160	10	1.0011	1.0009
6.	1,200	220	6	1.01267	1.001767
7.	1,800	100	10	1.0004	1.0009
8.	1,800	160	6	1.00067	1.0015
9.	1,800	220	8	1.01737	1.018

Using the ferret diameter concept, the diameter is found out at various orientations in a particular hole at both entry and exit. The diameter is found out at various orientations using Foxit Reader software after zooming the image of the holes. Circularity is found out using the following formula:

$$\text{Circularity} = D_{min}/D_{max} \tag{2}$$

Figure 3.4 shows the measurement of diameter only at four orientations, but actually at nearly eight orientations diameters were measured but due to unavailable of space only four orientations are shown, which consist of minimum diameter, maximum diameter, and two intermediate ones. The circularity values at both entry and exit for each of the responses are shown in Table 3.5.

After calculations of thrust, torque, D_f, and circularity for each response we had to find a procedure to determine optimal machining parameters for the multi-response process. So, the utility concept was decided to use for this purpose.

A product is being evaluated based on the number of qualities or characteristics it possesses. It is difficult for a product to have all the qualities which a customer wants to have in it. So, each of the products is represented by a term called 'overall utility' which describes the weighted average of all the qualities or characteristics it possesses. How good a product is related to a particular characteristic or quality is determined by the utility of the product on that particular characteristic. The overall utility of a product is the weighted sum of the utilities of each quality characteristic.

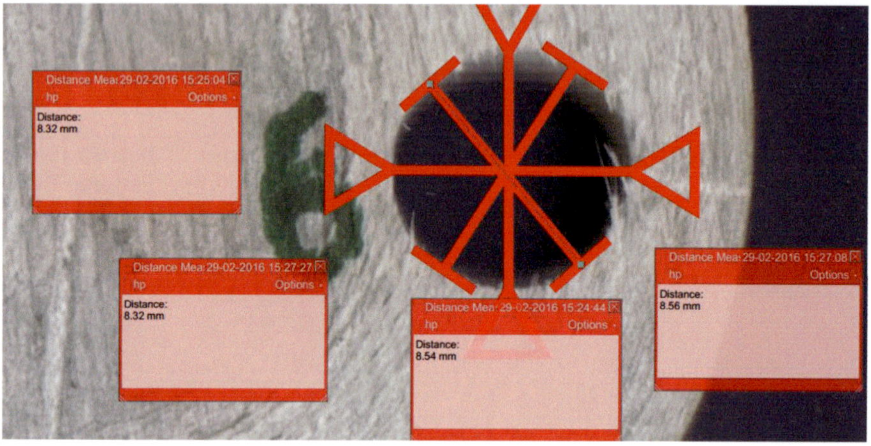

FIGURE 3.4 Measurement of diameter at various orientations at the entry of the hole.

TABLE 3.5 Calculation of Circularity at Entry and Exit

SL. No.	Spindle Speed (rpm)	Feed Rate (m/min)	Drill Diameter (mm)	Circularity at Entry	Circularity at Exit
1.	600	100	6	0.978793	0.986906
2.	600	160	8	0.964504	0.971779
3.	600	220	10	0.9638672	0.962174
4.	1,200	100	8	0.960526	0.978102
5.	1,200	160	10	0.9823702	0.982231
6.	1,200	220	6	0.990322	0.98697
7.	1,800	100	10	0.981427	0.9889669
8.	1,800	160	6	0.986688	0.985074
9.	1,800	220	8	0.971962	0.9642

In order to determine the utility of a particular characteristic, a preference scale (normally a logarithmic one) is chosen. The minimum acceptable level of a quality or characteristic is given a preference number 0 while the maximum acceptable level of a characteristic is given a preference number 9.

The preference number is given by the following formula:

$$\text{Preference number } P_i = Z^* \log (Y_i/Y_i') \qquad (3)$$

where; Y_i is the value of chosen characteristic 'i;' and Y_i' is the acceptable value of chosen characteristic 'i.'

$$Z = 9/(\log (Y_i^*/Y_i')) \tag{4}$$

where; Y_i^* is the best value of chosen characteristic 'i.'

$$\text{Subject to condition: } \sum W_i = 1$$

So, the overall utility is given by:

$$U = \sum W_i P_i \tag{5}$$

The acceptable value and best value are found in Table 3.6. The acceptable value for the D_f is the maximum of all experimental runs, while that of circularity is the minimum of all experimental runs. The best value for the D_f is the minimum of all experimental runs, while that of circularity is the maximum of all experimental runs. Then the utility values of each characteristic corresponding to all experimental runs are found out, and finally, the overall utility of each experimental run is found out which is shown in Table 3.7.

TABLE 3.6 Acceptable and Best Values for Each Characteristic

Characteristic	Acceptable Value (Y_i')	Best Value (Y_i^*)	Weights (W_i)
Thrust force	0.423	0.072	0.167
Torque	1.92	0.2	0.167
Delamination factor at entry	1.017375	1.0004	0.167
Delamination factor at exit	1.018	1.0009	0.167
Circularity at entry	0.96052	0.99032	0.167
Circularity at exit	0.962174	0.98896	0.167

In the present investigation, we have used the utility concept and Taguchi design method which uses S/N (signal to noise) ratios to find the optimal parametric setting for minimum damage of holes. S/N ratio gives us an idea about the deviation of characteristics from the mean values (desired values). There are three categories of S/N ratios, i.e., smaller the better, nominal the better, and larger the better. The S/N ratios are calculated using loss functions which are given below:

$$\text{Smaller the better: S/N} = -10 \times \text{Log} ((\sum z^2)/n) \tag{6}$$

$$\text{Larger the better: S/N} = -10 \times \text{Log} ((\sum 1/z^2)/n) \tag{7}$$

where; n is the number of observations, and z is the observed data.

A greater S/N ratio gives better performance while a lower S/N ratio suggests a lower performance.

TABLE 3.7 Overall Utility Calculation for Each Experimental Run

SL. No.	P1	P2	P3	P4	P5	P6	Overall Utility
1.	7.69916	4.3715	7.45	5.8616	5.55	8.3165	6.54146
2.	3.936	2.3265	1.724	1.515	1.218	3.256	2.33
3.	0	0	7.238	7.304	1.022	0	2.594
4.	5.27	3.5337	5.283	4.981	0	5.38	4.0746
5.	2.454	2.9275	8.625	9	6.625	6.685	6.05275
6.	8.528	9	2.479	8.54	9	8.34	7.6478
7.	3.6817	3.6878	9	9	6.34	9	6.785
8.	9	7.3865	8.855	8.48	7.915	7.685	8.22
9.	6.754	3.15	0	0	3.487	0.689	2.3467

Note: Notations P1, P2, P3, P4, P5, and P6 refer to preference number of thrust, torque, delamination factor at entry, delamination factor at the exit, circularity at entry, and circularity at exit, respectively.

3.3 RESULTS AND DISCUSSION

The analysis of variance (ANOVA) analysis is generally performed to know the statistical significance of the cutting parameters like spindle speed, feed rate, and drill bit diameter on the output characteristics such as thrust force, torque, D_f at entry, D_f at the exit, circularity at entry and circularity at the exit which is shown in Tables 3.9–14, respectively for GFRP. The coefficient of determination (R2) for thrust, torque, delamination at entry, delamination at the exit, circularity at entry, and circularity at exit is 98.3%, 92.7%, 87.7%, 84.1%, 86.4%, and 89.6% respectively.

From Table 3.8, it can be seen that D_d is the cutting parameter which is significantly affecting the thrust force, i.e., 83.657%, and spindle speed is the parameter next to D_d which is significantly affecting thrust force, i.e., 14.12%.

TABLE 3.8 Analysis of Variance for SN Ratios of Thrust

Source	DF	Seq SS	Adj SS	Adj MS	F	Percentage Contribution
Spindle speed	2	29.953	29.953	14.9764	8.34	14.12%
Feed rate	2	1.126	1.126	0.563	0.31	0.53%
Drill diameter	2	177.458	177.458	88.7289	49.44	83.657%
Residual error	2	3.589	3.589	1.7947	–	–
Total	**8**	**212.126**	–	–	–	–

From Figure 3.5, it is observed that the optimum parameters for minimum thrust force are obtained at spindle speed at level 3 (1,800 rpm), feed at level 1 (100 mm/min), and D_d at level 1 (6 mm). It is also observed that with an increase in spindle speed thrust force decreases and with an increase of both feed and D_d thrust force increases.

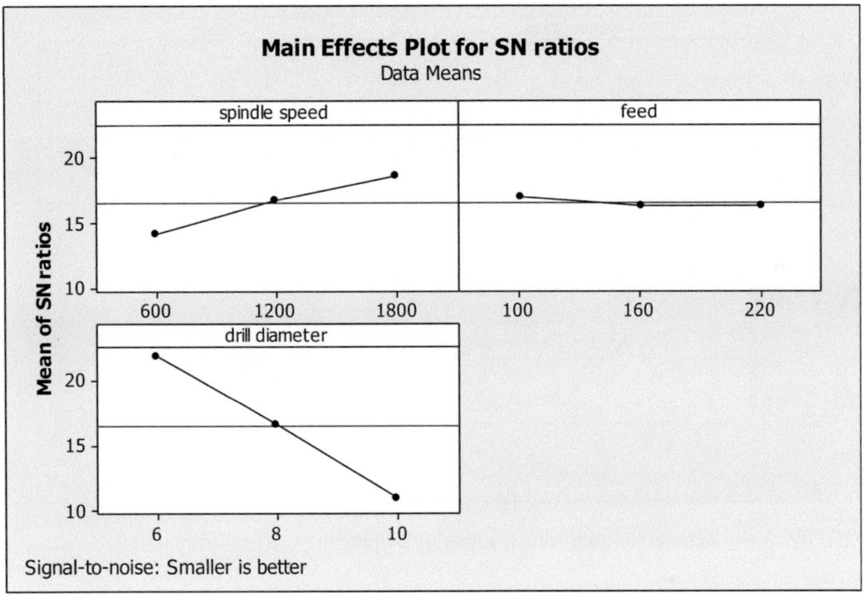

FIGURE 3.5 Main effect plot for SN ratios of thrust.

From Table 3.9, it is found that D_d is the cutting parameter which is significantly affecting torque, i.e., 66.338%.

TABLE 3.9 Analysis of Variance for SN Ratios of Torque

Source	DF	Seq SS	Adj SS	Adj MS	F	Percentage Contribution
Spindle speed	2	71.453	71.453	35.7263	3.57	26.044%
Feed rate	2	0.873	0.873	0.4363	0.04	0.318%
Drill diameter	2	181.998	181.998	90.9992	9.09	66.338%
Residual error	2	20.025	20.025	10.0124	–	–
Total	**8**	**274.348**	–	–	–	–

From Figure 3.6, it is found that the optimum parameters for minimum torque are obtained at spindle speed at level 2 (1,200 rpm), feed at level 2

(160 mm/min), and D_d at level 1 (6 mm). It is also observed that with an increase in D_d torque increases simultaneously.

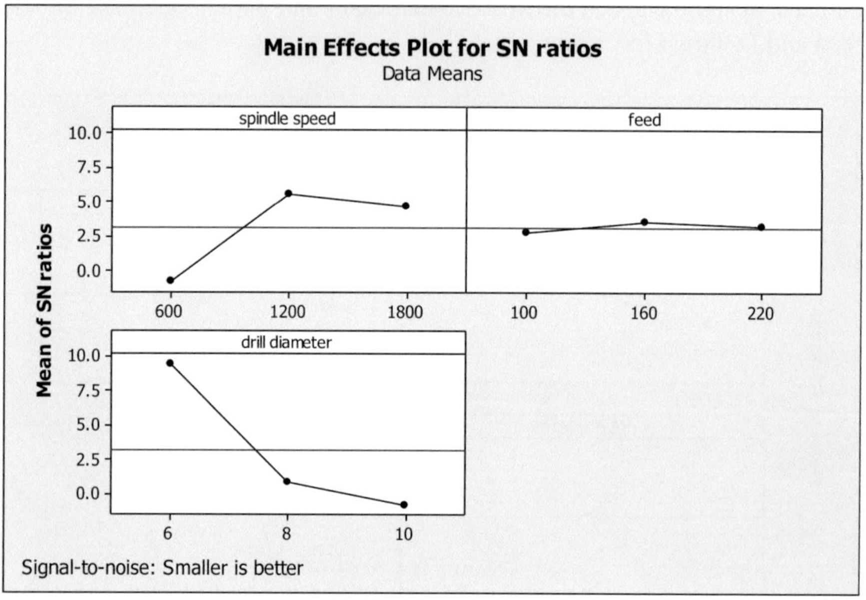

FIGURE 3.6 Main effect plot for SN ratios of torque.

From Table 3.10, it is found that D_d is the cutting parameter that is significantly affecting delamination at entry, i.e., 58.08% and feed is the parameter next to spindle speed which is affecting the D_f at entry, i.e., 28.18%.

TABLE 3.10 Analysis of Variance for SN Ratios of Delamination Factor at Entry

Source	DF	Seq SS	Adj SS	Adj MS	F	Percentage Contribution
Spindle speed	2	0.000127	0.000127	0.000063	0.04	0.508%
Feed rate	2	0.007035	0.007035	0.003518	2.32	28.18%
Drill diameter	2	0.014497	0.014497	0.007249	4.78	58.08%
Residual error	2	0.003031	0.003031	0.001516	–	–
Total	**8**	**0.02496**	–	–	–	–

From Figure 3.7, it is observed that the optimum parameters for minimum D_f at entry are obtained at spindle speed at level 3 (1,800 rpm), feed at level 1 (100 mm/min), and D_d at level 3 (10 mm). It is also observed that with an increase in spindle speed D_f decreases and with an increase in feed D_f increases.

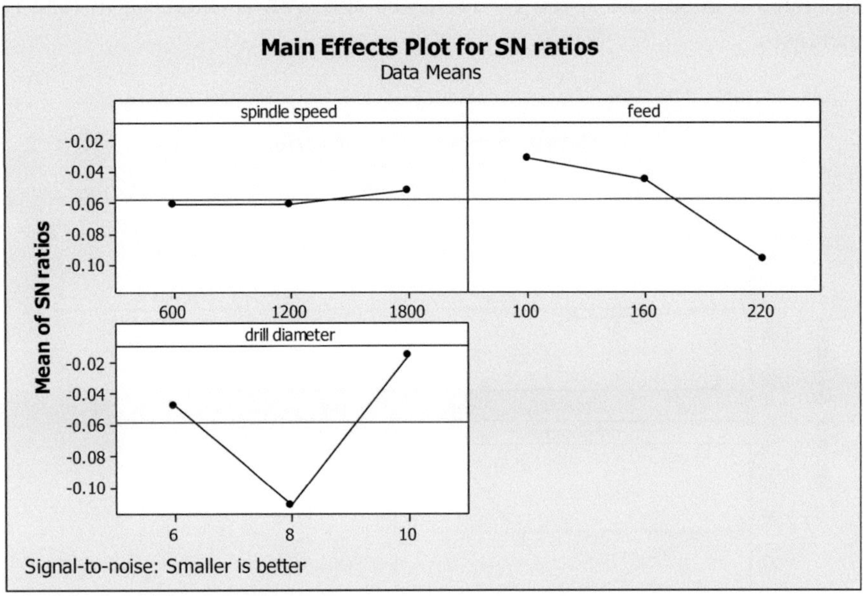

FIGURE 3.7 Main effect plot for SN ratios of delamination at entry.

Adsorption was rapid in the beginning up to the first 50 min and thereafter it is stabilized. Finally, it attained equilibrium at 180 min. There is no considerable change in percent adsorption at different temperatures, but it is slightly less at 50°C (96%) compared to other temperatures (98%).

From Table 3.11, it is found that D_d is the cutting parameter that is significantly affecting delamination at exit, i.e., 53.01% and feed is the parameter next to spindle speed which is affecting D_f at exit, i.e., 28.9%.

TABLE 3.11 Analysis of Variance for SN Ratios of Delamination Factor at Exit

Source	DF	Seq SS	Adj SS	Adj MS	F	Percentage Contribution
Spindle speed	2	0.000646	0.000646	0.000323	0.14	2.161%
Feed rate	2	0.008633	0.008633	0.004317	1.81	28.9%
Drill diameter	2	0.015841	0.015841	0.007921	3.33	53.01%
Residual error	2	0.004763	0.004763	0.002381	–	–
Total	**8**	**0.029882**	–	–	–	–

From Figure 3.8, it is found that the optimum parameters for minimum D_f at exit are obtained at spindle speed at level 3 (1,800 rpm), feed at level 1 (100 mm/min), and D_d at level 3 (10 mm). It is also observed that with an increase in spindle speed D_f decreases and with an increase in feed D_f increases.

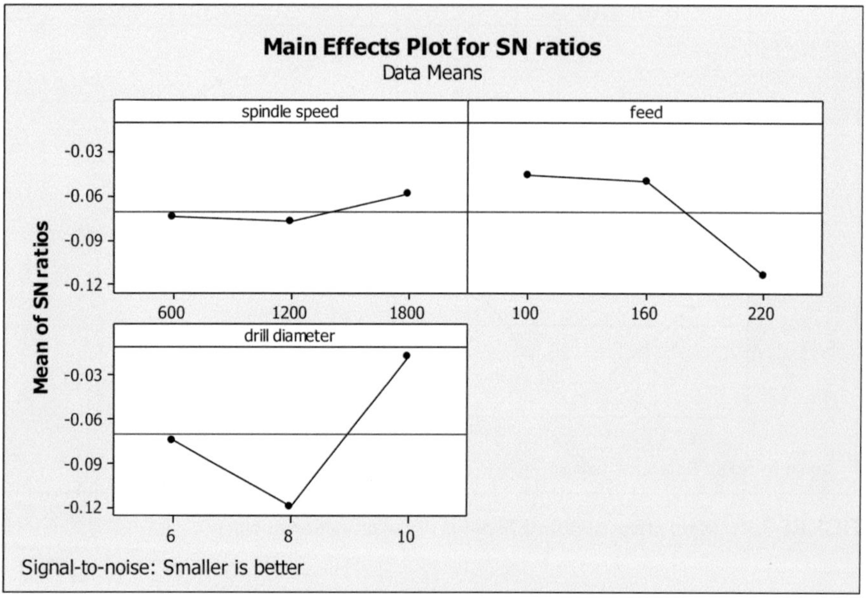

FIGURE 3.8 Main effect plot for SN ratios of delamination at exit.

From Table 3.12, it is found that D_d is the cutting parameter that is significantly affecting circularity at entry, i.e., 61.88%, and spindle speed is the parameter next to spindle speed which is affecting circularity at entry, i.e., 21.55%.

TABLE 3.12 Analysis of Variance for SN Ratios of Circularity at Entry

Source	DF	Seq SS	Adj SS	Adj MS	F	Percentage Contribution
Spindle speed	2	0.015933	0.015933	0.007967	1.59	21.55%
Feed rate	2	0.002186	0.002186	0.001093	0.22	2.957%
Drill diameter	2	0.045741	0.045741	0.022870	4.55	61.88%
Residual error	2	0.010049	0.010049	0.005025	–	–
Total	**8**	**0.073909**	–	–	–	–

From Figure 3.9, it is found that the optimum parameters for maximum circularity at entry are obtained at spindle speed at level 3 (1,800 rpm), feed at level 2 (160 mm/min), and D_d at level 1 (6 mm). It is also observed that with an increase in spindle speed circularity increases.

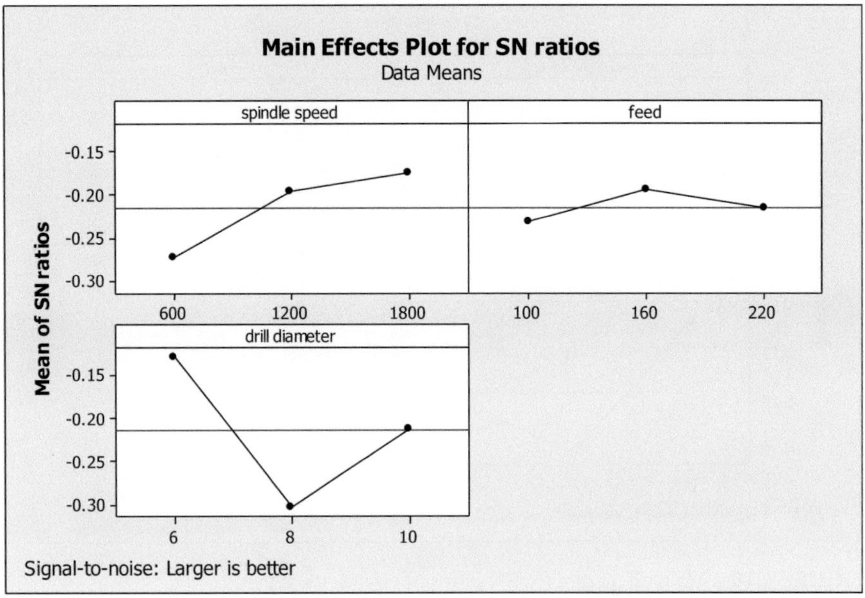

FIGURE 3.9 Main effect plot for SN ratios of circularity at entry.

From Table 3.13, it is found that D_d is the cutting parameter that is significantly affecting circularity at exit, i.e., 40.74% and feed is the parameter next to spindle speed which is affecting circularity at entry, i.e., 34.263%.

TABLE 3.13 Analysis of Variance for SN Ratios of Circularity at Exit

Source	DF	Seq SS	Adj SS	Adj MS	F	Percentage Contribution
Spindle speed	2	0.009584	0.009584	0.004792	1.41	14.64%
Feed rate	2	0.022423	0.022423	0.011211	3.31	34.263%
Drill diameter	2	0.026663	0.026663	0.013331	3.94	40.74%
Residual error	2	0.006776	0.006776	0.003388	–	–
Total	**8**	**0.065445**	–	–	–	–

From Figure 3.10, it is found that the optimum parameters for maximum circularity at exit are obtained at spindle speed at level 2 (1,200 rpm), feed at level 1 (100 mm/min), and D_d at level 1 (6 mm).

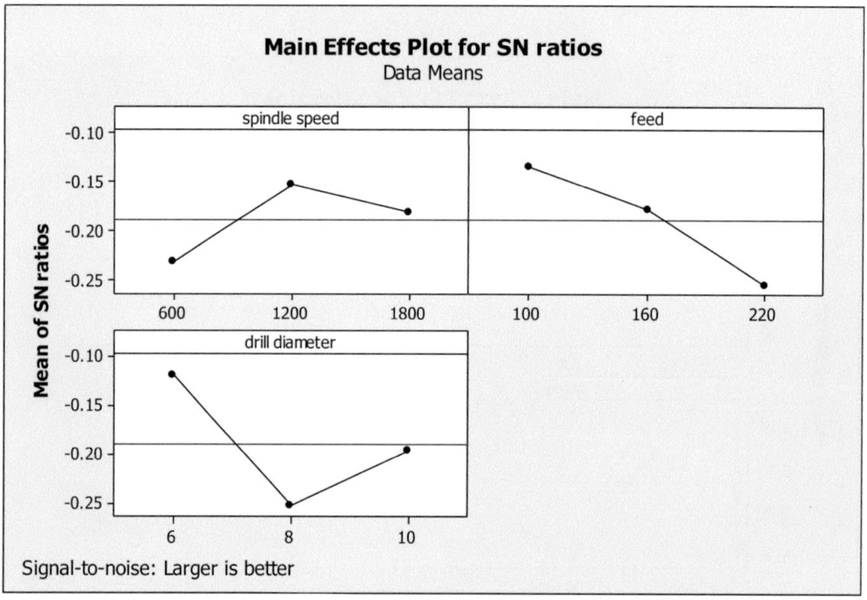

FIGURE 3.10 Main effect plot for SN ratios of circularity at exit.

Till now, we have found out the ANOVA and main plots for SN ratios of various output parameters like thrust force, torque, D_f at entry, D_f at the exit, circularity at entry, and circularity at the exit and based on the main plots for SN ratios we have found out the optimum parametric setting which must be given to get the desired individual output characteristics but, these optimum parametric settings vary from each other. As we know that while drilling, we cannot at once take all of these optimum parameters, so we have to find an optimum parameter that will give minimum thrust, torque, D_f, and maximum circularity. For such a case, we have used the utility concept. Now we can find the ANOVA and main plots for SN ratios of overall utility so that we can get the best parametric setting.

The coefficient of determination (R2) for overall utility is found out to be 99.1%. Table 3.14 shows the result of the ANOVA analysis of overall utility.

From Table 3.14, it is found that D_d is the cutting parameter that is significantly affecting overall utility, i.e., 64.4%, and spindle speed is the parameter next to spindle speed which is affecting circularity at entry, i.e., 20.205%.

TABLE 3.14 Analysis of Variance for SN Ratios of Overall Utility

Source	DF	Seq SS	Adj SS	Adj MS	F	Percentage Contribution
Spindle speed	2	33.596	33.596	16.7982	21.90	20.205%
Feed rate	2	24.072	24.072	12.0359	15.69	14.47%
Drill diameter	2	107.074	107.074	53.5368	69.78	64.4%
Residual error	2	1.534	1.534	0.7672	–	–
Total	**8**	**166.276**	–	–	–	–

It can be observed from Figure 3.11 that the optimum parameters for maximum overall utility are obtained at spindle speed at level 2 (1,200 rpm), feed at level 1 (100 mm/min), and D_d at level 1 (6 mm).

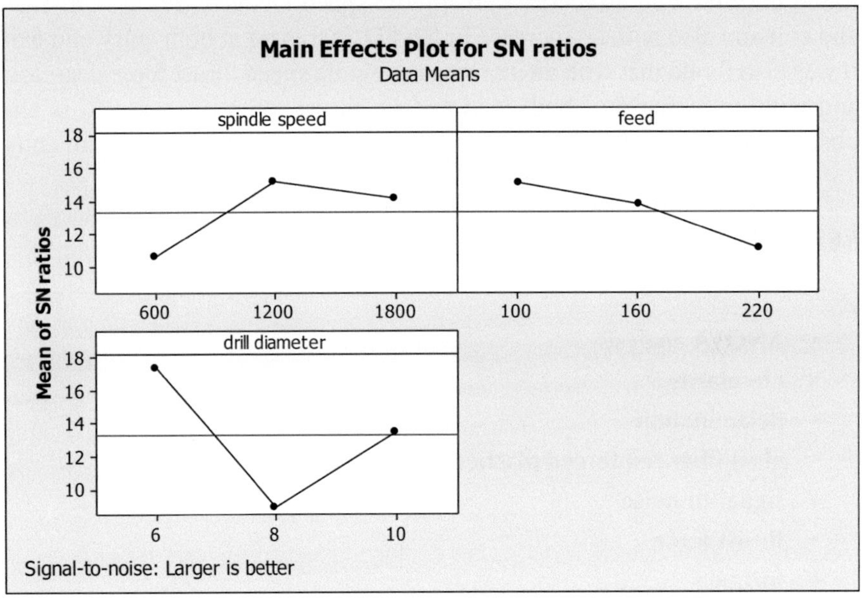

FIGURE 3.11 Main effect plot for SN ratios of overall utility.

3.4 CONFIRMATORY TEST

Finally, a confirmation test was conducted to test whether the obtained optimum parameters were correct or not. The results of the confirmation test are given in Table 3.15.

TABLE 3.15 Results of Confirmatory Test

Spindle Speed	Feed Rate	Drill Diameter	Thrust	Torque	Delamination at Entry	Delamination at Exit	Circularity at Entry	Circularity at Exit
1,200	100	6	0.074	0.17	1.0003	1.00087	0.9913	0.988

3.5 CONCLUSION

The results obtained from the confirmation test were found to give minimum thrust, minimum D_f, and maximum circularity when compared with the results of earlier responses. Hence the optimum parameters, i.e., spindle speed at level 2 (1,200 rpm), feed at level 1 (100 mm/rev), and D_d at level 1 (6 mm) obtained from the plot of SN ratios of overall utility is correct. It was found out that with an increase in spindle speed D_f decreases at both entry and exit and also with an increase in feed D_f increases at both entry and exit. It was also found that with an increase in spindle speed thrust force decreases and with an increase of both feed and D_d thrust force increases. It is also observed that with an increase in spindle speed circularity increases at entry.

KEYWORDS

- **ANOVA analysis**
- **circularity**
- **delamination**
- **glass fiber reinforced plastics**
- **signal-to-noise**
- **thrust force**
- **torque**
- **utility function**

REFERENCES

1. Abhishek, K., Sonkar, V., Datta, S., Mahapatra, S. S., Scholar, P. D., & Scholar, M. T., (2014). *Optimization in Drilling of MMC Composites: A Case Research, 1*(2), 79–83.

2. Abhishek, K., Datta, S., & Mahapatra, S. S., (2016). Multi-objective optimization in drilling of CFRP (polyester) composites: Application of a fuzzy embedded harmony search (HS) algorithm. *Meas. J. Int. Meas. Confed., 77*, 222–239. https://doi.org/10.1016/j.measurement.2015.09.015.
3. Mattson, M., (2009). *CNC Programming: Principles & Applications* (2nd edn.). Cengage Learning.
4. Sonkar, V., Abhishek, K., Datta, S., & Mahapatra, S. S., (2014). Multi-objective optimization in drilling of GFRP composites: A degree of similarity approach. In: *3rd International Conference on Materials Processing and Characterization (ICMPC 2014)* (Vol. 6, pp. 538–543). Elsevier BV. https://doi.org/10.1016/j.mspro.2014.07.068.
5. Abhishek, K., Datta, S., & Mahapatra, S. S., (2014). Optimization of thrust, torque, entry, and exist delamination factor during drilling of CFRP composites. *Int. J. Adv. Manuf. Technol., 76*(1–4), 401–416. https://doi.org/10.1007/s00170-014-6199-3.
6. Madhavan, S., & Prabu, S. B., (2012). Experimental investigation and analysis of thrust force in drilling of carbon fiber reinforced plastic composites using response surface methodology. *Int. J. Mod. Eng. Res., 2*(4), 2719–2723.
7. Shavali, G. S., Venugopal, G. E., Yadav, G. P. K., & Bandhu, D., (2016). Tensile and flexural characterization of Nomex and e-glass fiber reinforced epoxy composites. *Int. J. Sci. Res. Dev., 4*(1), 981–983.
8. Bandhu, D., Thakur, A., Purohit, R., Verma, R. K., & Abhishek, K., (2018). Characterization & evaluation of Al7075 MMCs reinforced with ceramic particulates and influence of age hardening on their tensile behavior. *J. Mech. Sci. Technol., 32*(7). https://doi.org/10.1007/s12206-018-0615-9.
9. Thakur, A., Purohit, R., Rana, R. S., & Bandhu, D., (2018). Characterization and evaluation of mechanical behavior of epoxy-CNT-bamboo matrix hybrid composites. In: *Materials Today: Proceedings* (Vol. 5). https://doi.org/10.1016/j.matpr.2017.11.655.
10. Tripathi, D. R., Vachhani, K. H., Bandhu, D., Kumari, S., Kumar, V. R., & Abhishek, K., (2021). Experimental investigation and optimization of abrasive waterjet machining parameters for GFRP composites using metaphor-less algorithms. *Mater. Manuf. Process., 36*(7), 803–813. https://doi.org/10.1080/10426914.2020.1866193.
11. Naga, P. S. M., Devaki, D. K., & Bandhu, D., (2016). Characterization of *Aegle marmelos* fiber reinforced composite. *Int. J. Eng. Res., 5*(Special 2), 345–349. https://doi.org/10.17950/ijer/v5i2/009.
12. Tripathi, D. R., Vachhani, K. H., Kumari, S., Dinbandhu, & Abhishek, K., (2020). Experimental investigation on material removal rate during abrasive water jet machining of GFRP composites. *Mater. Today Proc., 26*(Part 2), 1389–1392. https://doi.org/10.1016/j.matpr.2020.02.280.
13. Byrne, D., & Taguchi, S., (1987). Taguchi approach to parameter design. *Qual. Prog., 20*, 19–26.
14. Singh, I., Bhatnagar, N., & Viswanath, P., (2008). Drilling of uni-directional glass fiber reinforced plastics: Experimental and finite element study. *Mater. Des., 29*(2), 546–553. https://doi.org/10.1016/j.matdes.2007.01.029.
15. Mohan, N. S., Ramachandra, A., & Kulkarni, S. M., (2005). Influence of process parameters on cutting force and torque during drilling of glass-fiber polyester reinforced composites. *Compos. Struct., 71*(3, 4), 407–413. https://doi.org/10.1016/j.compstruct.2005.09.039.
16. Hocheng, H., & Tsao, C. C., (2003). Comprehensive analysis of delamination in drilling of composite materials with various drill bits. In: *Journal of Materials Processing*

Technology (Vol. 140, pp. 335–339). Elsevier. https://doi.org/10.1016/S0924-0136(03)00749-0.

17. Kilickap, E., (2010). Optimization of cutting parameters on delamination based on Taguchi method during drilling of GFRP composite. *Expert Systems with Applications* (pp. 6116–6122). Pergamon. https://doi.org/10.1016/j.eswa.2010.02.023.
18. Enemuoh, E. U., El-Gizawy, A. S., & Chukwujekwu, O. A., (2001). An approach for development of damage-free drilling of carbon fiber reinforced thermosets. *Int. J. Mach. Tools Manuf., 41*(12), 1795–1814. https://doi.org/10.1016/S0890-6955(01)00035-9.
19. Krishnaraj, V., Prabukarthi, A., Ramanathan, A., Elanghovan, N., Kumar, M. S., Zitoune, R., & Davim, J. P., (2012). Optimization of machining parameters at high speed drilling of CARBON FIBER REINFORCED PLASTIC (CFRP) laminates. *Compos. Part B Eng., 43*(4), 1791–1799. https://doi.org/10.1016/j.compositesb.2012.01.007.
20. Chen, W. C., (1997). Some experimental investigations in the drilling of carbon fiber-reinforced plastic (CFRP) composite laminates. *Int. J. Mach. Tools Manuf., 37*(8), 1097–1108. https://doi.org/10.1016/S0890-6955(96)00095-8.
21. Capello, E., (2004). Workpiece damping and its effect on delamination damage in drilling thin composite laminates. *J. Mater. Process. Technol., 148*(2), 186–195. https://doi.org/10.1016/S0924-0136(03)00812-4.
22. Dharan, C. K. H., & Won, M. S., (2000). Machining parameters for an intelligent machining system for composite laminates. *Int. J. Mach. Tools Manuf., 40*(3), 415–426. https://doi.org/10.1016/S0890-6955(99)00065-6.
23. Zitoune, R., & Collombet, F., (2007). Numerical prediction of the thrust force responsible of delamination during the drilling of the long-fiber composite structures. *Compos. Part A Appl. Sci. Manuf., 38*(3), 858–866. https://doi.org/10.1016/j.compositesa.2006.07.009.
24. Sharif, S., & Rahim, E. A., (2007). Performance of coated – and uncoated-carbide tools when drilling titanium alloy-Ti-6Al4V. *J. Mater. Process. Technol., 185*(1–3), 72–76. https://doi.org/10.1016/j.jmatprotec.2006.03.142.
25. Durão, L. M. P., Gonçalves, D. J. S., Tavares, J. M. R. S., De Albuquerque, V. H. C., Aguiar, V. A., & Torres, M. A., (2010). Drilling tool geometry evaluation for reinforced composite laminates. *Compos. Struct., 92*(7), 1545–1550. https://doi.org/10.1016/j.compstruct.2009.10.035.

CHAPTER 4

IDENTIFICATION OF PROBLEMS AND THEIR PRIORITIZATION IN THE TEXTILE INDUSTRIES USING MADM TECHNIQUES: FUZZY-TOPSIS AND FUZZY-VIKOR METHODS

JAY VAISHNANI[1], NITIN KUMAR LAXKAR[1], DIN BANDHU[2], KUMAR ABHISHEK[1], and SONI KUMARI[3]

[1]Department of Mechanical and Aero-Space Engineering, Institute of Infrastructure, Technology, Research, and Management (IITRAM), Ahmedabad, Gujarat, India

Department of Mechanical and Industrial Engineering, Manipal Institute of Technology Bengaluru, Manipal Academy of Higher Education, Manipal, Karnataka, India

[2]Department of Mechanical Engineering, GLA University, Mathura, Uttar Pradesh, India

ABSTRACT

This chapter deals with the identification of problems and their prioritization in the textile industries across Gujarat, India. The problems and parameters selected in this work are based on the primary data collected from the internet sources as well as from the textile industries in Gujarat through a questionnaire survey with complementary interviews. A total of seven problems have been identified and corresponding to these problems, six parameters have been selected. Weights to the parameters have been assigned using the modified digital logic (MDL) method followed by the implementation of

Decision-Making Models and Applications in Manufacturing Environments. Pushpdant Jain, Kumar Abhishek, & Prasenjit Chatterjee (Eds.)

the intended approaches of fuzzy-TOPSIS and fuzzy-VIKOR for prioritizing these problems. The fuzzy-TOPSIS approach proposes quality defects as the top-ranked problems among all the listed problems in the textile industries across Gujarat. Likewise, the fuzzy-VIKOR approach also advocates the same problem as the top raked among all the listed problems. Thus, both of these methodologies have ranked all the problems successfully while picking up the quality defects as the top-ranked problem. Among all the listed problems, the problem of quality defects has a larger potential to affect the efficiency and effectiveness of the textile industries in Gujarat.

4.1 INTRODUCTION

From the historical period to the present era, apparel has held a significant role in human life. It is known from the history's pages that the clothing has been produced since 2000 BC. Next to agriculture, apparel manufacturing became one of the large-scale economic activities that provide substantial employment. In today's world, textile and garment industries are important aspects of the global economy, employing tens of millions of people worldwide, predominantly women, in nearly 200 countries [1–3]. The textile industry in India is a major contributor to the country's GDP. It is one of India's most important export contributors, accounting for nearly 11.4% of the country's total export earnings in the fiscal year 2018–2019, totaling nearly USD 37.5 billion (INR 2,596 billion) and growing at a CAGR of 7% since 2004–2005. The industry is currently valued at USD 150 billion (INR 10,834 billion), with the domestic market accounting for 75% of total consumption and exports accounting for 25%. India came in second with 6% of global textile exports and fifth with 4% of global apparel exports. In employment terms, the Textile and apparel industry in India is the second largest after agriculture, employing nearly 100 million people directly and indirectly [4–6]. The Indian textile industry is generally categorized into three categories. The first one is the highly mechanized mill industry producing yarn and fabrics; the second one is the power loom factories making cloths from mill-made yarn, and the third category is the widely scattered handloom units making cloths from both mill-made and handspun yarn. Thus, the structure of this industry is very complex [7–9]. Gujarat is one of those states which is growing very fast with respect to time and leading industrial states in India and playing a big role in industrialization [10–12]. The state is well known for the development of hybrid cotton, ginning, compound mills, spinning units, and independent processing houses. Textile Industry, in general, has suffered a setback, in specific, in the organized sector, all over the country in the recent past

[13–15]. In this context, a study was done by Narayanan [16] that focuses on the structure, policy framework, and performance of the Indian textiles and garment industries. The study also explores the policy measures undertaken by the Indian government in the further uplifting of this sector in terms of production and trade. The study further narrates the difficulties faced by this sector and its competitiveness. Lal [17] discussed the export performance of the textile industry in developing nations. The purpose of the study was to identify and analyzation of the factors which have influenced the export performance in these countries. The study reveals that the current production of textile industries is highly influenced by some factors namely research and development, methods of farming, knowledge of soils, creation of software and the application of biotechnology, design capabilities, technical expertise, teaching, supervision, marketing, and management. Dixit & Lal [18] have quantitively analyzed the importance and contribution of the textile sector in inclusive development. The outcome of the work reveals that the textile industry leads to work growth and inclusion, however, it does not play a constructive position in terms of social justice due to its disorganized nature. Sakamoto et al. [19] investigated the real situation faced by the textile industry concerning environmental compliance. The obstacles and problems faced by factories in the construction and operation of effluent treatment plants (ETPs) are illustrated in this report. This is attributed to a lack of commitment on the part of corporations to comply with environmental legislation, as well as poor oversight and compliance by government officials. The setup's development and availability were also a problem as per the study. Franco [20] conducted a study and looked at the difficulties incumbent textile industries encountered in modifying their goods to adhere to the Cradle-to-Cradle system and circular economy concepts. The study's results provide a dynamic understanding of how mutual supplier-buyer creativity considerations (such as supply chain location, power balance, and a common vision) interact with complex aspects of product design, such as basic materials, layout, and versatility, to evaluate the production speed and quantity of circular products to be distributed, returned, and ultimately repurposed.

From the above-discussed literature reviews, it is clear that a lot of work has bees been done in the field of the textile and apparel industry by many academicians, scholars, policymakers, and manufacturing firms. Besides this, a few bits of literature also throw some light on the difficulties faced by textile industries in several developed and developing countries and also suggested a solution for them. In such firms, there is the existence several problems simultaneously, however, due to the lack of information or knowledge, it is quite difficult to prioritize these existing problems to find out their solution effectively. The implementation of several optimization algorithms, which have generally proven their viability in different areas of applications [21–30], has not yet been incorporated in the textile industries. In this regard where the difficulties must be prioritized for the smooth running of the firm, very little to no work has been attempted. Therefore, a case study has been done in which the problem identification and their prioritization in context with the textile industries in Gujarat, India has endeavored. To do so, an integrated approach of fuzzy-TOPSIS-VIKOR has been proposed here.

4.2 EXPERIMENTAL METHODS AND MATERIALS

Based on the primary data collected from the internet sources as well as from the textile industries in Gujarat through a questionnaire survey with complementary interviews, the most common problems along with their causes and effects have been identified. Table 4.1 represents all such common problems faced by the textile industries in Gujarat.

TABLE 4.1 Common Problems in the Textile Industry with Causes and Effects

SL. No.	Problems	Description of Causes	Effects
1.	Quality defects	Exposed raw edge, fabric defects, and defective snaps.	Low productivity, poor quality, low price of finished products.
2.	Shortage of raw material	Raw materials determine 35% of the total production cost. The country is short of cotton.	Low production and sickness to the mills.
3.	Raw material quality	The quality of raw material varies as the industry has to purchase the new cotton due to lack of storage area and huge demand for the clothes.	Low productivity, poor quality.
4.	Wastage of material	The use of conventional machinery and unskilled labor leads to the wastage of cotton.	High cost, low profit.
5.	Noise level	In SSIs, due to conventional machining, high level of noise causes.	Mental effects on labor, low productivity.
6.	Product fluctuation	Due to rapid change in the market, changing die cause more production time.	Low productivity.
7.	Competition in the foreign countries.	The Indian textile goods are facing stiff competition in foreign markets from Taiwan, South Korea, and Japan.	Losses of Indian textile industries.

Before the implementation of the proposed methodology, six parameters, based on the survey among the textile industries, have been selected. These parameters are independent of each other but have a very strong relationship with the problems listed in Table 4.1. Table 4.2 lists the selected parameters used in this work.

TABLE 4.2 List of Selected Parameters

SL. No.	Parameters	Code	Sub-Parameters
1.	Cost	Z1	Cost of poor quality, loss of brand value.
2.	Safety	Z2	The safety of the humans and assets involved.
3.	Maintenance	Z3	Maintainability, downtime.
4.	Environment	Z4	Health hazards, pollution.
5.	Automation	Z5	Degree of automation, human resistance to automation.
6.	Waste management	Z6	Chemical and textile waste management.

4.2.1 *FUZZY TOPSIS METHODOLOGY*

In small-scale industries, especially in the textile industry, the failure rate highly depends on the selection of the alternatives of the problems. Therefore, a Fuzzy TOPSIS approach has been used for problem selection in the textile industry. The method of fuzzy technique for order preference by similarity in ideal solution (Fuzzy TOPSIS) uses MDL method to obtain weightage of all basic parameters that have relation to the problem alternatives. Based on assigned weightage to the problems, a relative evaluation is done for obtaining the solution. The approach includes the following steps [31, 32]:

- **Step 1: Assigning Weightage to the Selected Parameters Using the MDL Method:** Here, the MDL method has been used for calculating the weightage of a particular parameter. Assigning weightage is very important in the TOPSIS method. This is due to the data collection which, generally, happens in the form of 1, 2, 3, where 1 means low, 2 means average, and 3 means high and after averaging these values, it will be possible to find the weightage for the variables. Therefore, initially, a weightage is assigned to all the selected parameter using the following formulae:

$$W_j = \frac{P_j}{\sum_{j=1}^{n} P_j} \quad (1)$$

where; P_j is the rating of the parameter; n is the number of parameters; and W_j is the final weightage.

- **Step 2: Defining Linguistic Variables (LVs):** These linguistic variables (LVs) will be useful in defining the relation between industrial problems and parameters (Eqn. (2)). For example, 'weight' is a LV whose values are 'very low,' 'low,' 'medium,' 'high,' 'very high,' etc. Fuzzy numbers can also represent these linguistic values.

$$\begin{cases} 0 & \text{if } x \leq a \\ \frac{(x-a)}{(b-a)} & \text{if } a \leq x \leq b \\ \frac{(c-x)}{(c-b)} & \text{if } b \leq x \leq c \\ 0 & \text{if } x \geq c \end{cases} \quad (2)$$

For each LV, define a trapezoidal fuzzy number as shown in Figure 4.1.

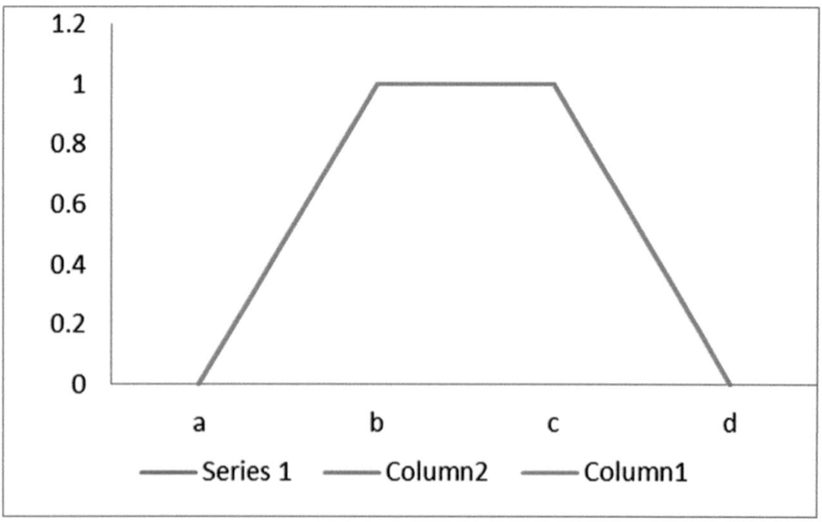

FIGURE 4.1 Trapezoidal fuzzy number.

μ is a membership function whose value lies between 0 and 1. One means all experts say that the given statement is fully true and the same follows for 0 too. Here, five linguistic terms have been defined in the range of very high (VH) to very low (VL) along with their corresponding trapezoidal fuzzy number as shown in Table 4.3.

TABLE 4.3 Linguistic Variables with Their Corresponding Trapezoidal Fuzzy Number

Linguistic Variables	Abbreviation	Fuzzy Number
Very high	VH	(7,8,9,9)
Moderate-high	MH	(5,6,7,8)
Average	A	(3,4,5,6)
Moderate-low	ML	(1,2,3,4)
Very low	VL	(0,0,1,2)

- **Step 3: Assigning Linguistic Variables and Constructing the Matrix:** Here, LVs are assigned to define the relationship between the problems and parameters. With this, a matrix is constructed for all defined problems.
- **Step 4: Defuzzification of the Trapezoidal Fuzzy Number and Matrix Construction with Them:** Here, the trapezoidal fuzzy number assigned to each LV is defuzzified using the centroid method (Eqn. (3)). With these defuzzified numbers, a matrix is constructed.

$$f_{ij} = Defuzzi\left(x_{ij}\right) = \frac{\int \mu\left(x\right) x dx}{\int \mu\left(x\right) dx} \tag{3}$$

- **Step 5: Calculating Normalized Fuzzy Decision Matrix (r_{ij}):** The normalized decision matrix is calculated using Eqn. (4):

$$r_{ij} = \frac{f_{ij}}{\sqrt{\sum_{i=1}^{m} (f_{ij})^2}} \tag{4}$$

- **Step 6: Calculating Weighted Decision Matrix (V_{ij}):** The weighted decision matrix is calculated using Eqn. (5):

$$V_{ij} = [r_{ij}]_{m\times n} \times [W_j]_{n\times m} \tag{5}$$

Here; $[r_{ij}]_{m\times n}$ is a normalized fuzzy decision matrix; and $[W_j]_{n\times m}$ is weightage matrix.

- **Step 7: Calculating Positive Ideal and Negative Ideal Solutions (NIS):** The positive ideal solution (PIS) (V_j^+) and NIS (V_j^-) are calculated using Eqns. (6) and (7), respectively:

$$V_j^+ = \{(max\ V_{ji}, j \in J_1), (min\ V_{ji}, j \in J_2)\}, \forall_j \tag{6}$$

$$V_j^- = \{(min\ V_{ji}, j \in J_1), (max\ V_{ji}, j \in J_2)\}, \forall_j \tag{7}$$

- **Step 8: Calculating Distances from the Positive Ideal Solution (PIS) and Negative Ideal Solution (NIS):** The distances from the PIS (d_i^+) and NIS (d_i^-) are calculated using Eqns. (8) and (9), respectively:

$$d_i^+ = \sqrt{\sum_{j=1}^{n} (V_{ij} - V_j^+)^2},\ i = 1,2,\ldots m \tag{8}$$

$$d_i^- = \sqrt{\sum_{j=1}^{n} (V_{ij} - V_j^-)^2},\ i = 1,2,\ldots m \tag{9}$$

- **Step 9: Calculating the TOPSIS Rank Index (C_i^+):** At last, the ranking to the problems is assigned using Eqn. (10).

$$C_i^+ = \frac{d_i^-}{(d_i^+) + (d_i^-)} \tag{10}$$

4.2.2 *FUZZY TOPSIS METHODOLOGY*

This proposed methodology is somewhat identical to the previous one where the initial four steps are the same. The variation comes in from the fifth step onwards, discussed as follows [33–37]:

- **Step 1:** Assigning weightage to the selected parameters using the MDL method.
- **Step 2:** Defining linguistic variables.
- **Step 3:** Assigning linguistic variables and constructing the matrix.
- **Step 4:** Defuzzification of the trapezoidal fuzzy number and matrix construction with them.

  Here, the trapezoidal fuzzy number assigned to each LV is defuzzified using the centroid method (Eqn. (11)). With these defuzzified numbers, a matrix is constructed.

$$Z_{ij} = Defuzzi(x_{ij}) = \frac{\int \mu(x) x dx}{\int \mu(x) dx} \tag{11}$$

- **Step 5: Calculating Positive Ideal and Negative Ideal Solutions (NIS):** Here, PIS (X_i^+) and NIS (X_i^-) for each defuzzified criteria using Eqns. (12) and (13).

$$X_i^+ = Min\,(Z_{ij}) \tag{12}$$

$$X_i^- = Min\,(Z_{ij}) \tag{13}$$

where; Z_{ij} is defuzzified value.

- **Step 6: Computing Fuzzy Weighted Sum (S_i):** Now the fuzzy weighted sum is calculated using Eqn. (14).

$$S_i = \sum W_j \times \frac{(X_i^+ - Z_{ij})}{(Z_{ij} - X_i^-)} \tag{14}$$

here; W_j is calculated weightage value; X_i^+ is the positive ideal value; X_i^- is the negative ideal value; and Z_{ij} is defuzzified value.

- **Step 7: Determining R_i:** Using Eqn. (15), the value of R_i is determined.

$$R_i = Max\,(S_i) \tag{15}$$

- **Step 8: Computing the VIKOR Rank Index (Q_i):** Using Eqn. (16), the VIKOR rank index (Q_i) is computed and on this basis, the ranking will be assigned.

$$Q_i = (1-\mu)\frac{(S_i - S^*)}{(S^- - S^*)} + (\mu)\frac{(R_i - R^*)}{(R^- - R^*)} \tag{16}$$

4.3 RESULTS AND DISCUSSION

Mathematical computations for obtaining the final result have been discussed in this section. Initially, the six selected parameters are assigned some weightage concerning the seven problems faced by textile industries in Gujarat. To do so, the MDL method is employed. Table 4.4 illustrates the weightage assigned to each parameter.

TABLE 4.4 The Weightage Assigned to Each Parameter Using the MDL Method

Parameters	S1	S2	S3	S4	S5	S6	S7	Weightage	Rank
Cost	2	3	2	1	2	3	2	0.208	3
Safety	3	2	3	3	3	2	2	0.250	1
Maintenance	2	1	1	1	2	1	2	0.139	5
Environment	3	1	2	2	3	2	3	0.222	2
Automation	1	2	1	1	1	1	1	0.111	6
Waste management	2	3	2	1	2	3	1	0.194	4

Note: Here; 1 is a relatively low weightage given by S's expert; 2 is a relatively medium weightage given by S's expert; 3 is a relatively high weightage given by S's expert; and S1, S2, S3, S4, S5, S6, S7, are different experts.

In the next phase, each of the seven alternatives is assigned the LVs as shown in Table 4.5.

TABLE 4.5 Linguistic Decision Matrix for Each Alternative

Alternatives	Z1	Z2	Z3	Z4	Z5	Z6
Quality defects	VH	ML	VH	VL	VH	A
Shortage of raw material	A	VL	ML	MH	VL	VL
Raw material quality	VH	VL	MH	VH	VL	VH
Wastage of material	ML	VL	A	VL	VH	VH
Noise level	VL	VL	A	A	MH	VL
Product fluctuation	VL	A	VH	VL	VH	ML
Competition in a foreign country	MH	VL	ML	A	MH	VL

Note: Here, VH (very high) describes a very high correlation or dependency with respect to parameters and problems. MH (moderate-high) describes moderate-high correlation or dependency with respect to parameters and problems. A (average) describes average correlation or dependency with respect to parameters and problems. ML (moderate-low) describes moderate-low correlation or dependency with respect to parameters and problems. VL (very low) describes a very low correlation or dependency with respect to parameters and problems.

A fuzzy number is assigned to each corresponding LV as shown in Table 4.6. The graphical representation of the fuzzy membership function is shown in Figure 4.2.

TABLE 4.6 Linguistic Variables with Their Corresponding Trapezoidal Fuzzy Number

Linguistic Variables	Abbreviation	Fuzzy Number
Very high	VH	(7,8,9,9)
Moderate-high	MH	(5,6,7,8)
Average	A	(3,4,5,6)
Moderate-low	ML	(1,2,3,4)
Very low	VL	(0,0,1,2)

The fuzzy values are defuzzified as presented in Table 4.7.

As discussed in the previous section that the initial four steps are common to both of the proposed methodologies. Therefore, the values listed in Tables 4.4–4.7 are utilized in both intended methods.

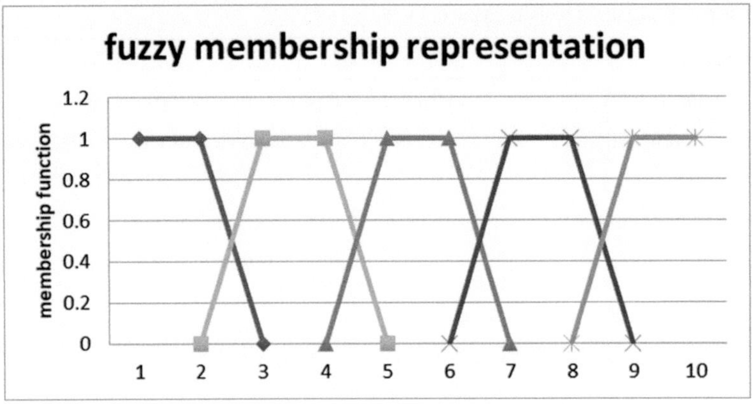

FIGURE 4.2 Representation of fuzzy membership function.

TABLE 4.7 Defuzzification of Fuzzy Values

Alternatives	Z1	Z2	Z3	Z4	Z5	Z6
Quality defects	8.222	2.500	8.222	0.778	8.222	4.500
Shortage of raw material	4.500	0.778	2.500	6.500	0.778	0.778
Raw material quality	8.222	0.778	2.500	8.222	0.778	8.222
Wastage of material	2.500	0.778	4.500	0.778	8.222	8.222
Noise level	0.778	0.778	4.500	4.500	6.500	0.778
Product fluctuation	0.778	4.500	8.222	0.778	8.222	2.500
Competition in foreign country	6.500	0.778	2.500	4.500	6.500	0.778

4.3.1 IMPLEMENTATION AND ANALYSIS OF FUZZY TOPSIS METHOD

Now, a normalized fuzzy decision matrix (r_{ij}) is calculated followed by a weighted decision matrix (V_{ij}) and tabulated in Tables 4.8 and 4.9, respectively.

TABLE 4.8 Normalized Fuzzy Decision Matrix (r_{ij})

Alternatives	Z1	Z2	Z3	Z4	Z5	Z6
Quality defects	0.542	0.165	0.542	0.051	0.542	0.296
Shortage of raw material	0.535	0.092	0.297	0.714	0.092	0.092
Raw material quality	0.567	0.053	0.172	0.567	0.053	0.567
Wastage of material	0.196	0.061	0.353	0.061	0.644	0.644
Noise level	0.084	0.084	0.489	0.489	0.707	0.084
Product fluctuation	0.061	0.353	0.645	0.061	0.645	0.196
Competition in foreign country	0.613	0.073	0.236	0.425	0.613	0.073

TABLE 4.9 Weighted Decision Matrix (V_{ij})

Alternatives	Z1	Z2	Z3	Z4	Z5	Z6
Quality defects	0.116	0.041	0.075	0.011	0.060	0.057
Shortage of raw material	0.111	0.023	0.041	0.158	0.010	0.017
Raw material quality	0.117	0.013	0.023	0.125	0.005	0.110
Wastage of material	0.040	0.015	0.049	0.013	0.071	0.124
Noise level	0.017	0.021	0.067	0.108	0.078	0.016
Product fluctuation	0.013	0.088	0.089	0.013	0.071	0.038
Competition in foreign country	0.127	0.018	0.032	0.094	0.068	0.014

TABLE 4.10 The Positive Ideal Solution (V_j^+) and Negative Ideal Solution (V_j^-)

	V_j^+	V_j^-
Z1	0.013	0.127
Z2	0.088	0.013
Z3	0.089	0.023
Z4	0.158	0.011
Z5	0.078	0.005
Z6	0.124	0.014

Next, the distances from the PIS (d_i^+) and NIS (d_i^-) are calculated along with the TOPSIS rank index (C_i^+). Based on the rank index calculations, the problems are prioritized (Table 4.11).

TABLE 4.11 Calculation of TOPSIS Rank Index (C_i^+) and rank Determination

Alternatives	D^i+	D^i-	C^i	Rank
Quality defects	0.1985	0.0921	0.8227	1
Shortage of raw material	0.1794	0.1496	0.4547	6
Raw material quality	0.1655	0.1493	0.4742	5
Wastage of material	0.1695	0.1571	0.4810	4
Noise level	0.1384	0.1698	0.5509	2
Product fluctuation	0.1564	0.1671	0.5165	3
Competition in a foreign country	0.1934	0.1047	0.3512	7

4.3.2 IMPLEMENTATION AND ANALYSIS OF FUZZY VIKOR METHOD

After the defuzzification of fuzzy numbers, a fuzzy weighted sum (S_i) is calculated. Using this calculation, the value R_i is determined as shown in Table 4.12.

TABLE 4.12 Calculation of S_j and R_j

Alternatives	Z1	Z2	Z3	Z4	Z5	Z6	S_i	R_i
Quality defects	0.208	0.134	0.00	0.222	0.00	0.097	0.661	0.222
Shortage of raw material	0.104	0.25	0.139	0.0510	0.111	0.194	0.604	0.250
Raw material quality	0.208	0.25	0.139	0.00	0.111	0.00	0.708	0.250
Wastage of material	0.048	0.25	0.090	0.222	0.00	0.00	0.61	0.250
Noise level	0.00	0.25	0.090	0.111	0.026	0.194	0.671	0.250
Product fluctuation	0.00	0.00	0.00	0.222	0.00	0.149	0.371	0.222
Competition in foreign country	0.160	0.250	0.139	0.111	0.026	0.194	0.88	0.250

These calculations help in determining the value of the VIKOR rank index (Q_i) which further leads to arranging the ranks of the problems (Table 4.13).

4.4 CONCLUSION

The textile industries in Gujarat are primarily affected by the selection of problem-solving techniques and the success rate of these techniques depends upon the accuracy, efficiency, and effectiveness of their respective project management teams. However, these firms still have some problems but lack of prioritization results in long-suffering with them. In this perspective, here, an intended methodology is proposed using fuzzy-TOPSIS and fuzzy-VIKOR approaches. The implementation of the fuzzy-TOPSIS approach suggests the quality defects as the top-ranked problems among all the listed problems in the textile industries across Gujarat. Likewise, the fuzzy-VIKOR approach represents the exact problem as the top-ranked challenge out of all the mentioned problems. This means that both approaches have correctly prioritized all the issues, with quality defects being identified as the most significant ones. The issue of quality defects has the greatest potential to impede the effectiveness and efficiency of Gujarat's textile businesses, although all of the problems outlined here are important.

TABLE 4.13 Calculation of VIKOR Rank Index (Q_i) and Rank Determination

Alternatives	S_i	R_i	Q_i	Rank
Quality defects	0.661	0.222	0.285	1
Shortage of raw material	0.604	0.250	0.729	4
Raw material quality	0.708	0.250	0.723	3
Wastage of material	0.61	0.250	0.734	5
Noise level	0.671	0.250	0.795	6
Product fluctuation	0.371	0.222	0.500	2
Competition in foreign country	0.88	0.250	1.00	7
S^*, R^*	0.371	0.222	–	–
S^-, R^-	0.88	0.250	–	–

KEYWORDS

- **fuzzy**
- **Gujarat**
- **multi-attributes decision-making**
- **textile**
- **Vlse Kriterijumska Optimizacija Kompromisno Resenje**

REFERENCES

1. Khude, P., (2017). A review on energy management in textile industry. *Innov. Energy Res., 06*(02), 1–5. https://doi.org/10.4172/2576-1463.1000169.
2. Bhaskar, M. S., Verma, P., & Kumar, A., (2013). Indian textile industries towards energy efficiency movement. *Int. J. Environ. Sci. Dev. Monit., 4*(3), 36–39.
3. Tandon, N., & Reddy, E. E., (2015). A study on emerging trends in e-commerce in India and its impact on the retail industry. *Int. J. Adv. Res. Technol., 2*(7), 267–276. https://doi.org/10.16962/elkapj/si.bm.icrbit-2015.40.
4. Raichurkar, P., & Ramachandran, M., (2015). Recent trends and developments in the textile industry in India. *Int. J. Text. Eng. Process., 1*(4), 47–50.
5. *Overview of Textiles Industry in India and Impact of COVID-19* – Blogs – Televisory https://www.televisory.com/blogs/-/blogs/overview-of-textiles-industry-in-india-and-impact-of-covid-19 (accessed on 17 February 2022).
6. *Textile Industry in India: Overview, Market Size, Exports, Growth.*| IBEF https://www.ibef.org/industry/textiles.aspx (accessed on 17 February 2022).
7. Wright, A. C., (1995). Strategy and structure in the textile industry: Spencer love and Burlington mills, 1923–1962. *Bus. Hist. Rev., 69*(1), 42–79. https://doi.org/10.2307/3117120.

8. Uddin, F., (2019). Introductory chapter: Textile manufacturing processes. In: *Textile Manufacturing Processes*. Intech Open. https://doi.org/10.5772/intechopen.87968.
9. Kozar, J. M., & Connell, K. Y. H., (2015). *Sustainability in the Apparel and Textiles Industry: A Conceptual Paper Addressing Previous Findings and Areas of Future Research* (pp. 229–237). Springer, Cham. https://doi.org/10.1007/978-3-319-10912-1_76.
10. Bagchi, A. K., Das, P., & Chattopadhyay, S. K., (2005). Growth and structural change in the economy of Gujarat, 1970–2000. *Econ. Polit. Wkly., 40*(28), 3039–3047.
11. Awasthi, D. N., (2000). Recent changes in Gujarat industry: Issues and evidence. *Econ. Polit. Wkly., 35*(35, 36), 3183–3187.
12. Dholakia, R. H., (2000). Liberalization in Gujarat: Review of recent experience. *Econ. Polit. Wkly., 35*(35, 36), 3121–3124.
13. Sharma, A., & Narula, S. A., (2020). What motivates and inhibits Indian textile firms to embrace sustainability? *Asian J. Sustain. Soc. Responsib., 5*(1), 6. https://doi.org/10.1186/s41180-020-0032-8.
14. Barwer, R., & Sharma, M. L. (2019). *International Journal of Basic and Applied Research Make in India: A Dream to Reality and Indian Textile Sector, 9*(2), 94–102.
15. Gulhane, S., & Turukmane, R., (2017). Effect of make in India on textile sector. *J. Text. Eng. Fash. Technol., 3*(1). https://doi.org/10.15406/jteft.2017.03.00084.
16. Narayanan, K., (2009). The Indian textiles and clothing industry and innovation policies. In: *Innovation Policies and International Trade Rules* (pp. 97–140). Palgrave Macmillan UK. https://doi.org/10.1057/9780230246201_4.
17. Lal, K., (2009). The textiles and clothing industry and economic development: A global perspective. In: *Innovation Policies and International Trade Rules* (pp. 10–36). Palgrave Macmillan UK. https://doi.org/10.1057/9780230246201_2.
18. Pankaj, D., & Lal, R. C., (2019). A critical analysis of Indian textile industry: An insight into inclusive growth and social responsibility. *Russ. J. Agric. Socio-Economic Sci., 88*(4), 53–61. https://doi.org/10.18551/rjoas.2019-04.08.
19. Sakamoto, M., Ahmed, T., Begum, S., & Huq, H., (2019). Water pollution and the textile industry in Bangladesh: Flawed corporate practices or restrictive opportunities? *Sustainability, 11* (7), 1951. https://doi.org/10.3390/su11071951.
20. Franco, M. A., (2017). Circular economy at the micro level: A dynamic view of incumbents' struggles and challenges in the textile industry. *J. Clean. Prod., 168*, 833–845. https://doi.org/10.1016/j.jclepro.2017.09.056.
21. Abhishek, K., Datta, S., Masanta, M., & Mahapatra, S. S., (2017). Fuzzy embedded imperialist competitive algorithm (ICA) for multi-response optimization during machining of CFRP (Epoxy) composites. In: *2017 Int. Conf. Adv. Mech. Ind. Autom. Manag. Syst. AMIAMS 2017 – Proc.* (pp. 100–103). https://doi.org/10.1109/AMIAMS.2017.8069196.
22. Abhishek, K., Chatterjee, S., Datta, S., & Mahapatra, S. S., (2017). Integrating principal component analysis, fuzzy linguistic reasoning and Taguchi philosophy for quality-productivity optimization. In: *Materials Today: Proceedings* (Vol. 4, pp. 1772–1777). Elsevier Ltd. https://doi.org/10.1016/j.matpr.2017.02.019.
23. Abhishek, K., Datta, S., & Mahapatra, S. S., (2017). Multi-objective optimization during drilling of CFRP composites: A PCA-fuzzy Taguchi integrated approach. *Int. J. Ind. Syst. Eng., 26*(2), 182–200. https://doi.org/10.1504/ijise.2017.10004386.
24. Bagada, C., Damor, H., Prajapati, V., & Abhishek, K., (2020). Utility function approach integrated with fuzzy for optimization in milling glass fiber reinforced epoxy composites. In: *Recent Advances in Mechanical Infrastructure* (pp. 85–91). Springer, Singapore. https://doi.org/10.1007/978-981-32-9971-9_10.

25. Tripathi, D. R., Vachhani, K. H., Bandhu, D., Kumari, S., Kumar, V. R., & Abhishek, K., (2021). Experimental investigation and optimization of abrasive waterjet machining parameters for GFRP composites using metaphor-less algorithms. *Mater. Manuf. Process.,* 1–11. https://doi.org/10.1080/10426914.2020.1866193.
26. Bandhu, D., Kumari, S., Prajapati, V., Saxena, K. K., & Abhishek, K., (2020). Experimental investigation and optimization of RMDTM welding parameters for ASTM A387 grade 11 steel. *Mater. Manuf. Process.* https://doi.org/10.1080/10426914.2020. 1854472.
27. Abhishek, K., Rakesh, K. V., Datta, S., & Mahapatra, S. S., (2019). An integrated multi-response optimization route combining principal component analysis, fuzzy inference system, nonlinear regression and JAYA algorithm: A case experimental study on machining of GFRP (epoxy) composites. *Int. J. Ind. Syst. Eng., 32*(4), 497–525. https://doi.org/10.1504/IJISE.2019.101334.
28. Dinbandhu, Prajapati, V., Vora, J. J., Das, S., & Abhishek, K., (2020). Experimental studies of regulated metal deposition (RMD™) on ASTM A387 (11) steel: Study of parametric influence and welding performance optimization. *J. Brazilian Soc. Mech. Sci. Eng., 42*(1), 78. https://doi.org/10.1007/s40430-019-2155-3.
29. Prajapati, V., Dinbandhu, Vora, J. J., Das, S., & Abhishek, K., (2020). Study of parametric influence and welding performance optimization during regulated metal deposition (RMD™) using grey integrated with fuzzy Taguchi approach. *J. Manuf. Process., 54,* 286–300. https://doi.org/10.1016/j.jmapro.2020.03.017.
30. Bandhu, D., & Abhishek, K., (2021). Assessment of weld bead geometry in modified shortcircuiting gas metal arc welding process for low alloy steel. *Mater. Manuf. Process.,* 1–19. https://doi.org/10.1080/10426914.2021.1906897.
31. Rathi, R., Khanduja, D., & Sharma, S. K., (2016). A fuzzy MADM approach for project selection: A six sigma case study. *Decis. Sci. Lett., 5*(2), 255–268. https://doi.org/10.5267/j.dsl.2015.11.002.
32. Rathi, R., Khanduja, D., & Sharma, S. K., (2015). Six sigma project selection using fuzzy TOPSIS decision making approach. *Manag. Sci. Lett., 5*(5), 447–456. https://doi.org/10.5267/j.msl.2015.3.009.
33. Liu, H. C., You, J. X., You, X. Y., & Shan, M. M., (2015). A novel approach for failure mode and effects analysis using combination weighting and fuzzy VIKOR method. *Appl. Soft Comput. J., 28*, 579–588. https://doi.org/10.1016/j.asoc.2014.11.036.
34. Sofiyabadi, J., Kolahi, B., & Valmohammadi, C., (2016). Key performance indicators measurement in service business: A fuzzy VIKOR approach. *Total Qual. Manag. Bus. Excell., 27*(9, 10), 1028–1042. https://doi.org/10.1080/14783363.2015.1059272.
35. Mandal, S., Singh, K., Behera, R. K., Sahu, S. K., Raj, N., & Maiti, J., (2015). Human error identification and risk prioritization in overhead crane operations using HTA, SHERPA and fuzzy VIKOR method. *Expert Syst. Appl., 42*(20), 7195–7206. https://doi.org/10.1016/j.eswa.2015.05.033.
36. Chatterjee, S., Abhishek, K., & Mahapatra, S. S., (2020). *Optimization of Quality Characteristics in Laser Drilling of Ti6Al4V Using VIKOR* (pp. 59–67). Springer, Singapore. https://doi.org/10.1007/978-981-32-9971-9_7.
37. Shah, B., Lakhani, H., Abhishek, K., & Kumari, S., (2020). Application of fuzzy linguistic modeling aggregated with VIKOR for optimal selection of solar power plant site: An empirical study. In: *Smart Innovation, Systems and Technologies* (Vol. 161, pp. 119–127). Springer Science and Business Media Deutschland GmbH. https://doi.org/10.1007/978-981-32-9578-0_11.

CHAPTER 5

SELECTION OF REFRIGERANTS FOR DOMESTIC APPLICATIONS USING MADM TECHNIQUES

ABHINEET SINGH[1], ABHEY GUPTA[1], DIN BANDHU[2], SONI KUMARI[3], DILEEP KUMAR GUPTA[1], and KUMAR ABHISHEK[1]

[1]*Department of Mechanical and Aero-Space Engineering, Institute of Infrastructure, Technology, Research, and Management (IITRAM), Ahmedabad, Gujarat, India*

[2]*Department of Mechanical and Industrial Engineering, Manipal Institute of Technology Bengaluru, Manipal Academy of Higher Education, Manipal, Karnataka, India*

[3]*Department of Mechanical and Aero-Space Engineering, Institute of Infrastructure, Technology, Research, and Management (IITRAM), Ahmedabad, Gujarat, India*

ABSTRACT

The factors that influence refrigerant production and selection are numerous and often contradictory. However, there are numerous options available today that take into account all aspects of refrigeration and air conditioning (RAC). The RAC industry is currently going through a critical period in which no new traditional refrigerants can be developed or consumed. To secure a brighter forthcoming time, safer and eco-friendly refrigerants will remain in demand. The worldwide community is currently exploring the refrigerants that work to save the ozone layer or contribute in the reduction of global warming. Natural refrigerants such as CO_2, NH_3, and hydrocarbons such as R290, R600, R600a,

Decision-Making Models and Applications in Manufacturing Environments. Pushpdant Jain, Kumar Abhishek, & Prasenjit Chatterjee (Eds.)

and hydrocarbon blends provide a viable solution to this difficulty, and they are still employed efficiently in certain systems. As there is a long list of such types of refrigerants, sometimes, it is confusing and tedious work to select the best refrigerant for home and domestic purposes. In this context, this work is an endeavor to choose the best available refrigerant using multi-attributes decision-making (MADM) techniques. Three MADM techniques namely simple additive weighting (SAW), gray relation analysis (GRA), and VIKOR have been implemented to select the ideal refrigerant from a prepared list of most common and usable refrigerants. All three methods R290 as the best refrigerant to be used for domestic purposes, keeping all the basic properties and environmental criteria in mind.

5.1 INTRODUCTION

RAC has now become an essential part of our daily lives. They are there to meet our heating and cooling requirements, i.e., from preserving the food to maintaining the thermal conditions at optimal comfort levels. They not only provide a healthy and comfortable living environment but also become a necessity in today's era as these things have raised the bar for the standard of living. RAC system contributes to a major part of industrial production as the equipment used in the system is used everywhere in our daily life, from our homes to our offices, from vehicles to buildings [1]. However, rapid growth in technology and the economy all over the world during the last few decades have given rise to many environmental issues, making us reconsider the cost we pay for our comfort. So an alternative approach in the selection of refrigerants was essential for balancing out the adverse effects of the RAC sector on the environment. The adjustments in the selection criteria of refrigerants because of environmental concerns and safety reasons have quickened during the most recent 30–40 years [2]. During the early days, the working fluid used for mechanical refrigeration was Air. Later on, the introduction of the vapor compression cycle leads to the development of small and efficient systems as compared to bulky and wasteful methods that were already in use. Initially, only Ammonia and CO_2 were used as the working fluids. The main reason behind using the refrigeration system during those days was to preserve the meat for the crew during the long sea voyages across the ocean. Since it was related to preserving food, ammonia because of its toxicity was not a suitable refrigerant and therefore CO_2 was preferred despite the higher-pressure requirement [3]. Midgely in the mid-1930s revolutionized the refrigeration industry with his invention of the chlorofluorocarbon R12. This led to the creation of the whole CFC family. Different members from this group appeared to have alluring properties like non-flammable, non-toxic, and good oil miscibility characteristics as well as great thermodynamic properties. This marked the beginning of the use of CFCs as refrigerants, allowing refrigeration to expand into the commercial and residential air conditioning segments [4]. The refrigerants before the use of CFCs were environmentally friendly, but the thirst of humans for an alternative with better chemical and thermo-physical properties led to the development of CFCs. Although it offered a significant stimulus to the RAC sector because of its enhanced performance, high reliability, and low development costs, it also had a disadvantage in the long term, which was eventually recognized, since it caused stratospheric ozone depletion. This layer of ozone in the stratosphere absorbs the majority of the dangerous UV-B radiation that strikes the planet, and its depletion causes a variety of problems. A few of those damages included eye cataracts, damages to the earth's food chain, suppression of immune system, skin cancers, etc. [5].

CFCs are one of the most dangerous gases affecting us indirectly because of this ozone layer depletion. The issue is not with the gases within the system, but with their release into the atmosphere. When CFCs undergo photolysis, it releases chlorine atoms. These atoms react at a very fast rate with ozone to form oxides and reduce the concentration of stratospheric ozone barrier formed there [6]. This was something that was not at all expected as this Ozone depletion can be mainly attributed due to exceptional stability, which was one of the desired properties in the refrigerants. Due to this property, CFCs survived in the atmosphere for decades. This enables them to diffuse to the stratospheric levels. CFCs were termed as a successful alternative because they fulfilled all the primary requirements of a refrigerant and brought a revolution in the RAC industry. But ultimately they were not a feasible solution in the long run [7]. During the era of the 1970s, awareness about environmental concerns was raised due to the effect that human activities were having on the Ozone levels in the stratosphere which further extended to the ozone hole formation discovery. This leads to the Montreal Protocol that sits under the Vienna Convention [8, 9].

The Montreal Protocol was signed on 16th September 1987 and came into effect on 1st January 1989, following to first meeting held in May 1989 at Helsinki, to avoid the creation and use of harmful toxic substances. Regular amendments have been made to the Montreal protocol to safeguard the deteriorating environmental condition as well as monitor the phasing out of harmful gases. During this period (1990–2000), developed and developing countries were given objectives and timelines to phase out dangerous refrigerants while manufacturers were required to introduce the first substitute refrigerant in 1989. The last meet was held in Kigali, Rwanda, in 2016, and many reforms have been made to the present scenario [2, 10–12]. After 1995, industrialized nations were forbidden from producing CFC refrigerants under the Montreal Protocol and other international accords. Table 5.1 lists such refrigerants.

TABLE 5.1 CFC Refrigerants Discontinued from Circulation under Montreal Protocol [2]

Refrigerant	Abbreviation	Structural Formulation	Meteorological Time (Years)	ODP Based on CFC-11	GWP Based on CFC in 100 Years
Dichlorodifluoromethane	CFC-12	CF_2Cl_2	103	1	8,510
Trichlorofluoromethane	CFC-11	$CFCl_3$	51	1	4,005
Chlorotrifluoromethane	CFC-13	$CClF_3$	–	–	–
Bromochlorodifluoromethane	CFC-12B_1	$CClBrF_2$	–	–	–
1,1,2-Trichloro-1,2,2-trifluoroethane	CFC-113	$C_2Cl_3F_3$	83	0.801	5,010
1,2-Dichloro-1,1,2,2-tetrafluoroethane	CFC-114	$C_2Cl_2F_4$	301	1.01	9,321
Chloropentafluoroethane	CFC-115	C_2ClF_5	1,702	0.61	9,310

Source: Adapted/Reprinted with permission from Ref. [2]. Copyright 2020. Elsevier.

The CFCs were replaced with HCFCs. Because the inclusion of a hydrogen atom rendered these chemicals significantly more vulnerable to assault by hydroxyl (OH) radicals, a considerable proportion of the HCFCs were annihilated when they entered the stratosphere. Despite being seen to be cleaner than CFCs, HCFCs have the propensity to damage the ozone layer [13, 14]. Table 5.2 lists such HCFC refrigerants which were phased out from circulation.

Later, in 1992 during the United nation framework convention, hydrofluorocarbons (HFCs) were approved as a cleaner refrigerant. These are the part of organic chemicals that lack chlorine atoms and contain just one or very few fluorine atoms, resulting in an ODP of 0. HFCs were used extensively as a replacement for CFCs and HCFCs, but it was soon discovered that the HFCs, although safe for ODP, are the major contributor to global warming as they release one of the major greenhouse gas and harm the environment at an accelerated pace. In accordance with the growing concerns over these issues, the Kyoto protocol came into the picture. Kyoto protocol, signed in December of 1997, was also a protocol based on environmental concerns that focus on minimizing greenhouse gases so as to reduce global warming. This protocol is primarily concerned with six carbon compounds including sulfur hexafluoride, CO_2, methane (CH_4), HFCs, nitrous oxide, and perfluorocarbons. These chemicals have been recognized as the most harmful possibilities for greenhouse gas production, and they serve a key position in worldwide environmental disruption. This leads to the phasing out of HFCs [14–19]. Such refrigerants are listed in Table 5.3.

TABLE 5.2 HCFC Refrigerants Discontinued from Circulation under Montreal Protocol [2]

Refrigerant	Abbreviation	Structural Formulation	Meteorological Time (Years)	ODP Based on CFC-11	GWP Based on CFC in 100 Years
Monochlorodifluoroethane	HCFC-22	$CHClF_2$	13.31	0.0551	1701
Monotetrafluoroethane	HCFC-124	$CHClF_2CF_3$	5.91	0.0221	481
Dichlorotrifluoroethane	HCFC-123	$CHCF_2Cl_3$	1.41	0.0201	94
chlorodifluoroethane	HCFC-142b	$C_2H_3ClF_2$	–	–	–
Dichlorofluoromethane	HCFC-21	$CHCl_2F$	–	–	–

Source: Adapted/Reprinted with permission from Ref. [2]. Copyright 2020. Elsevier.

The Montreal and Kyoto Protocols paved the way for finding a better alternative like HFOs. They have a limited atmospheric life and a very minimal global warming impact, but they are very combustible and expensive. Different blends and mixtures came into the picture. The introduction of F-gas regulations leads to producers and multinational manufacturers pioneering coolants that satisfy the GWP boundary of 150. These EU regulations 519/2014 went into effect in January 2015, with the main goal of reducing the number of obnoxious gas emissions and use to 79% of 1990 levels by 2030. F-gas allocation, leak seizure, and operating tasks such as refurbishing, deployment, and dismantling should only be undertaken by skilled personnel [22–24]. Table 5.4 shows the transition of refrigerants that can be summarized into three generations based on changing concerns and needs [25, 26].

TABLE 5.3 HFC Refrigerants Discontinued from Circulation under Kyoto Protocol [2, 20, 21]

Refrigerant	Abbreviation	Structural Formulation	Meteorological Time (Years)	ODP Based on CFC-11	GWP Based on CFC in 100 Years
1,1-difluoroethane	HFC-152a	CH_3CHF_2	1.5	0	140
1,1,2,2,3-pentafluoropropane	HFC-245ca	$CHF_2CF_2CH_2F$	6.6	0	560
Difluoromethane	HFC-32	CH_2F_2	5.6	0	650
1,1,1,2-tetrafluoroethane	HFC-134a	CH_2FCF_3	14.6	0	1,300
Pentafluoroethane	HFC-125	CHF_2CF_3	32.6	0	2,800
1,1,1,2,3,3,3-heptafluoropropane	HFC-227ea	CF_3CHFCF_3	36.5	0	2,900
1,1,1-trifluoroethane	HFC-143a	CH_3CF_3	48.3	0	3,800
1,1,1,3,3,3-hexafluoropropane	HFC-236fa	$CF_3CH_2CF_3$	209	0	6,300
1,1,1,3,3-pentafluoropropane	HFC-245fa	$CF_3CH_2CHF_2$	209	0	6,300
Octafluoropropane	PFC-218	C_3F_8	2,600	0	7,000
Trifluoromethane	HFC-23	CHF_3	264	0	11,700
1,1,1,2,3-pentafluoropropane	HFC-245eb	CF_3CHFCH_2F	–	0	–
HFO	R1234yf	HFO	0.029	0	–
Carbon dioxide	R744	CO_2	>50	0	–
Ammonia	R717	NH_3	<0.02	0	–

Source: Adapted from Ref. [2],[20],[21]

TABLE 5.4 Generation of Refrigerants based on Changing Needs [26]

Generation	Timeline	Aim	Examples
1st	1830s–1930s	Whatever worked	CO_2, NH_3, H_2O, CCl_4
2nd	1930s–1990s	Safety and durability	CFCs, HCFCs, NH_3, H_2O
3rd	1990s to Date	Environmental protection	HCFCs, HFCs, CO_2, NH_3, H_2O

Source: Adapted from Ref. [26]

These regulations, protocols, and environmental concerns all have a part in the refrigerant selections that are made today. While there is an increasing demand for energy in RAC applications, the environmental issues are in urgent need of being addressed, namely ozone depletion and global warming. These challenges although distinct are addressed together and can be addressed by using new working fluids or through an innovative thermodynamic cycle. The practical viability of the situation must also be kept in mind as of the currently available systems all over the world. The conditions and restrictions imposed on ideal refrigerants have increased a lot, and ODP, as well as GWP, are now essential for consideration in addition to the primary requirements [27, 28]. According to Calm et al. [29], the alternatives for CFCs are necessary as when Midgley and his associates invented the CFCs it was not with respect to environmental concerns and only focused on efficiency. As a result, novel refrigerants that meet today's standards are required [30].

The number of carbon atoms (m), hydrogen atoms (n), and fluorine atoms (p) contained in the molecule is used by ASHRAE to classify manmade refrigerants. As a result, ASHRAE recommends Eqn. (1) for designating refrigerants [24].

$$R\,(m-1)\,(n+1)\,(p) \tag{1}$$

Each refrigerant is allocated a number based on its chemical makeup, and the approach has been codified as an ASHRAE standard. The general numbering scheme is as follows:

Methane based:	000 Series
Ethane based:	100 Series
Propane based:	200 Series
Cyclic organic compounds:	300 Series
Zeotropes:	400 Series
Azeotropes:	500 Series
Organic compounds:	600 Series
Inorganic compounds:	700 Series
Unsaturated organic compounds:	1,000 Series

This gives a general idea about the refrigerant and what can be expected of its chemical properties as well as its reliability.

Refrigerants are also classified into two types based on their toxicity: A and B, i.e.,

$$\text{Class A} \geq 400 \text{ PPM and Class B} < 400 \text{ PPM}$$

Similarly, they are divided into three categories according to their flammability. During their test at 21°C:

- If the refrigerants show no flame, then they are in the Class 1 category.
- If the refrigerants show low flammability, then they are in the Class 2 category.
- If the refrigerants show high flammability, then they are in the Class 3 category.

These classifications help us to categorize the refrigerants based on the safety parameter [25]. In the search for a better alternative, a lot of research has been done to find an alternative that can be widely accepted all over the world. Singh [31] used the TOPSIS model on the thermo-physical properties of refrigerants to find an optimum solution for selecting a refrigerant. Similarly, Suriya [32] compared VCR systems for green refrigerants to find a better alternative. Dreepaul [33] revived the various refrigerants that are being used and what is the current state of the ones that are to be replaced in general. Benhadid-Dib [34] performed research on the environmental effect of refrigerants and the quest for a better replacement. Various other researchers and scientists have worked in this regard as environmental concern is a worldwide issue [35–44].

From this literature it can be seen that a lot of work has been done regarding refrigerants and how are they selected. This selection process is for different attributes or more or less based on general efficiency. There are various regulations as well as rules that affect the refrigerant that is in use. This work focuses on a selection of the best alternative out of the possible ones that are being in use currently based on the present norms and regulations as well as optimal usage benefits based on some of the most general properties of refrigerants. The work incorporates MADM techniques to get the desired results and compare different attributes that are sort of necessary in this present era of refrigeration and air-conditioning (AC).

5.2 EXPERIMENTAL METHODS AND MATERIALS

5.2.1 REFRIGERANT SELECTION CRITERIA

At present, the major refrigerants that are being used are not entirely perfect and possess one or more drawbacks. But still, for the lack of obvious choice, we have to settle for the best out of them or the one whose negative impact is the least. In order to find what the problems with the available alternative are, it is important to understand what is the present situation with respect to their usage [45, 46]. Table 5.5 presents the list of the refrigerants which are currently in use.

TABLE 5.5 Refrigerants Currently in Use [27]

Refrigerant	Present Situation
R-22	R22 is banned to use now, but it is still in use.
R-410A	R410A is being phased out due to its environmental effects.
R-32	R32 is highly toxic and highly flammable.
R-134A	R134A due to high GWP is in the process of being banned.
R-290 R-600A	The most eco-friendly refrigerants now easily accessible in the Indian market are "R-290" and "R-600A." However, because they are hydrocarbons, they are very combustible.

Source: Adapted from Ref. [27]

As it can be seen that these refrigerants, although in use, are not entirely feasible for long-term usage. This has set the search for the best one for long-term usage as a priority. There are various criteria, on which the selection process for refrigerant can be based, starting from physical properties to chemical ones, their economical evaluation to environmental concerns as shown in Figure 5.1.

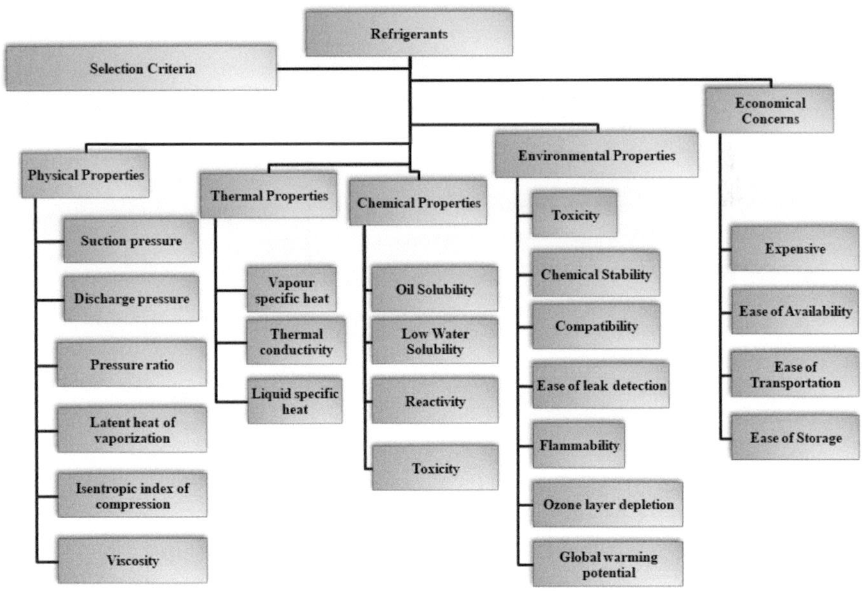

FIGURE 5.1 Criteria for refrigerant selection at a wide spectrum.

According to modern trends and requirements, the most essential criteria that design engineers would examine when choosing a refrigerant for a certain purpose are effectiveness, safety, durability, ecological friendliness, and simple affordability. Although these 5 factors are general, one's selection criteria should be in accordance with these factors as they affect the market the most.

Depending upon the aforementioned criteria, the factor to be assessed based on the user's understanding is the Mass flow rate. For excellent volumetric efficiency and minimal electricity utilization, the pressure ratio must be as reduced as feasible. The COP and ODP must follow the Montreal Protocol. The GWP number needs to be as low as feasible because greater GWP refrigerants are likely to be governed in the long term. Cost is one of the most important economic factors to be considered for refrigerant selection. Safety is decided based on ASHRAE standards and refrigerants should be selected only when they are safe for usage [47–50].

The refrigerants that are to be assessed were selected based on their present usage and prospects of being in use. Figure 5.2 depicts the criteria on which the refrigerants are selected as well as the list of refrigerants employed in this work.

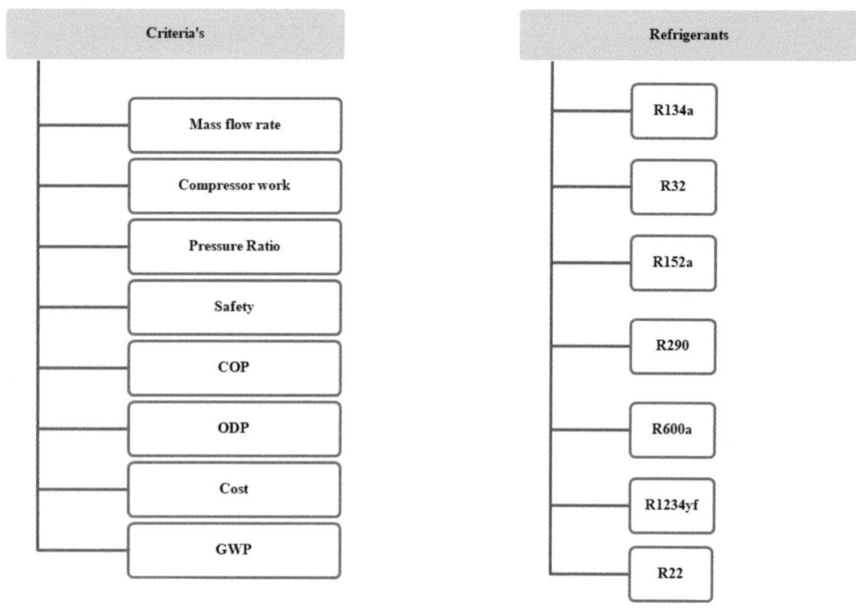

FIGURE 5.2 Carefully chosen criteria for refrigerant selection and the list of selected refrigerants.

5.2.2 DATA COLLECTION

REFPROP 9.0 was used to get the required enthalpies of the refrigerants, and then excel sheets were used to formulate the required data that has to be assessed for selection. Table 5.6 lists the finalized data for assessment.

TABLE 5.6 Data Used for Refrigerant Selection

Refrigerant	Mass Flow Rate (kg/s)	$W_{Compressor}$	Pressure Ratio	COP	Safety Group	ODP	GWP
R22	0.0492	3.5461	4.6130	1.9836	5	0.05	1810
R134a	0.0568	3.6905	4.7750	1.9060	5	0	1430
R32	0.0305	3.6937	4.0530	1.9043	4	0	675
R152a	0.0328	3.4293	4.7260	2.0511	3	0	138
R290	0.0303	3.7459	3.7840	1.8778	2	0	3
R600a	0.0322	3.6260	4.6230	1.9399	2	0	3
R1234yf	0.0289	3.7121	4.3510	1.8949	4	0	4

This data was composed for a general case where the domestic refrigerant cooling capacity is 2 tons, and the safety group is based on the increasing level of safety on a scale from 1 to 5 with B3 being 1 and A1 being 5 according to ASHRAE standards.

On the basis of importance, certain weightage was assigned to different categories. The attributes with their maximum and minimum values as well as their weightage are given in Table 5.7.

TABLE 5.7 Assigning Weightage to Attributes

Refrigerant	Mass Flow Rate	$W_{Compressor}$	Pressure Ratio	COP	Safety Group	ODP	GWP
Max.	0.057	3.75	4.775	2.05	5	0.05	1,810
Min.	0.029	3.43	3.784	1.88	2	0	3
Wt. taken	0.10	0.10	0.10	0.20	0.20	0.10	0.20

5.3 RESULTS AND DISCUSSION

Three different methods were used to determine the ranking of the refrigerants based on their attributes and applications. These methods and their respective outcomes are discussed as follows:

5.3.1 SAW METHOD

Under MADM, the simple additive weighting (SAW) method, alias weighted summation method is utilized. The SAW method's core idea is to obtain a weighted sum of each possibility on each characteristic. The SAW approach necessitates a procedure of standardizing the decision matrix (X) to a scale that may be equated to all current various rankings [51, 52].

If an attribute is a beneficial one, then the formula for normalizing is:

$$r_{ij} = \frac{x_{ij}}{\mathrm{Max}\left(x_{ij}\right)} \tag{2}$$

If the attribute is not beneficial, then the formula for normalizing is:

$$r_{ij} = \frac{\mathrm{Min}\left(x_{ij}\right)}{x_{ij}} \tag{3}$$

Weight is obtained by the level of importance of different attributes in selection as follows:

$$w = \frac{C1}{C1 + \ldots + Cn} \times 100\% \tag{4}$$

Table 5.8 shows the normalized data.

TABLE 5.8 Normalized Attributes

Refrigerant	N-Mass Flow Rate	N-$W_{Compressor}$	N-Pressure Ratio	N-COP	N-Safety Group	N-ODP	N-GWP
R22	0.588	0.967	0.820	0.947	0.400	0.000	0.002
R134a	0.588	0.929	0.792	0.985	0.400	1.000	0.002
R32	0.949	0.928	0.934	0.986	0.500	1.000	0.004
R152a	0.881	1.000	0.801	0.915	0.667	1.000	0.022
R290	0.956	0.915	1.000	1.000	1.000	1.000	1.000
R600a	0.897	0.946	0.819	0.968	1.000	1.000	1.000
R1234yf	1.000	0.924	0.870	0.991	0.500	1.000	0.750

Further, to decide the rank and to finalize the decision (Table 5.9), the sum of these weighted normalized data is done using Eqn. (5):

$$V_i = \sum_{j-1}^{n} = 1\ w_j\ r_{ij} \tag{5}$$

TABLE 5.9 Final SAW Table with Rank

Refrigerant	Final Weightage	Rank
R22	0.467	7
R134a	0.668	6
R32	0.729	4
R152a	0.722	5
R290	0.987	1
R600a	0.960	2
R1234yf	0.878	3

From Table 5.9, it can be articulated that the SAW method recommends R290 as the best refrigerant.

5.3.2 *GREY RELATION ANALYSIS (GRA) METHOD*

Similarly, one of the famous methods for multiple attribute selection is the gray relational analysis method. It works on a similar methodology of normalizing the decision matrix and finding the gray relational coefficient in order to obtain the gray grade of the decision matrix [53, 54].

The formula for normalizing the decision matrix is given as:

- **For Beneficial Attribute:**

$$x_i^*(k) = \frac{x_i^o(k) - \min x_i^o(k)}{\max x_i^o(k) - \min x_i^o(k)} \tag{6}$$

- **For Non-Beneficial Attribute:**

$$x_i^*(k) = \frac{\max x_i^o(k) - x_i^o(k)}{\max x_i^o(k) - \min x_i^o(k)} \tag{7}$$

Table 5.10 depicts the normalized attributes.

TABLE 5.10 Normalization of the Attributes

Refrigerant	N-Mass Flow Rate	N-$W_{Compressor}$	N-Pressure Ratio	N-COP	N-Safety Group	N-ODP	N-GWP
R22	0.272704	0.63106	0.163471	0.389724	0	0	0
R134a	0	0.174821	0	0.837553	0	1	0.210293
R32	0.94369	0.164768	0.728557	0.847026	0.333333	1	0.628113
R152a	0.859516	1	0.049445	0	0.666667	1	0.925291
R290	0.951891	0	1	1	1	1	1
R600a	0.881033	0.378789	0.15338	0.641754	1	1	1
R1234yf	1	0.106653	0.427851	0.901472	0.333333	1	0.999447

From the above data, we find the gray relational coefficient which is given by the formula:

$$\varepsilon_i(k) = \frac{\Delta_{min} + \epsilon\Delta_{max}}{\Delta_{oi}(k) + \epsilon\Delta_{max}} \tag{8}$$

For this, we find the deviation sequence matrix, i.e., $\Delta_{oi}(k)$ as shown in Table 5.11.

TABLE 5.11 Deviation Sequence Matrix

Refrigerant	D-Mass Flow Rate	D-$W_{Compressor}$	D-Pressure Ratio	D-COP	D-Safety Group	D-ODP	D-GWP
R22	0.727296	0.36894	0.836529	0.610276	1	1	1
R134a	1	0.825179	1	0.162447	1	0	0.789707
R32	0.05631	0.835232	0.271443	0.152974	0.666667	0	0.371887
R152a	0.140484	0	0.950555	1	0.333333	0	0.074709
R290	0.048109	1	0	0	0	0	0
R600a	0.118967	0.621211	0.84662	0.358246	0	0	0
R1234yf	0	0.893347	0.572149	0.098528	0.666667	0	0.000553

With the help of this matrix, the gray relational coefficient matrix was formulated as shown in Table 5.12.

Then, using Table 5.12, the gray relation grade was determined by means of Eqn. (9). Table 5.13 depicts the calculated gray relation grade along with the ranking of the refrigerants.

TABLE 5.12 Grey Relational Coefficient Matrix

Refrigerant	G-Mass Flow Rate	G-$W_{Compressor}$	G-Pressure Ratio	G-COP	G-Safety Group	G-ODP	G-GWP
R22	0.407	0.575	0.374	0.450	0.333	0.333	0.333
R134a	0.333	0.377	0.333	0.755	0.333	1.000	0.388
R32	0.899	0.374	0.648	0.766	0.429	1.000	0.573
R152a	0.781	1.000	0.345	0.333	0.600	1.000	0.870
R290	0.912	0.333	1.000	1.000	1.000	1.000	1.000
R600a	0.808	0.446	0.371	0.583	1.000	1.000	1.000
R1234yf	1.000	0.359	0.466	0.835	0.429	1.000	0.999

$$\gamma_i = \frac{1}{n}\sum\nolimits_{k=1}^{n} \omega_k (k) \varepsilon_i (k) \tag{9}$$

TABLE 5.13 Final GRA Table with Rank

Refrigerant	Final Weightage	Rank
R22	0.056061	7
R134a	0.071365	6
R32	0.092242	5
R152a	0.096172	4
R290	0.132079	1
R600a	0.111289	2
R1234yf	0.105013	3

From Table 5.13, it can be articulated that the GRA method recommends R290 as the best refrigerant.

5.3.3 *VIKOR METHOD*

VIKOR is another MADM method developed in 1998. The main idea of the VIKOR procedure lies in utilizing the solution space to highlight the positive and the negative ideal points. It also follows a similar methodology of normalizing the decision matrix, followed by the Utility measure (S_i) and Regret measure (R_f), after which Q_i calculated from which the rank is decided [55–58].

The normalized decision matrix (Table 5.14) is formed with the use of the following formula:

$$f_{ij} = \frac{I_i^j}{\sqrt{\sum_{i=1}^{m}\left(I_i^j\right)^2}} \tag{10}$$

TABLE 5.14 Data Normalization

Refrigerant	N-Mass Flow Rate	N-$W_{Compressor}$	N-Pressure Ratio	N-COP	N-Safety Group	N-ODP	N-GWP
R22	0.04293	2.23814	3.96031	3.10438	4.29256	0.04293	1553.90634
R134a	0.04550	2.00720	4.09939	4.54668	4.29256	0.00000	1227.67186
R32	0.01176	0.91174	3.47955	4.29256	3.43405	0.00000	579.49546
R152a	0.02704	1.89044	4.05733	4.62738	2.57554	0.00000	118.47463
R290	0.02438	2.03811	3.24861	3.41001	1.71702	0.00000	2.57554
R600a	0.02533	1.95140	3.96890	4.58188	1.71702	0.00000	2.57554
R1234yf	0.06173	2.15572	3.73538	3.22371	3.43405	0.00000	3.43405

With this normalized decision matrix, Utility measure S_i and Regret measure R_f (Table 5.15) are calculated using the following formulas:

- **For Beneficial Attribute:**

$$S_i = \sum_{i=1}^{n} w_i \left[\frac{\left(f_{ij}\right)_{max} - \left(f_{ij}\right)}{\left(f_{ij}\right)_{max} - \left(f_{ij}\right)_{min}}\right] \tag{11}$$

$$R_f = \text{maximum of} \left(\{w_i\}\frac{\left(f_{ij}\right)_{max} - \left(f_{ij}\right)}{\left(f_{ij}\right)_{max} - \left(f_{ij}\right)_{min}}\right) \tag{12}$$

- **For Non-Beneficial Attribute:**

$$S_i = \sum_{i=1}^{n} w_i \left[\frac{\left(f_{ij}\right) - \left(f_{ij}\right)_{min}}{\left(f_{ij}\right)_{max} - \left(f_{ij}\right)_{min}}\right] \tag{14}$$

$$R_f = \text{maximum of} \left(\{w_i\}\frac{\left(f_{ij}\right) - \left(f_{ij}\right)_{min}}{\left(f_{ij}\right)_{max} - \left(f_{ij}\right)_{min}}\right) \tag{15}$$

Using the utility measure and regret measure matrix, Q_i (Table 5.16) was calculated by means of Eqn. (16).

TABLE 5.15 Utility Measure and Regret Measure Matrix

Refrigerant	S_i	R_f
R22	0.746024	0.2
R134a	0.797459	0.2
R32	0.390887	0.156032
R152a	0.481035	0.2
R290	0.150312	0.084919
R600a	0.384216	0.194025
R1234yf	0.400116	0.133333

$$Q_i = v\left[\frac{(S_i)-(S_i)_{min}}{(S_i)_{max}-(S_i)_{min}}\right]+(1-v)\left[\frac{(R_i)-(R_i)_{min}}{(R_i)_{max}-(R_i)_{min}}\right] \tag{16}$$

TABLE 5.16 Final VIKOR Table with Rank

Refrigerant	Q_i	Rank
R22	0.976156	6
R134a	1	7
R32	0.544078	3
R152a	0.853314	5
R290	0	1
R600a	0.772087	4
R1234yf	0.410291	2

From Table 5.16, it can be articulated that the VIKOR method recommends R290 as the best refrigerant.

5.4 CONCLUSION

The work concludes with the selection of the best refrigerant using three different MADM techniques. The obtained outcomes are compared as tabulated in Table 5.17.

From this, it can be observed that the best suitable refrigerant for domestic purposes based on the data selection is R290. It is the refrigerant that should be adopted in the long run until a better alternative comes into the picture. But for the time being the refrigerant R290 is the optimal choice to be used for domestic purposes, with keeping all the basic properties and environmental criteria in mind.

TABLE 5.17 Final Comparison of the Outcomes and Ranking of the Refrigerants

Refrigerant	SAW	GREY	VIKOR	Rank
R22	7	7	6	7
R134a	6	6	7	6
R32	5	5	3	5
R152a	4	4	5	4
R290	1	1	1	1
R600a	2	2	4	2
R1234yf	3	3	2	3

KEYWORDS

- **grey relation analysis (GRA) method**
- **MADM techniques**
- **refrigerants**
- **selection and ranking**
- **simple additive weighting (SAW) method**
- **VIKOR method**

REFERENCES

1. Tassou, S. A., Ge, Y., Hadawey, A., & Marriott, D., (2011). Energy consumption and conservation in food retailing. *Appl. Therm. Eng., 31*(2,3), 147–156. https://doi.org/10.1016/j.applthermaleng.2010.08.023.
2. Roy, Z., & Halder, G., (2020). Replacement of halogenated refrigerants towards sustainable cooling system: A review. *Chem. Eng. J. Adv., 3*, 100027. https://doi.org/10.1016/j.ceja.2020.100027.
3. Sawalha, S., (2005). Using CO_2 in supermarket refrigeration. *ASHRAE J., 47*(8), 26–30.
4. Dreepaul, R. K., (2016). An assessment of hydrocarbons as an alternative refrigerant: The Mauritian context. *Int. J. Glob. Energy Issues, 39*(3, 4), 222–240. https://doi.org/10.1504/IJGEI.2016.076346.
5. United Nations, (2012). *Handbook for the Montreal Protocol on Substances That Deplete the Ozone Layer* (9th edn.).
6. UNEP, (2000). United nations environment programme annual report. *Glob. Environ. Outlook,* 1–36.
7. Bhatkar, V. W., Kriplani, V. M., & Awari, G. K., (2013). Alternative refrigerants in vapour compression refrigeration cycle for sustainable environment: A review of

recent research. *International Journal of Environmental Science and Technology* (pp. 871–880). Springer. https://doi.org/10.1007/s13762-013-0202-7.

8. Baran, B., Pal, A., Uddin, K., Thu, K., & Baran, S. B., (2018). Environmental assessment and characteristics of next generation refrigerants. *Green Asia Educ. Cent. Evergr. Jt. J. Nov. Carbon Resour. Sci. Green Asia Strateg., 05*(2), 58–66. https://doi.org/10.5109/1936218.
9. Peyyala, A., & Sudheer, N. V. V. S., (2017). Experimental investigation of COP using hydro carbon refrigerant in a domestic refrigerator. In: *IOP Conference Series: Materials Science and Engineering* (Vol. 225, p 012236). Institute of Physics Publishing. https://doi.org/10.1088/1757-899X/225/1/012236.
10. Morgenstern, O., Braesicke, P., Hurwitz, M. M., O'Connor, F. M., Bushell, A. C., Johnson, C. E., & Pyle, J. A., (2008). The world avoided by the Montreal protocol. *Geophys. Res. Lett., 35*(16). https://doi.org/10.1029/2008GL034590.
11. Velders, G. J. M., Ravishankara, A. R., Miller, M. K., Molina, M. J., Alcamo, J., Daniel, J. S., Fahey, D. W., et al., (2012). Preserving Montreal protocol climate benefits by limiting HFCs. *Science* (pp. 922–923). American Association for the Advancement of Science. https://doi.org/10.1126/science.1216414.
12. Chipperfield, M. P., Dhomse, S. S., Feng, W., McKenzie, R. L., Velders, G. J. M., & Pyle, J. A., (2015). Quantifying the ozone and ultraviolet benefits already achieved by the Montreal protocol. *Nat. Commun., 6*. https://doi.org/10.1038/ncomms8233.
13. Bolaji, B. O., & Huan, Z., (2013). Ozone depletion and global warming: Case for the use of natural refrigerant–a review. *Renewable and Sustainable Energy Reviews* (pp. 49–54). Pergamon. https://doi.org/10.1016/j.rser.2012.10.008.
14. Nyemba, W. R., Chinguwa, S., Marango, B. L., & Mbohwa, C., (2019). Evaluation and feasibility assessment of the sustainability of refrigeration systems devoid of harmful refrigerants for storage of vaccines. In: *Procedia Manufacturing* (Vol. 35, pp. 291–297). Elsevier B.V. https://doi.org/10.1016/j.promfg.2019.05.042.
15. Böhringer, C., (2003). The Kyoto protocol: A review and perspectives. *Oxford Rev. Econ. Policy, 19*(3), 451–466. https://doi.org/10.1093/oxrep/19.3.451.
16. Oberthür, S., & Ott, H. E., (1999). *The Kyoto Protocol*. Springer Berlin Heidelberg: Berlin, Heidelberg. https://doi.org/10.1007/978-3-662-03925-0.
17. Reilly, J., Prinn, R., Harnisch, J., Fitzmaurice, J., Jacoby, H., Kicklighter, D., Melillo, J., et al., (1999). Multi-gas assessment of the Kyoto protocol. *Nature, 401*(6753), 549–555. https://doi.org/10.1038/44069.
18. Manne, A. S., & Richels, R. G., (2000). The Kyoto protocol: A cost-effective strategy for meeting environmental objectives? In: *Efficiency and Equity of Climate Change Policy* (pp. 43–61). Springer Netherlands. https://doi.org/10.1007/978-94-015-9484-4_3.
19. Banjo, S. O., Bolaji, B. O., Ajayi, O. O., Olufemi, B. P., Osagie, I., & Onokwai, A. O., (2019). Performance enhancement using appropriate mass charge of R600a in a developed domestic refrigerator. In: *IOP Conference Series: Earth and Environmental Science* (Vol. 331, p 012025). Institute of Physics Publishing. https://doi.org/10.1088/1755-1315/331/1/012025.
20. MacCracken, C. N., Edmonds, J. A., Kim, S. H., & Sands, R. D., (1999). The economics of the Kyoto protocol. *Energy J., 20*, 25–71. https://doi.org/10.5547/issn0195-6574-ej-vol20-nosi-3.
21. Nordhaus, W. D., & Boyer, J. G., (1998). *Requiem for Kyoto: An Economic Analysis of the Kyoto Protocol.* Cowles Found. Discuss. Pap.

22. O'Neill, B. C., & Oppenheimer, M., (2002). Dangerous climate impacts and the Kyoto protocol. *Science,* 1971, 1972. https://doi.org/10.1126/science.1071238.
23. Aprea, C., Greco, A., & Maiorino, A., (2018). HFOs and their binary mixtures with HFC134a working as drop-in refrigerant in a household refrigerator: Energy analysis and environmental impact assessment. *Appl. Therm. Eng., 141*, 226–233. https://doi.org/10.1016/j.applthermaleng.2018.02.072.
24. Mota-Babiloni, A., Navarro-Esbrí, J., Makhnatch, P., & Molés, F., (2017). Refrigerant R32 as lower GWP working fluid in residential air conditioning systems in Europe and the USA. *Renewable and Sustainable Energy Reviews* (pp. 1031–1042). Elsevier Ltd. https://doi.org/10.1016/j.rser.2017.05.216.
25. Bolaji, B. O., (2020). Theoretical assessment of new low global warming potential refrigerant mixtures as eco-friendly alternatives in domestic refrigeration systems. *Sci. African, 10*, e00632. https://doi.org/10.1016/j.sciaf.2020.e00632.
26. McLinden, M. O., & Huber, M. L., (2020). (R)evolution of refrigerants. *Journal of Chemical and Engineering Data* (pp. 4176–4193). American Chemical Society. https://doi.org/10.1021/acs.jced.0c00338.
27. Vinoth, K. I., & Devaraj, A., (2019). Discussion of past, present and future perspectives of refrigerants and its future scope. In: *Lecture Notes in Mechanical Engineering* (pp. 461–471). Pleiades Publishing. https://doi.org/10.1007/978-981-13-2697-4_50.
28. Grzebielec, A., (2020). New refrigerants used by heat pumps. *Mod. Eng., 3*, 100–106.
29. Calm, J. M., Wuebbles, D. J., & Jain, A. K., (1999). Impacts on global ozone and climate from use and emission of 2,2-dichloro-1,1,1-trifluoroethane (HCFC-123). *Clim. Change, 42*(2), 439–474. https://doi.org/10.1023/A:1005496900220.
30. Calm, B. Y. J. M., & Hourahan, G. C., (2001). Refrigerant data summary. *Eng. Syst., 18*(11), 74–88.
31. Singh, S., & Das, G. M. S., (2015). Evaluation of refrigerant options using TOPSIS-MADM technique. *Int. J. Sci. Tech. Adv., 1*(3), 147–150.
32. Suriya, K. T., Velraj, M. R., Vidyasagar, M., Vignesh, M., Student, B. E., & Kannan, G. R., (2019). Comparative analysis of vapor compression refrigeration system using various green refrigerants. *Int. J. Eng. Res. Technol., 7*(6), 1–3.
33. Dreepaul, R. K., (2017). A study of alternative refrigerants for the refrigeration and air conditioning sector in Mauritius. In: *IOP Conference Series: Earth and Environmental Science* (Vol. 93, pp. 1–14). Institute of Physics Publishing. https://doi.org/10.1088/1755-1315/93/1/012054.
34. Benhadid-Dib, S., & Benzaoui, A., (2012). Refrigerants and their environmental impact substitution of hydro chlorofluorocarbon HCFC and HFC hydro fluorocarbon. Search for an adequate refrigerant. In: *Energy Procedia* (Vol. 18, pp. 807–816). Elsevier BV. https://doi.org/10.1016/j.egypro.2012.05.096.
35. Manju, L., Yadav, A. K., & Gupta, D. K., (2020). Thermodynamic analysis of transcritical CO_2 booster systems with flooded evaporator for supermarket application. In: *Smart Innovation, Systems and Technologies* (Vol. 161, pp. 293–304). Springer Science and Business Media Deutschland GmbH. https://doi.org/10.1007/978-981-32-9578-0_27.
36. Gupta, D. K., & Dasgupta, M. S., (2014). Simulation and performance optimization of finned tube gas cooler for trans-critical CO_2 refrigeration system in Indian context. *Int. J. Refrig., 38*(1), 153–167. https://doi.org/10.1016/j.ijrefrig.2013.09.041.
37. Purohit, N., Gupta, D. K., & Dasgupta, M. S., (2018). Experimental investigation of a CO_2 trans-critical cycle with IHX for chiller application and its energetic and exergetic

evaluation in warm climate. *Appl. Therm. Eng., 136*, 617–632. https://doi.org/10.1016/j.applthermaleng.2018.03.044.

38. Lata, M., & Gupta, D. K., (2021). Simulation and performance evaluation of trans-critical CO_2 refrigeration system with modified evaporative cooled finned tube gas cooler in Indian context. *Appl. Therm. Eng., 186*, 116500. https://doi.org/10.1016/j.applthermaleng.2020.116500.
39. Lata, M., & Gupta, D. K., (2021). Experimental investigation and simulation of modified evaporative cooling system. In: *Lecture Notes in Mechanical Engineering* (pp. 441–448). Springer. https://doi.org/10.1007/978-981-15-3639-7_52.
40. Lata, M., & Gupta, D. K., (2020). Performance evaluation and comparative analysis of trans-critical CO_2 booster refrigeration systems with modified evaporative cooled gas cooler for supermarket application in Indian context. *Int. J. Refrig., 120*, 248–259. https://doi.org/10.1016/j.ijrefrig.2020.08.004.
41. Lata, M., Yadav, R., & Gupta, D. K., (2020). Simulation and performance evaluation of CO_2 booster system integrated with modified evaporative cooling for supermarket application in India. *Jordan J. Mech. Ind. Eng., 14*(2), 183–193.
42. Sarkar, J., Bhattacharyya, S., & Lal, A., (2013). Selection of suitable natural refrigerants pairs for cascade refrigeration system. *Proc. Inst. Mech. Eng. Part A J. Power Energy, 227*(5), 612–622. https://doi.org/10.1177/0957650913487730.
43. Tashtoush, B., & Bani, Y. M., (2019). Comparative thermodynamic study of refrigerants to select the best environment-friendly refrigerant for use in a solar ejector cooling system. *Arab. J. Sci. Eng., 44*(2), 1165–1184. https://doi.org/10.1007/s13369-018-3427-4.
44. Brignoli, R., Brown, J. S., Skye, H. M., & Domanski, P. A., (2017). Refrigerant performance evaluation including effects of transport properties and optimized heat exchangers. *Int. J. Refrig., 80*, 52–65. https://doi.org/10.1016/j.ijrefrig.2017.05.014.
45. Kim, W. H., & Park, T. S., (2018). Non-premixed lean flame characteristics depending on air hole positions in a baffled micro combustor. *Appl. Therm. Eng., 129*, 431–445. https://doi.org/10.1016/j.applthermaleng.2017.10.064.
46. Ciconkov, R., (2018). Refrigerants: There is still no vision for sustainable solutions. *International Journal of Refrigeration* (pp. 441–448). Elsevier Ltd. https://doi.org/10.1016/j.ijrefrig.2017.12.006.
47. Hashimoto, M., Otsuka, T., Fukushima, M., Okamoto, H., Hayamizu, H., Ueno, K., & Akasaka, R., (2019). Development of new low-GWP refrigerants–refrigerant mixtures including HFO-1123. *Sci. Technol. Built Environ., 25*(6), 776–783. https://doi.org/10.1080/23744731.2019.1603779.
48. Cao, J., Chen, C., Gao, G., Yang, H., Su, Y., Bottarelli, M., Cannistraro, M., & Pei, G., (2019). Preliminary evaluation of the energy-saving behavior of a novel household refrigerator. *J. Renew. Sustain. Energy, 11*(1), 015102. https://doi.org/10.1063/1.5054868.
49. Senthilkumar, G., Saravanan, K., & Ajay, K. K., (2017). Investigation on nano refrigeration. In: *IOP Conference Series: Materials Science and Engineering* (Vol. 197, p. 012028). Institute of Physics Publishing. https://doi.org/10.1088/1757-899X/197/1/012028.
50. Belman-Flores, J. M., Pardo-Cely, D., Gómez-Martínez, M. A., Hernández-Pérez, I., Rodríguez-Valderrama, D. A., & Heredia-Aricapa, Y., (2019). Thermal and energy evaluation of a domestic refrigerator under the influence of the thermal load. *Energies, 12*(3), 400. https://doi.org/10.3390/en12030400.
51. Sri, D., & Tamando, S. H., (2019). Decision support systems assessment of the best village in Perbaungan sub-district with the simple additive weighting (SAW) method. *J. Mantik, 3*, 31–38.

52. Kumar, S. M., Jayaswal, P., & Kushwah, K., (2013). *Exploring Fuzzy SAW Method for Maintenance Strategy Selection Problem of Material Handling Equipment* (Vol. 3).
53. Wang, Z. X., (2013). A genetic algorithm-based grey method for forecasting food demand after snow disasters: An empirical study. *Nat. Hazards, 68*(2), 675–686. https://doi.org/10.1007/s11069-013-0644-8.
54. Prajapati, V., Dinbandhu, Vora, J. J., Das, S., & Abhishek, K., (2020). Study of parametric influence and welding performance optimization during regulated metal deposition (RMD™) using grey integrated with fuzzy Taguchi approach. *J. Manuf. Process., 54*, 286–300. https://doi.org/10.1016/j.jmapro.2020.03.017.
55. Opricovic, S., & Tzeng, G. H., (2007). Extended VIKOR method in comparison with outranking methods. *Eur. J. Oper. Res., 178*(2), 514–529. https://doi.org/10.1016/j.ejor.2006.01.020.
56. Vahdani, B., Hadipour, H., Sadaghiani, J. S., & Amiri, M., (2010). Extension of VIKOR method based on interval-valued fuzzy sets. *Int. J. Adv. Manuf. Technol., 47*(9–12), 1231–1239. https://doi.org/10.1007/s00170-009-2241-2.
57. Chatterjee, S., Abhishek, K., & Mahapatra, S. S., (2020). *Optimization of Quality Characteristics in Laser Drilling of Ti6Al4V Using VIKOR* (pp. 59–67). Springer, Singapore. https://doi.org/10.1007/978-981-32-9971-9_7.
58. Chang, C. L., (2010). A modified VIKOR method for multiple criteria analysis. *Environ. Monit. Assess., 168*(1–4), 339–344. https://doi.org/10.1007/s10661-009-1117-0.

CHAPTER 6

COMPARATIVE STUDY OF MCDM TECHNIQUES: TOPSIS, VIKOR, AND MOORA METHODS INTEGRATED WITH EWM METHOD FOR VENDOR SELECTION FOR MANUFACTURING INDUSTRY

JOSY GEORGE, SANDEEP KUMAR TEMBHARE, and PUSHPDANT JAIN

School of Mechanical Engineering, VIT Bhopal University, Sehore, Madhya Pradesh, India

ABSTRACT

In today's competitive world, vendor selection has become a very important function for any manufacturing firm, and the selection of a suitable method for vendor selection is immensely important. Many models and techniques have been developed to deal with selecting and evaluating vendors. In the present work, four different criteria are taken into consideration for vendor selection among ten alternatives. This work proposes a multicriteria decision-making (MCDM) based framework that is used to evaluate the vendors by using an entropy weight method (EWM) for the calculation of the weightage of each criterion. Once the weightage is calculated, then EWM is combined with Technique for Order Preference by Similarity to Ideal Solution (TOPSIS), Vlse Kriterijumska Optimizacija Kompromisno Resenje (VIKOR), multiobjective optimization on the basis of ratio analysis

Decision-Making Models and Applications in Manufacturing Environments. Pushpdant Jain, Kumar Abhishek, & Prasenjit Chatterjee (Eds.)

(MOORA) methods. A numerical example is proposed to understand the effectiveness, and comparison analysis is provided to test the feasibility of the method using EWM is used for weight calculation for selected criteria, and then these weights are used for further calculation of the rank of different alternatives using proposed methods.

6.1 INTRODUCTION

MCDM alludes to making the decision of the easiest different from among a limited arrangement of call options regarding various, in some cases clashing rules. Multi-standards dynamic methods are useful instruments to help leaders(s) to choose decisions inside the instance of cutting-edge issues. Especially, with the assistance of laptop applications, those ways became easier for the users, so they have found nice acceptance in several areas of decision-making processes in economy or management. MCDM could also be a sub-discipline of operation analysis that involves the analysis of a finite set of alternatives delineated inside the terms of evaluating criteria supported by the values and preferences of the selection maker. There's not a novel best resolution to associate MCDM downside which will be obtained while not incorporating preference info. The conception of the associate best resolution is sometimes replaced by the set of non-dominated solutions. A non-dominated resolution has the property that it's not possible to maneuver far off from it to the other resolution while not sacrificing a minimum of one criterion. Therefore, it is smart for the decision-maker (DM) to decide on a solution from the non-dominated set. Otherwise, she/he might do higher in terms of some or all of the standards, and not do worse in any of them. Generally, however, the set of non-dominated solutions is simply overlarge to be bestowed to the DM for his final alternative. Thus, we'd like tools that facilitate the DM to focus on his most popular solutions (or alternatives). A significant distinction between MCDM issues relies on whether or not the solutions are expressly or implicitly outlined.

The decision of the trafficker choice depends upon different kinds of criteria. Mainly, the value is that the foremost criteria thought-about whereas selecting a trafficker, others like product quality of the fabric, delivery time, and repair quality additionally play a significant role whereas choosing an acceptable vendor. Selecting the simplest trafficker isn't simple for a DM WHO invariably satisfies the whole needs of the consumers. Trafficker choice may be a multicriteria decision-making downside that features each qualitative and quantitative factor, a number of that conflict with one another. A multicriteria decision-making technique helps the DMs to measure a collection of alternatives supported by totally different criteria. The higher cognitive process is one among the foremost vital topics in numerous areas of the act and shows the way to create the proper choices to succeed in the tip goal. Call manufacturers over the past few decades have with success used multicriteria decision-making (MCDM) approaches to unravel advanced decision-making issues in an exceeding form of fields, like political economy, finance, logistics, environmental correction, business, engineering, medicine, law, and more.

Multicriteria higher cognitive process (MCDM) is one of the foremost well-known branches of the higher cognitive process [1]. MCDM is split into the multiobjective higher cognitive process (MODM) and multi-attribute higher cognitive process (MADM). However, the terms MADM and MCDM are accustomed to mean the identical category of models. MODM peruses call issues during which the choice house is continuous, whereas MCDM/MADM focuses on issues with distinct call areas and takes a look at to search out the best various amongst the present call alternatives regarding many criteria [2]. The various approaches that are investigated enclosed the EWM, TOPSIS, MOORA, and VIKOR: stepwise procedure.

6.2 LITERATURE SURVEY

The use of the MCDM ways for finding issues in numerous fields is rife and helps in creating important choices. MCDM is a good systematic and quantitative thanks to influence problems within the presence of many alternatives and several other typically totally different criteria. Deciding to relate to terribly advanced issues, as well as business-related choices and real-life choices, needs an associate degree applicable and reliable call network. Improvement could also be thought of as a decision-making method to urge the foremost out of obtainable resources for the most effective gettable results. Several real-world issues square measure multifarious or multiattribute issues that naturally involve multiple competitor goals that require to be optimized at a similar time among bound constraints or by selecting from the on-the-market distinct alternatives. In distinction to single-goal improvement, solutions to multiobjective and multiattribute issues correspond to a collection of solutions with tradeoffs, every expressing a peculiar tradeoff between totally different goals or attributes.

Improvement is often thought of as a decision-making method to maximize the effectiveness of the on-the-market resources wont to reach the most effective results doable. Multicriteria decision-making (MCDM) started within the Nineteen Seventies and was thought of as a stimulating space for several researchers. From the start hitherto, over seventy multicriteria decision-making techniques are discovered [3], wherever the MCDM are often divided into two different classifications: multiobjective decision-making (MODM) and multiattribute decision-making (MADM).

The MODM aims to style alternatives that optimize the multiple objectives of the administrator [4]. Two explicit important problems square measure thought of to pick the most effective marketer. One is what criteria ought to be used, and therefore the difference is relating to the strategy wont to compare vendors. The choice of a marketer is sophisticated by the various criteria that ought to be thought of. Meanwhile, varied approaches are often utilized to try and do a range [5]. Specific attention was given to the factors and analytical ways utilized in the seller choice method. In response to the hyperbolic interest in just-in-time (JIT) producing methods and analysis of JIT's impact on the seller, the choice was additionally mentioned by the authors.

A fuzzy dynamic methodology is made sense of Ref. [6] to manage the merchant choice issue in the store network framework. In this work, etymological qualities are utilized to survey the appraisals and loads for different elements. These phonetic evaluations can be communicated concerning trapezoidal or three-sided fuzzy numbers. Then, an order different measures independent direction (MCDM) model in light of fluffy sets hypothesis is proposed to manage the merchant choice issues in the production network the executive's framework. As per the idea of the TOPSIS method [7], a closeness coefficient is characterized to decide the positioning request of all merchants by computing the distances to both fuzzy positive-ideal arrangement (FPIA) and fuzzy negative-ideal arrangement (FNIA) all the while.

There are countless dynamic strategies that have been created to aid the merchant determination process. By taking special care of the changing necessities of merchant determination measures, choice techniques are made do occasionally. A few strategies have been famous for a really long time, while different techniques have been presented as of late. Because of the different models of merchant determination [8], a dynamic instrument obliging both the quantitative and the subjective rules is liked.

Merchant determination strategies are the arrangements or approaches used to direct the choice cycle. The strategies are picked are critical to the general determination process and impact the choice outcomes. It is essential to comprehend the reason why a firm picks one technique (or a blend of various strategies) over another. A few notable determination techniques have been created and ordered by various analysts throughout the long term. Certain strategies have been famous determination decisions for leaders for a really long time, while different techniques have just arisen as of late [9].

The entropy weight method (EWM) is utilized not exclusively to quantitatively gauge information sum yet conjointly to work out impartially the overall load of information. Entropy was initially authorized to infer a characteristic peculiarity of profit choppiness degree or the probability scale underneath a decent condition. In the event that entropy values region unit lower, the profit degrees region unit extra corresponding, inferring as on the mark of good entropy as feasible [10]. Alternately, assuming entropy values region units higher, the profit degrees have an extra sporadic expression.

6.3 METHODOLOGY ADOPTED

6.3.1 ENTROPY WEIGHT METHOD (EWM)

Using this method will decide loads alone in accordance with the inclination or decisions of chiefs, and furthermore, the expected vulnerability during this strategy is its primary weakness. The objective weight strategy works out loads upheld genuine perceptions, with none thought about the chief's inclinations, and doesn't have the advantage of the talented information or aptitude of the leaders. Hence, this section utilizes loads decided exploitation the EWM.

The entropy procedure might be alive of the vulnerability created with regards to applied science. Its acclimated portray irreversible advancement including movement or a technique and was brought into logical hypothesis by Shannon [27]. These days, this strategy has been broadly applied to shifted examination fields to take care of problem loads [11–16]. Information entropy is the mensuration of the issue level of a framework. It estimates the amount of accommodating data upheld by the information given. When the qualification in worth among assessed objects of a given marker is huge, and furthermore the entropy is pretty much nothing, this outlines that this pointer gives accommodating information, and its weight should be high. In differentiation, in the event that the qualification is close to nothing and furthermore the entropy is gigantic, the general weight is going to underneath. Consequently, entropy hypothesis is a goal approach of urgent loads [17].

The entropy weight estimation technique decides the heaviness of every reaction boundary with no thought of the chief. The fundamental idea of entropy weight estimation is that a higher weight list esteem is more helpful than a more modest file esteem. In this paper, the accompanying advances are utilized to assess the weight list of every reaction [18].

- **Step 1:** Development of a choice framework (X). A bunch of options (A = {Ai, I = 1, 2…., n}) is to be contrasted with concerning a bunch of rules (C = {C_j, j = 1, 2, …, m}). In this way, an n×m execution lattice (the choice network; X) can be acquired as:

$$X = \begin{bmatrix} x_{12} & x_{12} & \cdots & x_{1n} \\ x_{21} & x_{22} & \cdots & x_{2n} \\ \cdots & \cdots & \cdots & \cdots \\ \cdots & \cdots & \cdots & \cdots \\ \cdots & \cdots & \cdots & \cdots \\ x_{m1} & x_{m2} & \cdots & x_{mn} \end{bmatrix}$$

  where; x_{ij} is a crisp value indicating the performance rating of each alternative A_i concerning each criterion, C_j.
- **Step 2:** To discover objective weight by the entropy measure, the choice framework in Eqn. (1) should be standardized for every model C_j (j = 1, 2, …, m)

$$p_{ij} = \frac{x_{ij}}{\sum_{i=1}^{m} x_{ij}}$$

  The normalized decision matrix is obtained as a result of the process:

$$P = \begin{bmatrix} p_{12} & p_{12} & \cdots & p_{1n} \\ p_{21} & p_{22} & \cdots & p_{2n} \\ \cdots & \cdots & \cdots & \cdots \\ \cdots & \cdots & \cdots & \cdots \\ \cdots & \cdots & \cdots & \cdots \\ p_{m1} & p_{m2} & \cdots & p_{mn} \end{bmatrix}$$

- **Step 3:** Work out the entropy proportion of each file utilizing the accompanying condition:

$$e_j = -k \sum_{i=1}^{m} p_{ij} . \ln p_{ij}$$

  where; $k = \frac{1}{\ln(n)}$ is a constant which guarantees $0 < e_j < 1$

➢ **Step 4:** The level of divergence (d_j) of the normal inborn data contained by every basis C_j (j = 1, 2, …, m) can be determined a:

$$d_j = 1 - e_j$$

➢ **Step 5:** The goal weight for every standard C_j (j = 1, 2, …, m) is subsequently given by:

$$w_j = \frac{1-e_j}{\sum_{j=1}^{n} 1-e_j}$$

6.3.2 *TOPSIS METHOD*

TOPSIS (a technique for order preference by similarity to an ideal solution) method is presented by Chen & Hwang [19] concerning Hwang & Yoon [20]. The primary principle is that the chosen various ought to have the shortest distance from the best resolution and therefore the farthest distance from the negative ideal resolution. Within the TOPSIS technique, the proximity needed for the best (best) resolution is found, thus, each the positive-ideal resolution distance and therefore the negative-ideal resolution distance square measure thought-about. The most effective resolution that is expressed as a perfect or positive ideal resolution may be a resolution that maximizes the profit criterion and minimizes the value criterion. On the opposite hand, the negative or anti-ideal resolution may be a resolution that maximizes the value criterion and minimizes the profit criterion. Within the technique, all alternatives may be hierarchical by comparison the relative distances, since it's assumed that the closest various to the positive-ideal resolution is that the farthest various to the negative ideal resolution additionally.

The TOPSIS procedure consists of the following steps:

➢ **Step 1:** The structure of the matrix:

$$X = \begin{bmatrix} x_{12} & x_{12} & \cdots & x_{1n} \\ x_{21} & x_{22} & \cdots & x_{2n} \\ . & . & . & . \\ . & . & . & . \\ . & . & . & . \\ x_{m1} & x_{m2} & \cdots & x_{mn} \end{bmatrix}$$

- **Step 2:** Calculate the normalized the matrix D by using the following formula:

$$r_{ij} = \frac{x_{ij}}{\sqrt{\sum_{j=1}^{J} x_{ij}^{2}}}$$

- **Step 3:** Construct the weighted normalized decision matrix by multiplying:

$$V_{ij} = w_{ij}.r_{ij}$$

- **Step 4:** Determine the PIS and NIS:

$$A^{*} = \{(max\, v_{ij} \mid j \in J), (min\, v_{ij} \mid j \in J')\}$$

$$A^{-} = \{(min\, v_{ij} \mid j \in J), (max\, v_{ij} \mid j \in J')\}$$

- **Step 5:** Calculate the separation measure:

$$S_i^{*} = \sqrt{\sum_{j=1}^{n} \left(v_{ij} - v_j^{*}\right)^{2}}$$

$$S_i^{-} = \sqrt{\sum_{j=1}^{n} \left(v_{ij} - v_j^{-}\right)^{2}}$$

- **Step 6:** Calculate the relative closeness to the ideal solution:

$$C_i^{*} = S_i^{-} \Big/ S_i^{*} + S_i^{-},\ 0 \leq C_i^{*} \leq 1$$

- **Step 7:** Calculate the total score and select the alternative closest to 1.

6.3.3 *VIKOR METHOD*

The VIKOR procedure was acquainted as a pertinent strategy with carry out inside MCDM [21]. This procedure leans back unimportant the positive and negative ideal goal. The positive ideal goal shows the choices with the best cost though the negative ideal goal proposes the choices of the base qualities. It gives the most group utility to the mass and in this manner the base of individual lament for the adversary. The tradeoff goal could be a reasonable goal that is most elevated to the positive ideal goal and farthest from the negative ideal goal.

The main procedure of the VIKOR method [21–23] is described below:

- **Step 1:** The structure of the matrix:

$$X = \begin{bmatrix} x_{12} & x_{12} & \cdots & x_{1n} \\ x_{21} & x_{22} & \cdots & x_{2n} \\ . & . & . & . \\ . & . & . & . \\ . & . & . & . \\ x_{m1} & x_{m2} & \cdots & x_{mn} \end{bmatrix}$$

- **Step 2:** Calculate the weights to each indicator to entropy method which is introduced each indicators weight.
- **Step 3:** Make indicators value being dimensionless, establish a decision-making matrix.

The cost type indicators: $V_{ij} = \left(\dfrac{\min x_{ij}}{x_{ij}} \right)$

The benefit type indicators: $V_{ij} = \left(\dfrac{x_{ij}}{\max x_{ij}} \right)$

- **Step 4:** To calculate indicator PIS and NIS each criterion.

$$f_i^* = \left[\left(\max f_{ij} \mid i \epsilon I_2 \right), \left(\min f_{ij} \mid i \epsilon I_2 \right) \right],$$

$$f_i^- = \left[\left(\min f_{ij} \mid i \epsilon I_1 \right), \left(\max f_{ij} \mid i \epsilon I_2 \right) \right]$$

where; I_1 is a benefit type indicator set; and I_2 is a cost type indicator set.

- **Step 5:** To calculate the utility measure and regret measure for all the alternatives.
  a. Utility measure $S_i = \sum_{i=1}^{n} \left(w_i \left(f_i^* - f_{ij} \right) / \left(f_i^* - f_i^- \right) \right)$
  b. Regret measure $R_i = max \left[\left(w_i \left(f_i^* - f_{ij} \right) / \left(f_i^* - f_i^- \right) \right) \right]$

- **Step 6:** To calculate the value of the VIKOR index for each alternative:

$$Q_j = v\left(S_i - S^*\right) / \left(S^- - S^*\right) + (1 - v)\left(R_j - R^*\right) / \left(R^- - R^*\right)$$

where; $S^* = \min S_j$, $S^- = \max S_j$, $R^* = \min R_j$, $R^- = \max R_j$. Usually, the value of v is taken as 0.5. However, v can take any value from 0 to 1.

- **Step 7:** To compute the worth of the VIKOR file for every other option, as indicated by S, R, and Q independently to rank the plans, we get 3 position tables.

  Excused characteristic loads, propose a tradeoff arrangement, elective A1, which is the best positioned by the action Q assuming that the accompanying two circumstances are fulfilled:

  - **Condition 1:** 'OK benefit' Q (A2) – Q (A1) ≥ (1/(N–1)). A2 is the second-best option in the positioning by Q.
  - **Condition 2:** Acceptable dependability in navigation. Elective A1 should likewise be the best positioned by S as well as R. This compromise arrangement is steady inside a dynamic cycle, which could be: casting a ballot by greater part rule.

    (When v> 0.5 is required) or "by agreement" (when v = 0.5) or "with blackball" (when v< 0.5). In the event that one of the circumstances isn't fulfilled, then, at that point, a bunch of give and take arrangements is proposed, which comprises of; Alternatives A1 and A2 if by some stroke of good luck condition 2 isn't fulfilled.

    Choices A_1, A_2, A_m in the event that condition 1 isn't fulfilled. Not set in stone by the connection Q (Am) – Q (A1) ≥ (1/(N – 1)) for greatest M (the places of these options are "in closeness").

6.3.4 MOORA METHOD

MOORA, created by, is one of the MCDM strategy comprising of two stages, in particular, the reference approach and in this manner the greatness connection framework approach, and allows measure the supportive and non-helpful rules during a technique for picking one more from a gathering of choices. The MOORA approach, first used by Brauers [12], perhaps a similarly new multi-rules dynamic philosophy. The procedure depends on a greatness connection framework and dimensionless action [13]. MOORA philosophy comprises of 5 significant advances [14].

- **Step 1: Creating the Decision Matrix:** The technique begins with a choice grid of reactions of various options in contrast to assessment standards. The information given is addressed as network X_{mxn}. Where x_{ij} is the exhibition proportion of ith elective on j^{th} property, m is the quantity of options, and n is the quantity of traits.

$$X = \begin{bmatrix} x_{12} & x_{12} & \cdots & x_{1n} \\ x_{21} & x_{22} & \cdots & x_{2n} \\ \cdot & \cdot & \cdot & \cdot \\ \cdot & \cdot & \cdot & \cdot \\ \cdot & \cdot & \cdot & \cdot \\ x_{m1} & x_{m2} & \cdots & x_{mn} \end{bmatrix}$$

where; x_{ij} is the performance measure of ith alternative on j^{th} criteria, m is the number of alternatives, and n is the number of criteria.

- **Step 2: Normalization Procedure:** MOORA alludes to a proportion framework where every reaction of an option on measures is contrasted with a denominator, which is delegate for all choices worried that goal. This ratio can be expressed as below:

$$x^*_{ij} = \frac{x_{ij}}{\sqrt{\sum_{i=1}^{m} x^2_{ij}}}$$

where; x^*_{ij} is a dimensionless number which belongs to the interval [0,1] representing the normalized performance of ith alternative on j^{th} criteria.

- **Step 3: For Multiobjective Optimization:** These standardized exhibitions are included the situation of augmentation (for useful traits) and deducted on account of minimization (for non-gainful characteristics). Then, at that point, the enhancement issue becomes:

$$y_i = \sum_{j=1}^{g} x^*_{ij} - \sum_{j=g+1}^{n} x^*_{ij}$$

where; g is the quantity of qualities to be augmented; (n–g) is the quantity of characteristics to be limited; and y_i is the standardized appraisal worth of ith elective for every one of the properties. At times, it isn't unexpected to see that a few credits are a higher priority than others. To give more significance to a trait, it very well may be increased with its comparing weight (importance coefficient) [12].

- **Step 4: Determine the Weighted Assessment Value:** By and large, it isn't unexpected to see that some choice models are a higher priority than others. To build the need of rules, it very well may be increased by its weight. Whenever these rules loads are thought about, becomes as follows:

$$y_i = \sum_{j=1}^{g} w_j . x_{ij}^{*} - \sum_{j=g+1}^{n} w_j . x_{ij}^{*}$$

where; w_j is the priority of jth criteria, which can be assigned using different multicriteria decision-making methods.

- **Step 5: Ranking of Alternatives:** Choice options should be positioned in the inclination request as indicated by diminishing upsides of y_i*. Appraisal worth can be positive or negative contingent upon the rules circumstance and need values.

6.4 NUMERICAL ILLUSTRATION

A numerical example is illustrated in this section, a manufacturing industry that wants to select a suitable vendor for raw material supply. The numerical is solved with the help of three different methods and then results are compared. In the present problem, there are ten vendors (V1 to V10) and four criteria (C1, C2, C3, C4) in which C2 is cost attribute and remaining is benefit attribute. Initial information about the vendor and criteria is given in Table 6.1.

TABLE 6.1 Initial Parameters for Vendor Selection

	C1	C2	C3	C4
V1	75	120	5	80
V2	88	212	4	74
V3	68	225	6	84
V4	56	180	5	66
V5	66	230	7	58
V6	81	130	3	94
V7	94	275	2	54
V8	86	195	3	62
V9	71	315	5	43
V10	79	380	9	87

6.4.1 *WEIGHT CALCULATION: IMPLEMENTATION OF ENTROPY WEIGHT METHOD (EWM)*

Finding the applicable weight for each criterion is one of the main points in multiattribute decision-making (MADM) problems. The EWM is one of the various approaches for finding weights (Table 6.2).

TABLE 6.2 Normalized Matrix with Entropy-Based Weights

	C1	C2	C3	C4
V1	0.0982	0.0531	0.102	0.114
V2	0.1152	0.0937	0.0816	0.1054
V3	0.089	0.0995	0.1224	0.1197
V4	0.0733	0.0796	0.102	0.094
V5	0.0864	0.1017	0.1429	0.0826
V6	0.106	0.0575	0.0612	0.1339
V7	0.123	0.1216	0.0408	0.0769
V8	0.1126	0.0862	0.0612	0.0883
V9	0.0929	0.1393	0.102	0.0613
V10	0.1034	0.168	0.1837	0.1239
Weightage	0.0605	0.325	0.4694	0.1451

6.4.2 *VENDOR SELECTION BASED ON TOPSIS METHOD*

It's an approach of compensatory aggregation that compares a set of choices by relating weights for each criterion, homogenizing scores for each criterion, and calculating the geometric distance between each choice and the ideal choice, which is the best score in each criterion (Tables 6.3 and 6.4).

TABLE 6.3 Normalized Matrix, TOPSIS Method

	C1	C2	C3	C4
V1	0.3073	0.1590	0.2993	0.3520
V2	0.3606	0.2809	0.2395	0.3256
V3	0.2786	0.2982	0.3592	0.3696
V4	0.2295	0.2385	0.2993	0.2904
V5	0.2704	0.3048	0.4191	0.2552
V6	0.3319	0.1723	0.1796	0.4135
V7	0.3852	0.3644	0.1197	0.2376
V8	0.3524	0.2584	0.1796	0.2728
V9	0.2909	0.4174	0.2993	0.1892
V10	0.3237	0.5036	0.5388	0.3828
Weightage	0.0605	0.3250	0.4694	0.1451

TABLE 6.4 TOPSIS, Closeness Coefficient Table

	S+	S–	C*	Rank
V1	0.1129	0.1422	0.5576	4
V2	0.1465	0.0941	0.3909	7
V3	0.0961	0.1334	0.5812	3
V4	0.1171	0.1214	0.5091	5
V5	0.0773	0.1550	0.6671	1
V6	0.1687	0.1161	0.4077	6
V7	0.2093	0.0467	0.1825	10
V8	0.1729	0.0857	0.3313	9
V9	0.1441	0.0889	0.3815	8
V10	0.1121	0.1988	0.6394	2

6.4.3 VENDOR SELECTION BASED ON VIKOR METHOD

To work out choice issues with clashing and non-commensurable (various units) standards, expecting that exchange is decent for compromise, the chief needs an outcome that is the nearest to the ideal, and the decisions are assessed by totally settled measures. VIKOR positions decisions and decides the outcome named arrangement that is the nearest to the ideal (Tables 6.5 and 6.6).

TABLE 6.5 Normalized Matrix, VIKOR Method

	C1	C2	C3	C4
V1	0.7979	1.0000	0.5556	0.8511
V2	0.9362	0.5660	0.4444	0.7872
V3	0.7234	0.5333	0.6667	0.8936
V4	0.5957	0.6667	0.5556	0.7021
V5	0.7021	0.5217	0.7778	0.6170
V6	0.8617	0.9231	0.3333	1.0000
V7	1.0000	0.4364	0.2222	0.5745
V8	0.9149	0.6154	0.3333	0.6596
V9	0.7553	0.3810	0.5556	0.4574
V10	0.8404	0.3158	1.0000	0.9255
Weightage	0.0605	0.3250	0.4694	0.1451

TABLE 6.6 Ranking of Alternatives in VIKOR Method

S_i	R_i	Q_i	Rank
0.3383	0.2682	0.0940	1
0.6078	0.3353	0.4922	7
0.4927	0.2217	0.1506	2
0.5667	0.2682	0.3168	5
0.5083	0.2272	0.1769	3
0.4595	0.4023	0.4829	6
0.8509	0.4694	1.0000	10
0.6888	0.4023	0.7065	9
0.7440	0.2940	0.5418	8
0.3688	0.3250	0.2383	4

6.4.4 VENDOR SELECTION BASED ON MOORA METHOD

The MOORA strategy is a multi-¬objective improvement procedure that can be effectively applied to address different kinds of perplexing dynamic issues in the assembling climate (Tables 6.7 and 6.8).

TABLE 6.7 Normalized Matrix, MOORA Method

	C1	C2	C3	C4
V1	0.3073	0.1590	0.2993	0.3520
V2	0.3606	0.2809	0.2395	0.3256
V3	0.2786	0.2982	0.3592	0.3696
V4	0.2295	0.2385	0.2993	0.2904
V5	0.2704	0.3048	0.4191	0.2552
V6	0.3319	0.1723	0.1796	0.4135
V7	0.3852	0.3644	0.1197	0.2376
V8	0.3524	0.2584	0.1796	0.2728
V9	0.2909	0.4174	0.2993	0.1892
V10	0.3237	0.5036	0.5388	0.3828
Weightage	0.0605	0.3250	0.4694	0.1451

TABLE 6.8 MOORA, Max-Min Data with Ranking

	Maximum	Minimum	Y = (Max-Min)	Rank
V1	0.2102	0.0517	0.1585	2
V2	0.1815	0.0913	0.0902	7
V3	0.2391	0.0969	0.1422	4
V4	0.1965	0.0775	0.1190	5
V5	0.2501	0.0991	0.1510	3
V6	0.1644	0.0560	0.1084	6
V7	0.1140	0.1184	–0.0044	10
V8	0.1452	0.0840	0.0612	8
V9	0.1856	0.1357	0.0499	9
V10	0.3280	0.1637	0.1644	1

6.5 RESULT AND DISCUSSION

Vendor determination choices are a vital piece of any company's generally well-defined course of action. One might say that seller determination is one movement in which immense conceptualizing and preliminaries are done. This MCDM issue assists the selector with settling on the most appropriate choice in regard to the merchant choice interaction. According to numerous writing audits, it is demonstrated that this is perhaps the main movement is merchant choice. Consequently, the merchant determination strategy and, most importantly, the choice of a sufficient technique is huge, particularly from the part of the end-product.

In this section, an illustrative model was effectively directed pointed toward showing that multi-rules navigation can be effectively applied for the seller determination, or we can say a multi-rules choice model for assessing choices for merchant choice issues. For this reason, a two-venture philosophy is presented, in which the entropy weight technique decides the heaviness of rules by means of the aptitude of dynamic colleagues. Then, at that point, TOPSIS VIKOR and MOORA techniques apply EWM' loads as info loads. By the correlation of the positioning technique, the outcome is resolved which would be generally adequate to the choice. At the point when all techniques were looked at, it was observed that the distinctions saw which were addressed with the assistance of charts (Figure 6.1).

In view of the four measures of manageability, a sum of ten sellers in an assembling firm were thought of, where it was inferred that merchant 5 with TOPSIS strategy, vendor 1 with VIKOR Method, and merchant 10 with MOORA Method was determined and supposed to be the best arrangement or most reasonable merchant for additional business with an assembling firm, which has been affirmed by examination with another vendor.

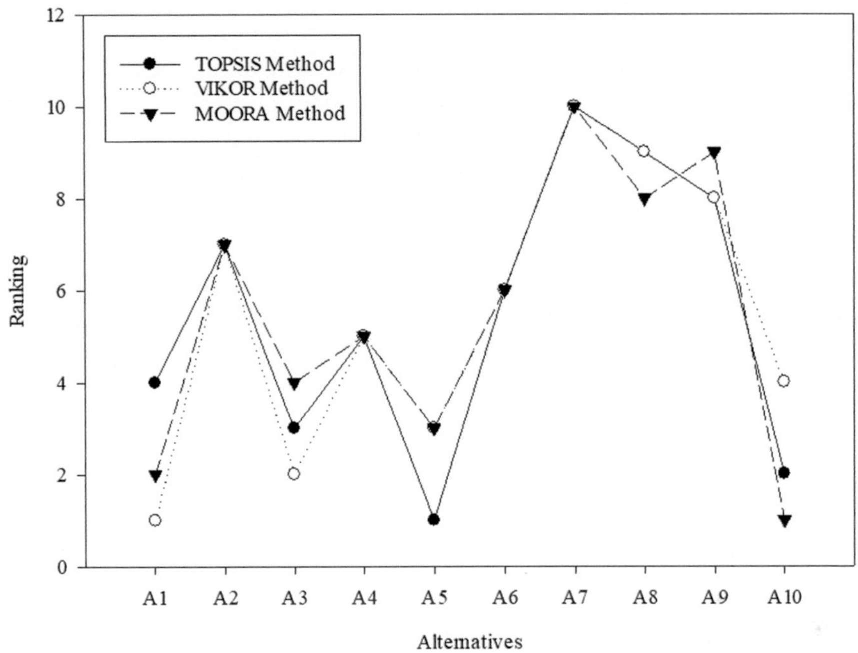

FIGURE 6.1 Rank comparison of TOPSIS, VIKOR, and MOORA methods.

6.6 CONCLUSION

Some of the most common strategies for solving an MCDM problem have been used in this chapter. The methods of selection for vendors are picked by applying a decision-making approach to discover the best accessible vendors on the market. After inspection of the aggregations on multiple process parameters in diverse scenarios, the presented models are easy to use and meaningful to aggregate process data. To get the final ranking preferences in decreasing order, TOPSIS, VIKOR, and MOORA methods are performed; this enables a comparison of relative performance. From the data accessible for the above example through the end solution processing. Since the comparison analysis given in this chapter, we have concluded that there may not be a single MCDM method that can always ensure the best decision, so this is recommended to the DM should always be vigilant before accepting the results of an MCDM method.

6.7 FUTURE SCOPE

In many sorts of issues, the multi-measures dynamic technique is utilized for choosing the best option among the arrangement of choices. Further work should be possible by involving more than one technique for positioning and contrasting the outcomes with a concentrate on the productivity, unwavering quality, and awareness of those strategies. Multi-measures dynamic techniques can be utilized for enhancements in the options by changing the upsides of options and investigating the adjustment of rankings. This examination can give a method for working on in an enormous number of fields.

KEYWORDS

- **entropy weight method**
- **multicriteria decision making**
- **multiobjective optimization**
- **technique for order performance**
- **vendor selection**
- **Vlse Kriterijumska Optimizacija Kompromisno Resenje**

REFERENCES

1. Panapakidis, I. P., & Christoforidis, G. C., (2018). Optimal selection of clustering algorithm via multicriteria decision analysis (MCDA) for load profiling applications. *Applied Sciences, 8*(2).
2. Triantaphyllou, E., (2000). *Multicriteria Decision-Making Methods: A Comparative Study* (p. 290). Springer US.
3. Alias, M. A., Hashim, S. Z. M., & Samsudin, S., (2008). Multi criteria decision making and its applications: A literature review. *Jurnal Teknologi Maklumat, 20*(2), 129–152.
4. Singh, A., & Malik, S. K., (2014). Major MCDM techniques and their application-A review. *IOSR Journal of Engineering, 4*(5), 15–25.

5. Weber, C. A., Current, J. R., & Benton, W., (1991). Vendor selection criteria and methods. *European Journal of Operation Research, 50*, 2–18.
6. Chen-Tung, C., & Ching-Torng, L., (2006). A fuzzy approach for vendor evaluation and selection in supply chain management. *Production Economics, 102*, 289–301.
7. Chunhui, L., & Aizhen, L., (1999). The application of TOPSIS method to comprehensive assessment of environmental quality. *Journal of Geological Hazard and Environmental Preservation, 10*(2), 9–13.
8. Boran, F. E., Genç, S., Kurt, M., & Akay, D., (2009). A multicriteria intuitionistic fuzzy group decision making for vendor selection with TOPSIS method. *Expert Systems with Applications, 36*, 11363–11368.
9. Li, C. C., & Fun, Y. P., (1997). A new measure for vendor performance evaluation. *IIE Transactions, 29*(1), 753–758.
10. Yanfu, Z., (2015). TOPSIS method based on entropy weight for vendor evaluation of power grid enterprise. In: *2nd International Conference on Education Reform and Modern Management (ERMM 2015)* (pp. 334–337).
11. Brunsell, N. A., Schymanski, S. J., & Kleidon, A., (2011). Quantifying the thermodynamic entropy budget of the land surface: Is this useful? *Earth System Dynamics, 2*(1), 87–103. https://doi.org/10.5194/esd-2-87-2011.
12. Singh, V. P., (2011). Hydrologic synthesis using entropy theory: Review. *Journal of Hydrologic Engineering, 16*(5), 421–433. https://doi.org/10.1061/(asce)he.1943-5584.0000332.
13. Ruddell, B. L., Brunsell, N. A., & Stoy, P., (2013). Applying information theory in the geosciences to quantify process uncertainty, feedback, scale. *Eos, Transactions American Geophysical Union, 94*(5), 56. http://doi.org/10.1002/2013EO050007.
14. Islamoglu, M., Apan, M., & Oztel, A., (2015). An evaluation of the financial performance of REITs in Borsa Istanbul: A case study using the entropy-based TOPSIS method. *International Journal of Financial Research, 6*(2), 124–138. http://doi.org/10.5430/ijfr.v6n2p124.
15. Xu, J., Feng, P., & Yang, P., (2016). Research of development strategy on China's rural drinking water supply based on SWOT-TOPSIS method combined with AHP-entropy: A case in Hebei province. *Environmental Earth Sciences, 75*(1), 58. https://doi.org/10.1007/s12665-015-48856.
16. Işık, A. T., & Adalı, E. A., (2017). The decision-making approach based on the combination of entropy and ROV methods for the apple selection problem. *European Journal of Interdisciplinary Studies, 3*(3). http://doi.org/10.26417/ejis.v3i3.p81-6.
17. Zou, Z., Yun, Y., & Sun, J., (2006). Entropy method for determination of weight of evaluating indicators in fuzzy synthetic evaluation for water quality assessment. *Journal of Environmental Sciences, 18*(5), 1020–1023. https://doi.org/10.1016/s1001-0742(06)60032-6.
18. Deng, H., Yeh, C. H., & Willis, R. J., (2000). Inter-company comparison using modified TOPSIS with objective weights. *Computers & Operations Research, 27*(10), 963–973.
19. Chen, S. J., & Hwang, C. L., (1992). *Fuzzy Multiple Attribute Decision Making: Methods and Applications.* Berlin: Springer-Verlag.
20. Hwang, C. L., & Yoon, K. P., (1981). *Multiple Attribute Decision Making: Methods and Applications Survey.* New York: Springer-Verlag.

21. Musfirah, A. H., & Jaharah, A. G., (2012). Magnesium and aluminum alloys in the automotive industry. *Journal of Applied Sciences Research* (Vol. 8, No. 9, pp. 4865–4875). ISSN 1819-544X.
22. Chang, C. L., & Hsu, C. H., (2009). Multicriteria analysis via the VIKOR method for prioritizing land-use restraint strategies in the Tseng-Wen reservoir watershed. *Journal of Environmental Management, 90*(11), 3226–3230.
23. Radhika, S., Sameer, K. D., & Swapna, D., (2013). VIKOR method for multi criteria decision making in academic staff selection. *Journal of Production Research and Management* (Vol. 3, No. 2, pp. 30–35). ISSN: 2249–4766.
24. Brauers, W. K. M., (2003). Optimization methods for a stakeholder society: A revolution in economic thinking by multiobjective optimization. *Nonconvex Optimization and Its Applications.* Kluwer Academic Publishers.
25. Chakraborty, S., (2011). Applications of the MOORA method for decision making in a manufacturing environment. *The International Journal of Advanced Manufacturing Technology, 54*(9–12), 1155–1166.
26. Brauers, W. K. M., Zavadskas, E. K., Turskis, Z., & Vilutiene, T., (2008). Multiobjective contractors' ranking by applying the MOORA method. *Journal of Business Economics and Management, 9*(4), 245–255.
27. Shannon, C. E. (1948). A Mathematical Theory of Communication. *Bell System Technical Journal, 27*, 379–423, 623–656, July & October.

CHAPTER 7

SELECTION OF CUTTING FLUIDS FOR MACHINING TITANIUM ALLOYS USING MCDM METHODS

MEET SHAH[1], YUG MODI[1], DIN BANDHU[2], SONI KUMARI[3] and KUMAR ABHISHEK[1]

[1]*Department of Mechanical and Aerospace Engineering, Institute of Infrastructure, Technology, Research, and Management (IITRAM), Ahmedabad, Gujarat, India*

[2]*Department of Mechanical and Industrial Engineering, Manipal Institute of Technology Bengaluru, Manipal Academy of Higher Education, Manipal, Karnataka, India*

[3]*Department of Mechanical Engineering, GLA University, Mathura, Uttar Pradesh, India*

ABSTRACT

Titanium and its alloys are in high demand across a wide range of industries. Owing to their high strength-to-weight ratio, they are also useful in producing armored vehicles and the superstructure of surface ships, in addition to their extensive uses in the aerospace and petrochemical industries. Despite this, machining titanium and its alloys are problematic due to elevated temperatures and high tool wear rates (TWRs). Since the range of allowable cutting conditions for dry machining is restricted, it is not ideal for industrial production. Flood cooling with cutting fluids is a popular solution for lowering the cutting temperatures. Meanwhile, cutting tools and cutting fluids are constantly being improved in order to increase the machining of these alloys. Cutting fluids get rid of heat and metal chips from the cutting

Decision-Making Models and Applications in Manufacturing Environments. Pushpdant Jain, Kumar Abhishek, & Prasenjit Chatterjee (Eds.)

zone and deliver lubrication. Cutting fluid eliminates certain health and safety concerns when introducing new ones. Long-term exposure to cutting fluid and mist can be hazardous to one's skin. Researchers have been developing various cutting fluids and techniques for applying these fluids in order to achieve sustainable machining. But ambiguity about the best available cutting fluid remains the same. Since there are so many different kinds of cutting fluids, choosing the right one for machining can be difficult and time-consuming. In this context, this work is an attempt to use multi-criteria decision-making (MCDM) methods to choose the best possible cutting fluid. Three MCDM techniques, namely analytic hierarchy process (AHP), simple additive weighting (SAW), and TOPSIS, have been implemented to select the ideal cutting fluid from a prepared list of the most common and usable cutting fluids. All three methods suggest LB2000 vegetable-based oil as the best-cutting fluid to be used for machining titanium alloys, with keeping all the basic properties and environmental criteria in mind.

7.1 INTRODUCTION

Titanium and its alloys are commonly used in aerospace due to their outstanding combination of high specific strength that can be sustained at high temperatures, fracture tolerance, and corrosion resistance. Other commercial and industrial, such as petroleum refining, oil engineering, surgical implantation, emission control, radioactive waste disposal, petrochemical, and maritime applications, are also seeing an increase in their use. They have developed themselves as reliable engineering materials, available in a variety of alloys and wrought types, namely bar, billet, sheet, plate, wire, and so on [1–5]. Despite expanded use and processing, titanium and its alloys remain costlier than many other metals due to the complexities of the extraction process, the difficulty of melting, and fabrication and machining issues. To reduce the expense of titanium-made parts, near-net-shape approaches, namely casting, forging, and powder metallurgy have been incorporated. Most titanium-made components, however, are still made using traditional machining techniques, namely turning, drilling, milling, tapping, reaming, and grinding [6–8]. Owing to the intrinsic properties of the materials, machining titanium and its alloys is still a difficult process. Since titanium is chemically responsive, it has a propensity to weld to the cutting tool while machining, resulting in chipping and early failure of the cutting tool. Because of its poor thermal conductivity, the temperature at the interface of the workpiece and cutting tool rises, reducing tool life. Furthermore, its machinability is hampered by its high resistance at high temperatures and poor modulus of elasticity [9–13]. According to their unique metallurgical features, titanium alloys can be classified into four categories: near α alloys, α-β alloys, and β alloys. α alloys have an α-phase microstructure as they comprise α stabilizers, which are often combined with neutral elements. Ti-5Al-2.5Sn is a single-phase α-alloy that has exceptional tensile properties and creep stability at room and elevated temperatures up to 300°C. As a result, it's mostly used in cryogenic and corrosion resistance applications. Near α alloys have a high degree of α-stabilization along with a small amount of β-stabilizers. Their microstructural appearance displays mostly of α-phase with only minor amounts of β-phase. Examples of near α alloys are Ti-8Al-1Mo-1V and Ti-65A1-5Zr-0.5Mo-0.25Si. They are similar to α alloys with a higher temperature operating range of 400 to 520°C. α-β alloys contain α and β-stabilizers and their microstructures show a combination of α and β-phases. The most common examples of these alloys are Ti-6Al-4V and Ti-4Al-2Sn-4Mo-0.5Si. These alloys are capable of maintaining their strength at elevated temperatures. Therefore, they are often employed in high-strength applications at a higher temperature range of 350–400°C. β alloys consist of β-stabilizers and their microstructure shows the presence of β-phases. These alloys are highly dense and possess excellent formability, high hardenability, and enhanced forgeability due to their BCC microstructure. At room temperature, these alloys display similar strength as α-β alloys. However, they are inferior to α-β alloys in terms of maintaining their properties at elevated temperatures. Ti-10V-2Fe-3Al is an example of β alloy [14–19]. When machining titanium alloys, the main issues are high temperatures and high stresses produced at the tool's cutting edge. As a general rule, a cutting fluid must be used to reduce the issue. The cutting fluid not only serves as a coolant but also as a lubricant, lowering tool-tip temperatures and reducing cutting forces and chip welding, which are popular with titanium alloys, thereby extending tool life. The right selection of cutting fluid has a substantial impact on the life of the cutting tool. Since the evolution in machining processes, lots of cutting fluids have been developed along with different strategies for their application during the process. Most of the cutting fluids and their implementation strategies are being used by several industries, nowadays [20–22]. According to Baradie [23], proper application of cutting fluid increases the process's productivity besides reducing its cost. It further increases tool life, decreases surface roughness (SR), increases dimensional accuracy (A), and decreases the amount of power consumed. Cutting fluids can be classified in different ways, and one of their classifications is shown in Figure 7.1.

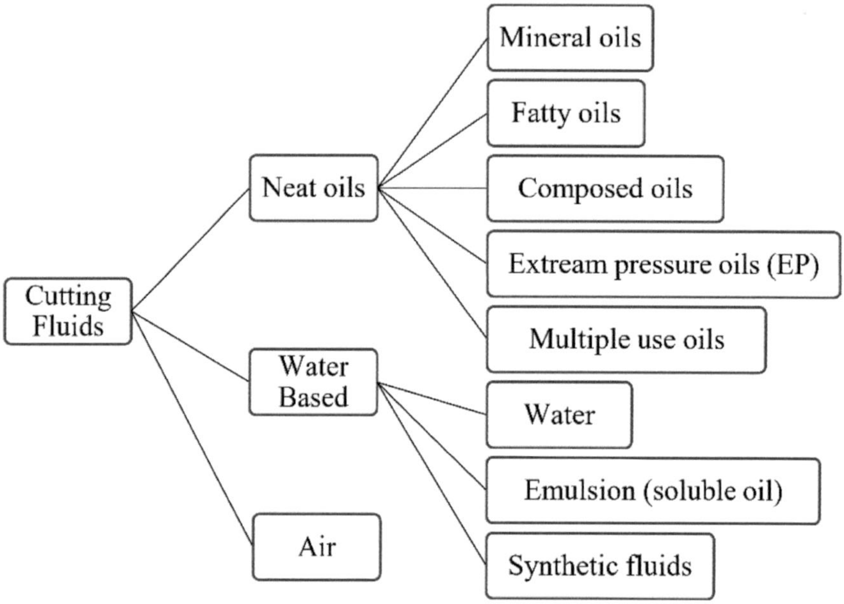

FIGURE 7.1 Classification of cutting fluids.

To develop methods during the selection of cutting fluids for alloys, Kisel et al. [24] studied four different water-based cutting fluids and their effectiveness on machining different alloys such as steel 45, VT3 alloy, D16 alloy, and KhN77TYuR alloy. A mathematical formula was also derived for the optimal selection of cutting fluids. Petkovic et al. [25] noticed that there are various MCDM methods for ranking the cutting fluid out of several cutting fluids. But in that, different attributes have diverse significance. Hence, a novel method was developed, known as the performance selection index (PSI) method for ranking the available cutting fluids. In this method, weightage distribution is not important for finding the best-cutting fluid. In a study, Cakir et al. [26] found that the use of cutting fluid reduces SR, decreases heat generation, and improves tool life. Based on numerous attributes, selection criteria for the best available cutting fluid were proposed for several machining processes. Chiffre & Belluco [27] carried out a comprehensive investigation for machining different workpiece materials namely stainless steels, carbon steels, and aluminum alloys using different cutting fluids such as vegetable oil-based oils, water-based, and straight mineral oils. The study was done in terms of tool life, surface finish, and cutting forces. Holmes [28] described the nature of cutting fluid and their mode of action along with the factors that have an impact on their selection. The study details the general characteristics of oil and their impact on machining various alloys, namely magnesium alloy, steel alloy, aluminum alloys, etc. It also provides guidelines for the selection of cutting fluid for various materials. Okafor & Nwoguh [29] investigated three vegetable-based oil, namely low oleic soybean oil, high oleic soybean oil, and acculube LB2000, and compared them with emulsion-based cutting fluid. Tiwari [30] implemented the MADM methods for selecting cutting fluids and ranked them accordingly. SAW and WPM methods were incorporated for selecting the best-cutting fluid during turning AISI304L. Prasad & Chakraborty [31] explored one of the multi-criterion decision-making tools for the selection of cutting fluid. In it, the potential of similarity-based method was examined for selecting a particular cutting fluid based on several attributes.

The aforementioned pieces of literature reveal that most of the researchers have taken a single or few base cutting fluids and carried out their study to observe the cutting fluid's effect on machining different alloys. But while doing so, the cutting fluid properties were not kept in focus, or if considered most of the focus was kept on viscosity and thermal conductivity, and the rest of the properties were ignored. Also, the literature shows the abundance of work on cutting fluids selection for materials like steel alloys, aluminum alloys, etc., but the literature on cutting fluid selection for titanium alloy machining is scarce. Besides this, prior-state-of-art reveals that several cutting fluids are chosen and analyzed by researchers and academicians. But there remains always confusion about which cutting fluid is best because every cutting fluid has a different property and each of them has a different impact on machining titanium alloys. In this context, this work is an attempt to use multi-criteria decision-making (MCDM) methods to choose the best possible cutting fluid.

7.2 EXPERIMENTAL METHODS AND MATERIALS

7.2.1 CUTTING FLUIDS AND THEIR ATTRIBUTES

Based on the different attributes of cutting fluids, five base cutting fluids that are commonly used in the machining of titanium alloys are taken as shown in Table 7.1. These cutting fluids are arranged in a ranking system using MCDM techniques. The flowchart of the selection of best cutting is shown in Figure 7.2.

TABLE 7.1 Cutting Fluids with Attributes Data

Cutting Fluids	Density (gm/ml)	Viscosity, 40°C (mm^2/s)	Flash Point (°C)	Thermal Conductivity (W/mK)	Specific Heat (J/kgK)
Coconut oil	0.962	27	219	0.154	3,500
Canola oil	0.965	45.6	229	0.168	1,938
LB2000 vegetable-based oil	0.92	35	320	0.22	1,957
Palm oil	0.92	39.4	252	0.17	1,902
Synthetic ester (MIDEL 7131)	0.95	28	270	0.143	1,933

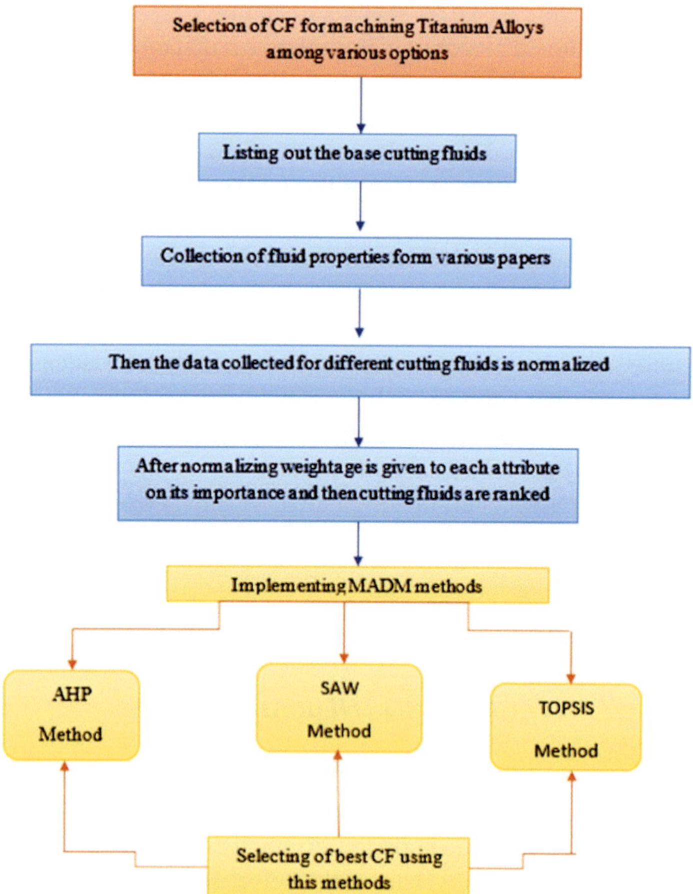

FIGURE 7.2 Cutting fluids and their attributes.

7.2.2 ANALYTICAL HIERARCHY PROCESS (AHP) METHOD

It is a method in which a comparison of attributes is done in pairs. There is a scale that was proposed by Saaty & Vargas [32] for comparing one attribute over another. After scaling attributes, the weightage of the attribute is decided, and the data is normalized. Then the row of normalized values of attributes is added and criteria/attribute weightage is obtained. Then there is also a procedure for checking the obtained attribute weightage whether it is correct or not. Thereafter, the normalized value is multiplied with the corresponding weightage and the whole row is added to obtain the result. Let's, understand this as follows (Table 7.2):

In the following example, A, B, C, and D are the attributes. CF1 and CF2 are the two cutting fluids. w_1, w_2, w_3, and w_4 are the weights for attributes A, B, C, and D, respectively. Here x_{ij} represents the values that are obtained after the normalization of the attribute data [33–35].

TABLE 7.2 AHP Method

Cutting Fluids	A	B	C	D	Preference Score	Rank
CF1	x_{11}	x_{12}	x_{13}	x_{14}	0.8	1
CF2	x_{21}	x_{22}	x_{23}	X24	0.7	2
Weights	w_1	w_2	w_3	w_4	–	–

The formula for prioritizing value calculation for cutting fluid CF1 is shown in Eqn. (1):

$$\text{Preference Score (CF1)} = (x_{11} \times w_1) + (x_{12} \times w_2) + (x_{13} \times w_3) + (x_{14} \times w_4) \quad (1)$$

To obtain the preference score for other cutting fluids, similar kinds of calculations will be repeated.

7.2.3 SIMPLE ADDITIVE WEIGHTING (SAW) METHOD

The weighted summation method is another name for the simple additive weighting (SAW) method. The multi-attribute technique SAW is based on the idea of a weighted summation. The process would seek a weighted summation of each alternative's results on each attribute. The choice with the highest score will be the best and will be recommended. This approach can be used to help overlay operations in a Geographic Information System. This method is suggested for completing a settlement in some multi-attribute decision problems. The SAW method's simple principle of determining the amount of weighted output scores on each option on all attributes is extremely helpful [36–38].

If an attribute is a beneficial one, then the formula for normalizing is:

$$r_{ij} = \frac{x_{ij}}{\text{Max}(x_{ij})} \tag{2}$$

If the attribute is not beneficial, then the formula for normalizing is:

$$r_{ij} = \frac{\text{Min}(x_{ij})}{x_{ij}} \tag{3}$$

Weight is obtained by the level of importance of different attributes in selection as follows:

$$w = \frac{C1}{C1 + \ldots + Cn} \times 100\% \tag{4}$$

7.2.3 TOPSIS METHOD

Technique for order preference by similarity to ideal solution, abbreviated as TOPSIS, is a MADM technique used to find the best substitutes in the solution space for a problem. Ching-Lai Hwang and Kwangsun Yoon proposed it in 1981 [39, 40]. The straightforward logic of the approach makes it easier for the user to understand and implement. It measures the degree of closeness to the ideal solution where the chosen/best substitutes should have the minimum separation distance from the positive ideal solution (PIS). Meanwhile, it must be too far from the negative ideal solution (NIS) [41, 42]. The PIS consists of the best performance values shown by any substitute for respective attributes. At the same time, the NIS is a set of worst performance values [41, 43]. This method has substantial application areas such as design and engineering, logistics, supply chain management (SCM), marketing, manufacturing, etc., due to its several advantageous features. It is simple to understand, program, and use. It maintains the same number of steps, regardless of problem size [44, 45]. The approach involves the following steps for converting multiple attributes into a single response [41–43]:

➢ **Step 1: Formation of a Decision Matrix:**

$$D = \begin{matrix} A_1 \\ A_2 \\ . \\ A_i \\ . \\ A_m \end{matrix} \begin{bmatrix} x_{11} & x_{12} & . & x_{1j} & x_{1n} \\ x_{21} & x_{22} & . & x_{2j} & x_{2n} \\ . & . & . & . & . \\ x_{i1} & x_{i2} & . & x_{ij} & . \\ . & . & . & . & . \\ x_{m1} & x_{m2} & . & x_{mj} & x_{mn} \end{bmatrix} \tag{5}$$

where; $A_i (i = 1, 2, \ldots, m)$ denotes the promising substitutes; $x_j (j = 1, 2, \ldots, n)$ signifies the attributes relating to the performance of the substitute, $j = 1, 2, \ldots, n$ and x_{ij} signify the performance of A_i for attribute X_j.

➢ **Step 2: Normalization of Decision Matrix:**

$$r_{ij} = \frac{x_{ij}}{\sqrt{\sum_{i=1}^{m} x_{ij}^2}} \tag{6}$$

where; r_{ij} depicts the normalized performance of A_i for attribute X_j

➢ **Step 3: Formation of a Weighted Normalized Decision Matrix:**

$$V = [v_{ij}] \; V = w_j r_{ij} \tag{7}$$

$$D = \begin{bmatrix} y_{11} & y_{12} & . & y_{1j} & y_{1n} \\ y_{21} & y_{22} & . & y_{2j} & y_{2n} \\ . & . & . & . & . \\ y_{i1} & y_{i2} & . & y_{ij} & . \\ . & . & . & . & . \\ y_{m1} & y_{m2} & . & y_{mj} & y_{mn} \end{bmatrix} \tag{8}$$

where; $\sum_{j=1}^{n} w_j = 1$

➢ **Step 4: Finding the Positive Ideal (Best) and the Negative Ideal (Worst) Solutions:**

- **The Positive Ideal (Best) Solution:**

$$A^+ = \left\{ \left(\max_i y_{ij} \middle| j \in J \right), \left(\min_i y_{ij} \middle| j \in J' \middle| i = 1, 2, \ldots, m \right) \right\}$$
$$= \left\{ y_1^+, y_2^+, \ldots, y_j^+, \ldots y_n^+ \right\} \tag{9}$$

- **The Negative Ideal (Worst) Solution:**

$$A^- = \left\{ \left(\min_i y_{ij} \middle| j \in J \right), \left(\max_i y_{ij} \middle| j \in J' \middle| i = 1,2,\ldots,m \right) \right\}$$
$$= \left\{ y_1^-, y_2^-, \ldots, y_j^-, \ldots y_n^- \right\} \quad (10)$$

where; $J = \{j = 1, 2, \ldots, n | j\}$: allied with the advantageous attributes. $J' = \{j = 1, 2, \ldots, n | j\}$: allied with nonadvantageous attributes.

➢ **Step 5: Distance Measurement of Substitutes from Ideal Solutions:** The separation distance of each substitute from the ideal solutions is measured by n-dimensional Euclidean distance using the following equations:

$$S_i^+ = \sqrt{\sum_{j=1}^{n} \left(y_{ij} - y_j^+ \right)^2} \; i = 1,2,\ldots,m \quad (11)$$

$$S_i^- = \sqrt{\sum_{j=1}^{n} \left(y_{ij} - y_j^- \right)^2} \; i = 1,2,\ldots,m \quad (12)$$

➢ **Step 6: Evaluation of Overall Performance Coefficient Nearest to the Ideal Solution:**

$$C_i^+ = \frac{S_i^-}{S_i^+ + S_i^-}, i = 1,2,\ldots,m; 0 \le C_i^+ \le 1 \quad (13)$$

7.3 RESULTS AND DISCUSSION

Three different methods were used to determine the ranking of the cutting fluid based on their attributes and applications. These methods and their respective outcomes are discussed as follows:

7.3.1 AHP METHOD

A pair-wise comparison of the attributes is in Table 7.3. This comparison is done to assign weightage to the attributes.

The pair-wise compared data is normalized. The normalized data with obtained attribute weights are shown in Table 7.4. Normalization of pair-wise compared attribute values was done by dividing those values by the sum of the column as calculated in Table 7.3. Then attributes weights were calculated by dividing the summed total of the attribute row by the number of attributes (i.e., 5 in this work).

TABLE 7.3 Pairwise Comparison of Attributes (AHP)

	Density (gm/ml)	Viscosity, 40°C (mm²/s)	Flash Point (°C)	Thermal Conductivity (W/mK)	Specific Heat (J/kgK)
Density (gm/ml)	1	0.142857	0.25	0.142857	0.333333
Viscosity, 40°C (mm²/s)	7	1	5	2	5
Flash Point (°C)	4	0.2	1	0.2	2
Thermal Conductivity (W/mK)	7	0.5	5	1	5
Specific Heat (J/kgK)	3	0.2	0.5	0.2	1
Sum	22	2.042857	11.75	3.542857	13.33333

TABLE 7.4 Normalized Data with Obtained Attribute Weights

	Density (gm/ml)	Viscosity, 40°C (mm²/s)	Flash Point (°C)	Thermal Conductivity (W/mK)	Specific Heat (J/kgK)	Attribute Weights
Density (gm/ml)	0.045455	0.06993	0.021277	0.040323	0.025	0.040397
Viscosity, 40°C (mm²/s)	0.318182	0.48951	0.425532	0.564516	0.375	0.434548
Flash Point (°C)	0.181818	0.097902	0.085106	0.056452	0.15	0.114256
Thermal Conductivity (W/mK)	0.318182	0.244755	0.425532	0.282258	0.375	0.329145
Specific Heat (J/kgK)	0.136364	0.097902	0.042553	0.056452	0.075	0.081654

Now, attribute weights obtained in Table 7.4 were verified. To do so, pairwise compared values (Table 7.3) were multiplied to the attribute weights (Table 7.4). Then weightage sum value was calculated by simply adding the values of the row as shown in Table 7.5. Thereafter the ratio of weighted sum value and attribute weight was calculated (Table 7.5).

Using Eqns. (14) and (15), consistency index and consistency ratio (CR) were calculated and tabulated as shown in Table 7.6. The CR was less than 0.1, which indicates that the obtained attribute weights (in Table 7.4) are correct.

$$\text{C.I.} = \frac{\lambda\text{max} - n}{n-1} \tag{14}$$

$$\text{Consistency Ratio} = \text{C.I./R.I.} \tag{15}$$

TABLE 7.5 Verification of Attribute Weights

	Density (gm/ml)	Viscosity, 40°C (mm^2/s)	Flash Point (°C)	Thermal Conductivity (W/mK)	Specific Heat (J/kgK)	Weightage Sum Value	The Ratio of Weighted Sum Value and Attribute Weight
Density (gm/ml)	0.040397	0.062078	0.028564	0.047021	0.027218	0.205278	5.081548
Viscosity, 40°C (mm^2/s)	0.282779	0.434548	0.57128	0.65829	0.40827	2.355167	5.419808
Flash Point (°C)	0.161588	0.08691	0.114256	0.065829	0.163308	0.591891	5.180405
Thermal Conductivity (W/mK)	0.282779	0.217274	0.57128	0.329145	0.40827	1.808748	5.495286
Specific Heat (J/kgK)	0.121191	0.08691	0.057128	0.065829	0.081654	0.412712	5.054389
Attribute Weights	0.040397	0.434548	0.114256	0.329145	0.081654	–	–

TABLE 7.6 Calculation of Consistency Index and Consistency Ratio

Number of Attributes (n)	λ_{max} (Average of Ratio Obtained)	Consistency Index $(C.I.) = \frac{\lambda_{max} - n}{n-1}$	Consistency Ratio = C.I./R.I. (where random index R.I. = 1.12)
5	5.246287	0.061572	0.054975

After that, the preference number was calculated using Eqn. (16). Based on the calculated preference score, the cutting fluids were ranked as shown in Table 7.7.

$$Pi = \sum_{j=1}^{N} w_{ij} \left(x_{ij}\right)_{normalized} \tag{16}$$

TABLE 7.7 Calculation of Preference Score and Ranking of Cutting Fluids

Cutting Fluids Used in Machining Titanium Alloy (Ti-6Al-4V)	Density (gm/ml)	Viscosity, 40°C (mm^2/s)	Flash Point (°C)	Thermal Conductivity (W/mK)	Specific Heat (J/kgK)	Preference Score	Rank
Weights	0.040397	0.434548	0.114256	0.329145	0.081654	–	–
Coconut oil	0.956341	1	0.684375	0.7	1	0.863431	2
Canola oil	0.953368	0.592105	0.715625	0.763636	0.553714	0.674136	5
LB2000 vegetable-based oil	1	0.771429	1	1	0.559143	0.864677	1
Palm oil	1	0.685279	0.7875	0.772727	0.543429	0.726873	4
Synthetic ester (MIDEL 7131)	0.968421	0.964286	0.846875	0.65	0.552286	0.813951	3

7.3.2 SAW METHOD

Initially, higher-beneficial and lower-beneficial attributes were arranged as shown in Table 7.8. Here, L represents the lower beneficial attribute and H represents the higher beneficial attributes.

Thereafter, the value in Table 7.8 were normalized as shown in Table 7.9. Weights were assigned to each attribute as shown in Table 7.10.

TABLE 7.8 Arrangement of Higher and Lower Values of the Attributes

	L	L	H	H	H
Cutting Fluids Used in Machining Titanium Alloy (Ti-6Al-4V)	**Density (gm/ml)**	**Viscosity, 40°C (mm²/s)**	**Flash Point (°C)**	**Thermal Conductivity (W/mK)**	**Specific Heat (J/kgK)**
Coconut oil	0.962	27	219	0.154	3,500
Canola oil	0.965	45.6	229	0.168	1,938
LB2000 vegetable-based oil	0.92	35	320	0.22	1,957
Palm oil	0.92	39.4	252	0.17	1,902
Synthetic ester (MIDEL 7131)	0.95	28	270	0.143	1,933
Higher or lower value as per required	0.92	27	320	0.22	3,500

TABLE 7.9 Normalized Decision Matrix

Cutting Fluids Used in Machining Titanium Alloy (Ti-6Al-4V)	Density (gm/ml)	Viscosity, 40°C (mm²/s)	Flash Point (°C)	Thermal Conductivity (W/mK)	Specific Heat (J/kgK)
Coconut oil	0.956341	1	0.684375	0.7	1
Canola oil	0.953368	0.592105	0.715625	0.763636	0.553714
LB2000 vegetable-based oil	1	0.771429	1	1	0.559143
Palm oil	1	0.685279	0.7875	0.772727	0.543429
Synthetic ester (MIDEL 7131)	0.968421	0.964286	0.84375	0.65	0.552286

TABLE 7.10 Assigning Weights to the Attributes

Attributes	Density (gm/ml)	Viscosity, 40°C (mm²/s)	Flash Point (°C)	Thermal Conductivity (W/mK)	Specific Heat (J/kgK)
Weights	0.040397	0.434548	0.114256	0.329145	0.081654

Further to decide the rank and to finalize the decision (Table 7.11), the sum of these weighted normalized data is done using Eqn. (17).

$$V_i = \sum_{j-1}^{n} = 1\ w_j\ r_{ij} \tag{17}$$

7.3.2 *TOPSIS METHOD*

Initially, the decision matrix has been formed, and the weights (as calculated in the SAW method) were assigned to each attribute as shown in Table 7.12.

TABLE 7.11 Calculation of Preference Score and Ranking of the Cutting Fluids

Cutting Fluids Used in Machining Titanium Alloy (Ti-6Al-4V)	Preference Score	Rank
Coconut oil	0.863431	2
Canola oil	0.674136	5
LB2000 vegetable-based oil	0.864677	1
Palm oil	0.726873	4
Synthetic ester (MIDEL 7131)	0.813594	3

TABLE 7.12 Formation of Decision Matrix

Cutting Fluids Used in Machining Titanium Alloy (Ti-6Al-4V)	Density (gm/ml)	Viscosity, 40°C (mm^2/s)	Flash Point (°C)	Thermal Conductivity (W/mK)	Specific Heat (J/kgK)
Coconut oil	0.962	27	219	0.154	3,500
Canola oil	0.965	45.6	229	0.168	1,938
LB2000 vegetable-based oil	0.92	35	320	0.22	1,957
Palm oil	0.92	39.4	252	0.17	1,902
Synthetic ester (MIDEL 7131)	0.95	28	270	0.143	1,933
Weights	0.040397	0.43458	0.114256	0.329145	0.081654

Normalization was done using Eqn. (6) and the values were tabulated (Table 7.13).

TABLE 7.13 Normalized Decision Matrix

Cutting Fluids Used in Machining Titanium Alloy (Ti-6Al-4V)	Density (gm/ml)	Viscosity, 40°C (mm^2/s)	Flash Point (°C)	Thermal Conductivity (W/mK)	Specific Heat (J/kgK)
Coconut oil	0.455931	0.338301	0.376021	0.398039	0.67122
Canola oil	0.457353	0.571353	0.393191	0.434225	0.371664
LB2000 vegetable-based oil	0.436025	0.438539	0.549437	0.568627	0.375308
Palm oil	0.436025	0.493669	0.432682	0.439394	0.36476
Synthetic ester (MIDEL 7131)	0.450244	0.350831	0.463588	0.369608	0.370705

Thereafter, weightage for the attributes was calculated using Eqn. (7) and based on this a weighted normalized decision matrix was formed as shown in Table 7.14.

TABLE 7.14 Formation of Weighted Normalized Decision Matrix

Cutting Fluids Used in Machining Titanium Alloy (Ti-6Al-4V)	Density (gm/ml)	Viscosity, 40°C (mm²/s)	Flash Point (°C)	Thermal Conductivity (W/mK)	Specific Heat (J/kgK)
Coconut oil	0.018418	0.147008	0.042963	0.131013	0.054808
Canola oil	0.018476	0.24828	0.044924	0.142923	0.030348
LB2000 vegetable-based oil	0.017614	0.190566	0.062777	0.187161	0.030645
Palm oil	0.017614	0.214523	0.049437	0.144624	0.029784
Synthetic ester (MIDEL 7131)	0.018188	0.152453	0.052968	0.121655	0.03027

The positive ideal (best) and the negative ideal (worst) solutions were computed using Eqns. (9) and (10), respectively. These values are presented in Table 7.15.

TABLE 7.15 Ideal Best and Ideal Worst Values

	Density (gm/ml)	Viscosity, 40°C (mm²/s)	Flash Point (°C)	Thermal Conductivity (W/mK)	Specific Heat (J/kgK)
V+	0.017614	0.147008	0.062777	0.187161	0.054808
V–	0.018476	0.24828	0.042963	0.121655	0.029784

Euclidian distance measurement of substitutes was accomplished from the obtained ideal solutions using Eqns. (11) and (12). Thereafter, using Eqn. (13), the preference number for each attribute was calculated and the rank was assigned to the cutting fluids accordingly. Table 7.16 represents the values of Euclidian distance along with the preference number and the ranking of the cutting fluids.

TABLE 7.16 Distance Measurement Form the Ideal, Preference Number, and Ranks of the Cutting Fluids

Cutting Fluids Used in Machining Titanium Alloy (Ti-6Al-4V)	Si+	Si–	Pi	Rank
Coconut oil	0.059547	0.104737	0.637536	2
Canola oil	0.11459	0.021366	0.157154	5
LB2000 vegetable-based oil	0.049811	0.089533	0.642532	1
Palm oil	0.084686	0.04135	0.328081	4
Synthetic ester (MIDEL 7131)	0.070848	0.09635	0.576264	3

7.4 CONCLUSION

The work concludes with the selection of the best-cutting fluid using three different MCDM techniques. The obtained outcomes are compared and charted in Table 7.17.

TABLE 7.17 Correlation of Three Methods for Final Ranking of the Cutting Fluids

Cutting Fluids Used in Machining Titanium Alloy (Ti-6Al-4V)	AHP	SAW	TOPSIS	Rank
Coconut oil	2	2	2	2
Canola oil	5	5	5	5
LB2000 vegetable-based oil	1	1	1	1
Palm oil	4	4	4	4
Synthetic ester (MIDEL 7131)	3	3	3	3

From this, it can be observed that the best suitable cutting fluid for machining difficult-to-cut titanium alloys based on the selected data is LB2000 vegetable-based oil. It is the cutting fluid that should be adopted for machining titanium alloys because it provides high thermal conductivity besides reducing the temperature at the junction of the tooltip and work-piece. Its low viscosity allows it to flow into intricate parts, which effectively provides lubrication and cooling effects during the machining operation. Hence, for the time being, the cutting fluid LB2000 vegetable-based oil is the optimal choice to be used for titanium's machining, with keeping all the basic properties and environmental criteria's in mind.

KEYWORDS

- **analytic hierarchy process**
- **cutting fluids**
- **lubrication**
- **multi-criteria decision-making**
- **simple additive weighting**
- **technique for order preference**
- **titanium alloys**

REFERENCES

1. Chatterjee, S., Mahapatra, S. S., & Abhishek, K., (2020). Experimental evaluation for improving multiple quality characteristics of drilled holes: A case study on drilling of Ti6Al4V. *J. Adv. Manuf. Syst., 19*(1), 189–214. https://doi.org/10.1142/S0219686720500109.
2. Jaiswal, A. P., Khanna, N., & Bajpai, V., (2020). Orthogonal machining of heat-treated Ti-10-2-3: FE and experimental. *Mater. Manuf. Process., 35*(16), 1822–1831. https://doi.org/10.1080/10426914.2020.1802039.
3. Khanna, N., & Agrawal, C., (2020). *Titanium Machining Using Indigenously Developed Sustainable Cryogenic Machining Facility* (pp. 183–205). Springer, Cham. https://doi.org/10.1007/978-3-030-18854-2_8.
4. Chatterjee, S., Mahapatra, S. S., & Abhishek, K., (2016). Simulation and optimization of machining parameters in drilling of titanium alloys. *Simul. Model. Pract. Theory, 62*, 31–48. https://doi.org/10.1016/j.simpat.2015.12.004.
5. Chatterjee, S., Abhishek, K., & Mahapatra, S. S., (2020). *Optimization of Quality Characteristics in Laser Drilling of Ti6Al4V Using VIKOR* (pp. 59–67). Springer, Singapore. https://doi.org/10.1007/978-981-32-9971-9_7.
6. Ezugwu, E. O., & Wang, Z. M., (1997). *Titanium Alloys and Their Machinability: A Review, 68*.
7. Veiga, C., Davim, J. P., & Loureiro, A. J. R., (2013). Review on machinability of titanium alloys: The process perspective. *Rev. Adv. Mater. Sci., 34*(2), 148–164.
8. Chatterjee, S., Mahapatra, S. S., Mondal, A., & Abhishek, K., (2018). An experimental study on drilling of titanium alloy using CO_2 laser. *Sadhana – Acad. Proc. Eng. Sci., 43*(8), 1–14. https://doi.org/10.1007/s12046-018-0903-1.
9. Rahman, M., Wang, Z. G., & Wong, Y. S., (2006). A review on high-speed machining of titanium alloys. *JSME International Journal, Series C: Mechanical Systems, Machine Elements and Manufacturing* (pp. 11–20). The Japan Society of Mechanical Engineers. https://doi.org/10.1299/jsmec.49.11.
10. Modgil, A., (2003). *Effects of High-Speed Machining on Surface Topography of Titanium Alloy (TI6AL4V)*. University of Florida (2003). http://ufdcimages.uflib.ufl.edu/UF/E0/00/28/46/00001/modgil_a.pdf.
11. Arrazola, P. J., Garay, A., María, I. L., Armendia, M., Marya, S., & Le Maître, F., (2009). Machinability of titanium alloys (Ti6Al4V and Ti555.3). *J. Mater. Process. Technol., 209*(5). https://doi.org/10.1016/j.jmatprotec.2008.06.020ï.
12. Kikuchi, M., & Okuno, O., (2004). Machinability evaluation of titanium alloys. *Dent. Mater. J., 23*(1), 37–45. https://doi.org/10.4012/dmj.23.37.
13. Khanna, N., & Davim, J. P., (2015). Design-of-experiments application in machining titanium alloys for aerospace structural components. *Meas. J. Int. Meas. Confed., 61*, 280–290. https://doi.org/10.1016/j.measurement.2014.10.059.
14. Ahmed, Y. M., Salleh, K., Sahari, M., Ishak, M., & Khidhir, B. A., (2014). Titanium and its alloy. *Int. J. Sci. Res., 3* (10), 1351–1361.
15. Guillemot, F., (2005). Recent advances in the design of titanium alloys for orthopedic applications. *Expert Review of Medical Devices* (pp. 741–748). Future Drugs Ltd. https://doi.org/10.1586/17434440.2.6.741.

16. Bieler, T. R., Trevino, R. M., & Zeng, L., (2005). Alloys: Titanium. In: *Encyclopedia of Condensed Matter Physics* (pp. 65–76). Elsevier Inc. https://doi.org/10.1016/B0-12-369401-9/00536-2.
17. Froes, F. H., & Whittaker, M. T., (2020). Titanium and its alloys. In: *Reference Module in Materials Science and Materials Engineering.* Elsevier. https://doi.org/10.1016/b978-0-12-819726-4.00066-1.
18. Polmear, I., StJohn, D., Nie, J. F., & Qian, M., (2017). Titanium alloys. In: *Light Alloys* (pp. 369–460). Elsevier. https://doi.org/10.1016/b978-0-08-099431-4.00007-5.
19. Sankaran, K. K., & Mishra, R. S., (2017). Titanium alloys. In: *Metallurgy and Design of Alloys with Hierarchical Microstructures* (pp. 177–288). Elsevier. https://doi.org/10.1016/b978-0-12-812068-2.00005-9.
20. Taylor, C. M., Hernandez, S. G. A., Marshall, M., & Broderick, M., (2018). Cutting fluid application for titanium alloys Ti-6Al-4V and Ti-10V-2Fe-3Al in a finish turning process. In: *Procedia CIRP* (Vol. 77, pp. 441–444). Elsevier B.V. https://doi.org/10.1016/j.procir.2018.08.279.
21. García-Martínez, E., Miguel, V., Martínez-Martínez, A., Manjabacas, M. C., & Coello, J., (2019). Sustainable lubrication methods for the machining of titanium alloys: An overview. *Materials (Basel), 12*(23), 3852. https://doi.org/10.3390/ma12233852.
22. Revuru, R. S., Posinasetti, N. R., Vsn, V. R., & Amrita, M., (2017). Application of cutting fluids in machining of titanium alloys—A review. *Int. J. Adv. Manuf. Technol., 91*(5–8), 2477–2498. https://doi.org/10.1007/s00170-016-9883-7.
23. El Baradie, M. A., (1996). Cutting fluids: Part I. Characterization. *J. Mater. Process. Technol., 56* (1–4), 786–797. https://doi.org/10.1016/0924-0136(95)01892-1.
24. Kisel, A. G., Makashin, D. S., Averkov, K. V., & Razhkovskii, A. A., (2018). Effectiveness and physical characteristics of machining fluid. *Russ. Eng. Res., 38*(7), 508–512. https://doi.org/10.3103/S1068798X18070092.
25. Petković, D., Madić, M., Radovanović, M., & Gečevska, V., (2017). Application of the performance selection index method for solving machining MCDM problems. *Facta Univ. Ser. Mech. Eng., 15*(1), 97–106. https://doi.org/10.22190/FUME151120001P.
26. Çakīr, O., Yardimeden, A., Ozben, T., & Kilickap, E., (2007). *Selection of Cutting Fluids in Machining Processes Manufacturing and Processing.*
27. De Chiffre, L., & Belluco, W., (2000). Comparison of methods for cutting fluid performance testing. *CIRP Ann.–Manuf. Technol., 49* (1), 57–60. https://doi.org/10.1016/S0007-8506(07)62895-9.
28. Holmes, P. M., (1971). Factors affecting the selection of cutting fluids. *Industrial Lubrication and Tribology* (pp. 47–55). MCB UP Ltd. https://doi.org/10.1108/eb052964.
29. Okafor, A. C., & Nwoguh, T. O., (2020). A study of viscosity and thermal conductivity of vegetable oils as base cutting fluids for minimum quantity lubrication machining of difficult-to-cut metals. *Int. J. Adv. Manuf. Technol., 106*(3, 4), 1121–1131. https://doi.org/10.1007/s00170-019-04611-3.
30. Tiwari, V. V., (2015). MADM for selection of vegetable based cutting fluids by saw method and Wpm. *Int. J. Res. Technol. Manag., 1*(1), 16–28.
31. Prasad, K., & Chakraborty, S., (2018). Application of the modified similarity-based method for cutting fluid selection. *Decis. Sci. Lett., 7*(3), 273–286. https://doi.org/10.5267/j.dsl.2017.8.002.
32. Saaty, T., & Vargas, L., (2012). *Models, Methods, Concepts & Applications of the Analytic Hierarchy Process*. Springer. https://doi.org/10.1007/978-1-4614-3597-6.

33. Jurenka, R., Cagáňová, D., & Špirková, D., (2019). Application of AHP method in decision-making process. In: *EAI/Springer Innovations in Communication and Computing* (pp. 3–15). Springer Science and Business Media Deutschland GmbH. https://doi.org/10.1007/978-3-319–76998-1_1.
34. Triantaphyllou, E., (2000). *Multi-Criteria Decision-Making Methods* (pp. 5–21). Springer, Boston, MA. https://doi.org/10.1007/978-1-4757-3157-6_2.
35. Aziz, N. F., Sorooshian, S., & Mahmud, F., (2016). MCDM-AHP method in decision makings. *ARPN J. Eng. Appl. Sci., 11*(11), 7217–7220. https://doi.org/10.1109/TIE.2013.2297315.
36. Wira, T. P. D., & Agustian, P. A., (2018). Comparison analysis of simple additive weighting (SAW) and weighted product (WP) in decision support systems. *MATEC Web Conf., 215*, 1–5. https://doi.org/10.1051/matecconf/201821501003.
37. Ibrahim, A., & Surya, R. A., (2019). The implementation of simple additive weighting (SAW) method in decision support system for the best school selection in Jambi. *J. Phys. Conf. Ser., 1338*(1), 1–8. https://doi.org/10.1088/1742-6596/1338/1/012054.
38. Widianta, M. M. D., Rizaldi, T., Setyohadi, D. P. S., & Riskiawan, H. Y., (2018). Comparison of multi-criteria decision support methods (AHP, TOPSIS, SAW & PROMENTHEE) for employee placement. In: *Journal of Physics: Conference Series* (Vol. 953, p 12116). Institute of Physics Publishing. https://doi.org/10.1088/1742-6596/953/1/012116.
39. Vommi, V. B., (2017). TOPSIS with statistical distances: A new approach to MADM. *Decis. Sci. Lett., 6*(1), 49–66. https://doi.org/10.5267/j.dsl.2016.8.001.
40. Panda, S. N., Bagal, D. K., Pattanaik, A. K., Patnaik, D., Barua, A., Jeet, S., Parida, B., & Naik, B., (2020). Comparative evaluation for studying the parametric influences on quality of electrode using Taguchi method coupled with MOORA, DFA, and TOPSIS method for electrochemical machining. In: *Recent Advances in Mechanical Infrastructure* (pp. 115–129). https://doi.org/10.1007/978-981-32-9971-9_13.
41. Sonkar, V., Abhishek, K., Datta, S., & Mahapatra, S. S., (2014). Multi-objective optimization in drilling of GFRP composites: A degree of similarity approach. In: *3rd International Conference on Materials Processing and Characterization (ICMPC 2014)* (Vol. 6, pp. 538–543). Elsevier BV. https://doi.org/10.1016/j.mspro.2014.07.068.
42. Kamaraj, M., Santhanakrishnan, R., & Muthu, E., (2018). Investigation of surface roughness and MRR in drilling of Al_2O_3 particle and sisal fiber reinforced epoxy composites using TOPSIS based Taguchi method. In: *IOP Conference Series: Materials Science and Engineering* (Vol. 402). Institute of Physics Publishing. https://doi.org/10.1088/1757-899X/402/1/012095.
43. Abhishek, K., (2015). *Experimental Investigations on Machining of CFRP Composites: Study of Parametric Influence and Machining Performance Optimization* (pp. 1–277). Department of Mechanical Engineering, National Institute of Technology, Rourkela. https://doi.org/10.1007/s11390-005-0552-9.
44. Sun, C. C., & Lin, G. T. R., (2009). Using fuzzy TOPSIS method for evaluating the competitive advantages of shopping websites. *Expert Syst. Appl., 36*(9), 11764–11771. https://doi.org/10.1016/j.eswa.2009.04.017.
45. Velasquez, M., & Hester, P. T., (2013). *An Analysis of Multi-Criteria Decision-Making Methods, 10.*

CHAPTER 8

DECISION-MAKING FRAMEWORK FOR SUSTAINABILITY ASSESSMENT OF MANUFACTURING

BRAJESH K. DWIVEDI[1] and ASHUTOSH K. SRIVASTAVA[2]

[1]*Environmental Sciences, Department of Botany, University of Allahabad, Allahabad, Uttar Pradesh, India*

[2]*Dr. Shyama Prasad Mukherjee Government PG College, Bhadohi, Uttar Pradesh, India*

ABSTRACT

Manufacturing sectors have a major shared impact on the basic issues, i.e., economic, social, and environmental aspects of the sustainability dimensions. These aspects have relevant indicators on the issues of quality and quantity and are framed to rank of sustainability indicators rank of nation. Besides this, we suggest that the following important points: (i) basic approach for environmental impact assessment (EIA) study; (ii) basic occupational and emission standard for the proposal of industry and also old industries be reviewed; (iii) to develop directory of environmental expert; (iv) that the competent authority responsibility specially Ministry of Environment, Forest and Climate Change for conducting EIA studies by forming a study team/expert for every EIA; and (v) concern have been raised on issues such as reduction in the time period allotted to public hearings, the extended time give to submission of compliance reports, and regularization of manufacturing projects through 'post-fact' clearance, it is expected to incorporate the perspectives of multiple stakeholders in a balanced manner. A model of framework performance elements has also been using interpretive structure

Decision-Making Models and Applications in Manufacturing Environments. Pushpdant Jain, Kumar Abhishek, & Prasenjit Chatterjee (Eds.)

modeling (ISM). In this connection, in India, legislation exists already that makes EIA mandatory for every new manufacturing project proposal.

8.1 INTRODUCTION

Sustainable development (SD) refers to the under control imposed by the current scale of technology and the socio-economic structures on the ability of the ecosphere for safe mode specially for coming generation. The processes of development and technology over the five decades has profound to the increase in chemicals piles (general to carcinogenic) ultimately damage natural system on a grand scale. The concept of SD also needs to reconcile economic development, the quality of life, and the environment, within various political frameworks and educational contribution [1–3], which interrelate at the global levels. For all its inadequacies, the concept of SD promotes the message that resources are not infinite and should be conserved. This is, after all the essence of good housekeeping and the old dictum of waste not, want not, still holds true at all spatial and temporal scales [4]. The problem such as environmental degradation are often perceived as someone else's responsibility, because of the tiered way that society organizes itself. But generally human activities not being undertaken in a sustainable way. Either the resources itself is being rapidly depleted, or the production and/or consumption of the resources jeopardizing the wider environment, e.g., global warming, soil degradation, habit alteration, etc. [5, 6]. The ecosystem's intrinsic values and rights, considering it a resource to be exploited for human settlement, food, and energy production, SD [7–12]. Now this concern also as a pattern of social and structural economic transformation [13–17].

Thus, the doctrine of sustainability provides a mechanism for development to occur in harmony with environmental care with active actions [18, 19]. Some of the following approaches are framed to achieve sustainability:

- Problem-solving approach – this refers to a series of logic steps to identify problems, needs, and solutions.
- Environmental system approach:
- Bio-geophysical units' approach:
- Specific situation approach.
- Professional discipline approach refers to professional like air, water, land, noise, waste (liquid, solid, gaseous, and particulate), environmental health and urban management.
- Voluntary sectors approach – this represents the approach NGOs encourage and/or support.

8.1.1 ENVIRONMENTAL MANAGEMENT (SUSTAINABLE) TOOLS

The sustainability science has a major focus on management strategies for the global sustainability, appropriate methodologies, and institutional needs; political agenda of SD and the nature society interactions [20, 21]. This led the public to demand that environmental issues be incorporated into the decision-making processes. In recent times, several management tools are being applied for SD actions. EIA (environmental impact assessment) and SEA (Strategies environmental assessment) are among the available tools for environmental assessment [22, 23]. The following tools considered as major outcome in this course include:

- Environment assessment (EA);
- Economic assessment, usually through cost-benefit analysis (CBA);
- Environment impact statement (EIS);
- Environmental audits;
- Waste minimization program and environmental management system.
- Life cycle assessment (LCA) and environmental design (ED).

8.1.2 THE EMERGENCE OF SUSTAINABILITY ASSESSMENT (ENVIRONMENTAL IMPACT ASSESSMENT)

Environmental impact assessment (EIA) is an anticipatory and participatory management tools for informed decision making on optional use of natural resources and SD [5, 24–35].

Some approaches commonly used by proponents to evaluate the sustainability of a proposal include sustainability oriented multi-criteria analysis (MCA) [36] or forms of cost-benefit analysis (CBA) [37]. Now, more than 150 developed countries have incorporated EIA either with in the legislature mechanism or as guidelines in decision making processes. The basic aspects, applications, and categories on EIA are summarized in Table 8.1.

8.2 IMPACT ON DECISION

Decision-making serves as an instrument for SD at different scales. Any department/agency may make a decision on the direction of subjective policy and on the resultant sectors' programmers. Various combinations of stakeholders (e.g., private sector and public sector) can also take decision at different scales in the planning and development process [38, 39]. Figure 8.1 represent a sound integrated decision-making process.

TABLE 8.1 Aspects, Applications, and Categories of EIA

Aspects	EIA Category
Application:	
• General	• Impact assessment (IA)
• Global strategic (to policies, plans)	• Environmental Assessment (EA)
• Local specific (to individual projects, sites)	• Environmental impact assessment (EIA)
	• Strategic environmental assessment (SEA)
	• Project EA or EIA
Source of Impact:	
• Specific projects	• Project EA or EIA
• Policies	• SEA
• Development aid	• Policy assessment
• New policy	
• Change in policy	
Impact Receptors:	
• The whole environment	• EA or EIA
• Ecosystem and their components	• Ecological impact assessment
• People (individuals)	• Health impact assessment
• Population communities	• Demographic impact assessment
Impact Categories:	
• Economic	• Economic impact assessment
• Ecological	• Ecological impact assessment
• Cumulative	• Cumulative effects assessment
• Social and culture	• Social impact assessment
• Health	• Health impact assessment
• Climate	• Climate impact assessment
• Noise	• Noise impact assessment

In this connection, there is a need to classify most approached criteria, i.e., socioeconomic and bio-physical environmental categories with examples of negotiable and non-negotiable impacts (Table 8.2).

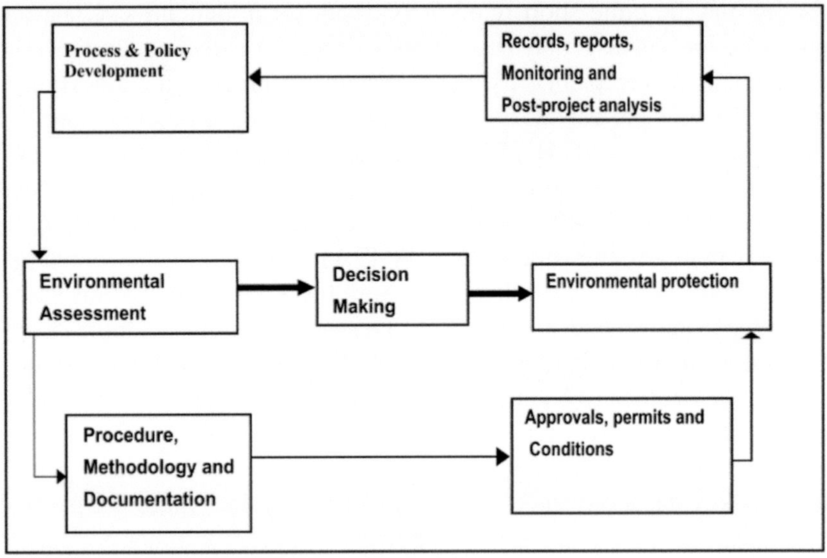

FIGURE 8.1 Sustainable consideration in decision-making.

TABLE 8.2 Judging Environmental Acceptability

	Non-Negotiable Impact	Negotiable Impact
Ecological (physical and biological components).	• Degrades essential life support system. • Degrade conservation. • Estate. • Adversely affects ecological integrity. • Loss of biodiversity.	• No degradation beyond carrying capacity. • No degradation of productive systems. • Wise use of natural resources.
Social (human as individuals or in social groupings).	• Loss of human life. • Reduce public health and safety unacceptably. • Unreasonably degrades quality of life where people live.	• Community benefits and cost and where they are borne. • Reasonable appointment of intergenerational equity. • Compatibility with defined environmental policy goals.

8.3 DECISION PLANNING AND DEVELOPMENT PROCESS

Development Process involves major projects have a life cycle, including a number of key stages from planning to project closedown. For some projects, this life may be quite short (e.g., 5–10 years for a medium-size/fast-track opencast); for others, it may be very long (e.g., 50 years plus for a reservoir or a nuclear power plant). EIA is a systematic process for assessing the impact of major projects as an aid to improved decision-making and environmental outcome [40, 41]. EIA processes are normally located within the EIA system. Laws and regulations normally relate to planning and consents/licensing framework [42]. The level of integration between the EIA system and the planning framework varies. The EIA may have an impact on a planning official's report, on a planning committee's decision, and on pre-and post-submission project modification. Decision is rarely a simple 'yes' or 'no,' and are more likely to be conditional [43]. Decision, for example, are to be made on whether EIA is needed at all (the screening stage); the scope of the EIA; the alternatives under considerations; project design and redesign; the range of mitigation measures and implementation and monitoring during the 'post-key-decision' stage of the project life cycle. Figure 8.2 gives the key stage in the planning and development life cycle for major projects and the environmental assessment processes.

8.3.1 CRITERIA FOR ASSESSING THE IMPACT OF ENVIRONMENTAL IMPACT ASSESSMENT (EIA) ON DECISION

Synthesizing the various criteria, the following are identified as particularly important in assessing the impact of EIA on decision:

- Centrality in decision-making (i.e., to what extent does EIA make a difference?).
- Integration in the project cycle and approval processes (i.e., to what extent is EIA easily by passed?).
- Multi-choice in way of their application to various stages in the EIA processes (e.g., from considerations of alternatives to implementation of projects modification.
- Large optimizing range of environmental factors both biophysical and socio-economic.

8.4 ENVIRONMENTAL DESIGN (ED)

The manufacturing sectors has been undergoing change to achieve different scales that include maintaining high quality and low cost, staying competitive in an international marketplace, and meeting consumer preference and regulatory demand for reduced environmental impacts [44]. Designing for the environment or environmental design (ED) also contribute to productivity. It is a strategy for effectively organizing environmental issues [45].

ED or designing for environment is a perspective in which environmental aspects of an existing or new product, processes, or facility, design/redesign are optimized systematically and continuously for minimal adverse impact on the environment [46, 47]. To elaborate, the environmental issues relating to processes and product are addressed, generally, after their implementation, and this involve redesigning of the processes and products to the extent possible [48, 49]. However, such tasks are expensive and time consuming and rarely attain the required levels of environmental improvements. It is, therefore, necessary to evaluate potential environmental impacts at the designing stage, and it is this principle that underlies the concept of ED [50, 51]. In other words, there is a need to shift the conventional values. This shift in focus results in processes and products that are environmentally sound as well as utilization oriented. The principle of ED as below:

- The design of project, process or product must incorporate environmental objectives, in addition to other objectives such as economic efficiency and practically.
- The core environmental objectives are the overall sustainability of;
- While sustainability is required all details of Annexure-I, II, III by MEFCC. [Annexure-I, List of projects/Industries listed by Ministry of Environment, Forest and Climate change (MEFCC), Govt of India, they have necessary EIA before starting of project. Annexure-II, It is all legal proceedings to adopt properly EIA process as per as MEFCC rule. Annexure-III, It is a competent authority co to establishing performing team for particular project/ Industry as per as MEFCC rule.]
- Environmental objectives refer specially to a projects Annexure-I. These objectives are to be agreed upon at the outset by administrators, sponsors, and designers.
- Multi-sectoral and multi-disciplinary, design for environmental objectives is inevitable.

ED can also provide a long-term strategic folio of a company's future production and consumption [51]. In this context, consider the following as: (i) increased innovation; (ii) high ability to compete, add value and attract customers; (iii) become more cost-effective; (iv) reduce environmental impacts and liability; and (v) gain systems perspective.

8.4.1 ENVIRONMENTAL DESIGN (ED) FOR MANUFACTURED PRODUCTS

Industrial production is responsible for a great deal of the adverse environmental impact that we experience in the world today [32]. A product may cause adverse environmental impacts during its production, use, and beyond. This point to the fact that the development of a few industrial products involves many decision-making situations where alternative solution to a given design problem must be ranked, and priorities given, according to relevant criteria. Many different and very criteria may be applied. Among the important demands on the new product may be:

- Ease of manufacture;
- Performance;
- Appearance;
- Manufacturing cost;
- Durability;
- Safety;
- Services.

The management of trade-offs in sustainability assessment requires good processes as discussed earlier authors [52–54] for socio-economic benefit. The following nutshell are essential for tradeoffs, i.e., net gains, Burden of argument, Avoidance of significant adverse effects, Protection of the future, Explicit justification, Open process and also participatory processes.

8.5 SUSTAINABLE APPROACH: MANUFACTURING'S

The dynamic of relationship between manufacturing activity and nature can be better understood by positioning the circular flow model with in a larger framework to generate the materials balance model [44], this model illustrate the link between economic activity and the natural environment, this relationship is defined by the flow of resources from nature to an economic system and the return flow of residuals from economic activity back to the environment [55]. Throughout environmental economics, the primary focus has been on the flow of residuals and the potential damages associated with their release to nature. Indeed, the conventional policy approach uses end-of-pipe, command-and-control instruments aimed at limiting the amount and toxicity of residuals released to the environment after they are generated. Implicit in this approach is a perception that the flow of materials through the economy is linear and open, a linear flow of materials assumes that materials run in one direction, entering an economic system as input and exiting as wastes or residuals. This so-called "cradle-to-grave" flow emphasizes use, waste generation, and disposal, which coincides with policies aimed at abating contaminated residuals only at the end of the flow. Although such policy efforts can and do improve environmental quality in the short run, they do not adequately address the materials balance model in the context of policy initiatives [57].

Part of the solution is to recognize that the flow of materials through an economy need not be linear. Instead, product design, manufacturing processes, and energy use can be modified to achieve a cyclic, or closed, flow of materials, as shown in Figure 8.2. Notice that this ISM model, considered a "cradle-to-cradle" flow, suggest a much broader approach to environmental protection than controlling residuals after they have been generated. Waste prevention, energy, and resource efficiency, and design for reuse or recycling are among the options suggested by this paradigm. In sum, the key implication of a cyclic system is that economic activity can be altered throughout the cycle of production and consumption to achieve a reduction in the associated environmental impacts [44].

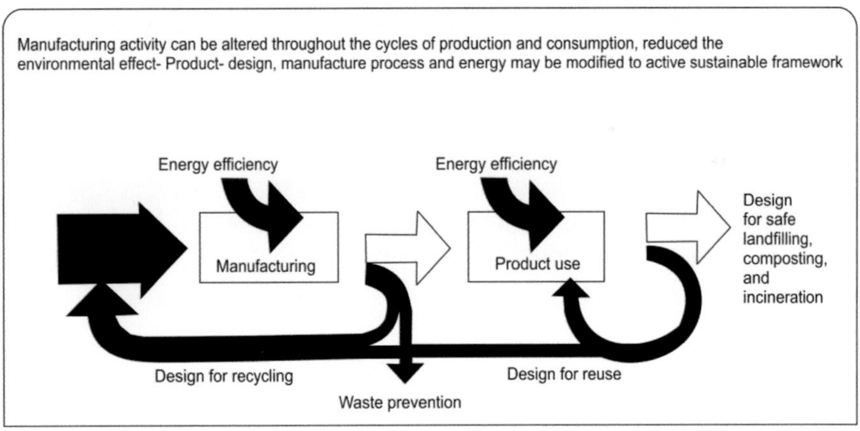

FIGURE 8.2 Sustainable flow of materials in manufacturing operation.

To put the notion of a cyclic materials flow into practices, a life cycle assessment (LCA) can be undertaken. This is an analytical tool that examines the environmental impacts of a manufacturing product by evaluating all stages of a product or process from raw materials extraction to final disposal. For these reasons, there is a real temptation to position pollution prevention (P2) under the broad umbrella of industrial ecology specially manufacturing sectors [58]. P2 promotes risk reduction through minimizing or eliminating wastes, while industrial (manufacturing) ecology argues in favor of using wastes as inputs in other production processes. In this same regard, P2 solution are aimed mainly at the single firm (though other sectors participate), while industrial ecology is aimed at a network of businesses [59]. This also means, from a benefit-cost perspective, that the potential gains of manufacturing ecology have to be shared. Other approach, viz., source reduction, source segregation, raw material substitution, changes in manufacturing processes and product substitution refers to the selection of environmentally safe over their potentially polluting substitutes as a sustained framework. Extended product responsibility (EPR) is closely linked to SD and is preventive in orientation. It is that in the product chain-designers, manufactures, distributors, consumers, recyclers, remanufacturers, disposers, and decision-making framework for market decision when choosing a product [60]. Design for the environment (DfE) is an initiative that promotes the use of environmental considerations along with cost and performance in product design and development. The need for setting a global environmental agenda and implementing it is more than apparent. Identifying an appropriate course of action is not easy, and it takes time. This is a call upon time for cooperation from every market sector (manufacturer) to make a decision framework for competent green favors in the way of sustainability.

ACKNOWLEDGMENT

The authors are extremely grateful to the Coordinator, Environmental Sciences, Department of Botany, University of Allahabad, for the support and encouragement.

KEYWORDS

- **design for the environment**
- **environmental evaluation expert**
- **extended product responsibility**
- **interpretive structure modeling**
- **manufacturing environment**
- **sustainability assessment**

REFERENCES

1. Acher, (2007). Policy science contribution to analysis to promote sustainability. *Sustain. Sci., 2*(2), 141–149.
2. Kajikawa, Y., (2008). Research core and framework of sustainability science. *Sustain. Sci.,* 3(2), 221–231.
3. Barth, M., & Michelsen, G. W., (2013). The concept of sustainable economic development. *Sustain Sci, 8*, 103–119.
4. Quental, N., Lourenco, J., & Silva, A., (2011). Sustainability: Characteristics and scientific roots. *Environ Dev. Sustain., 13*, 1257–1276.
5. Dwivedi, B. K., & Tripathi, G. C., (2015). Sustainability: The concept and its economic approach. *The Indian Journal of Economics., XCVI, 380*, 175–187.
6. Gibson, R. B., (2006). Sustainability assessment: Basic components of a practical approach. *Impact Assessment and Project Appraisal., 24*, 170–182.
7. Dwivedi, B. K., (2016). Ecological integrity of river water quality: In relation to primary productivity and cyanotoxin occurrence. *Int J Adv Biol Sci., 3*, 200–207.
8. Dwivedi, B. K., & Srivastava, A. K., (2017). Diatoms as indicator of pollution gradients of the River Ganga, Allahabad. *Int. J. Curr. Microbiol. App. Sci., 6*, 4323–4334.
9. Dwivedi, B. K., (2018). Potential risk assessment of cyanotoxin for water resource. *Int. J. Curr. Microbiol. App. Sci., 7*, 3604–3616.
10. Dwivedi, B. K., & Srivastava, A. K., (2018). An approach to assess potential heavy metal in the sediments transport of river Ganga channel. *Int. J. Curr. Res. in Biosci. Plant. Biol., 5*(1), 67–76.
11. Dwivedi, B. K., (2021). *Energy Crises: Energy within Environmental Constraints in the Book Artificial Intelligence for Renewable Energy System*. Elsevier Publication (accepted).
12. Dwivedi, B. K., (2021). Challenges of urban water surveillance and its sustainable contrivance. In: *Urban Water Crises and Management Strategies for Sustainable Development*. Elsevier *Publication* (accepted).
13. Feldmann, L., Vanderhaegen, M., & Pirotte, C., (2001). The EU's SEA directive: Status and links to integration and sustainable development. *Environmental Impact Assessment Review, 21*, 203–222.
14. Dwivedi, B. K., (2021a). Aquifer (hydrogeological) study of southern areas of Prayagraj, India. *Asian Journal of Water, Environment and Pollution, 18*, 85–89.
15. Dwivedi, B. K., (2021b). River water quality in relation to alluvial deposition and irreparable ecological approach. *Asian Journal of Water, Environment and Pollution.*
16. Dwivedi, B. K., (2021c). Fine – grained sediments dynamics of river Ganga and its technical contrivance. *Journal of Applied Water Engineering and Research.*
17. Dwivedi, B. K., & Tripathi, G. C., (2021). Economic activity in river system: A challenges of hydrology. *The Indian Journal of Economics.*
18. Pope, J., Annandale, D., & Morrison, S. A., (2004). Conceptualizing sustainability assessment. *Environmental Impact Assessment Review, 24*, 595–616.
19. Gibson, R. B., (2011). Application of a contribution to sustain-ability test by the joint review panel for the Canadian Mackenzie gas project. *Impact Assessment and Project Appraisal, 29*(3), 231–244.
20. Hansmann, R., (2010). Sustainable learning, an introduction to the concept and its motivational aspects. *Sustainability, 2*, 2873–2897.

21. Esteves, A. M., Franks, D., & Vanclay, F., (2012). Social impact assessment: The state of the art. *Impact Assessment and Project Appraisal*, 30. doi: 10.1080/14615517.2012.660356.
22. Kain, J. H., & Soderberg, H., (2008). Management of complex knowledge in planning for sustainable development: The use of multi-criteria decision aids. *Environmental Impact Assessment Review, 28*, 7–21.
23. Lee, N., & Kirkpatrick, C., (2001). Methodologies for sustainability impact assessments of proposals for new trade agreements. *Journal of Environmental Assessment, Policy and Management, 3*, 395–412.
24. Morgan, R., (2012). Environmental impact assessment: The state of the art. *Impact Assessment and Project Appraisal, 30*(1), doi: 10.1080/14615517.2012.661557.
25. Bond, A. J., & Morrison, S. A., (2011). Re-evaluating sustainability assessment: Aligning the vision and the practice. *Environmental Impact Assessment Review, 31*, 1–7.
26. Hacking, T., & Guthrie, P., (2008). A framework for clarifying the meaning of triple bottom-line, integrated, and sustain-ability assessment. *Environmental Impact Assessment Review, 28*, 73–89.
27. Gasparatos, A., Elharam, M., & Horner, M., (2008). A critical review of reductionist approaches for assessing the progress towards sustainability. *Environmental Impact Assessment Review, 28*, 286–311.
28. Gaudreau, K., & Gibson, R. B., (2010). Illustrating integrated sustainability and resilience-based assessments: A small-scale biodiesel project in Barbados. *Impact Assessment and Project Appraisal, 28*, 233–243.
29. Galbraith, L., Bradshaw, B., & Rutherford, M. B., (2007). Towards a new supra regulatory approach to environmental assessment in Northern Canada. *Impact Assessment and Project Appraisal, 25*, 27–41.
30. Harris, R. B., (2012). Health impact assessment: The state of the art. *Impact Assessment and Project Appraisal,* 30. doi: 10.1080/14615517.2012.666035.
31. Thomas, B. F., Urmila, J. T., & Samuel, H., (2015). Environmental impact assessment and strategic environmental assessment research in the UK. *Journal of Environmental Assessment Policy and Management, 17*, 1–12.
32. Fundingsland, T. M., & Hanusch, M., (2012). Strategic environmental assessment: The state of the art. *Impact Assessment and Project Appraisal,* 30. doi: 10.1080/14615517.2012.666400.
33. Hacking, T., & Guthrie, P., (2006). Sustainable development objectives in impact assessment: Why are they needed and where do they come from? *Journal of Environmental Assessment Policy and Management, 8*, 341–371.
34. Bebbington, J., Brown, J., & Frame, B., (2007). Accounting technologies and sustainability assessment models. *Ecological Economics, 61*, 224–236.
35. Ness, B., (2007). Categorizing tools for sustainability assessment. *Ecological Economics, 60*, 498–508.
36. Chanchitpricha, C., & Bond, A., (2013). Conceptualizing the effectiveness of impact assessment processes. *Environmental Impact Assessment Review, 43*, 65–72.
37. Ekins, P., & Vanner, R., (2007). Sectoral sustainability and sustainability assessment methodologies: A review of methodology in light of collaboration with the UK oil and gas sector. *Journal of Environmental Planning and Management, 50*, 87–111.
38. Morrison, S. A., & Fischer, T. B., (2006). What is wrong with EIA and SEA anyway? A sceptic's perspective on sustainability assessment. *Journal of Environmental Assessment Policy and Management, 8*, 19–39.

39. Noble, B. F., (2009). Promise and dismay: The state of strategic environmental assessment systems and practices in Canada. *Environmental Impact Assessment Review, 29*, 66–75.
40. Pahl, W. C., (2007). Transitions towards adaptive management of water facing climate and global change. *Water Resour. Manage.*, 49–62.
41. Bond, A. J., & Morrison, S. A., (2009). Sustainability appraisal: Jack of all trades, master of none? *Impact Assessment and Project Appraisal., 27*, 321–329.
42. Berkes, F., (2009). Evolution of co-management: Role of knowledge generation, bridging organizations and social learning. *Journal of Environmental Management, 90*, 1692–1702.
43. Garrick, D., & Hall, J. W., (2014). Water security and society: Risks, metrics, and pathways *Annu. Rev. Environ. Resour., 39*, 611–639.
44. Thomas, J. M., & Callan, S. J., (2007). *Environmental Economics*. South-Western, a part of Cengage Learning, Canada.
45. Pope, J., Morrison, S. A., & Annandale, D., (2005). Applying sustainability assessment models. *Impact Assessment and Project Appraisal, 23*, 293–302.
46. Therivel, R., & Minas, P., (2002). Ensuring effective sustainability appraisal. *Impact Assessment and Project Appraisal, 20*, 81–91.
47. Therivel, R., (2009). Sustainability-focused impact assessment: English experiences. *Impact Assessment and Project Appraisal, 27*, 155–168.
48. Faircheallaigh, C., (2009). Effectiveness in social impact assessment: Aboriginal peoples and resource development in Australia. *Impact Assessment and Project Appraisal, 27*, 95–110.
49. Faircheallaigh, C., (2010). Public participation and environmental impact assessment: Purposes, implications, and lessons for public policy making. *Environmental Impact Assessment Review, 30*, 19–27.
50. Sinclair, A. J., Diduck, A., & Fitzpatrick, P., (2008). Conceptualizing learning for sustainability through environmental assessment: Critical reflections on 15 years of research. *Environmental Impact Assessment Review, 28*, 415–428.
51. Slootweg, R., & Jones, M., (2011). Resilience thinking improves SEA: A discussion paper. *Impact Assessment and Project Appraisal, 29*, 263–276.
52. Shields, D. J., Blengini, G. A., & Solar, S. V., (2011). Integrating life cycle assessment and other tools for ex ante integrated sustainability assessment in the minerals industry. *American Journal of Applied Sciences, 8*, 1214–1227.
53. Sadler, B., (1999). A framework for environmental sustainability assessment and assurance. In: Petts, J., (ed.), *Handbook of Environmental Impact Assessment* (pp. 12–32). Oxford: Blackwell.
54. Govender, K., Hounsome, R., & Weaver, A., (2006). Sustain-ability assessment: Dressing up SEA? – experiences from South Africa. *Journal of Environmental Assessment Policy and Management, 8*, 321–340.
55. Zhou, Z., Mo, Q., & Qin, L., (2012). A new indicator for life cycle sustainability assessment of fuels. *Advanced Materials Research, 347*, 360–363.
56. Deng, Y., Cai, L., & Gong, Z., (2012). Non-weighting assessment model: A new tool for sustainable urban landscape planning. *Advanced Materials Research*, 361–363.
57. Pope, J., & Grace, W., (2006). Sustainability assessment in context: Issues of process, policy and governance. *Journal of Environmental Assessment Policy and Management, 8*, 373–398.

58. Van Den, B., (2005). *Economics of Industrial Ecology*. Cambridge, MA: MIT Press.
59. Sanchez, L. E., & Morrison, S. A., (2011). Learning about knowledge management for improving environmental impact assessment in a government agency: The Western Australian experience. *Journal of Environmental Management, 92*, 2260–2271.
60. Kibert, N. C., (2004). Extended producer responsibility: A tool for achieving sustainable development. *Journal of Land Use, 19*, 503–523.

CHAPTER 9

ARAS-BASED SELECTION OF OPTIMAL SPECIMEN AMONG PRPS AND ITS VALIDATION USING MVGFD

P. V. RAJESH and M. SRIRAM PRASANTH

Department of Mechanical Engineering, Saranathan College of Engineering, Tiruchirappalli, Tamil Nadu, India

ABSTRACT

Composites result from joining two or more engineering materials with different physical and chemical properties. When combined, these individual materials can produce a resultant composite with superior mechanical and tribological properties compared to the unit materials. In a composite, typically, there are two entities. One is a matrix, and the other is reinforcement. The composite materials are heterogeneous when viewed on micro-scale but homogeneous, when magnified. A polymer matrix composite (PMC) is made up of various short or continuous fibers bound together by an organic polymer resin. PMCs are designed to transfer loads between fibers through the matrix. Some of the advantages with PMCs include their light weight, high stiffness, and their high strength along the direction of their reinforcements. When the reinforcement inside the polymer matrix is necessarily a powder made up of fine particles, then the composite structure is known as Particle/Particulate Reinforced Plastic (PRP). In the ensuing work, hybrid PRP composites samples are created through a novel technique called Stir Moulding. Several experiments are carried out to gauge their relevant properties. The best composite specimen is then chosen using Additive Ratio ASsessment (ARAS), a quantitative type Multi-Criteria Decision Making

Decision-Making Models and Applications in Manufacturing Environments. Pushpdant Jain, Kumar Abhishek, & Prasenjit Chatterjee (Eds.)

(MCDM) method. The result thus obtained is validated using multi-variant general factorial design (MVGFD) methodology.

9.1 INTRODUCTION

Composites are formed, when: (i) a metal and a non-metal is joined; (ii) a metal and two or more non-metals are joined; and (iii) two or more similar/dissimilar non-metals are joined. Composites are heterogeneous in nature (i.e., the constituents exist in different distinctive phases). When two different materials are joined together to form a composite by mechanical means with the application of heat and pressure, the final product retains the chemical properties of the parent materials separately but differs in mechanical strength. Generally, composites are selected for specific applications based on either their choice of matrix material or shapes and sizes of reinforcements [1]. In modern times, Composites are preferred over individual materials worldwide due to their better strength-to-weight ratio, improved hardness, enhanced stiffness, and excellent resistance to wear and corrosion. If more than one reinforcement is used to fabricate composite specimens, then they are called hybrid composites [2].

In polymer matrix composites (PMCs), the matrix is a polymer (non-metal), and reinforcements are either metals or non-metals. Generally, polymers are classified into two categories: Thermoplastics and Thermosets. Thermoplastics do not undergo any chemical change in their composition when heated. Due to additive polymerization, they can be reheated, remolded, and recycled a number of times. Thermosetting plastics can be heated and shaped only once. Due to condensation polymerization, the chemical reaction in the thermosetting process is irreversible. Polymers and their associated compounds are fast replacing metals and ceramics almost everywhere. Owing to their low cost, relative ease of transportation and usage, flexibility, and robustness in production, they are largely preferred in structural, medical, and biological applications. Zainab S. Al-Khafaji, in his research article, proposed a novel method to fabricate Polymer composites reinforced with alumina and their possible usage as implant materials in the human body [3]. Some of the typical applications include automobile body covers, panels, interior cabins in houses, aircrafts, and ships, vehicle bumpers, decorative tabletops, etc. P. K. Mallick explained in detail about the various processing techniques used for producing PMCs in one of his book chapters. He also presented the different process parameters, their range and their influence over physical, mechanical, and tribological properties of polymer-based composites [4].

Muti-criteria decision making (MCDM) is a calculative optimization method based on the selection of the best possible combination of analyzed criteria/properties from a given set of different combinational alternatives through ranking. T. Martin Aruldoss and his associates conducted a detailed survey on various MCDM methods and their applications. They attempted to compare different MCDM techniques based on their methodology and found out the differences prevailing among them [5]. Additive ratio assessment (ARAS) is one of the latest MCDM techniques proposed by Zavadskas in which the priorities of various alternatives are determined according to a utility function value. The utility function value used for the determination of the efficiency of a feasible alternative is directly proportional to the relative weights of each criterion involved in the process [6].

Multi-variant general factorial design (MVGFD) is a software-based methodology to create an experimental design for the research work. The experimental design consisting of input factors and output responses are evaluated to derive a regression equation for each response. The regression equation establishes a relationship between independent factors and responses depending upon them. Finally, an optimal combination is selected with the help of analysis of variance (ANOVA) among different factors and responses. The advantage of MVGFD lies in its flexibility of choosing varying levels among different factors. S. Jayabal and his colleagues performed optimization on the machinability behavior of polyester hybrid composites using General Factorial Design methodology to find out the factors that influence machining. Based on the inferences obtained from the research study, they came to a conclusion that the percentage of reinforcement plays a pivotal role in deciding the surface morphology [7].

9.2 EXPERIMENTAL METHODS AND MATERIALS

9.2.1 SELECTION OF MATERIALS

Material selection plays a crucial role in the formation and utility of PRPs. It is important that the constituents should have good bonding with each other. The matrix and reinforcements should exhibit good flexibility, reasonable hardness, and low porosity together as a composite.

9.2.1.1 MATRIX

In PRPs, matrix acts as a binder that houses reinforcement particles. The matrix material chosen for this research work is PER. Ester is a monomer formed from the condensation of a dibasic organic acid and a polyol. When ester functionally repeats itself to form a continuous chain, it is called polyester. Unsaturated polyester resin is a thermoset. Normally polyester resins are commercially available as solutions (i.e., polyester granules dissolved in reactive diluents like styrene) [8]. They cure at glass transition temperature. These resins offer the following merits:

- Resistance to chemical attacks;
- Resistance to aging;
- Low cost;
- Low shrinkage during solidification;
- Reasonable temperature withstanding capacity.

9.2.1.2 REINFORCEMENTS

Reinforcements are also called fillers in PRPs. Two different fine particle reinforcements are selected for this work: Alumina and rice husk powder (RHP). They provide strength and hardness to the composites.

Alumina, also known as Aluminum Oxide, is one of the most commonly used powdered ceramic with high hardness, high compressive strength and excellent resistance to corrosion [9].

Rice Husk, a by-product of paddy is an agricultural waste rich in lignin and silica. This light weight fibrous material is removed from rice using winnowing [10]. In this research work, finely ground RHP along with Alumina is used to strengthen the resin matrix.

9.2.1.3 INITIATOR

Initiator starts the chemical reaction by reacting with a monomer to form an intermediate component called free radical that links other monomers successively. This continuous association of a large number of monomers leads to the creation of a polymeric compound. Methyl ethyl ketone peroxide (MEKP), a colorless and oily liquid is used as initiator in this work.

9.2.1.4 PROMOTER

The purpose of catalyst is to speed up/accelerate a chemical reaction without taking part in that reaction. Catalyst can also be called promoter. For the formation of composite and hardening of resin, the role of initiator is vital [11]. Cobalt salt solution is the commonly used catalyst.

9.2.2 FABRICATION METHOD

Stir molding process is used to fabricate composite specimens. Required quantity of alumina and RHPs are initially weighed and pulverized separately in a ball mill for about 2 hours to obtain micron-sized fine particulates. They are then passed together through a set of sieves to obtain uniform particle size. Both particles are taken collectively in a mortar, crushed, and mixed using a pestle to avoid clustering. After obtaining an even mixture, both the powders are preheated at 200°C for about 15 minutes in a muffle furnace shown in Figure 9.1 to remove moisture and enhance miscibility within the resin matrix. About 50 ml of resin is taken in a beaker. The powder mixture is added to the resin as per mass-based ratio [12]. The beaker is then placed in a magnetic stirrer setup shown in Figure 9.2.

FIGURE 9.1 Muffle furnace.

FIGURE 9.2 Magnetic stirrer.

The entire composition is mixed thoroughly for about 15 minutes at a speed of 250–300 rpm with the help of rotating stir bars/stir pellets shown in Figure 9.3, dropped inside the beaker. At 8th minute, 4–5 drops of initiator are poured into the mixture. Catalyst is added to the mixture 2 minute before the end of the process. Finally, the semi-solid solution is poured in the mold cavity of the die, prepared as per ASTM (American Society for Testing and Materials) dimensions shown in Figure 9.4. The mold cavity is enclosed by a top plate and base plate, so that it is sandwiched in between. The top plate, mold cavity and base plate are fastened together tightly with the help of bolt and nut assembly at all the four corners. The semi-solid solution starts to solidify gradually. The curing time/setting time of the composite is one day. After that, they can be removed from the mold cavity.

The composition of different PRP specimens to be fabricated by varying the proportion of reinforcements using stir molding is determined from design of experiments (DoEs) in general factorial design using design expert software v.10 and is tabulated in Table 9.1.

FIGURE 9.3 Stir bars.

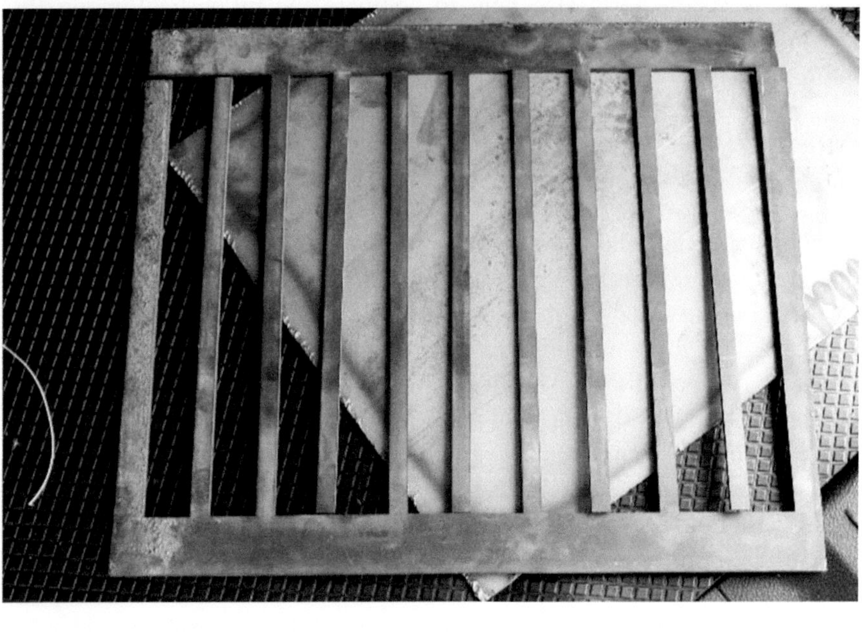

FIGURE 9.4 EDM machined mold cavity for PRPs.

TABLE 9.1 Composition of Composite Specimens

Specimen No.	Poly Ester Resin		Alumina Powder		Rice Husk Powder	
	ml	%	g	%	g	%
1	50	100	0	0	0	0
2	50	100	1.25	2.5	0	0
3	50	1,000	0	0	1.25	2.5
4	50	100	2.5	5	0	0
5	50	100	0	0	2.5	5
6	50	100	1.25	2.5	1.25	2.5
7	50	100	2.5	5	1.25	2.5
8	50	100	1.25	2.5	2.5	5
9	50	100	2.5	5	2.5	5

9.3 TESTS CONDUCTED AND RESULTS

After the fabrication of specimens as per the experimental design, appropriate tests are conducted to evaluate the mechanical properties of composite specimens. These tests also help to compare the properties among PRP specimens without reinforcement, with single reinforcement and with hybrid reinforcements.

9.3.1 TENSILE TEST

The tensile test is conducted to measure the ultimate tensile strength (UTS) of a specimen. UTS is the extent up to which the material can be pulled apart [13]. As the universal testing machine (UTM) in the institution is having a capacity of 40 T, it cannot measure the tensile parameters of PRP, which are very low compared to metal alloys. Hence, spring testing machine (STM) (range: 0–2.5 KN) shown in Figure 9.5 is used for measuring the tensile stress of the PRP composite specimens in this research work.

Two fixtures are designed in such a way that they can be fixed in STM and used to hold the specimens in the top end as well as bottom end. The 3D view of the fixtures is given as in Figure 9.6.

The specimens are fixed between the gripping portions of the top and bottom fixtures. The STM is made to move apart so that the specimen can be pulled in opposite direction and elongated till the permissible level. After withstanding a particular peak load, it broke into two pieces. This maximum load is called ultimate tensile load [14]. The UTS is obtained by dividing the ultimate load with an area of cross-section of the specimen. The test samples are created, and the test is done as per ASTM standards for tension testing of polymer materials 'D3039' [15]. The same procedure is repeated for all the remaining specimens, and the results are tabulated in Table 9.2.

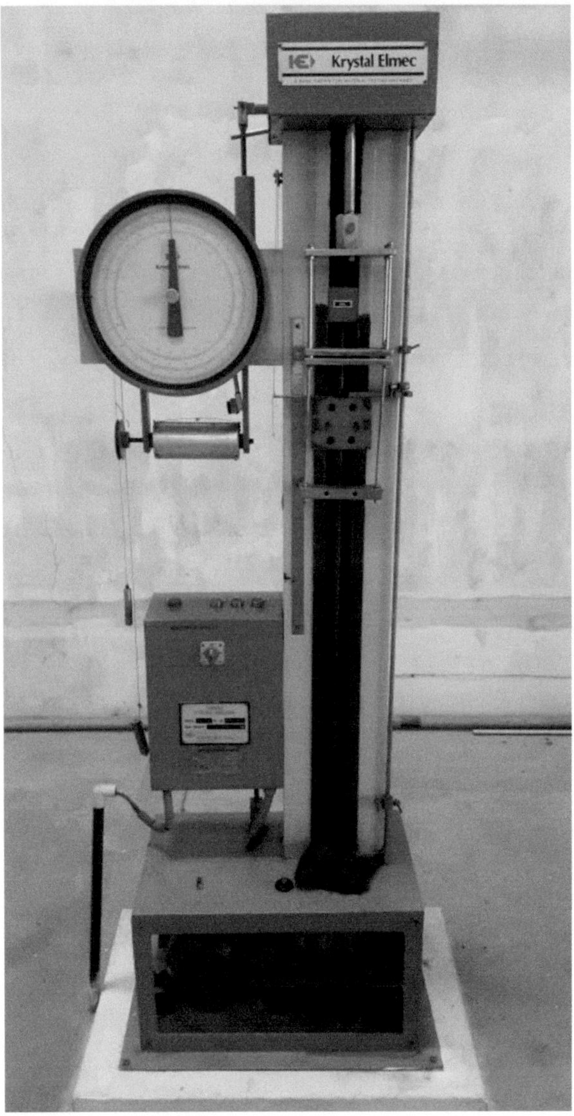

FIGURE 9.5 Spring testing machine.

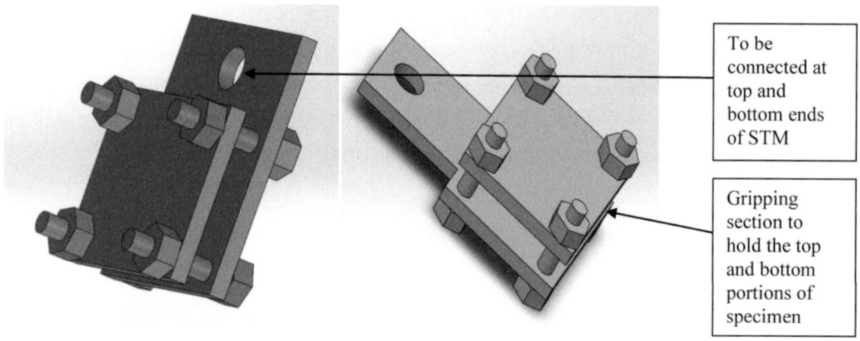

FIGURE 9.6 Fixtures for tensile test.

TABLE 9.2 Tensile Test Results

Specimen No.	Area of Cross Section 'A_0'	Maximum Load 'P_{max}'	Ultimate Tensile Strength $S = P_{max}/A_0$
	mm²	N	MPa
1	75	2,300	30.67
2	75	1,620	21.60
3	75	1,800	24.00
4	75	1,520	20.27
5	75	1,580	21.07
6	75	1,450	19.33
7	75	1,420	18.93
8	75	1,440	19.20
9	75	1,370	18.27

The results indicated that the unreinforced specimen with 100% resin has the highest tensile strength. The addition of reinforcement reduces the UTS. Particularly, the specimens containing alumina powder have less tensile strength when compared to the specimens containing RHP. Hybrid reinforcements are insignificant in determining UTS.

The tensile test results are represented as a graph in Figure 9.7.

9.3.2 *FLEXURAL TEST*

Flexural test, also known as three-point bending test is conducted to measure the flexural strength (FS) or flexibility of a material. FS is a defined as the tendency of a material that deforms under a continuously increasing load till it fractures. In other words, the higher the FS, lower the stiffness [16]. This test is performed in the compression chamber (bottom section) of the UTM shown in Figure 9.8.

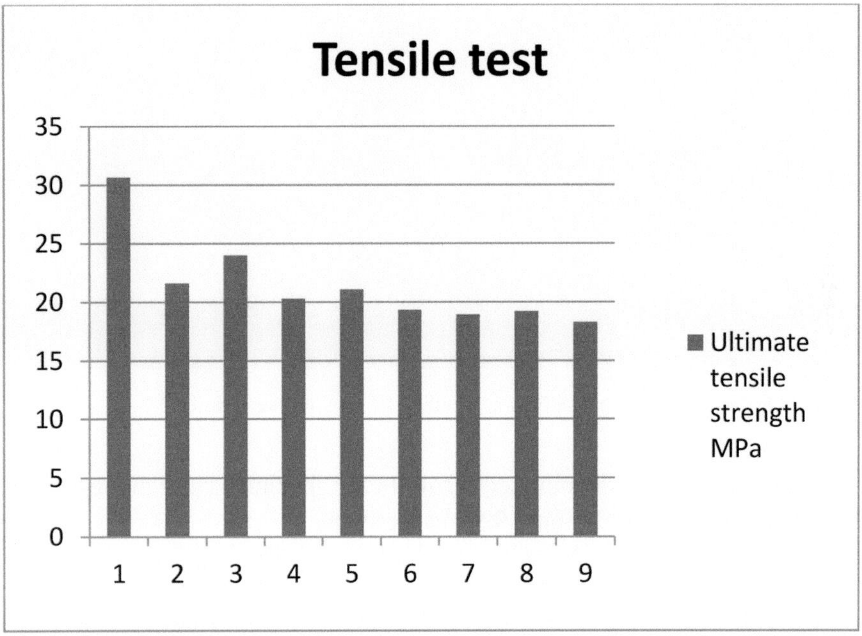

FIGURE 9.7 Tensile test results.

The composite specimens are prepared as per dimensions given in ASTM Standard test methods for flexural properties of unreinforced and reinforced plastics 'D790.' The specimen is placed above simply supported fixtures. Load is applied using a plunger at the exact middle of the specimen. Load is increased till the deflecting specimen breaks. The maximum/highest load absorbed by the specimen before it breaks is flexural load [17]. The FS is calculated from the formula given as in Eqn. (1):

$$\sigma_f = 3PL/(2bt^2) \quad (1)$$

where; σ_f is the FS/stress at yield in MPa; P is the ultimate load given at the center-point/middle of the specimen, before it ruptures in N; L is the span length between two simply supported fixtures in mm; b is the width of the specimen in mm; and t is the thickness of the specimen in mm.

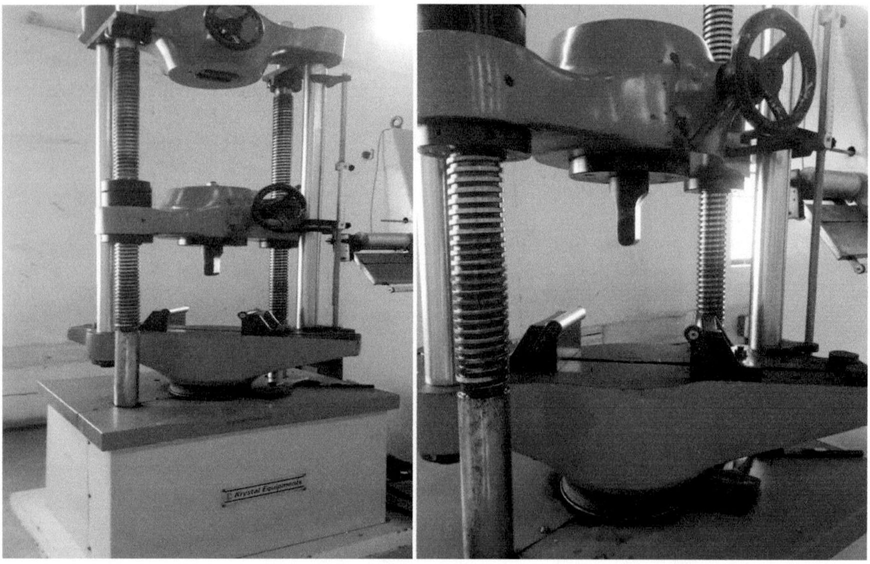

FIGURE 9.8 (a) UTM; and (b) 3-point bending flexural test setup in UTM.

The flexural test is conducted on all 9 PRP specimens and the results are shown in Table 9.3 and Figure 9.9, respectively.

TABLE 9.3 Flexural Test Results

Specimen No.	Flexural Strength (FS)
	MPa
1	62.50
2	58.72
3	69.24
4	48.43
5	53.08
6	45.25
7	42.36
8	36.65
9	40.47

The results indicated that the composite specimen containing 2.5% of RHP alone as reinforcement shows the highest FS. Other specimens follow the trend as the percentage of reinforcement increases, FS decreases. The unreinforced specimen has the second highest FS. Hybrid reinforcements have no significant influence over FS.

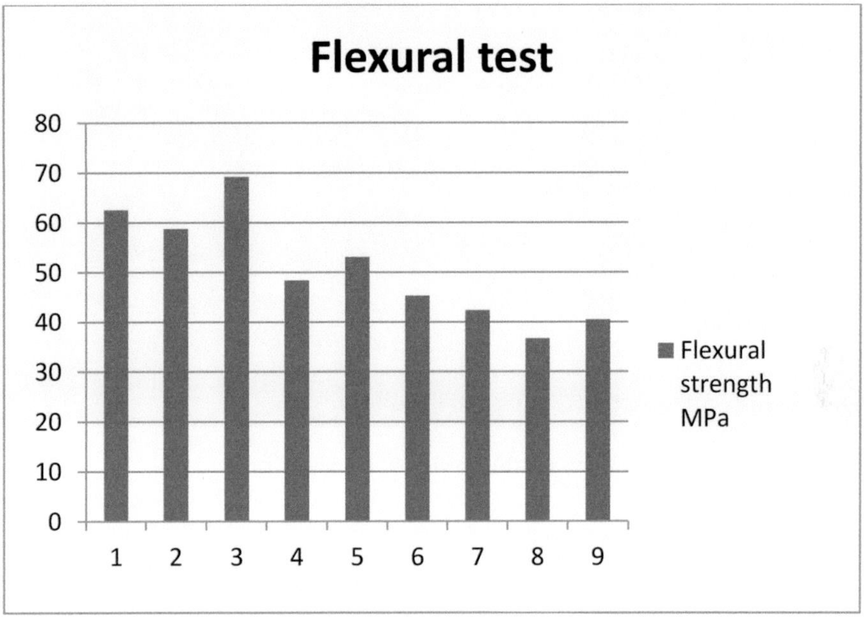

FIGURE 9.9 Flexural test results.

9.3.3 *HARDNESS TEST*

Hardness is the property of a material to resist wear and indentation. For polymers and their associated composites, Rockwell Hardness test is conducted. Rockwell Hardness Tester is shown in Figure 9.10.

The Rockwell test determines the hardness by measuring the depth of impression created by a penetrating indenter under a large load compared to the penetration made by a pre-load. The test procedure for Rockwell hardness of a material involves the application of a minor load followed by a major load [18]. The minor load fixes the zero position at graduated B scale. The major load of 100 kgf is applied, held constant for a few seconds and then removed, while the minor load is still retained. The depth of penetration from the zero datum is measured from the dial indicator. The penetration depth and hardness are inversely proportional [19]. The hardness test results are represented in Table 9.4.

FIGURE 9.10 (a) Front view; and (b) top view of Rockwell hardness tester.

TABLE 9.4 Hardness Test Results

Specimen No.	Rockwell Hardness 'HR_B'
1	64
2	105
3	97
4	124
5	101
6	112
7	118
8	113
9	120

The results infer that the unreinforced composite specimen has the lowest hardness. Hardness values increase when reinforcements are added. It is also noted that the addition of alumina powder enhances hardness more when compared to RHP. Hybrid reinforcements give marginal increase to hardness when compared to individual reinforcements.

The pictorial representation of hardness test results is indicated in Figure 9.11.

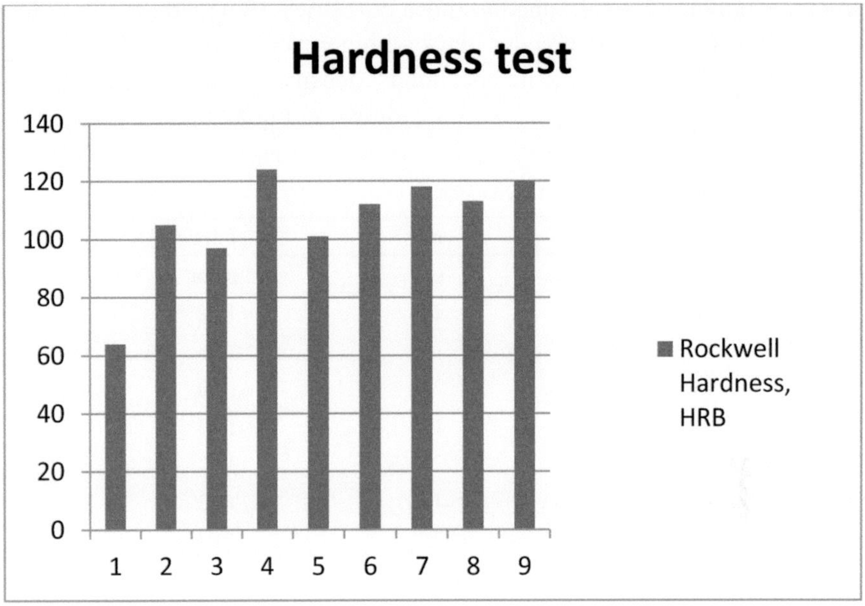

FIGURE 9.11 Hardness test results.

9.3.4 DETERMINATION OF PARTICLE DENSITY

Density of the material is the mass per unit volume. The density of the composite specimens should be as low as possible so that they can have high strength-to-weight ratio [9].

Particle density is the true density of a material, in which air bubbles filling porosity are removed when immersed in water medium. It is measured using Archimedes' principle mentioned in Eqn. (2):

$$\rho_{PMC} = (m_A/(m_A - m_B)) \text{ x } \rho_{H2O} \quad (2)$$

where; ρ_{PMC} is the density of the composite in g/cm^3; m_A is the mass of the composite sample in air in g; m_B is the apparent mass of the same composite suspended in distilled water in g; and ρ_{H2O} is the density of distilled water (1 g/cm^3).

Based on the above-mentioned formula, the particle densities of all the 9 PRP specimens are calculated and presented in Table 9.5.

From the results, it is evident that the unreinforced specimen has the highest density. When making composites, the chance of porosity is high and small air bubbles get entrapped in those porosities. Hence, there is a reduction in density. Alumina powder plays a bigger role in reducing the density than RHP. In hybrid composite specimens' density is nominal/intermediate. The densities of all the 9 specimens are shown in Figure 9.12.

TABLE 9.5 Measurement of Particle Density

Specimen No.	m_A	m_B	Particle Density
	g	g	g/cm³
1	4.6	2.2	1.92
2	4.4	1.9	1.76
3	4.5	2.0	1.80
4	3.9	1.7	1.77
5	4.3	2.0	1.87
6	4.0	1.8	1.82
7	4.4	2.1	1.91
8	4.2	1.9	1.83
9	3.8	1.7	1.81

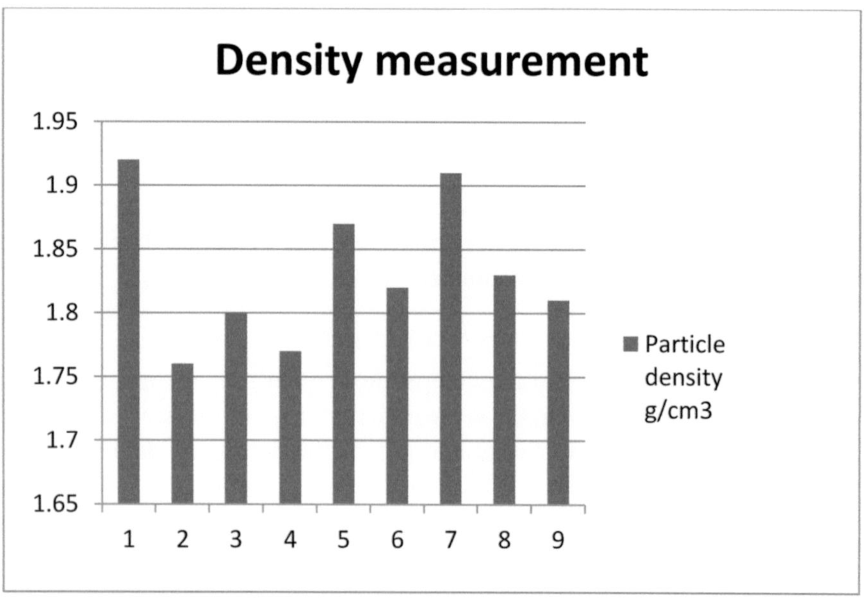

FIGURE 9.12 Particle density measurement results.

Finally, the results obtained by performing corresponding tests are summarized and presented in Table 9.6.

TABLE 9.6 Input Process Parameters and Output Variables

Specimen No.	Composition of Al_2O_3	Composition of RHP	Ultimate Tensile Strength	Flexural Strength	Rockwell Hardness HR_B	Particle Density
	%	%	MPa	MPa	–	g/cm^3
1	0	0	30.67	62.50	64	1.92
2	2.5	0	21.60	58.72	105	1.76
3	0	2.5	24.00	69.24	97	1.80
4	5	0	20.27	48.43	124	1.77
5	0	5	21.07	53.08	101	1.87
6	2.5	2.5	19.33	45.25	112	1.82
7	5	2.5	18.93	42.36	118	1.91
8	2.5	5	19.20	36.65	113	1.83
9	5	5	18.27	40.47	120	1.81

9.4 NUMERICAL OPTIMIZATION USING ARAS

Additive Ratio ASsessment (ARAS) is one of the least complicated and most recently developed MCDM technique to find out the optimal specimen combination from a set of different alternatives by ranking the alternatives. Like most of the MCDM techniques, this was also first proposed by E. Zavadskas in his research article to solve complex problems arising in the field of material selection and supplier selection [20]. This method converts the non-beneficiary criteria or cost criteria into beneficiary criteria by finding out the inverse of criteria elements (x_{ij}). Procedural steps of optimization for the above-mentioned research work:

9.4.1 CONSTRUCTION OF DECISION MATRIX WITH 'M' ALTERNATIVES AND 'N' CRITERIA

The different specimen combinations are considered as alternatives, which are presented row-wise (i). The different properties are considered as criteria and the corresponding test results are arranged column-wise (j) [21]. The Decision Matrix thus constructed is represented in Table 9.7.

TABLE 9.7 Decision Matrix

Alternatives	Criteria			
	UTS	FS	RH	PD
	MPa	MPa	–	g/cm^3
1	30.67	62.50	64	1.92
2	21.60	58.72	105	1.76
3	24.00	69.24	97	1.80
4	20.27	48.43	124	1.77
5	21.07	53.08	101	1.87
6	19.33	45.25	112	1.82
7	18.93	42.36	118	1.91
8	19.20	36.65	113	1.83
9	18.27	40.47	120	1.81

9.4.2 CLASSIFICATION OF CRITERIA AND ASSIGNING WEIGHT FACTORS FOR EACH CRITERION BASED ON THE RELATIVE IMPORTANCE

Table 9.8 presents the classification of different criteria and the corresponding weightage allocated to each criterion.

TABLE 9.8 Criteria and Their Weight Factors

SL. No.	Criteria	Category	Objective	Weightage (w)
1.	Ultimate tensile strength	Beneficiary (larger-the-better)	Maximization	0.35
2.	Flexural strength	Beneficiary (larger-the-better)	Maximization	0.35
3.	Rockwell hardness	Beneficiary (larger-the-better)	Maximization	0.20
4.	Particle density	Cost/non-beneficiary (smaller-the-better)	Minimization	0.10

9.4.3 STANDARDIZATION OF DECISION MATRIX

Beneficiary criteria such as UTS, flexural strength (FS), and Rockwell Hardness can be proceeded further with linear normalization strategies. In contrast, non-beneficiary criteria, such as Particle Density, should be converted into beneficiary category by calculating the reciprocals of the values at the particular column [20] as given in Eqn. (3):

$$x_{ij}' = 1/x_{ij} \tag{3}$$

The modified decision matrix is represented as Table 9.9.

TABLE 9.9 Modified Decision Matrix

Alternatives	Criteria			
	UTS	**FS**	**RH**	**1/PD**
1	30.67	62.50	64	0.521
2	21.60	58.72	105	0.568
3	24.00	69.24	97	0.556
4	20.27	48.43	124	0.565
5	21.07	53.08	101	0.535
6	19.33	45.25	112	0.549
7	18.93	42.36	118	0.524
8	19.20	36.65	113	0.546
9	18.27	40.47	120	0.552

9.4.4 *NORMALIZATION OF CRITERIA ELEMENTS*

The normalized values (r_{ij}) for each criterion are calculated as per the following Eqn. (4) and the normalized Decision Matrix, R is shown in Table 9.10.

$$r_{ij} = (x_{ij}' - \min_i (x_{ij}'))/(\max_i (x_{ij}') - \min_i (x_{ij}')) \tag{4}$$

TABLE 9.10 Normalized Decision Matrix

Alternatives	Criteria			
	UTS	**FS**	**RH**	**1/PD**
1	1	0.79	0	0
2	0.269	0.68	0.683	1
3	0.462	1	0.550	0.73
4	0.161	0.36	1	0.93
5	0.226	0.5	0.617	0.29
6	0.085	0.26	0.800	0.6
7	0.053	0.18	0.900	0.06
8	0.075	0	0.817	0.54
9	0	0.12	0.933	0.67

9.4.5 FORMATION OF WEIGHT NORMALIZED DECISION MATRIX

Weight normalized decision matrix table; W is formed based on the following Eqn. (5):

$$w_{ij} = r_{ij} \times w \tag{5}$$

where; normalized criteria elements (r_{ij}) are multiplied with allotted weight factors to form weight normalized values (w_{ij}).

Table 9.11 indicates the weight normalized decision matrix.

TABLE 9.11 Weight Normalized Decision Matrix

Alternatives	Criteria			
	UTS	**FS**	**RH**	**1/PD**
1	0.35	0.278	0	0
2	0.094	0.237	0.137	0.1
3	0.162	0.35	0.11	0.073
4	0.056	0.127	0.2	0.093
5	0.079	0.176	0.123	0.029
6	0.03	0.092	0.16	0.06
7	0.019	0.061	0.18	0.006
8	0.026	0	0.163	0.054
9	0	0.041	0.187	0.067

9.4.6 DETERMINATION OF OPTIMALITY FUNCTION FOR I^{TH} ALTERNATIVE

The Optimality function (O_i) for each alternative is determined by adding the weight normalized criteria elements together in each row [22]. The same has been shown in Table 9.12.

9.4.7 CALCULATION OF THE DEGREE OF UTILITY FOR EACH ALTERNATIVE

The degree of utility (U_i) is calculated from the ratio between row-wise optimality functions and the highest optimality function represented as Eqn. (6).

$$U_i = O_i/O_{max(i)} \tag{6}$$

TABLE 9.12 Optimality Function

O_i
0.628
0.568
0.695
0.476
0.408
0.343
0.266
0.244
0.295

Table 9.13 represents the utility degree of each alternative.

TABLE 9.13 Degree of Utility

Alternatives/Specimen No.	U_i	U_i in Percentage
1	0.903	90.3
2	0.817	81.7
3	1.000	100
4	0.685	68.5
5	0.587	58.7
6	0.493	49.3
7	0.382	38.2
8	0.351	35.1
9	0.424	42.4

The 3rd alternative with the highest utility value is the best choice among the set of different alternatives. The optimized set of parameters and results are shown in Table 9.14.

9.5 VALIDATION OF ARAS USING MVGFD

MVGFD is one of the categories in Full Factorial Methodology. The experimental design for determining the factors and their levels in fabricating composite specimens is suggested by MVGFD. It is one of the most flexible techniques where different number of levels can be chosen among factors, and a number of factors need not be equal to a number of levels [23]. Table 9.15 shows the factors and their levels in MVGFD.

TABLE 9.14 Optimized Solutions by ARAS

Specimen No.	Composition of Al_2O_3	Composition of RHP	Ultimate Tensile Strength	Flexural Strength	Rockwell Hardness HR_B	Particle Density
	%	%	MPa	MPa	–	g/cm³
3	0	2.5	24.00	69.24	97	1.80

TABLE 9.15 Factors and Their Levels

Symbols	Factors/Process Parameters	Experimental Levels		
		Low	Medium	High
A	Composition of Al_2O_3 (wt.%)	0	2.5	5
B	Composition of RHP (wt.%)	0	2.5	5

The DoEs, which indicates the factors and responses, is shown in Table 9.16.

TABLE 9.16 Input Factors and Output Responses

Specimen No.	Factors		Responses			
	A: Composition of Al_2O_3	B: Composition of RHP	Ultimate Tensile Strength	Flexural Strength	Rockwell Hardness HR_B	Particle Density
	%	%	MPa	MPa	–	g/cm³
1	0	0	30.67	62.50	64	1.92
2	2.5	0	21.60	58.72	105	1.76
3	0	2.5	24.00	69.24	97	1.80
4	5	0	20.27	48.43	124	1.77
5	0	5	21.07	53.08	101	1.87
6	2.5	2.5	19.33	45.25	112	1.82
7	5	2.5	18.93	42.36	118	1.91
8	2.5	5	19.20	36.65	113	1.83
9	5	5	18.27	40.47	120	1.81

9.5.1 CREATION OF ANOVA TABLE FOR EACH RESPONSE

Tables 9.17–9.20 indicate the ANOVA results of UTS, FS, Rockwell hardness and particle density, respectively.

TABLE 9.17 ANOVA for Response 1: Ultimate Tensile Strength

Source	Sum of Squares	DOF	Mean Square	F Value	p-Value Prob > F	Inference
Model	114.4464	5	22.88928	14.81769	0.0252	Significant
A	55.63215	1	55.63215	36.01425	0.0093	–
B	32.66667	1	32.66667	21.14722	0.0193	–
AB	14.44	1	14.44	9.347937	0.0551	–
A^2	9.316806	1	9.316806	6.031365	0.0912	–
B^2	2.390756	1	2.390756	1.547689	0.3018	–
Residual	4.634178	3	1.544726	–	–	–
Cor total	119.0806	8	–	–	–	–

Std. Dev.	1.24287	**R^2**	0.961084
Mean	21.48222	**Adj R^2**	0.896223
C.V. %	5.785575	**Pred R^2**	0.52558
PRESS	56.49419	**Adeq Precision**	11.59835

TABLE 9.18 ANOVA for Response 2: Flexural Strength

Source	Sum of Squares	DOF	Mean Square	F Value	p-Value Prob > F	Inference
Model	738.0289	3	246.0096	5.556373	0.0476	Significant
A	478.1123	1	478.1123	10.79864	0.0218	–
B	259.3838	1	259.3838	5.858441	0.0601	–
AB	0.5329	1	0.5329	0.012036	0.9169	–
Residual	221.3761	5	44.27522	–	–	–
Cor total	959.405	8	–	–	–	–

Std. Dev.	6.653963	**R^2**	0.769257
Mean	50.74444	**Adj R^2**	0.630811
C.V. %	13.11269	**Pred R^2**	0.303933
PRESS	763.7506	**Adeq Precision**	6.989068

TABLE 9.19 ANOVA for Response 3: Rockwell Hardness

Source	Sum of Squares	DOF	Mean Square	F Value	p-Value Prob > F	Inference
Model	2479.583	5	495.9167	10.59525	0.0401	Significant
A	1666.667	1	1666.667	35.60831	0.0094	–
B	280.1667	1	280.1667	5.985757	0.0920	–
AB	420.25	1	420.25	8.978635	0.0578	–
A^2	72	1	72	1.538279	0.3030	–
B^2	40.5	1	40.5	0.865282	0.4209	–
Residual	140.4167	3	46.80556	–	–	–
Cor total	2620	8	–	–	–	–

Std. Dev.	6.841459	$\mathbf{R^2}$	0.946406
Mean	106	**Adj** $\mathbf{R^2}$	0.857082
C.V. %	6.454206	**Pred** $\mathbf{R^2}$	0.350698
PRESS	1701.172	**Adeq Precision**	9.831077

TABLE 9.20 ANOVA for Response 4: Particle Density

Source	Sum of Squares	DOF	Mean Square	F Value	p-Value Prob > F	Inference
Model	0.008603	5	0.001721	0.297455	0.8876	Not significant
A	0.001667	1	0.001667	0.288138	0.6287	–
B	0.0006	1	0.0006	0.10373	0.7685	–
AB	0.002025	1	0.002025	0.350088	0.5957	–
A^2	0.003756	1	0.003756	0.649272	0.4793	–
B^2	0.000556	1	0.000556	0.096046	0.7769	–
Residual	0.017353	3	0.005784	–	–	–
Cor total	0.025956	8	–	–	–	–

Std. Dev.	0.076054	$\mathbf{R^2}$	0.331443
Mean	1.832222	**Adj** $\mathbf{R^2}$	–0.78282
C.V. %	4.150933	**Pred** $\mathbf{R^2}$	–7.13789
PRESS	0.211223	**Adeq Precision**	1.395641

9.5.2 FORMATION OF REGRESSION EQUATION IN TERMS OF CODED FACTORS FOR EACH RESPONSE

Eqns. (7)–(10) represent the regression equations for UTS, FS, Rockwell hardness and particle density, respectively.

$$\text{Ultimate Tensile Strength} = +19.31 - 3.05*A - 2.33*B + 1.90*A*B + 2.16*A^2 + 1.09*B^2 \tag{7}$$

$$\text{Flexural Strength} = +50.74 - 8.93*A - 6.58*B + 0.37*A*B \tag{8}$$

$$\text{Rockwell Hardness} = +113.00 + 16.67*A + 6.83*B - 10.25*A*B - 6.00*A^2 - 4.50*B^2 \tag{9}$$

$$\text{Particle Density} = +1.81 - 0.017*A + 0.010*B + 0.022*A*B + 0.043*A^2 - 0.017*B^2 \tag{10}$$

9.5.3 3D INTERACTION PLOTS FOR EACH RESPONSE

Figures 9.13–9.16 represent the 3D plots showing the interaction between two factors and their effect on various responses (i.e., UTS, FS, Rockwell Hardness, and Particle Density, respectively).

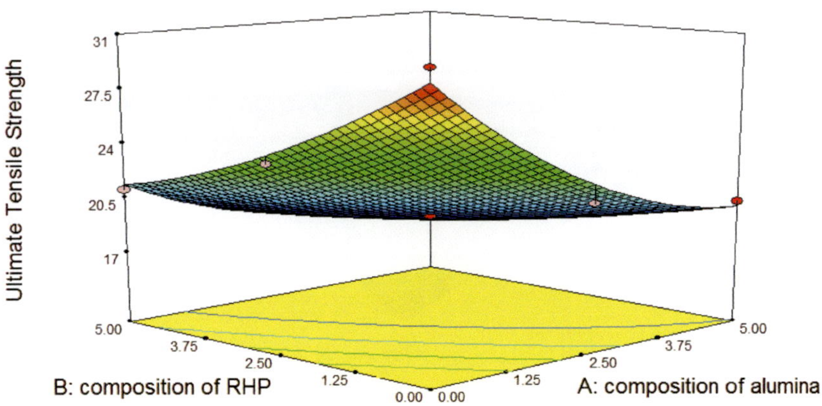

FIGURE 9.13 3D interaction plot showing the effect of factors A and B on UTS.

From the above plot, it is clear that UTS values decrease with increase in the addition of reinforcements. An increase in the weight percentage of RHP reduces the UTS gradually, whereas an increase in the proportion of alumina powder reduces the UTS in a parabolic manner.

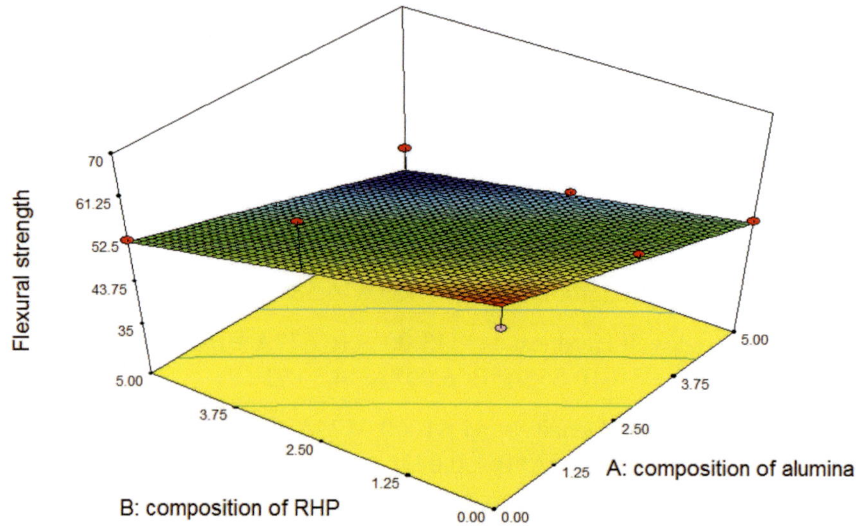

FIGURE 9.14 3D interaction plot showing the effect of factors A and B on FS.

From the above plot, it is clear that FS values decrease with increase in the addition of reinforcements. An increase in the weight percentage of RHP reduces the FS gradually, whereas an increase in the proportion of alumina powder reduces the FS more steeply.

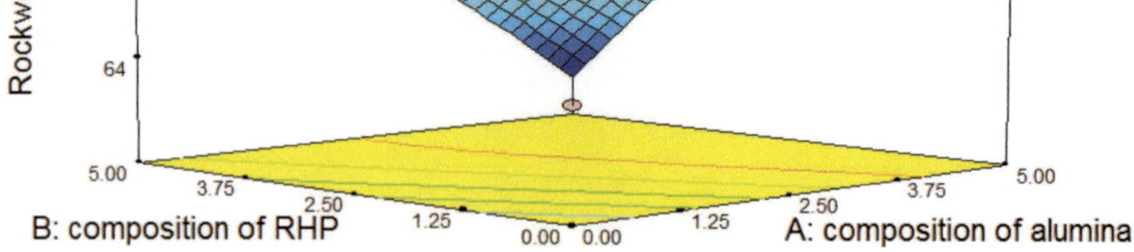

FIGURE 9.15 3D interaction plot showing the effect of factors A and B on RH.

From the above plot, it is clear that hardness values increase with increase in the addition of reinforcements. There is a sudden increase in hardness, when more amount of RHP is present in the composite. Increasing the percentage of alumina also has the same effect.

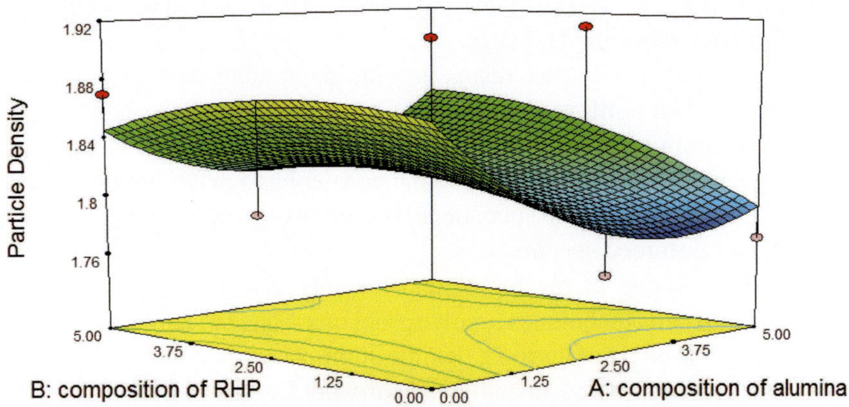

FIGURE 9.16 3D interaction plot showing the effect of factors A and B on PD.

Particle density initially reduces and then increases slightly with an increase in the composition of alumina. But the addition of RHP has an inconclusive effect over particle density, as the particle density is almost stagnant with the increase in the proportion of RHP. Table 9.21 shows the optimized results proposed by MVGFD.

TABLE 9.21 Optimized Solutions by MVGFD

Composition of Al_2O_3	Composition of RHP	Ultimate Tensile Strength	Flexural Strength	Rockwell Hardness HR_B	Particle Density	Desirability
%	%	MPa	MPa	–	g/cm³	%
0	2.21	24.0027	64.0907	96.4001	1.79918	74.8

The above-mentioned optimal solutions predicted by MVGFD are identically very close to the input and output values of specimen 3, which is proposed as the best specimen by ARAS.

9.6 CONCLUSION

The above detailed research work leads to the following inferences:

- PRP specimens containing single reinforcements provide substantial enhancement of mechanical and physical properties rather than hybrid reinforcements.
- Stir molding is one of the easy and least complicated methods to produce reinforced plastics.
- Unsaturated polyester resins provide good adhesiveness and strong bond when reinforced with ceramic and agricultural waste powders.
- Non-reinforced resin specimen has high UTS and FS, whereas the specimen reinforced with alumina powder has a better hardness value.
- Non-reinforced resin specimens have relatively higher particle density than reinforced specimens.
- ARAS is one of the robust and recently discovered MCDM-based optimization techniques to solve problems related to material selection.
- MVGFD is one of the most flexible optimization techniques performed conveniently using an analytical software Design Expert v.10.
- When the optimized results obtained from ARAS are validated using MVGFD, it proposes identical results.

KEYWORDS

- **additive ratio assessment**
- **alumina**
- **composites**
- **multi-variant general factorial design**
- **particle reinforced plastic**
- **rice husk powder**

REFERENCES

1. Mark, J. E. (1996). Ceramic-reinforced polymers and polymer-modified ceramics. *Polymer Engineering & Science, 36*(24), 2905–2920. doi: 10.1002/pen.10692.
2. Sinha, A. K., Narang, H. K., & Bhattacharya, S., (2020). Mechanical properties of hybrid polymer composites: A review. *Journal of the Brazilian Society of Mechanical Sciences and Engineering, 42*(8), 50–62. doi: 10.1007/s40430-020-02517-w.
3. Al-Khafaji, Z. S., (2018). Preparation and modelling of composite materials (polyester-alumina) as implant in the human body. *International Journal of Mechanical Engineering and Technology, 9*(4), 468–478.

4. Mallick, P. K., (2017). *Processing of Polymer Matrix Composites* (1st edn., pp. 1–356). CRC Press. doi: 10.1201/9781315157252. ISBN: 9781315157252.
5. Martin, A. T., Miranda, L., & Prasanna, V. V., (2013). A survey on multi criteria decision making methods and its applications. *American Journal of Information Systems, 1*(1), 31–43. doi: 10.12691/ajis-1-1-5.
6. Zavadskas, E. K., & Turskis, Z., (2010). A new additive ratio assessment (ARAS) method in multicriteria decision-making. *Technological and Economic Development of Economy, 16*(2), 159–172. doi: 10.3846/tede.2010.10.
7. Jayabal, S., Natarajan, U., & Sekar, U., (2010). Regression modeling and optimization of machinability behavior of glass-coir-polyester hybrid composite using factorial design methodology. *The International Journal of Advanced Manufacturing Technology, 55*(1–4), 263–273. doi: 10.1007/s00170-010-3030-7.
8. Selvam, R., Ravi, S., & Raja, R., (2017). Fabrication of SiC particulate reinforced polyester matrix composite and investigation. *IOP Conference Series: Materials Science and Engineering, 197,* 42–52. doi: 10.1088/1757-899x/197/1/012052.
9. Latief, F. H., Chafidz, A., Junaedi, H., Alfozan, A., & Khan, R., (2019). Effect of alumina contents on the physicomechanical properties of alumina (Al_2O_3) reinforced polyester composites. *Advances in Polymer Technology,* 1–9. doi: 10.1155/2019/5173537.
10. Neeraj, B., Prakash, C. G., & Nisha, R., (2020). Rice husk as a fiber in composites: A review. *Journal of the Mechanical Behavior of Materials, 29,* 147–162. doi: 10.1515/jmbm-2020-0015.
11. Baskaran, R., Sarojadevi, M., & Vijayakumar, C. T., (2011). Unsaturated polyester nano-composites filled with nano-alumina. *Journal of Materials Science, 46*(14), 4864–4871. doi: 10.1007/s10853-011-5398-7.
12. Suhot, M. A., Hassan, M. Z., Aziz, S. A., & Md. Daud, M. Y., (2021). Recent progress of rice husk reinforced polymer composites: A review. *Polymers, 2391*(13), 1–19. doi: 10.3390/polym13152391.
13. Channabasavaraju, S., Shivanand, H. K., & Santhosh, K. S., (2013). Evaluation of tensile and flexural properties of polymer matrix composites. *International Journal of Modern Engineering Research, 3*(5), 3177–3180.
14. Singh, I. V., Shedbale, A., & Mishra, B. (2015). Material property evaluation of particle reinforced composites using finite element approach. *Journal of Composite Materials, 50*(20), 2757–2771. doi: 10.1177/0021998315612539.
15. Arjmandi, R., Hassan, A., Majeed, K., & Zakaria, Z., (2015). Rice husk filled polymer composites. *International Journal of Polymer Science*, 1–32. doi: 10.1155/2015/501471.
16. Gundogdu, M., (2014). The evaluation of flexural strength of composite resin materials with and without fiber. *Dentistry, 04*(09), 33–39. doi: 10.4172/2161-1122.1000259.
17. Khan, M. Z., Srivastava, S. K., & Gupta, M., (2018). Tensile and flexural properties of natural fiber reinforced polymer composites: A review. *Journal of Reinforced Plastics and Composites*, 44–68. doi: 10.1177/0731684418799528.
18. Shaker, J., Eddres, A., Hubi, H., Alaa, R., & Omran, N. M., (2012). Effects of chopped alumina fibers on hardness and impact strength of unsaturated polyester. *Journal of Babylon University/Engineering Sciences, 20*(4), 1–12.
19. Akindapo, J. O., Agov, E. T., Garba, D. K., & Ogabi, R. O., (2017). Comparative assessment of mechanical properties of groundnut shell and rice husk reinforced epoxy composites. *American Journal of Mechanical Engineering, 5*(3), 76–86. doi: 10.12691/ajme-5-3-2.

20. Chatterjee, P., & Chakraborty, S., (2013). Gear material selection using complex proportional assessment and additive ratio assessment-based approaches: A comparative study. *International Journal of Materials Science and Engineering, 1*(2), 104–111. doi: 10.12720/ijmse.1.2.
21. Athawale, V. M., & Chakraborty, S., (2012). Material selection using multi-criteria decision-making methods: A comparative study. *Proceedings of the Institution of Mechanical Engineers, Part L: Journal of Materials: Design and Applications, 226*(4), 266–285. doi: 10.1177/1464420712448979.
22. Singaravel, B., Shankar, D. P., & Prasanna, L., (2018). Application of MCDM method for the selection of optimum process parameters in turning process. *Materials Today: Proceedings, 5*(5), 13464–13471. doi: 10.1016/j.matpr.2018.02.341.
23. Alavi, F., & Jahan, M., (2017). Optimization of process parameters in micro-EDM of Ti-6Al-4V based on full factorial design. *International Journal of Advanced Manufacturing Technology, 92*(1–4), 167–187.

CHAPTER 10

OPTIMIZATION OF PMEDM PROCESS PARAMETERS FOR MRR, TWR, RA, AND HV USING TAGUCHI METHOD AND GREY RELATIONAL ANALYSIS FOR DIE STEEL MATERIALS

NGUYEN HUU PHAN, BANH TIEN LONG, and NGUYEN VAN THIEN

Hanoi University of Industry, Bac Tu Liem, Hanoi, Vietnam

ABSTRACT

Mixing powder into dielectric fluid in electrical discharge machining (PMEDM) is a very interesting technological solution in current research. This method has the highest efficiency in simultaneously improving the productivity and quality of a machined surface. At present, the primary solution for improving the processing efficiency of PMEDM is the optimization of process parameters. In this study, the process parameters in PMEDM using micro-titanium powder for die-steels have been optimized for single and multi-objective problems. Material removal rate (MRR), tool wear rate (TWR), surface roughness (SR), and the micro–hardness of a machined surface (HV) in PMEDM were quality indicators of optimization problems. The process parameters used in the study included workpiece materials (SKD61, SKD11, SKT4), electrode materials (copper, graphite), electrode polarity, pulse–on time, pulse–off time, current, and titanium powder concentration. Some interaction pairs among the process parameters

Decision-Making Models and Applications in Manufacturing Environments. Pushpdant Jain, Kumar Abhishek, & Prasenjit Chatterjee (Eds.)

were also used to evaluate the effect on the optimal results. Taguchi and grey relational analysis (GRA) method have been used to solve single and multi-objective optimization problems. The results showed that the optimal parameters were verified by experiment, and its accuracy was good. The finished machined surface under optimum conditions was also analyzed. The machined surface quality under optimum conditions was good.

10.1 INTRODUCTION

Past research into powder-mixed electric discharge machining (PMEDM) methods have proven promising to improve both the productivity and quality of electric discharge machining (EDM). A suitable powder is mixed into the dielectric fluid used in EDM, which can lead to both an increased material removal rate (MRR) and tool wear rate (TWR). In addition, surface roughness (SR) can be reduced and the micro-hardness (HV) of the surface machining can be greatly increased. Furthermore, increasing the concentration of powder 2–6 g/l can lead to an increased MRR, while concurrently reducing the TWR [1]. Graphite powder mixed in the dielectric fluid can be beneficial to the influence of EDM and has increased the MRR by 68%, decreased the TWR by 28%, and reduced the breakdown voltage by 30% [2]. The quality of the surface layer of steel H13 after machining by PMEDM, has also been improved significantly [3]. Results have demonstrated that an increase in the electrode size can lead to an increase in both SR and thickness. When white powder is mixed into the dielectric fluid, there is a strong increase in MRR, further roughing the EDM [4]. However, the TWR and chip mixed with powder has still not been addressed. Research results of the influence of some kinds of powder materials on the SR, MRR, and TWR in PMEDM have shown that graphite powder is more influential than silicon powder [5]. An increase in concentration has led to increases in MRR increases and decreases in SR. The Productivity and quality surface machining of EDM can be increased with Al powder mixed into the dielectric fluid [6]. Results have shown that the MRR increases, while both the TWR and SR decrease. Results obtained are similar for silicon powder as well as other powders with an increased concentration, all of which led to an increase in the MRR [7].

Taguchi method has been widely used to optimize quality characteristics in the field of EDM [8]. The effect of current, pulse-on time, and pulse-off time on the TWR of the copper electrode during machining of EN31 was assessed using the Taguchi method [9]. The results showed that using powder reduces the TWR. Conversely, an increase in current and pulse on time increased the TWR. Taguchi's method was used to evaluate the level of influence of aluminum powder on SR during machining of a H13 workpiece [10]. Negative electrode polarity and Al powder mixed into the dielectric fluid helped to reduce SR. The SR increased due to an increase in the current and duration of the pulses. An optimal value of MRR was determined by Taguchi's method [11]. Powder mixed in the dielectric fluid led to an increased MRR. The maximum value of the MRR obtained was 12.47 mm^3/min with powder concentrations of 6 g/L. During the machining of EN31 steel using a PMEDM process, MRR and SR were optimized. The results showed that the MRR and SR were strongly influenced by the concentration of powder and the intensity of electrical discharge [12]. The PMEDM process efficiency was better than that of the EDM process [13]. This contributed to the effectiveness of the PMEDM method. Three different powder materials were used, namely Gr powder, SiC, and Al_2O_3, in the dielectric media. The Gr powder helped to increase the MRR, while SiC powder helped to reduce the TWR [14]. By using SiC powder, the productivity of the EDM process improved significantly during the machining of WC [15]. The MRR was increased by 90% in comparison to the EDM process. The maximum value of the MRR (1.419 mm^3/min) was obtained using Taguchi's method with a powder concentration of 8 g/L. Singh et al. studied the effect of Al powder size and concentration on the MRR, TWR, and SR. The results indicated that the size and concentration of powder, whether too small or too high, led to an increased MRR. The TWR and SR were reduced when mixed powder was added to the dielectric fluid [16]. Ganachari et al. used Taguchi's method, combined with the GRA method to optimize the parameters of PMEDM [17]. The Al and SiC powders were used in the dielectric fluid, and SR were selected as the response. The best quality of machining surface was obtained when a PMEDM process was used. The relationship between MRR and EDM process parameters, including time pulse generator, current and solvent flow, were shown during the drilling of the EDM hole [18]. The increase in the duration of pulses, as well as the intensity of electrical discharge led to an increased MRR. However, an increase in the value of feed rate reduced the MRR. One of the disadvantages of the Taguchi method was that it could only optimize one of the quality characteristics of PMEDM. It was effective only for individual cases where the productivity was the highest or where the SR or TWR were the smallest. Thus, the results processed by Taguchi's method can only be used to solve problems and are not able to evaluate the overall effectiveness of PMEDM. In practice, manufacturers encourage researchers to develop manufacturing processes that optimize both productivity and quality. Recent studies have investigated successful combinations of the Taguchi method with other optimization techniques, therefore optimizing multiple characteristics of the EDM, including a combination of the MRR and SR, or even the MRR, TWR, and SR. The SR and width of the cut are two characteristics used to optimize machining processes simultaneously for stainless steel SS304 with a wire EDM [19]. The optimal values were obtained with a current of 3A current, and a pulse on-time and pulse-off time of 15μs and 4 μs, respectively. These results demonstrated that the time-pulse generator included process parameters that strongly affected both of the indicators studied. The effect of combining the Taguchi method with GRA was given when this combination was optimized simultaneously with an electrode-wear ratio (EWR), the MRR and SR Ti-6Al-machining in 4 V by the EDM [20]. The results showed that the MRR was increased by 87.3%, the EWR decreased by 25.7%, and the SR was reduced by 18.9%, as compared with experimental results in a matrix of the Taguchi method. The machining productivity increased and TWR decreased in the EDM at optimized conditions [21]. The optimization method included a combination of Taguchi's method and GRA. Accuracy optimization calculations were verified experimentally with the largest deviations of 8.8%. SR and kerf width have been optimized simultaneously in WEDM using Taguchi–GRA [22]. The results have shown that ton is the most influential process parameter on wire feed and gap voltage, respectively, and it reduced the effect to SR and left width. Taguchi–GRA was used to simultaneously optimize MRR, SR, recast layer thickness (RLT) and micro-hardness (HV) in PMEDM of H11 die steel [23]. The results show that the quality of the work surface improved significantly (gray relational grade increased by 0.1055). Many quality indicators have been optimized by Taguchi–GRA in dry EDM using a Cu electrode of AISI D2 steel [24, 25]. The machined surface morphology and microstructure under optimum conditions was analyzed, and its results were good. Chromium and SiC powders mixed into the dielectric of EDM have significantly improved dry EDM. In addition, the surface topography of H11 steel was significantly improved [26, 27]. Indicators including MRR, TWR, EWR, and SR in PMEDM were optimized simultaneously by TOPSIS and GRA. The results showed that both methods in combination are a solution for multi-objective optimization in this field. Surface quality at optimum conditions has also been analyzed and evaluated, and the results have shown that the surface quality improved. Powder mixed into dielectric fluid in the EDM of plastic mold steel prevented an increase in the length, width, and depth of cracks on the machined surface [28, 29]. However, the study of simultaneous optimization of MRR and SR parameters in EDM by the Taguchi–TOPSIS method was more effective than Taguchi–GRA [30]. Some quality indicators in EDM for AISI D2 tool steel with CNT as dielectric fluid by Taguchi–TOPSIS was presented in Ref. [31]. Optimized quality indicators including MRR, SR, and fractal dimension were studied. The results showed that of the three surveyed parameters (current (I), voltage (V), and ton), V was the most influential (42.42%), and ton was the least effective (11.13%). EDM machining for AISI D2 tools using steel nanofluids has been optimized with multi-targeting by Taguchi–GRA. SR and fractal dimensions were used as the optimal quality indicators. The results of the optimization process was good, and the obtained accuracy was 6.02%. In optimizing multiple targets, the TOPSIS method is simple, easily comprehensible, and robust [32]. In addition, this method allows the consideration of a number of quantitative and qualitative selection attributes. Therefore, it is a solution that allows us to approach the multi-objective optimization problem more objectively. By combining the concepts between the Taguchi method and GRA, the productivity was improved and the machining electrode wear was reduced in the machining of a super-hard alloy (Ti-6A1-4V) by EDM [33]. The success of using a combined Taguchi method with GRA contributed significantly in improving the machining efficiency of EDM [34].

In this study, we investigated the single-parameter optimization characteristics of the PMEDM quality, including MRR, TWR, SR, and HV, as well as the multi-parameter optimization of all four characteristics. The combination of both the Taguchi method and GRA was used to optimize problems related to the quality characteristics in PMEDM. Furthermore, PMEDM was used for machining steel titanium powder molding technology with the parameters used by other researchers, including workpiece material, tool material, polarity, pulse-on time, current, pulse-off time, and powder concentration. We further considered the influence of interactions to optimize the results, including the interactions of the workpiece and tool materials, the interactions of workpiece materials and the powder concentrations, and the interaction of tool materials and powder concentrations.

10.2 EXPERIMENTAL PROCEDURE

An electrical discharge machine, AG40L (Sodick, Inc. USA), was used. A schematic experiment is shown in Figure 10.1. The tank was made of CT3 steel with size 330×180×320 and motor shafts were fitted by stirring at 100 rev/min with titanium powder and were mixed with the dielectric fluid (oil HD-1) during the experiment (*see* Figure 10.1). The workpiece materials included SKD61, SKD11, and SKT4 mound steel, where the common type used in industry standards were selected for this study. Samples measured 45×27×10 mm. Furthermore, Cu, Gr are among the two materials most commonly used, and have been a focus of recent research. The electrode was shaped into a circular cylinder and it had a diameter measuring 23 mm. The size of the particle of titanium powder measuring 45 μm were selected and mixed into the dielectric fluid.

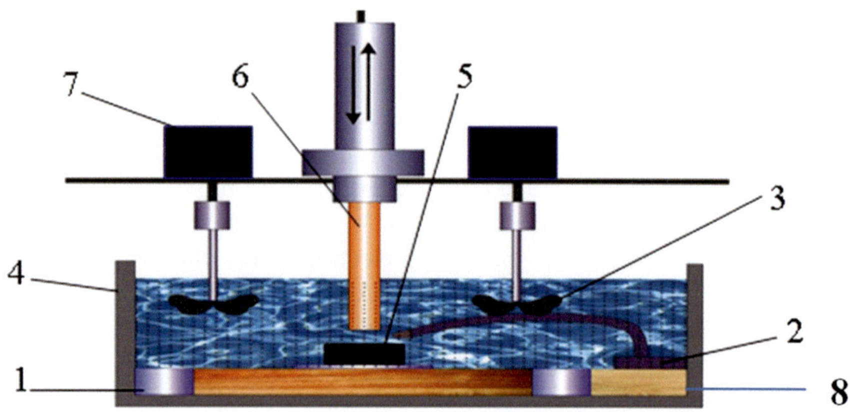

(1) Magnets; (2) pump; (3) nozzle; (4) machining tank; (5) workpiece; (6) electrode; (7) stirring; and (8) wood panels.

FIGURE 10.1 Schematic line diagram.

The mass before and after processing was measured with an electronic scale AJ 203 (Shinko Denshi Co. LTD–Japan), where the largest mass measured 200 g, with an accuracy of 0.001 g. The SR was measured using a strain gauge transducer contact, SJ-301 (Mitutoyo–Japan). The surface hardness (HV) was measured by a 1,106 Met Indenta (airline Buehler–USA). The surface morphology was characterized by scanning electron microscope (SEM) JEOL 6490 JED2300 (JEOL–Japan).

10.3 EXPERIMENTAL METHODS

10.3.1 TAGUCHI METHOD

The overall objective was to create a product with the highest quality and lowest price. These methods were based on traditional experimental design, following the Taguchi method, which can examine the influence of technological parameters, specific to the quality of the production process and cost. In other words, the method is frequently used to determine the best practices. The Taguchi method is a tool used to design experiments based on orthogonal matrix, specific to Taguchi, and is used to assess the process parameters. This method is often expensive. The experimental parameters could receive more than two levels, including a communication between the different possibilities that exist in an experimental design. This method included a simple experimental design and standardization. The experimental design of Taguchi method was implemented by the orthogonal matrix (table) for placement of the technological parameters, which were examined by their levels with the smallest number of experiments during the time as well as the least expensive. The selection of tables was based on the number of parameters and their change rates. The analysis of variance (ANOVA) was based on data obtained from Taguchi's experimental design and was used to select new parameter values to optimize the quality characteristics. Data from the table was analyzed by using charts, pictures, ANOVA, and the Fisher ratio test (F).

To analyze the results of experiments, Taguchi used a coefficient, S/N, to evaluate the impact of interference. The ratio, S/N, has a greater value for input parameters and was minimally impacted by noise. In experimental studies, the valuation ratio, S/N, was the highest possible for the results. The value obtained for S/N for the quality characteristics of a product under the conditions examined were much higher than those of the random effects of interference. A technology designed with the highest S/N will always give optimum results and the most change. The S/N coefficient was calculated to determine the effects of these parameters as well as their interactions to the quality characteristics in processing. The Optimal regime of the technological parameters was determined by the characteristics of the coefficient S/N [35].

The higher-the-better:

$$(S/N)_{HB} = -10 \log(MSD_{HB}) \tag{1}$$

In Eqn. (1), MSD_{HB} is defined through the following expression:

$$MSD_{HB} = \frac{1}{r}\sum_{i=1}^{r}\left(\frac{1}{y_i^2}\right)$$

In this case, MSD_{HB} represents the mean square deviation, r represents some the number of repetitions, and y_i is the value of experiment results.

The *normal-the-best:*

$$(S/N)_{NB} = -10\log(MSD_{NB}) \tag{2}$$

In Eqn. (2), MSD_{NB} is defined as:

$$MSD_{NB} = \frac{1}{r}\sum_{i=1}^{r}(y_i - y_0)^2$$

In this expression, MSD_{NB} is the mean square deviation, and y_0 represents the standard values or target values.

The *lower-the-better:*

$$(S/N)_{LB} = -10\log(MSD_{LB}) \tag{3}$$

In Eqn. (3), MSD_{LB} is defined as:

$$MSD_{LB} = \frac{1}{r}\sum_{i=1}^{r}y_i^2$$

In this equation, MSD_{LB} represents the mean square deviation, and $\sum_{i=1}^{r}y_i^2$ is the overall typical value of each experiment. In this study, the S/N ratio of the output parameters of the experiment are identified in Table 10.1.

TABLE 10.1 Characteristics of Result

SL. No.	Characteristics	Type of Characteristic	Unit
1.	MRRz	Higher-the-better	mm^3/min
2.	TWR	Lower-is-better	mm^3/min
3.	HV	Higher-the-better	HV
4.	Ra	Lower-is-better	μm

In the current study, the interaction effect of the input parameters was considered, as shown in Table 10.2. In the field of PMEDM, researchers have studied the effect of powder size, workpiece material, electrode material, current, pulse-on time, and pulse-off time. In this study, the interaction terms were considered, specifically workpiece material, x-electrode material (AXB), workpiece material, x-powder concentration (AxG), and electrode material x-powder concentration (BxG).

Taguchi's orthogonal arrays were used for designing the experiments. There are many orthogonal array's available in the Taguchi's method, therefore selection depended upon the number of factors and degrees of freedom of each factor. In this study, seven main factors were considered, out of which, two factors were at two levels, each having one degree of freedom. Five of the main factors had three levels, with each having two degrees of freedom. Also, the study considered three interaction terms. Thus, the total sum of degrees of freedom, including the main factors as well as the interaction terms, was 20. Therefore, based on the 20 degrees of freedom, the L_{27} orthogonal array suited the present requirements as it had 26 degrees of freedom. The remaining 6 degrees of freedom were assigned as random error. The L_{27} orthogonal array had 13 columns, each with 2 degrees of freedom. Coefficient A was assigned to the column 1, B to column 2, G to column 5, C to column 9, D to column 10, E to column 12, and F to column 13, as shown in Table 10.3.

TABLE 10.2 Input Parameters and its Levels

SL. No.	Factors	Symbols	Level			DOF
			Level 1	Level 2	Level 3	
1.	Workpiece material	A	SKD61	SKD11	SKT4	2
2.	Electrode material	B	Cu	Cu*	Gr	1
3.	Polarity	C	–	+	–*	1
4.	Pulse-on time (μs)	D	5	10	20	2
5.	Current (A)	E	8	4	6	2
6.	Pulse-off time (μs)	F	38	57	85	2
7.	Powder concentration Ti (g/l)	G	0	10	20	2
8.	Interaction of workpiece material and tool material	A×B	–	–	–	2
9.	Interaction of workpiece material and powder concentration	A×G	–	–	–	4
10.	Interaction of tool material and powder concentration	B×G	–	–	–	2
11.	Total					20

10.3.2 GREY RELATIONAL ANALYSIS

The Grey relational theory is effective in studying subjects with limited information, whether not enough or uncertain. The gray relational analysis (GRA) can effectively determine the complex relationship between the quality characteristics specified. Recently, Grey relational analysis has become a powerful tool in analyzing the process of considering multiple quality characteristics. To optimize the process parameters for multiple performance characteristics in PMEDM, such as MRR, TWR, surface roughness (Ra), and hardness of surface layer (HV), Grey relational analysis was used along with the Taguchi method, following the steps provided.

TABLE 10.3 Experimental Design based on L_{27} Orthogonal Array

Exp. No.	Workpiece Material	Electrode Material	Electrode Polarity	Pulse on Time (μs)	Pulse Current (A)	Pulse of Time (μs)	Powder Concentration (g/l)
1	SKD61	Cu	–	5	8	38	0
2	SKD61	Cu	+	10	4	57	10
3	SKD61	Cu	–*	20	6	85	20
4	SKD61	Cu*	+	10	6	85	0
5	SKD61	Cu*	–*	20	8	38	10
6	SKD61	Cu*	–	5	4	57	20
7	SKD61	Gr	–*	20	4	57	0
8	SKD61	Gr	–	5	6	85	10
9	SKD61	Gr	+	10	8	38	20
10	SKD11	Cu	+	20	4	85	0
11	SKD11	Cu	–*	5	6	38	10
12	SKD11	Cu	–	10	8	57	20
13	SKD11	Cu*	–*	5	8	57	0
14	SKD11	Cu*	–	10	4	85	10
15	SKD11	Cu*	+	20	6	38	20
16	SKD11	Gr	–	10	6	38	0
17	SKD11	Gr	+	20	8	57	10
18	SKD11	Gr	–*	5	4	85	20
19	SKT4	Cu	–*	10	6	57	0
20	SKT4	Cu	–	20	8	85	10
21	SKT4	Cu	+	5	4	38	20
22	SKT4	Cu*	–	20	4	38	0
23	SKT4	Cu*	+	5	6	57	10
24	SKT4	Cu*	–*	10	8	85	20
25	SKT4	Gr	+	5	8	85	0
26	SKT4	Gr	–*	10	4	38	10
27	SKT4	Gr	–	20	6	57	20

* *Dummy treated.*

➢ **Step 1: Pre-processing Data:** Data pre-processing involved transforming an original sequence into a comparable sequence. A series of various units were transformed to dimensionless quantities. Experimental results were thus normalized in a range of 0–1. Usually, each series would be normalized by dividing the data in the original series by their average.

The original reference sequence and sequence for comparison were represented as $xo(k)$ and $xi(k)$, $i = 1, 2, \ldots, m$; $k = 1, 2, \ldots, n$, respectively, where m was the total number of experiments to be considered, and n was the total number of observed data. Data pre-processing converted the original sequence to a comparable sequence. If the target value of the original sequence was "higher-the-better," then the original sequence was normalized as follows:

$$x_i^*(\mathrm{k}) = \frac{x_i^0(\mathrm{k}) - \min x_i^0(\mathrm{k})}{max\, x_i^0(\mathrm{k}) - \min x_i^0(\mathrm{k})} \quad (4)$$

In Eqn. (4), $x_i^*(\mathrm{k})$ represents the sequence data after handling; x_i^0 represents the original sequence of data; $max\ x_i^0(\mathrm{k})$ represents the largest value of $x_i^0(\mathrm{k})$; and min $x_i^0(\mathrm{k})$ represents the smallest value of $x_i^0(\mathrm{k})$.

If the purpose was "lower-the-better," then the original sequence was normalized as follows:

$$x_i^*(\mathrm{k}) = \frac{max\, x_i^0(\mathrm{k}) - x_i^0(\mathrm{k})}{max\, x_i^0(\mathrm{k}) - \min x_i^0(\mathrm{k})} \quad (5)$$

Furthermore, if there was a specific target value, then the original sequence was normalized using (6):

$$x_i^*(\mathrm{k}) = \frac{\left|x_i^0(\mathrm{k}) - OB\right|}{\max.\left\{max\, x_i^0(\mathrm{k}) - \mathrm{OB}; \mathrm{OB} - \min x_i^0(\mathrm{k})\right\}} \quad (6)$$

In Eqn. (6), OB represents the target value.

- **Step 2: Grey Relational Coefficient (GRC):** It was calculated to express the relationship between the best, reference, and actual normalized S/N ratio. The gray relational coefficient was expressed using equation as given:

$$\gamma_{0,i}(\mathrm{k}) = \frac{\Delta_{\min} - \zeta.\Delta_{max}}{\Delta_{0i}(\mathrm{k}) + \zeta.\Delta_{max}} \qquad 0 \le \gamma_{0,i}(\mathrm{k}) \le 1 \quad (7)$$

In Eqn. (7), $\Delta_{0i}(\mathrm{k})$ represents the deviation sequence of reference sequence $x_0^*(\mathrm{k})$ and comparability sequence $x_i^*(\mathrm{k})$, $\Delta_{0i}(\mathrm{k}) = |x_0^*(\mathrm{k}) - x_i^*(\mathrm{k})|$ represents the absolute value of the difference between $x_0^*(\mathrm{k})$

and $x_i^*(k)$, Δ_{min} = min.min $\Delta_{0i}(k)$, Δ_{max} = max.max $\Delta_{0i}(k)$, ζ is the distinguishing coefficients $\zeta \in |0, 1, i = 1, 2, \ldots, m$ where; 'm' is the total number of experiment to be considered; and k = 1, 2, …, n where; 'n' is the total number of observation data.

- **Step 3: Grey Relational Grade:** This was a weighted sum of the grey relational coefficient (GRC). The quality characteristics were evaluated on the basis of overall gray level relationship and were determined by (8):

$$\Gamma_{0,i} = \sum_{k=1}^{n} W_k \cdot \gamma_{0,i}(k) \tag{8}$$

In this equation, W_k represents the weights of the kth machining characteristic, and $\sum_{k=1}^{n} W_k = 1$. Furthermore, $\gamma_{0,i}(k)$ represents the GRC of the kth machining characteristic.

- **Step 4: Optimal Grey Relational Grade Value:** In order to predict the optimal values of the Grey relational grade, which provides an index of multimachining characteristics, only significant parameters were included. These were found by utilizing an ANOVA. The optimal grey relational grade, (η_{op}) was predicted using Eqn. (9) as described below:

$$\eta_{op} = \eta_m + \sum_{i=1}^{q} (\eta_i - \eta_m) \tag{9}$$

where; η_m is the average grey relational grade; η_i represents the average gray relational grade at the optimum level; and q is the number of parameters that significantly affects the machining characteristics.

- **Step 5: Confirming Experiments:** The results of this step were used to evaluate the exact results of the optimization model.

10.4 RESULTS AND DISCUSSION

10.4.1 RESULT OF EXPERIMENTS

Each experiment was repeated three times, and Minitab software was used to analyze the experimental results. The results were processed by Minitab 17 to determine the mean value of the machining characteristics as well as the coefficient, S/N. The results are given in Table 10.4.

TABLE 10.4 Results of Experiments

Exp.	MRR		TWR		SR		HV	
No.	$\overline{MRR}$	S/N	$\overline{TWR}$	S/N	$\overline{R}_a$	S/N	$\overline{HV}$	S/N
1	10.487	20.385	1.95	–5.878	3.35	–10.50	506.7	54.03
2	8.169	17.650	2.011	–6.859	3.21	–10.15	658.96	56.34
3	3.152	9.859	1.495	–3.572	2.56	–8.16	581.6	55.27
4	10.239	20.186	4.426	–12.922	3.55	–11.00	496.68	53.84
5	14.304	22.957	4.364	–12.806	3.61	–11.14	828.92	58.29
6	0.089	–25.225	0.054	24.565	1.45	–3.27	629.84	55.94
7	37.466	31.465	11.499	–21.216	4.78	–13.60	544.58	54.70
8	23.575	27.376	9.935	–19.951	3.24	–10.20	748.42	57.46
9	38.843	31.637	19.626	–25.857	4.35	–12.78	626.18	55.91
10	18.882	25.511	2.01	–61.877	4.16	–12.38	509.72	54.12
11	3.857	11.538	1.179	–14.855	2.05	–6.24	679.54	56.33
12	14.496	22.263	3.56	–11.032	3.20	–10.10	664.2	56.43
13	10.608	20.505	2.25	–7.341	3.35	–10.51	546.02	54.70
14	0.320	–11.925	0.132	16.988	2.04	–6.21	679.2	56.54
15	23.577	26.797	1.495	–3.831	4.57	–13.20	655.18	56.29
16	23.885	25.272	7.439	–17.677	4.57	–13.20	469.82	53.41
17	59.669	35.498	14.073	–22.98	4.45	–12.97	907.64	59.12
18	17.159	24.683	5.491	–14.818	2.74	–8.77	683.52	56.68
19	1.252	1.623	0.587	43.491	2.55	–8.12	530.72	54.47
20	20.745	26.301	5.078	–14.114	4.31	–12.70	624.58	55.78
21	4.374	12.707	2.902	–9.298	2.46	–7.86	631.68	55.93
22	0.198	–14.342	0.277	10.402	2.26	–7.09	468.04	53.35
23	6.782	16.544	4.715	–13.492	2.89	–9.23	544.38	54.69
24	19.682	25.866	4.413	–12.923	3.50	–10.89	613.84	55.72
25	10.649	20.526	4.537	–13.135	3.23	–10.19	445.44	52.90
26	25.970	28.281	9.041	–19.202	3.24	–10.20	681.22	56.6233
27	54.360	34.705	14.581	–23.35	5.65	–15.05	832.66	58.2904

10.4.2 OPTIMUM MACHINING CHARACTERISTICS CALCULATED FROM THE TAGUCHI METHOD

10.4.2.1 ANOVA FOR S/N RATIO

An ANOVA was verified for the signal-to-noise ratio of both the main and interaction terms. Table 10.5 illustrates the ANOVA results for S/N for the

MRR, TWR, SR, and HV with 90% confidence. The F values of the parameters have shown which parameters were the most influential on providing the optimal values of the MRR, TWR, SR, and HV.

TABLE 10.5 ANOVA for S/N Ratio for MRR, TWR, SR, and HV

Source	F			
	MRR	TWR	SR	HV
Material workpiece (A)	–	–	–	–
Material electrode (B)	16.39	25.48	55.47	6.71
Polarity (C)	–	–	15.1	–
Pulse on time (D)	–	–	35.72	3.56
Current (E)	5.62	4.71	20.97	–
Pulse of time (F)	–	–	–	–
Concentration (G)	–	–	–	50.82
A×B	–	–	–	–
A×G	4.88	3.51	18.12	3.44
B×G	–	–	–	4.51

10.4.2.2 OPTIMUM MACHINING CHARACTERISTICS

10.4.2.2.1 Optimum Material Removal Rate (MRR)

Figure 10.2 describes the influence of each parameter of the coefficient S/N of the MRR. These results indicated that the SKD11 workpiece material (A2), the Gr electrode material (B2), the positive polarity of the electrodes (C2), the pulse generator at a time of 20μs (D3), the current 8A (E3), the time stopped pulse of 85μs (F3), and the starch concentration of 10 g/L (G2), were the parameters that contributed to the largest coefficient value S/N. The image shown in Figure 10.3 describes the influence of the interaction between the parameters of the factor S/N of the MRR. The results indicated that the interaction between the steel electrode (SKD61) and Gr (A1×B2), the interaction of the SKD11 steel with titanium powder with a concentration of 20 g/L (A3×G3), the Gr electrode which interacted with the titanium powder with a concentration of 20 g/L (B2×G3), were the interactions which had the greatest impact on values of the largest S/N. This resulted in a good influence on the optimum value of the MRR.

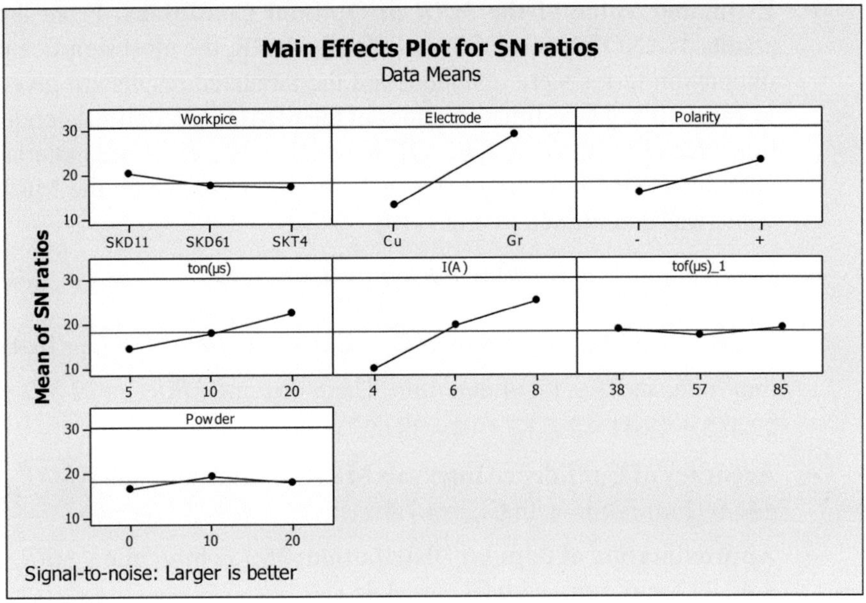

FIGURE 10.2 Main effects plot for S/N of MRR.

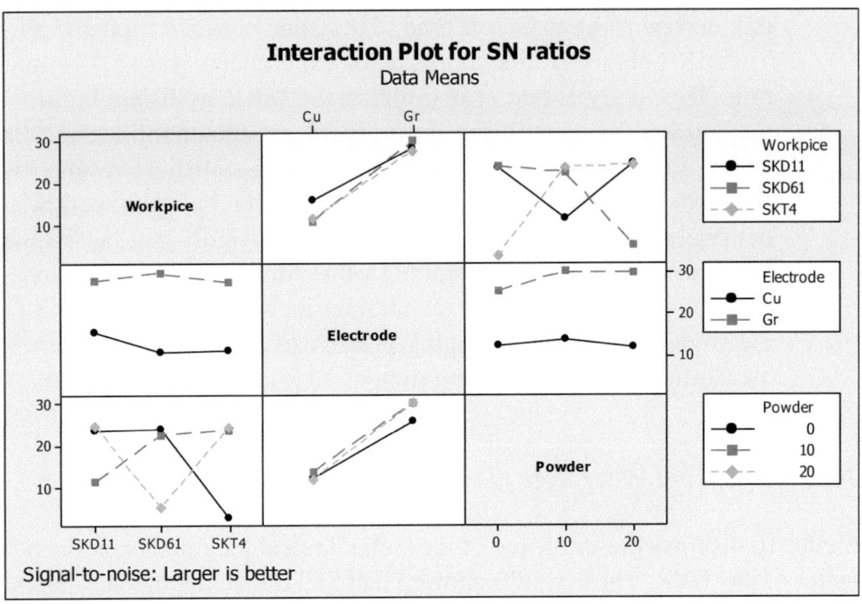

FIGURE 10.3 Interaction plot for S/N of MRR.

- **Estimated Value of the $\overline{MRR}$ at Optimal Conditions:** From the results of ANOVA on the S/N ratio for the MRR, the most significant interaction terms were obtained, and the tabulated results are given in Table 10.5. The estimated values of the MRR under optimal conditions, A2, B2, C2, D3, E3, F3, G2, in which only 2 electrode material parameters (B) and current (E) had a strong influence on the MRR value was determined by Eqn. (10):

$$\mu_{B_2,E_3,A_2\times G_3} = \overline{MRR_{OP}} = \overline{B_2} + \overline{E_3} + \overline{A_2 \times G_3} - 2\times\overline{T} \tag{10}$$

In this equation, $\overline{B_2}$ mm³/min, $\overline{E_3}$ = 22.165 mm³/min, $\overline{A_2 \times G_3}$ =18.41 mm³/min, and $\overline{T}$ = 17.14 mm³/min. Therefore, and $\overline{MRR}_{OP}$ = 32.397 + 22.165 + 18.41 – 2 × 17 × 14 = 38.692 mm³/min.

- **Accuracy of Confidence Intervals MRR:** 31.292 mm³/min $\leq \overline{MRR}_{OP}$ ≤46.092 mm³/min with CI_{CE}= ±7.4.
- **Approximation of Popular Distribution:** 35.152 mm³/min $\leq \overline{MRR}_{OP}$ ≤ 42.232 mm³/min, with CI_{POP}=±3.45.
- **Confirmation Experiments:** These were carried out using the optimal process parameters as used in the material workpiece (SKD11), material electrode (Gr), reverse polarity, pulse-on time (20 µs), current (8A), pulse-off time (85µs), and concentration (10 g/L). The MRR obtained after confirmation experiments was 45.734 mm³/min. This study aimed at optimizing the MRR by using Taguchi's method and to identify the most significant factor influencing the MRR. Based on both the experimental and statistical results, the electrode materials, current, and the interaction between workpiece materials and titanium powder were the most significant factors that influenced the MRR. The optimal value for the MRR was obtained at 45.734 mm³/min, where conditions included the use of the Gr electrode during the machining process of SKD11 steel, a current measuring 8A, and a concentration of 10 g/L of titanium powder.

10.4.2.2.2 Tool Wear Rate (TWR)

Figure 10.4 shows the influence of the technological parameters on the S/N factor of the TWR. These results showed that SKD11 steel materials, copper electrode materials, cathode polarization, time pulse of 5µs, current of 4A, pulse of 57 µs downtime, and concentration of titanium powder of 0.0 g/L have a positive influence on the S/N factor of the TWR. These parameters greatly impacted the optimal results of the TWR. Figure 10.5 illustrates the influence of the interaction between the parameters on the S/N of the TWR. Results indicated that interaction of SKD11 steel with Cu electrode materials, the interaction of SKT4 steel with titanium powder concentration of 0 g/L, the interaction of the Cu electrode materials with titanium powder concentration of 20 g/L were the interactions with the strongest influence on the S/N factor of the TWR.

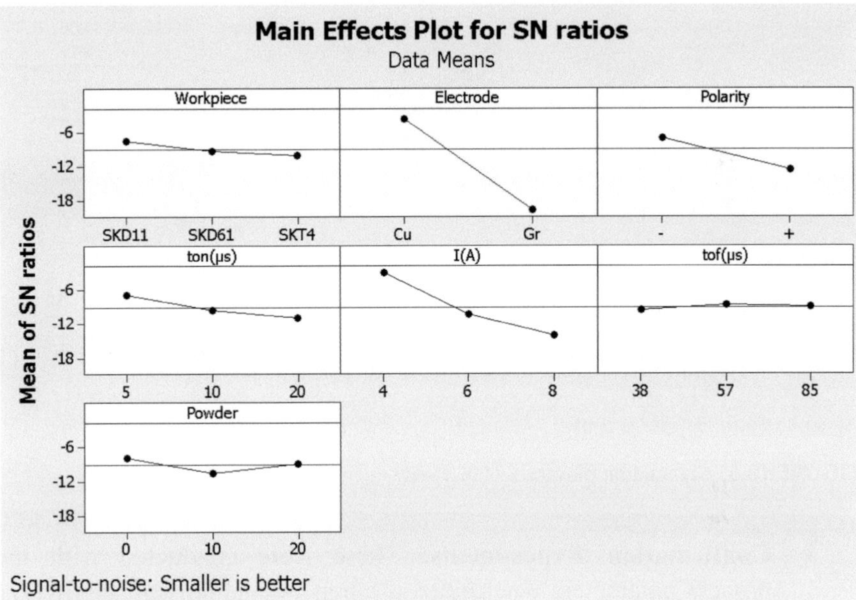

FIGURE 10.4 Main effects plot for S/N of TWR.

- **Estimated Value of the $\overline{TWR}$ at Optimal Conditions:** Reasonable process parameters of the MRR consisted of the following: A2, B1, C1, D2, E1, F2, and G2, of which two electrode material parameters (B) and current (E) had a strong influence on the TWR. The TWR's value was determined by the formula given in Eqn. (11):

$$\mu_{B_2,E_3,G_1,B_2\times G_1} = \overline{TWR}_{op} = \overline{B_1} + \overline{E_1} + \overline{A_3 \times G_3} - 2\times\overline{T} \tag{11}$$

In this equation, $\overline{B_1}$ = 2.383 mm³/min; $\overline{E_1}$ = 3.713 mm³/min; $\overline{A_3 \times G_3}$ = 7.30 mm³/min; $\overline{T}$ = 5.152 mm³/min. Consequently, $\overline{TWR}_{op}$ = 2.383 + 3.713 + 7.30 – 2 × 5.152 = 3.092 mm³/min. Furthermore, the predicted confidence interval for the confirmation experiments was 0.69 mm³/min $\leq \overline{TWR}_{op} \leq$ 5.492 mm³/min with $CI_{CE} = \pm 2.4$. Additionally, the 90% confidence interval of the population was 2.542 mm³/min $\leq \overline{TWR}_{op} \leq$ 3.642 mm³/min with $CI_{POP} = \pm 0.55$.

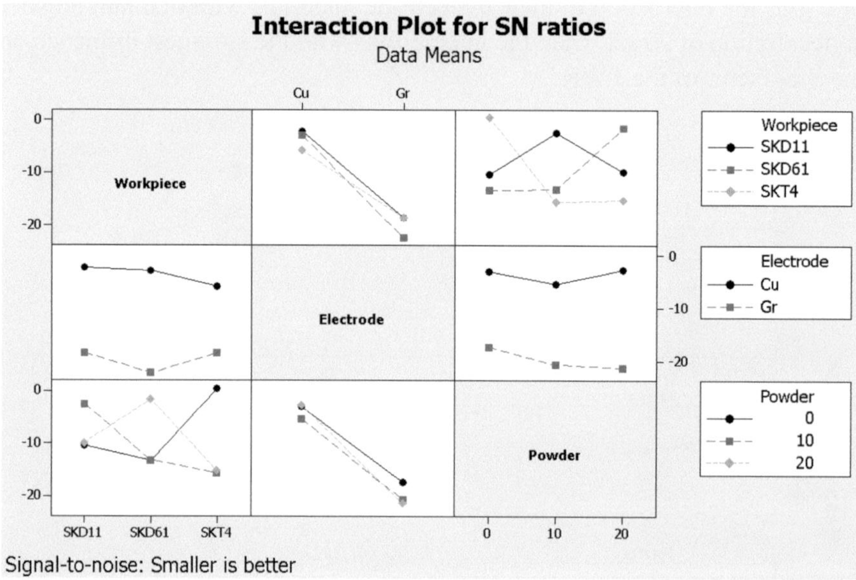

FIGURE 10.5 Interaction plot for S/N of TWR.

- **Confirmation Experiments:** These were conducted with the process parameters determined through calculations of the SKD11 workpiece material, the Cu electrode, the electrode polarization agreement, pulse duration of 10 μs, a current of 4A, the horizontal development pulse of 57 μs, and powder concentration of 10 g/L. The results of the TWR = 2.93 mm³/min, and the difference between the calculated results and the experimental results was 4.1%.

10.4.2.2.3 Surface Roughness (SR)

The greater S/N factor showed that research results were less affected by noise. The S/N factor value for Ra was affected by the parameters shown in Figure 10.6. The results showed that the steel materials, SKT4 (A3), the Cu electrode material (B1), the electric polarization negative (C1), with a time pulse of 5μs (D1), the current flow of 4A (E1), a downtime pulse of 85μs (F3), and a titanium powder concentration of 10 g/L (G2), positively influenced the S/N factor for Ra. These process parameters decreased the degree to which the Ra was affected by the noise, and consequently resulted in the most optimal value for Ra with the least amount of alteration. Figure 10.7 demonstrates the influence of the interaction between the parameters to S/N factor for Ra. The results indicated that the interactions between the SKT4 steel and the copper electrode materials (A3×B1), the interactions between the SKD61 steel and the titanium powder concentration of 20 g/L (A1×G3), and the interactions between the Cu electrode materials and the titanium powder concentration of 20 g/L (B1×G3), included the pairs of interactions that influenced the S/N factors for Ra.

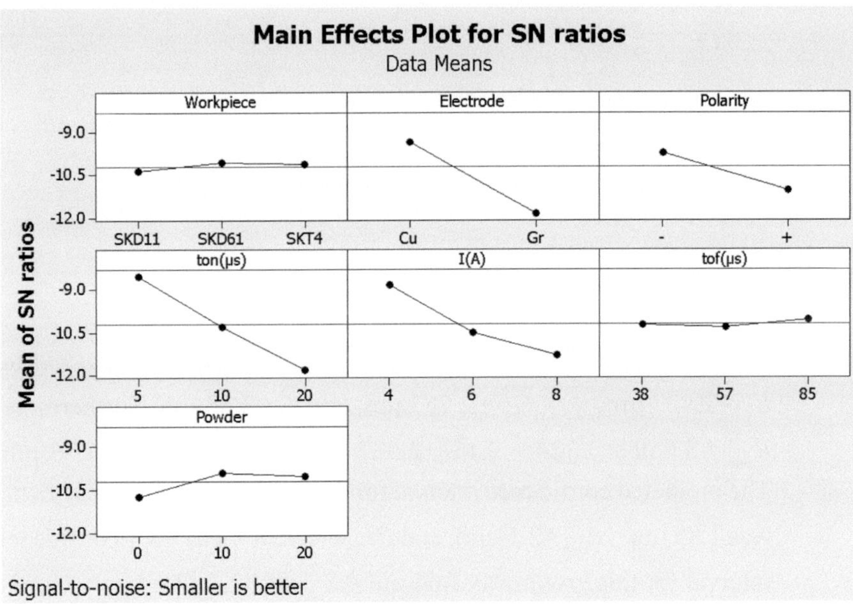

FIGURE 10.6 Main effects plot for S/N of SR.

➢ **Estimated Value of $\overline{R}_a$ under Optimal Conditions:** Process parameters for Ra included the following: A1, B1, C1, D1, E1, F2, and G1. Under these conditions, the electrode material parameters (B), the electrode polarization (C), the time pulse generator, and the current flow (E) had strong influences on Ra. Therefore, the Ra value was determined by the formula (12):

$$\mu_{B_2,E_3,G_1,B_2\times G_1} = \overline{R}_{aop} = \overline{B_1} + \overline{C_1} + \overline{D_1} + \overline{E_1} + \overline{A_1 \times G_1} - 4\times\overline{T} \quad (12)$$

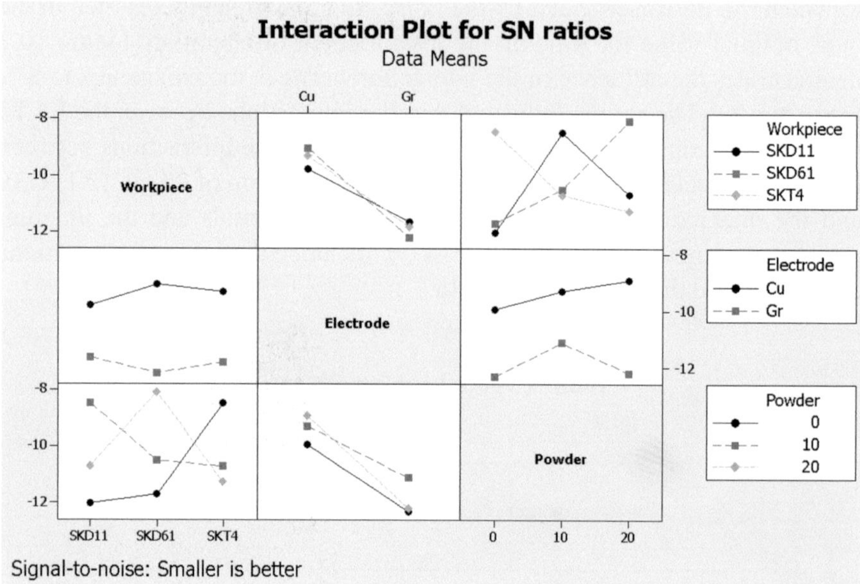

FIGURE 10.7 Interaction plot for S/N of SR.

In this expression, $\overline{B_1}$ = 3.059 μm, $\overline{C_1}$ = 3.324 μm, $\overline{D_1}$ = 3.153 μm, $\overline{E_1}$ = 2.927 μm, $\overline{A_1 \times G_3}$ = 2.787 μm, and $\overline{T}$ = 3.38 μm. Furthermore, $\overline{R}_{aop}$ = 3.059 + 3.324 + 3.153+2.927 + 2.787 – 4 × 3.38 = 1.73 μm. The predicted confidence interval for these confirmation experiments was 1.35 μm $\leq \overline{R}_{aop} \leq$2.11 μm, and $CI_{CE} = \pm 0.387$. The 90% confidence interval for this group was 1.64 μm $\leq \overline{R}_{aop} \leq$1.82, and $CI_{POP} = \pm 0.09$.

- **Confirmation Experiments:** These were conducted where the process parameters were determined through calculations of the SKD61 workpiece material, the Cu electrode, the electrode polarization upon, a pulse generator time of 5 μs, a current 4A, the horizontal development pulse of 85 μs, and powder concentration of 10 g/L. The result of Out = 1.47 μm demonstrated high accuracy, with discrepancies between the calculated and experimental results of 1.5%.

10.4.2.2.4 Micro-Hardness of Machined Surface

The S/N factor of the HV which was affected by the parameters is given in Figure 10.8. The results showed that the SKD11 steel (A2), the Gr electrode material (B2), the negative electrode polarization (C1), a time pulse of 20μs (D3), a current flow of 8A (E3), a pulse-off time of 57μs (F2), and a titanium powder concentration of 10 g/L (G2) had a positive influence on the S/N factor for the HV. Figure 10.9 shows the influence of the interaction between the parameters of the S/N factor for the HV. The results indicated that the interaction between the SKD11 steel and the Gr electrode materials (A2×B2), the interaction between the SKD61 steel and the titanium powder concentration of 10 g/L (A1×G2), the interaction between the electrode material and the titanium powder concentration of 10 g/L (B2×G2), were the pairs of interactions that influenced the S/N factor of the HV.

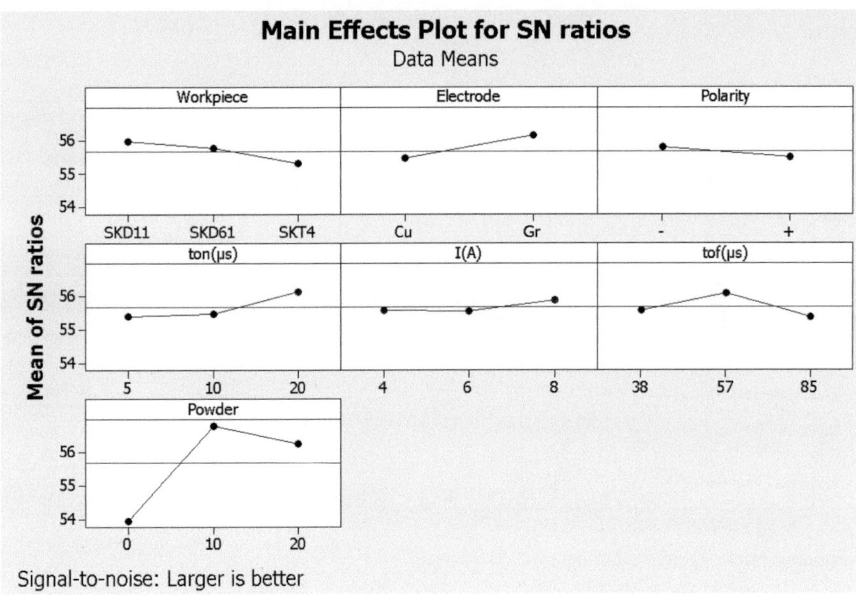

FIGURE 10.8 Main effects plot for S/N of HV.

- **Estimated Value of $\overline{R_a}$ at Optimal Conditions:** Reasonable process parameters of the MRR included A1, B2, C1, D3, E3, F2, and G2. It was determined that three parameters, including the electrode materials (B), the time pulse generator (D), and the Ti powder concentration (G), had a strong influence on the HV. The value was determined by the Eqn. (13):

$$\mu_{B_2,E_3,G_1,B_2\times G_1} = \overline{HV}_{op} = \overline{B_2} + \overline{D_3} + \overline{G_2} + \overline{A_1 \times G_2} + \overline{B_2 \times G_2} - 4\times\overline{T} \tag{13}$$

In this equation, $\overline{B_2}$ = 659.9 HV, $\overline{D_3}$ = 661.4 HV, $\overline{G_2}$ = 705.9 HV, $\overline{A_1 \times G_2}$ = 645.433 HV $\overline{B_2 \times G_2}$ = 679.03 HV, $\overline{T}$ = 621.825 HV. Additionally, $\overline{HV}_{toiuu}$ =659.9 + 661.4 + 705.9 +645.433+679.03 – 4 × 621.825 = 864.363 HV. The predicted confidence interval for the confirmation experiments was 789.81 HV ≤ $\overline{HV}_{op}$ ≤ 938.913 HV and CI_{CE} = ±74.55. The 90% confidence interval for the group was 830.14 HV ≤ $\overline{HV}_{op}$ ≤898.58 HV, and CI_{POP} = ±34.22.

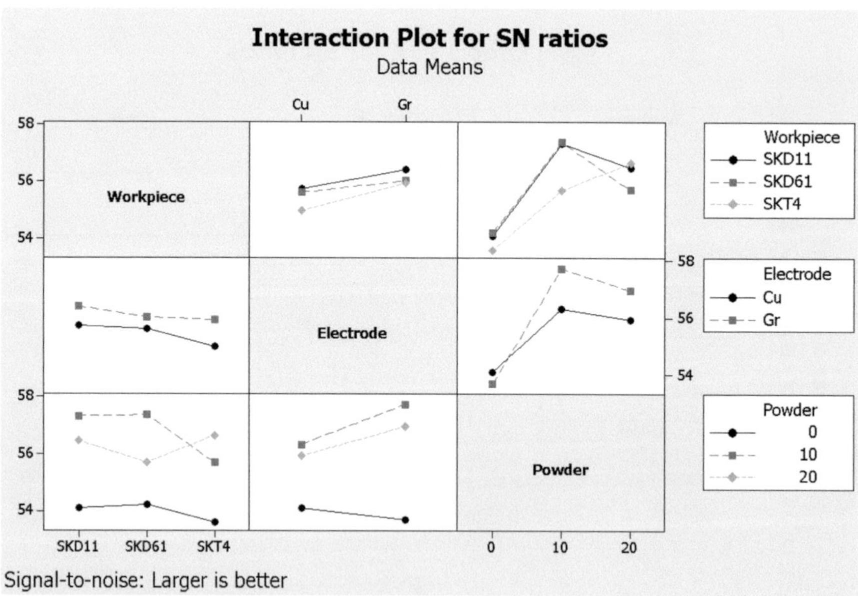

FIGURE 10.9 Interaction plot for S/N of HV.

- **Confirmation Experiments:** These of microscopic hardness surface layer processing were conducted with the process parameters as determined through calculations of the SKD11 workpiece material, the Gr electrode, the electrode polarization, a pulse generator time of 20 μs, a current of 8A, a horizontal pulse duration of 57 μs, and a powder concentration of 10 g/L. The results for HV were HV = 917.87 and discrepancies between the calculated and experimental results were 5.8%.

10.4.2.3 MULTI-CHARACTERISTIC OPTIMIZATION USING GRA

- **Step 1: Connecting the Taguchi Method with GRA:** In the present work, the S/N ratios of the machining characteristics have been normalized with a "larger-the-better" type characteristic. Therefore, Eqn. (1) was utilized in order to normalize the data. The S/N ratios of four machining characteristics for all 27 sequences are listed in Table 10.6. Using Eqn. (1) with m = 27 and n = 4, the S/N ratios have been normalized as shown in Table 10.6. Therefore, the larger normalized S/N ratio corresponds to a higher performance and the best-normalized S/N ratio was equal to unity.

TABLE 10.6 Sequence After Data Pre-Processing

Exp. No.	S/N Ratio Normalization			
	MRR	**TWR**	**SR**	**HV**
	Reference Sequence			
	1.0000	**1.0000**	**1.0000**	**1.0000**
1	0.7511	0.3962	0.3862	0.1811
2	0.7061	0.3768	0.4160	0.5536
3	0.5778	0.4420	0.5849	0.3802
4	0.7478	0.2565	0.3438	0.1513
5	0.7935	0.2588	0.3319	0.8673
6	0.0000	1.0000	1.0000	0.4879
7	0.9336	0.0920	0.1231	0.2898
8	0.8662	0.1171	0.4117	0.7334
9	0.9364	0.0000	0.1927	0.4834
10	0.8355	0.3901	0.2267	0.1958
11	0.6054	0.4834	0.7479	0.5512
12	0.7820	0.2940	0.4202	0.5676
13	0.7531	0.3672	0.3854	0.2885
14	0.2191	0.8497	0.7504	0.5848
15	0.8567	0.4368	0.1570	0.5444

TABLE 10.6 *(Continued)*

Exp. No.	S/N Ratio Normalization			
	MRR	TWR	SR	HV
	Reference Sequence			
	1.0000	**1.0000**	**1.0000**	**1.0000**
16	0.8316	0.1622	0.1570	0.0823
17	1.0000	0.0571	0.1766	1.0000
18	0.8219	0.2189	0.5331	0.6079
19	0.4422	0.5991	0.5883	0.2525
20	0.8486	0.2329	0.1995	0.4622
21	0.6247	0.3284	0.6104	0.4867
22	0.1792	0.7191	0.6757	0.0716
23	0.6879	0.2452	0.4941	0.2876
24	0.8414	0.2565	0.3531	0.4526
25	0.7534	0.2523	0.4126	0.0000
26	0.8811	0.1320	0.4117	0.5977
27	0.9869	0.0497	0.0000	0.8659

- **Step 2: Grey Relational Coefficient (GRC):** The factor "ζ" was set to 0.5 in this study, while m = 27, n = 4. Using the results of the S/N normalization in Table 10.3 to determine Δ_{0i}(k), the results of Δ_{0i}(k) are given in Table 10.7.
An analysis the results of Table 10.7 shows the following:

$$\Delta_{max} = \Delta_{06}(1) = \Delta_{09}(2) = \Delta_{25}(4) = \Delta_{27}(3) = 1,0000$$

$$\Delta_{min} = \Delta_{17}(1) = \Delta_{06}(2) = \Delta_{06}(4) = \Delta_{17}(3) = 0,0000$$

We replaced the value calculated in Eqn. (3) to determine the $\gamma_{0,i}$ (k). The results are included in Table 10.7. The GRC was determined by the average value of $\gamma_{0,i}$(k).
- **Step 3: Analysis of the Grey Relational Grade and a Selection of Optimal Level of Parameters:** Figure 10.10 illustrates the grey relational grade for MRR_{max}, TWR_{min}, Ra_{min}, HV_{max}, where the grey relational grade was larger for better quality characteristics. The Grey relational grade value was the largest for experiment 6, where GRG = 0.7068. Furthermore, 4 of the 27 experimental results provided the best quality characteristics: 6, 14, 17, and 27.

TABLE 10.7 The Deviation Sequences and Grey Relational Coefficients and Grade Values

Exp. No.	Deviation Sequences				Grey Relational Coefficients				$\bar{\gamma}$	Rank
	$\gamma_{0i}(1)$	$\gamma_{0i}(2)$	$\gamma_{0i}(3)$	$\gamma_{0i}(4)$	MRR	TWR	SR	HV		
1	0.2489	0.6038	0.6138	0.8189	0.6676	0.4530	0.4489	0.3791	0.4872	23
2	0.2939	0.6232	0.5840	0.4464	0.6298	0.4452	0.4613	0.5283	0.5161	14
3	0.4222	0.5580	0.4151	0.6198	0.5422	0.4726	0.5464	0.4465	0.5019	18
4	0.2522	0.7435	0.6562	0.8487	0.6647	0.4021	0.4325	0.3707	0.4675	25
5	0.2065	0.7412	0.6681	0.1327	0.7077	0.4028	0.4280	0.7903	0.5822	5
6	1.0000	0.0000	0.0000	0.5121	0.3333	1.0000	1.0000	0.4940	0.7068	1
7	0.0664	0.9080	0.8769	0.7102	0.8828	0.3551	0.3631	0.4132	0.5035	17
8	0.1338	0.8829	0.5883	0.2666	0.7889	0.3616	0.4594	0.6522	0.5655	6
9	0.0636	1.0000	0.8073	0.5166	0.8872	0.3333	0.3825	0.4918	0.5237	12
10	0.1645	0.6099	0.7733	0.8042	0.7524	0.4505	0.3927	0.3834	0.4947	19
11	0.3946	0.5166	0.2521	0.4488	0.5589	0.4918	0.6648	0.5270	0.5606	7
12	0.2180	0.7060	0.5798	0.4324	0.6964	0.4146	0.4630	0.5363	0.5276	11
13	0.2469	0.6328	0.6146	0.7115	0.6694	0.4414	0.4486	0.4127	0.4930	22
14	0.7809	0.1503	0.2496	0.4152	0.3904	0.7689	0.6670	0.5463	0.5931	4
15	0.1433	0.5632	0.8430	0.4556	0.7772	0.4703	0.3723	0.5232	0.5358	10
16	0.1684	0.8378	0.8430	0.9177	0.7481	0.3737	0.3723	0.3527	0.4617	27
17	0.0000	0.9429	0.8234	0.0000	1.0000	0.3465	0.3778	1.0000	0.6811	2
18	0.1781	0.7811	0.4669	0.3921	0.7374	0.3903	0.5171	0.5605	0.5513	8
19	0.5578	0.4009	0.4117	0.7475	0.4727	0.5550	0.5484	0.4008	0.4942	20
20	0.1514	0.7671	0.8005	0.5378	0.7676	0.3946	0.3845	0.4818	0.5071	16
21	0.3753	0.6716	0.3896	0.5133	0.5712	0.4268	0.5621	0.4934	0.5134	15
22	0.8208	0.2809	0.3243	0.9284	0.3786	0.6403	0.6066	0.3500	0.4939	21
23	0.3121	0.7548	0.5059	0.7124	0.6157	0.3985	0.4971	0.4124	0.4809	24
24	0.1586	0.7435	0.6469	0.5474	0.7592	0.4021	0.4360	0.4774	0.5187	13
25	0.2466	0.7477	0.5874	1.0000	0.6697	0.4007	0.4598	0.3333	0.4659	26
26	0.1189	0.8680	0.5883	0.4023	0.8079	0.3655	0.4594	0.5541	0.5467	9
27	0.0131	0.9503	1.0000	0.1341	0.9745	0.3448	0.3333	0.7885	0.6103	3

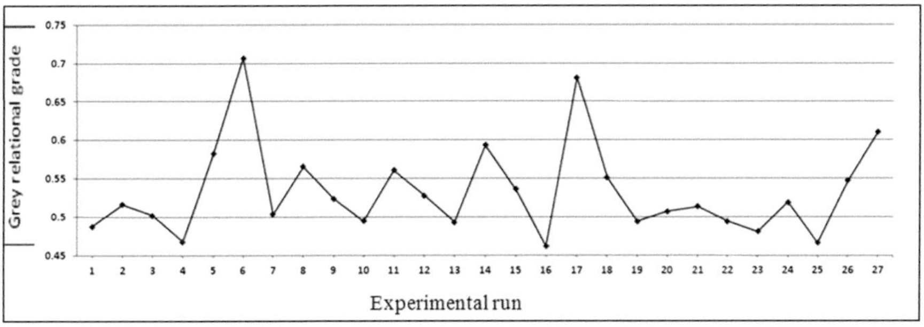

FIGURE 10.10 Grey relational grades for MRR_{max}, TWR_{min}, R_{amin}, HV_{max}.

While the relationship between the processing parameters and the quality characteristics is still unknown, the optimal combination of process parameters can be scaled more accurately. The ANOVA results of the GRC obtained through the Minitab 16 software was used to evaluate the influence of the processing parameters on the machining quality through PMEDM, including the maximum MRR, minimum TWR, minimum Ra, and maximum HV. The results of the ANOVA Grey relational grades were demonstrated with 90% confidence, and are given in Table 10.8. This includes the parameters which demonstrate the strongest influence on the maximum MRR, minimum TWR, minimum Ra, and maximum HV, including the powder concentration of powder (G). Furthermore, Table 10.9 ranks the influence of parameters related to the Grey relational grade, where the powder concentration was the strongest influence and the electrode materials were the smallest. In addition, Figure 10.11 shows the optimal combination of parameters: A2 (SKD11), B2 (Gr), C1(–), D3 (20 μs), E1 (4A), F2 (57 μs), and G2 (10 g/L).

TABLE 10.8 ANOVA for Gray Relational Grade

SL. No.	Parameter	DOF	SS	V	F	F_{table}	P (%)
1.	Material workpiece (A)	2	0.0046	0.0023	0.77	–	–
2.	Material electrode (B)	1	0.0022	0.0022	0.74	–	–
3.	Polarity (C)	1	0.0022	0.0022	0.74	–	–
4.	Pulse on time (D)	2	0.0039	0.0020	0.65	–	–
5.	Current (E)	2	0.0032	0.0016	0.54	–	–
6.	Pulse of time (F)	2	0.0080	0.0040	1.34	–	–
7.	Concentration (G)	2	0.0313	0.0157	5.25	3.463	0.04
8.	AxB	2	0.0031	0.0016	0.52	–	–
9.	AxG	4	0.0130	0.0033	1.33	–	–
10.	BxG	2	0.0049	0.0025	0.82	–	–
11.	Error	6	0.0179	0.0030	–	–	–
12.	Total	26	0.0947	–	–	–	–

➢ **Step 4: Optimal Grey Relational Grade Value:**

$$\eta_{op} = \eta_m + \sum_{i=1}^{1}\left(\eta_i - \eta_m\right) = 0.5327 + (0.5593 - 0.5327) = 0.5593$$

➢ **Step 5: Confirmation Experiments:**

TABLE 10.9 Response Table for Gray Relational Grade

Level	Parameters						
	A	B	C	D	E	F	G
1	0.5443	0.5264	0.5392	0.5361	0.5466	0.5228	0.4846
2	0.5394	0.5455	0.5199	0.5166	0.5198	0.5571	0.5593
3	0.5146	–	–	0.5456	0.5318	0.5184	0.5544
Delta	0.0298	0.0191	0.0193	0.0290	0.0268	0.0386	0.0746
Rank	3	7	6	4	5	2	1

Mean value of gray relation grade = 0.5327.

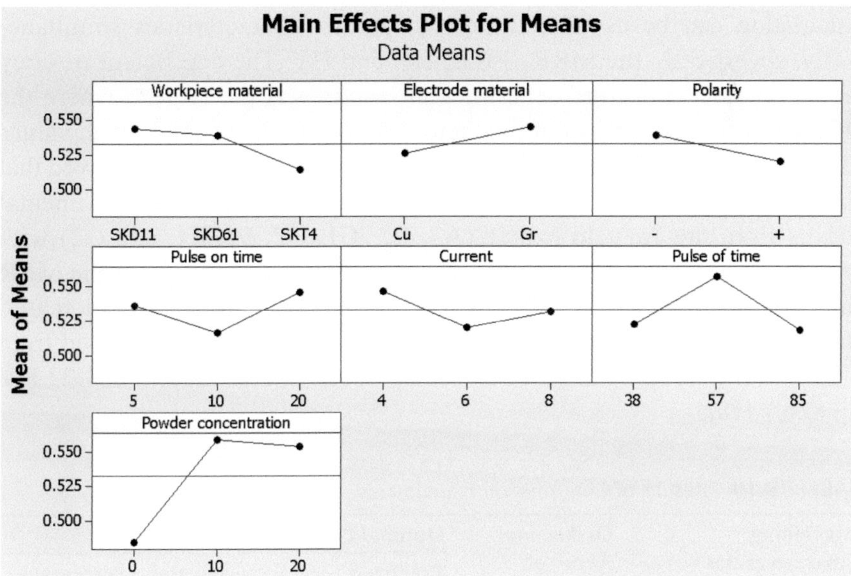

FIGURE 10.11 The main effects of the factors on the gray relational grade.

In order to achieve simultaneous, reasonable values for all four of the quality characteristics, MRR, TWR, Ra, HV, through the PMEDM using the titanium powder mixed in the dielectric fluid, the technological parameters were defined as the SKD11 material workpiece, the graphite electrode, the negative electrode polarity, the pulse-on time of 20 μs, a current of 4A, a pulse-off time of 57 μs, and a concentration of powder 10 g/L (Figure 10.11). The optimal values for the quality characteristics, MRR, TWR, Ra, and HV were defined by Eqn. (14):

$$(\text{MRR, TWR, R}_a\text{, HV})_{OP} = \overline{A}_2 + \overline{B}_2 + \overline{C}_1 + \overline{D}_3 + \overline{E}_1 + \overline{F}_2 + \overline{G}_2 - 6 \times \overline{Y} \quad (14)$$

Rather than using Eqn. (14), we have forecast the results as follows, where MRR = 42.281 mm^3/min, TWR = 9.951 mm^3/min, Ra = 2.903 µs, and HV =817.7 HV.

To evaluate the accuracy of the calculation, two iterative experiments were conducted with each. Comparative experimental results under optimal conditions, with the results in the experimental matrix and the optimization calculations by GRA, is shown in Table 10.10. The results showed that the difference between the values calculated by GRA and those from the experiments was not large, 5.9%. Furthermore, a comparison of the calculated results and the experimental treatment showed that a complete calculation can be used accurately to predict characteristics simultaneously, specifically the MRR, TWR, Ra, and HV. The coefficient of Grey levels obtained from the experimental results was the largest, where the smallest were obtained from the factors related to the characteristic values in the matrix experiments following the Taguchi method. This showed that the processing efficiency improved significantly. Comparing experimental results from the Taguchi matrix (A3, B2, C1, D2, E1, F1, and G2) with experimental results (A2, B2, C1, D3, E1, F2, G2), showed that the MRR increased from 25.970 mm^3/min up 44.784 mm^3/min, the TWR increased from 9.041 mm^3/minute down 9.637 mm^3/minute, the R was reduced from 3.24m down 2.81µm, and the surface hardness increased from 681.22 HV up 826.7 HV.

TABLE 10.10 Results of Confirmation Experiments

Machining Characteristics	Orthogonal Array of Taguchi	Optimal Parameters		
		Predicted Value	Experimental Value	Percentage Difference
Level	A_3, B_2, C_1, D_2, E_1, F_1, G_2	A_2, B_2, C_1, D_3, E_1, F_2, G_2	A_2, B_2, C_1, D_3, E_1, F_2, G_2	
MRR (mm^3/min)	25.970	42.281	44.784	–5.9
TWR (mm^3/min)	9.041	9.951	9.637	–3.2
SR (µs)	3.24	2.90	2.81	3.6
HV (HV)	681.22	817.7	826.7	–1.1
Grey relational grade	0.337	0.559	0.591	–

Figures 10.12–10.14 demonstrate that both the topography and white layer of the machined surface were poor within the optimal parameters. The results

showed that the EDM surface, using Ti powder mixed in the dielectric fluid under optimal parameters, resulted in a number of larger craters; however, the size and depth were both reduced. Debris particles also appeared to be less prevalent on the EDM surface. The number of cracks was reduced, and both the size and depth were smaller. The white layer had a high hardness and uniform thickness. This helped to improve the working of the processing surface. Figure 10.15 represents the electrode surface topography of graphite under both poor and optimal parameters. The Gr electrode surface included a collection of many craters. The topography of the electrode surface under optimum parameters was hypothesized to increase the number of dents, but instead resulted in both a reduced size and depth, which increased the reliability and accuracy of worn electrodes.

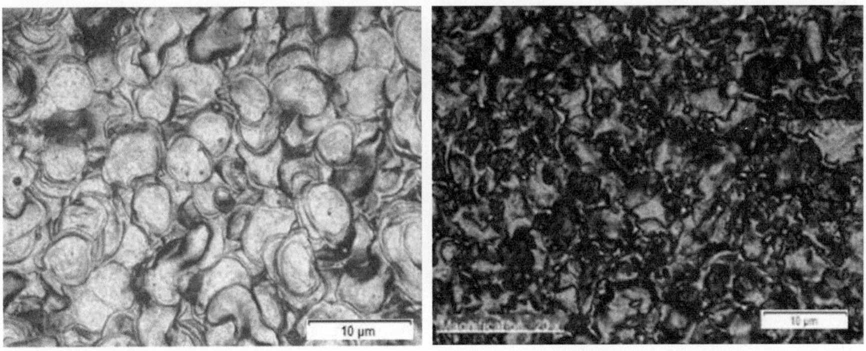

(a) A2, B2, C1, D1, E2, F1, G1; (b) Optimal parameters.

FIGURE 10.12 SEM of SKD11 workpiece surface.

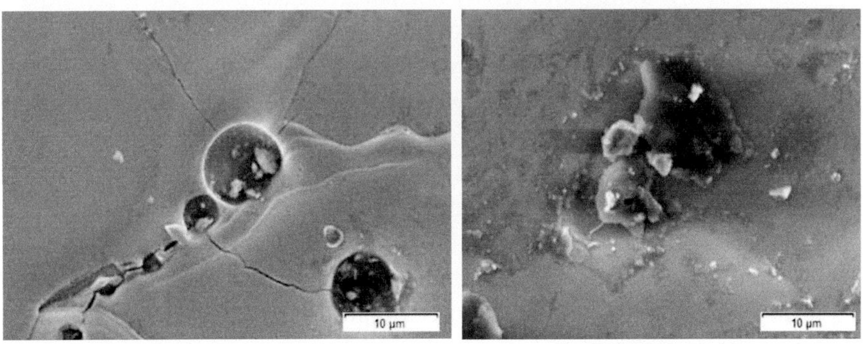

(a) A2, B2, C1, D1, E2, F1, G1; (b) Optimal parameters.

FIGURE 10.13 Micro-cracks on SKD11 workpiece surface.

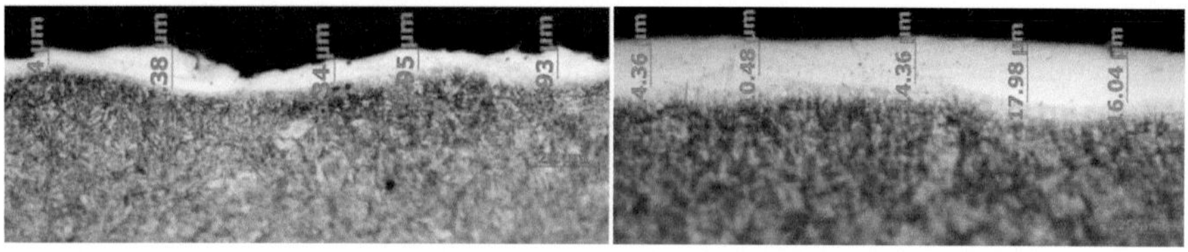

(a) A2, B2, C1, D1, E2, F1, G1; *(b) Optimal parameters.*

FIGURE 10.14 White layer on SKD11 workpiece surface.

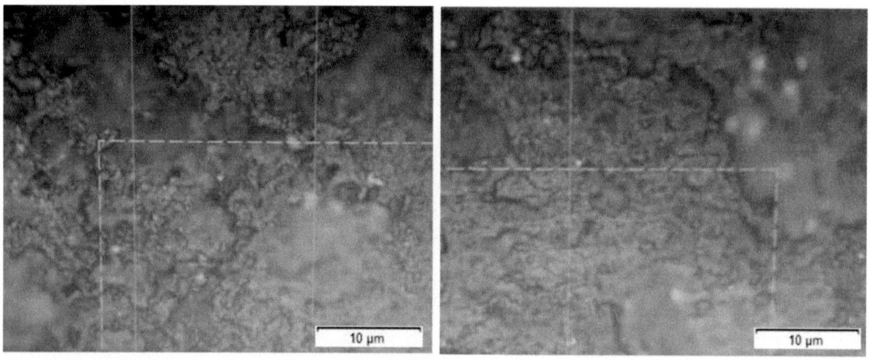

(a) A2, B2, C1, D1, E2, F1, G1; *(b) Optimal parameters.*

FIGURE 10.15 SEM of graphite electrode surface.

10.5 CONCLUSIONS

Research has shown that titanium powder mixed into the dielectric fluid in the EDM has improved the productivity and surface quality of fine machining.

The use of the Taguchi method to optimize individual quality indicators of machining process was refined by PMEDM (MRR, TWR, R_a, HV). Both B and AxG provided the most powerful influence on the S/N coefficient of the four characteristic qualities, including the MRR, Ra, HV, and TWR. The A×B on F had a low impact on all quality characteristics. Parameter E strongly affected the MRR, Ra, and TWR. Parameter C only had a strong influence on Ra, where parameter D influenced both the SR and HV, and parameters G and BxG influenced the HV. This was significant for the application of the PMEDM on the specific production requirements used in manufacturing practices.

The Grey relational analysis (GRA) was employed for simultaneous optimization of multi-machining characteristics in PMEDM as MRR, TWR,

R_a, HV. Using GRA, optimization of the multi-machining characteristics was converted into single characteristic, referred to as the Grey relational grade. Based on an ANOVA for the grey relational grade, values showed that the only parameters where the concentrations of titanium powder had a strong influence on the quality characteristics, were the MRR, TWR, Ra, and HV of the PMEDM. The optimal combination of the process parameters for multiple performance optimization was set at A_2 (SKD11), B_2 (Gr), C_1 (–), D_3 (20μs), E_1 (4A), F_2 (57μs), and G_2 (10 g/L). The surface quality after processing under optimal conditions was acceptable. The reliability trail of the electrode increased and was topography acceptable. This contributed to improving the quality and accuracy of the EDM. The Grey relational analysis was used simultaneously to optimize the characteristics of 4 parameters of PMEDM: MRR, TWR, Ra, and HV. Although the TWR increased, the increase was small. However, the number of grey level factors obtained from the optimal experimental results demonstrated the highest GRA, which showed that the processing efficiency improved significantly. Compared with the results received from the orthogonal array obtained through the Taguchi method, result under optimal conditions through the GRA showed that the MRR increased by 72.44%, the TWR decreased by 6.5%, the Ra decreased by 15.3%, and micro-hardness surface layer increased by 21.3%. This approach proved a successful method for improving the processing efficiency of study subjects.

Furthermore, the surface layer after using PMEDM with the titanium powder in optimal conditions through the GRA was of good quality, where the topographies had a good shape, the number and size of cracks were reduced, and the white layer was of uniform thickness. This contributed to a higher efficiency of titanium powder mixed into the dielectric fluid on both the productivity and quality obtained through the EDM machining process.

KEYWORDS

- **electric discharge machining**
- **material removal rate**
- **recast layer thickness**
- **surface roughness**
- **Taguchi method**
- **tool wear rate**

REFERENCES

1. Ojha, K., Garg, R. K., & Singh, K. K., (2011). Experimental investigation and modeling of PMEDM process with chromium powder suspended dielectric. *International Journal of Applied Science and Engineering.*
2. Jeswani, M. L., (1981). *Effect* of the *addition* of *graphite powder* to *kerosene used* as the *dielectric fluid* in *electrical discharge machining*, *Wear, 70*, 133–139.
3. Pec¸as, P., & Henriques, E., (2003). Influence of silicon powder-mixed dielectric on conventional electrical discharge machining, *International Journal of Machine Tools & Manufacture, 43*, 1465–1471.
4. Zhao, W. S., Meng, Q. G., & Wang, Z. L., (2002). The application of research on powder mixed EDM in rough machining. *Journal of Materials Processing Technology, 129*, 30–33.
5. Özerkan, B., & Çogun, C., (2005). Effect of mixed dielectric on machining performance in electric discharge machining. *Gazi University Journal of Science,* 18.
6. Singh, P., Kumar, A. L., Beri, N., & Kumar, V., (2010). Influence of electrical parameters in powder mixed electric discharge machining (PMEDM) of Hastelloy. *Journal of Engineering Research and Studies, 1*, 93–105.
7. Kansal, H. K., Singh, S., & Kumar, P., (2007). Effect of silicon powder mixed EDM on machining rate of AISI D2 die steel. *Journal of Manufacturing Processes, 9*(1).
8. Sanghani, C. R., & Acharya, G. D., (2014). A review of research on improvement and optimization of performance measures for electrical discharge machining. *Int. Journal of Engineering Research and Applications, 4*(1), 433–450.
9. Parkash, V., & Kumar, D., (2013). Effect of powder mixed dielectric medium on tool wear rate in EDM. *International Journal of Scientific Research (IJSR), 2*(2), 107–109.
10. Singh, G., Singh, P., Tejpal, G., & Singh, B., (2012). Effect of machining parameters on surface roughness of H13 Steel in EDM process using powder mixed fluid. *International Journal of Advanced Engineering Research and Studies, 2*(1), 148–150.
11. Rajendra, M., & Rao, G. K. M., (2014). Experimental evaluation of performance of electrical discharge machining of D3 die steel with Al_2O_3 abrasive mixed dielectric material by using design of experiments. *International Journal of Scientific Engineering and Technology, 3*, 599–606.
12. Kumar, V., Rajpal, & Singh, M., (2014). Experimental study of surface parameters of EN31 on powder mixed EDM using Taguchi methodology, *International Journal for Scientific Research & Development, 2*(07), 122–125.
13. Razak, M. A., Abdul-Rani, A. M., & Nanimina, A. M., (2015). Improving EDM efficiency with silicon carbide powder-mixed dielectric fluid. *International Journal of Materials, Mechanics and Manufacturing, 3*(1), 40–43.
14. Rathi, M. G., & Mane, D. V., (2014). Study on effect of powder mixed dielectric in EDM of inconel 718. *International Journal of Scientific and Research Publications, 4*(11), 1–7.
15. Kaldhone, S. Y., Kavade, M. V., & Rawat, U., (2014). Effect of powder mixed dielectric on performance measures of EDM for tungsten carbide. *International Journal of Innovative Research in Advanced Engineering (IJIRAE), 1*(10), 106–111.
16. Singh, P., Kumar, A., Beri, N., & Kumar, V., (2010). Some experimental investigation on aluminum powder mixed EDM on machining performance of Hastelloy steel. *International Journal of Advanced Engineering Technology, 1,* 28–45.

17. Ganachari, V. S., Kavade, M. V., & Mohite, S. S., (2013). Effect of mixture of Al and SiC powder on surface roughness in PMEDM using Taguchi method with GRA optimization. *Int. J. Adv. Engg. Res. Studies, 2(4), 04–07.*
18. Kumar, S., Kundu, S., & Chaudhary, R., (2014). Optimization of process parameter and experimental investigation of MRR on H-13 die tool steel using EDM with application of Taguchi technique. *International Journal of Engineering and Management Research, 4*(4), 162–167.
19. Zahid, A. K., Arshad, N. S., Noor, Z. K., Urfi, K., & Quadir, G. A., (2014). Multi response optimization of wire electrical discharge machining process parameters using Taguchi based grey relational analysis. *Procedia Materials Science 6*, 1683–1695.
20. Tang, L., & Du, Y. T., (2014). Experimental study on green electrical discharge machining in tap water of Ti–6Al–4V and parameters optimization. *Int. J. Adv. Manuf. Technol., 70*, 469–475.
21. Rajesh, K., Anish, K., Mohinder, P. G., Ajit, S., & Neeraj, S., (2015). Multiple performance characteristics optimization for Al 7075 on electric discharge drilling by Taguchi grey relational theory. *Journal of Industrial Engineering International*, 1–14.
22. Vijay, K. M., & Man, S. A., (2017). Micro-EDM multiple parameter optimization for Cp titanium. *The International Journal of Advanced Manufacturing Technology, 89*, 897–904.
23. Durairaj, M., Sudharsun, D., & Swamynathan, N., (2013). Analysis of process parameters in wire EDM with stainless steel using single objective Taguchi method and multi objective grey relational grade. *Procedia Engineering, 64*, 868–877.
24. Tripathy, S., & Tripathy, D. K., (2017). Surface characterization and multi-response optimization of EDM process parameters using powder mixed dielectric. *Materials Today: Proceedings, 4*, 2058–2067.
25. Rajiv, K. S., & Jagdeep, S., (2014). Determination of multi-performance characteristics for powder mixed electric discharge machining of tungsten carbide alloy. *Proc. IME Part B: Journal of Engineering Manufacture, 230*, 303–312.
26. Pragadish, N., & Kumar, M. P., (2016). Optimization of dry EDM process parameters using grey relational analysis. *Arabian Journal for Science and Engineering, 41*, 4383–4390.
27. Tripathy, S., & Tripathy, D. K., (2016). Multi-attribute optimization of machining process parameters in powder mixed electro-discharge machining using TOPSIS and grey relational analysis. *Engineering Science and Technology, an International Journal, 19*, 62–70.
28. Tripathy, S., & Tripathy, D. K., (2017). Multi-response optimization of machining process parameters for powder mixed electro-discharge machining of H-11 die steel using grey relational analysis and topsis. *Journal Machining Science and Technology An International Journal, 21,* 362–384.
29. Bülent, E., Fevzi, U., & Nihal, E., (2014). Suspended SiC particle deposition on plastic mold steel surfaces in powder-mixed electrical discharge machining. *Proc IME Part B: Journal of Engineering Manufacture, 229*, 475–486.
30. Mohammadreza, S., & Behnam, K., (2017). Investigation of carbon nanotube added dielectric on the surface characteristics and machining performance of Ti–6Al–4V alloy in EDM process. *Journal of Manufacturing Processes, 25*, 212–219.
31. Behnam, K., & Mohammadreza, S., (2017). Effects of hybrid electrical discharge machining processes on surface integrity and residual stresses of Ti-6Al-4V titanium alloy. *The International Journal of Advanced Manufacturing Technology, 93*, 1999–2011.

32. Dastagiri, M., Rao, P. S., & Valli, P. M., (2016). TOPSIS, GRA methods for parametric optimization on wire electrical discharge machining (WEDM) process. *All India Manufacturing Technology, Design and Research Conference.*
33. Prabhu, S., & Vinayagam, B. K., (2016). Multiresponse optimization of EDM process with nanofluids using TOPSIS method and genetic algorithm. *Archive of Mechanical Engineering, 63,* 45–71.
34. Gadakh, V. S., (2012). Parametric optimization of wire electrical discharge machining using Topsis method. *Advances in Production Engineering & Management, 7*, 157–164.
35. Laxman, J., & Guru, R. K., (2014). Optimization of EDM process parameters on titanium super alloys based on the grey relational analysis. *International Journal of Engineering Research, 3*(5), 344–348.
36. Lin, J. L., & Lin, C. L., (2002). The use of the orthogonal array with grey relational analysis to optimize the electrical discharge machining process with multiple performance characteristics. *International Journal of Machine Tools and Manufacture, 42*(2), 237–244.
37. Roy, R., (1990). *A Primer on the Taguchi Method*. New York: Van Nostrand Reinhold.

CHAPTER 11

SELECTION OF OPTIMAL RAPID PROTOTYPING PROCESS USING MULTI-VARIANT MCDM-BASED APPROACHES

JAGADISH[1], AMITAVA RAY[2], SURAJIT BARAD[1], and SHUBHAM B. PATIL[1]

[1]*Department of Mechanical Engineering, National Institute of Technology, Raipur, Chhattisgarh, India*

[2]*Principal Jalpaiguri Government Engineering College, Jalpaiguri, West Bengal, India*

ABSTRACT

Rapid prototyping (RP) is one of the important techniques for building a prototype for product development. These components can be used in assemblies, product testing, and tooling for short or medium-run production. Various RP processes are available, including SLS, SLA, IJM, FDM, LOM, SGC, and 3D plotting. Due to the rapid growth of RP technology, the selection of the most RP process to meet the user requirements from among a number of RP processes becomes increasingly important. Improper selection leads to the failure of the RP product. Selection of RP process is very difficult and is a multi-criteria decision-making problem. Here, three integrated MCDM-based methods like PCA-MOORA, PCA-VIKOR, and DEAR are employed for optimal RP process selection. PCA method is used for weights extraction while MOORA, VIKOR, and DEAR for ranking the RP process. Further, validation and comparative

Decision-Making Models and Applications in Manufacturing Environments. Pushpdant Jain, Kumar Abhishek, & Prasenjit Chatterjee (Eds.)

study of the proposed methods is done by considering the case study of RP process selection. The result shows that Z40 2 RP process yields optimal process among others in all three methods. Therefore, it is concluded that the proposed methods can be used for optimal selection of another manufacturing process.

11.1 INTRODUCTION

Earlier industries used various traditional manufacturing processes like casting and molding, machining, joining, and shear and forming to make a product. In the traditional processes of manufacturing, the final product is made by removing material from the unfinished product. The limitation of traditional manufacturing system is the high cost of manufacturing, less innovation in design and create more waste. In the 21st century, the demand of customer is increasing day by day, i.e., customer do not have time to wait for the product. The industries which fulfill the demand of customer will sustain the market. Industries need to use latest technology to survive in the competitive market. RP is one of the prominent techniques that can overcome the drawbacks of traditional manufacturing system and satisfy the customer. RP is capable of fabricating parts quickly with complex shapes easily, minimum material waste and reduction of design errors.

11.1.1 RP SYSTEMS AND ITS SELECTION

RP is a set of techniques used to quickly manufacture a physical part or assembly scale model using data from three-dimensional (3D) computer-aided design (CAD). Generally, all RP techniques involve five basic steps. The first stage is to create a CAD model of design. The object to be built is modeled using a CAD software package. Various professional software available in the market are AutoCAD, CATIA, and Solidworks. In the Second stage, CAD model is coveted into standard tessellation language (STL) format. STL format represents a 3D surface as an assembly of planar triangles. CAD model is converted into the form of millions of small triangles. STL file is split into thin cross-sectional layers in the third stage. Software slices the STL model into a number of layers from 0.01 to 0.7 mm thick depending upon the RP technology. The actual construction of part take place and model is built by adding layer upon layer of material is done in fourth stage. The final stage involves removing of model from the machine for post-processing which includes cleaning, finishing, polishing, etc.

The techniques used in rapid prototyping (RP) majorly classified into three groups. First is powder RP technique: selective laser sintering (SLS), in SLS technique focused laser move on the bed to sinter particular area of powder material to layer after layer creating a solid object. Surrounding powder near sinter area acts as support for the subsequent layer. In SLS material used are nylon and polyamide powder. SLS used in medical field to impart bone and aerospace hardware. Electron beam melting (EBM), parts are produced in layer with a high vacuum electron beam by melting powder. The material used in EBM is conductive metal instead of thermoplastic polymer. Second is liquid RP technique: stereolithography (SLA), it is a technology use to manufacture layer by layer models using photochemical processes that bind chemical monomer into polymers. Third is solid RP technique: laminated object manufacturing (LOM), In LOM laminates of plastic or metal are subsequently glued together and formed with a laser cutting on laminated objects. Fused deposition modeling (FDM), in FDM extrusion head is filled with filament material and it is heated to a semi-liquid state. The head is moved under computer control to define the printed shape. The materials used in FDM are polylactic acid (PLA), high-impact polystyrene (HIPS), and thermoplastic polyurethane (TPU).

There are a large number of different users for various RP techniques. Every year there are also a number of new technologies introduced. Various users of RP technology are automobile, aerospace, biomedical, construction industries, etc. Every user has unique expectation from RP technique. Each RP technology has its own strength, weakness, and application. RP machines vary in terms of application, quality, cost, accuracy, size, material used, etc. Most of the RP user are finding difficulty to quickly pick an appropriate RP technology to satisfy their need. Therefore, to select appropriate RP technology for one specific need is very difficult task. However, the selection of optimal RP process from different alternatives taking into account various criteria categorizes it as a multi-criteria decision making (MCDM) problem. So, in this chapter various MCDM techniques are used to select an optimal RP process.

11.1.2 LITERATURE REVIEW

A variety of research are being carried out to help RP clients choose the most suitable RP technology to take the best decision. Development of RP technologies has created the selection of RP system difficult from among the range of RP systems available worldwide. Earlier researcher compare different RP technology based on bench-marking studies, such as Mahesh et al. [1] integrate fuzzy decision and benchmarking to select RP system. Limitation of benchmark study is, it is time consuming and expensive. The first try to develop a computer-based program to select an RP system was done by Hornberger [2]. The program provides general information of RP system available at that time. Possibilities for Fast Retrieval of data with user-friendly and flexible design of a system is possible by the IRPS system [3]. Masood & Soo [3] implemented in the United States, Japan, Germany, and Israel a rule based expert system for choosing a program for RP from all commercial RP systems. Byun et al. [4] used a modified TOPSIS approach for selection of the RP system. Lan et al. [5] integrates the expert system and fuzzy synthetic evaluation (FSF). Under the expert system feasible alternative is generated. FSF is used to decide the ranking order and decide most suitable RP system. Both qualitative and quantitated criteria are included to get more accurate result. Lokesh et al. [6] uses AHP model to select RP system. Models consist of various issue and that issue structure into hierarchy. Top level issue in the hierarchy is a selection of RP technology. Second level consist of social and environmental issue, process requirement issue, product requirement issue and other company issue. The third level consists of sub-issue such as production cost, speed, variety, surface finish, part size, etc. Process rating is done as per the need of the decision maker. Researcher found FDM as appropriate technology. Shahrabi et al. [7] uses hybrid AHP and TOPSIS for selection of RP technology. AHP method is used for determining the weight of the criteria's and TOPSIS is used for ranking of the alternatives. According to his study, SLS technology is preferred one. Kumar et al. [8] uses analytical network process approach (ANP) in the selection of RP technology for university performing research work. Complex proportional assessment (COPRAS) method was used by author Mondal et al. [9] to selecting RP system. COPRAS is account for both positive and negative criteria which can be assessed separately within the evaluation process. The results obtained using COPRAS with those originating from previous researchers were same. From the above literature, researchers have used various hybrid MCDM techniques to select optimum RP system, but very few research works done on hybrid PCA-MOORA, PCA-VIKOR, and DEAR. In this book chapter, all three hybrid MCDM method will be used to select optimum RP system using a case study.

11.2 PROPOSED METHODOLOGY

11.2.1 PCA-MOORA

In this section, a hybrid decision-making model is proposed based on MOORA and PCA to solve a multi-criteria decision-making problem. The steps of the proposed methodology have been discussed below [10]:

- **Step 1: Formation of the Decision Matrix:** First, a decision matrix is created in which the number of alternatives and number of criteria are listed. The response of each alternative to each criterion is recorded in a matrix form. The matrix is helpful when examining a large number of decision factors and evaluating the relative importance of each factor.

$$D_{ij} = \begin{bmatrix} x_{11} & x_{12} & \cdots & \cdots & x_{1n} \\ x_{21} & x_{22} & \cdots & \cdots & x_{2n} \\ \vdots & \vdots & \vdots & \vdots & \vdots \\ \vdots & \vdots & \vdots & \vdots & \vdots \\ x_{m1} & x_{m2} & \cdots & \cdots & x_{mn} \end{bmatrix} \tag{1}$$

  where; D_{ij} is the decision matrix; x_{ij} is the response of ith alternative to j^{th} criteria; m represents the number of alternatives; and n represents the number of criteria.

- **Step 2: Normalization of the Decision Matrix:** Normalization of the decision matrix is done to make the responses non-dimensional so that they can be brought into a comparable space. Normalization is done by dividing each response of the alternative to a criterion by the square root of the sum of the squares of each response to the criterion.

$$N_{ij} = \frac{x_{ij}}{\sum_{i=1}^{n} \left(x_{ij}\right)^2} \tag{2}$$

  where; x_{ij} is the response of ith alternative to j^{th} criteria; and N_{ij} represents the normalized matrix.

- **Step 3: Calculating the Weights for each Criterion by Applying PCA Method:** The importance of each criteria affecting the objective is different. So, calculating weights for each criterion is an important task. Here PCA is used to calculate the weights. In this method, a mean-corrected normalized matrix (MCN_{ij}) is created. Then a correlation matrix is formed using the equation given below:

$$R_{jl} = \left[\frac{con(x_i(j), x_i(l))}{\sigma x_i(j) \times \sigma x_i(l)} \right] \quad (3)$$

where; $x_i(j)$ is the GRC of each response, $cov(x_i(j), x_i(l))$ is the covariance of response variables j and l, $\sigma x_i(j)$ and $\sigma x_i(l)$ are the standard deviation of response variables j and l.

The correlation matrix R_{jl} is used to calculate the eigenvalues using the equation given below:

$$(R - \lambda_x I_m)V_{ik} = 0 \quad (4)$$

where; λ_x eigenvalues, $\sum_{k=1}^{n} \lambda_k = n$; k = 1, 2, …, n, and $V_{ik}[a_{k1}, a_{k2}, \ldots, a_{km}]^T$ are the eigenvectors corresponding to the eigenvalue λ_k.

For a value λ there is a vector ***V*** that satisfies the equation:

$$S{*}V = \lambda{*}V. \quad (5)$$

where; V is the eigenvector; and λ is eigenvalue.

For each eigenvalue, there is an associated eigenvector. The principal component is obtained by multiplying the mean corrected normalization matrix with the eigenvectors.

$$PC_p = MCN_{ij} \times \begin{bmatrix} v_{1m} \\ v_{2pm} \\ \vdots \\ v_{pm} \end{bmatrix} \quad (6)$$

The proportions for principal components are evaluated by dividing the corresponding eigenvalue by the sum of all eigenvalues. The eigenvectors which has high proportion is used to evaluate the weights. The weight of each criterion is evaluated as the square of the corresponding eigenvector.

- **Step 4: Evaluating Normalized Assessment Values for each Criterion:** A normalized assessment value of each criterion is calculated by adding the normalized response multiplied with corresponding weights for beneficiary criteria and subtracting for non-beneficiary criteria as shown in the equation given below:

$$Ny_i = \sum_{j=1}^{g} w_i x_{ij} - \sum_{j=g+1}^{n} w_i x_{ij} \tag{7}$$

where; Ny_i is the normalized assessment value; w_i is the weight of each criterion; and x_{ij} is the response of ith alternatives to j^{th} criteria.

- **Step 5: Ranking of Alternatives:** Finally, the alternatives are arranged in descending order of their normalized assessment value. The alternative with the highest normalized assessment value is ranked first.

11.2.2 PCA-VIKOR

In these following steps, a hybrid MCDM approach PCA-VIKOR is described. PCA method is used to calculate the weights of each criterion. The weights are obtained in the similar procedure described in Steps 1–3 of PCA-MOORA method [10].

- **Step 1: Determination of the Ideal and NIS:** The ideal solution A^+ and the NIS A^- are computed as follows:

$$A^+ = \{\max X_{ij} \mid i = 1, 2, ..., m\} = X_1^+, X_2^+, ...X_j^+, ..., X_n^+ \tag{8}$$

$$A^- = \{\max X_{ij} \mid i = 1, 2, ..., m\} = X_1^-, X_2^-, ...X_j^-, ..., X_n^- \tag{9}$$

where; X_{ij} is the response in the normalized matrix.

- **Step 2: Calculating the Utility Measure and Regret Measure:** The utility measure (S_i) and regret measure (R_i) are calculated as follows:

$$S_i = \sum_{j=1}^{m} \left(W_j \times \frac{X_i^+ - X_{ij}}{X_i^+ - X_i^-} \right) \tag{10}$$

$$R_i = \max_j \left(W_j \times \frac{X_i^+ - X_{ij}}{X_i^+ - X_i^-} \right) \tag{11}$$

where; Xij is the response in the normalized matrix.

- **Step 3: Calculating the VIKOR Index:** The VIKOR index are calculated as follows:

$$Q_i = \alpha * \frac{S_i - S^*}{S^- - S^*} + (1 - \alpha) * \frac{R_i - R^*}{R^- - R^*} \tag{12}$$

where; α is the weight given to the utility measure; and (1–α) is given to the regret measure, where α Є [0,1].

$$S^* = \min_i S_i, S^- = \max_i S_i, R^* = \min_i R_i, \text{ and } R^- = \max_i R_i$$

- **Step 4: Ranking of the Alternatives:** In this method ranking of the alternative is done in the ascending order of the VIKOR index. The alternative with the lowest VIKOR index is ranked first and with highest VIKOR index is ranked last.

11.2.3 DEA METHOD

DEA is a widely accepted approach for assessing decision-making unit (DMU) efficiency. DEA has gained too much attention and extensive use by business and academic researchers because of its easy and successful application. Moreover, DEA has been used extensively for advanced manufacturing technologies selection [14].

- **Step 1:** Compute the weights for each output, i.e., the performance measurement ratio for each test to the sum of all performance measures, corresponds to all experiments. The calculation is based on the weight portion of each product calculated by using the following equations:

$$W_c = \frac{C_i}{\sum C_i} \tag{13}$$

where; W_c is the weight of each output, and C_i is the response of each alternative to each criterion.

- **Step 2:** Transform the output data into weighted data after multiplying the output data with the appropriate weight fractions.

$$WC = W_c \times C_i \tag{14}$$

where; WC is the weighted matrix; W_c is the weight of each output; and C_i is the response of each alternative to each criterion.

- **Step 3:** Calculate the multi-performance ranking index (MPRI) by dividing the larger the better performance characteristics with smaller-the-better performance characteristics using the following equation:

$$MPRI = \frac{B_1 + B_2 + B_3 + ...B_b}{NB_1 + NB_2 + NB_3 + ...NB_{nb}} \tag{15}$$

where; B_i is the weighted i^{th} response; NB_i is the weighted i^{th} response; b is the total number of beneficiary criteria; and nb is the total number of non-beneficiary criteria.

- **Step 4:** The MPRI obtained from above is arranged in descending order and is ranked in that order, i.e., the highest MPRI value is ranked first and the lowest MPRI value is ranked last.

11.3 CASE STUDY

In this case study, selection of RP process has been carried out. The study includes six RP systems namely SLA, SLS, FDM, LOM, Quadra T and Z40 2, out of which one optimal process is selected. To evaluate the selection, six attributes such as accuracy (A), tensile strength (S), elongation (E), surface roughness (R), cost of the part (C) and build time (B). The obtained results after the experimentation has been tabulated in Table 11.1.

TABLE 11.1 List of Different RP Processes and Properties

RP Process	Accuracy (A)	Tensile Strength (S)	Elongation (E)	Surface Roughness (R)	Cost of the Part (C)	Build Time (B)
SLA 3500	120.0	65.00	5.000	6.500	0.745	0.500
SLS 2500	150.0	40.00	8.500	12.50	0.745	0.500
FDM 8000	125.0	30.00	10.00	21.00	0.665	0.745
LOM 1015	185.0	25.00	10.00	20.00	0.590	0.410
Quadra T	95.00	30.00	6.000	3.500	0.745	0.410
Z40 2	600.0	5.000	1.000	15.50	0.135	0.255

The decision matrix is created by forming a matrix of responses of different RP process such as SLA, SLS, FDM, LOM, Quadra T and Z40 2 to various criteria like accuracy (A), tensile strength (S), elongation (E), surface roughness (R), cost of the part (C) and build time (B). Among the six criteria, accuracy (A), tensile strength (S) and elongation (E) are considered as beneficial criteria whereas surface roughness (R), cost of the part (C) and build time (B) are considered as non-beneficiary criteria. The decision matrix formed is as shown in Table 11.1.

Weight of each criterion has been calculated using principal component analysis. For the calculation of weight, the decision matrix is normalized using the Eqn. (2). This process is required to convert all the responses of different units into a comparable non-dimensional state. The normalized decision matrix is as tabulated in Table 11.2.

TABLE 11.2 Normalized Matrix

RP Process	A	S	E	R	C	B
SLA 3500	0.1777	0.7145	0.2735	0.1808	0.4737	0.4143
SLS 2500	0.2222	0.4397	0.4649	0.3478	0.4737	0.4143
FDM 8000	0.1852	0.3298	0.5470	0.5842	0.4228	0.6174
LOM 1015	0.2740	0.2748	0.5470	0.5564	0.3751	0.3398
Quadra T	0.1407	0.3298	0.3282	0.0974	0.4737	0.3398
Z40 2	0.8887	0.0550	0.0547	0.4312	0.0858	0.2113

First, the mean-corrected form of the normalized matrix is created. Then a correlation matrix is formed from the mean corrected, normalized matrix using Eqn. (3). The correlation matrix is used to calculate the eigenvalues and eigenvectors. The percentage contribution of the eigenvector is calculated by dividing the corresponding eigenvalue by the sum of all eigenvalues and is tabulated in Table 11.3.

TABLE 11.3 Eigenvalue and Proportion for Principal Components

Principal Component	Eigenvalues	Proportion (%)
First	3.6713	61.19
Second	1.7052	28.42
Third	0.4048	6.75
Fourth	0.2078	3.46
Fifth	0.0109	0.18
Sixth	0.0000	0.00

The square of the eigenvalues signifies the contribution of the corresponding response to the principal component. From Table 11.3, it is concluded that the first eigenvalue has the highest proportion percent, so the square of the corresponding eigenvectors becomes the weights. So, the weights, in this case, are 0.2594, 0.1598, 0.1548, 0.0077, 0.2556, and 0.1626.

The weights evaluated from the PCA method are applied in the MOORA (XXX) and VIKOR (XXX) methods. In MOORA method, the normalized assessment values for each RP process are calculated using Eqn. (7) and is tabulated in Table 11.4. In VIKOR method [10], the ideal and NIS are calculated using Eqns. (8) and (9).

$$A^{+} = \{0.887, 0.7145, 0.5470, 0.5842, 0.4737, 0.6174\};$$

$$A^{-} = \{0.1407, 0.0550, 0.0547, 0.0974, 0.0858, 0.2113\}.$$

The weights for each criterion are calculated using PCA method as described above and the weights evaluated are 0.2594, 0.1598, 0.1548, 0.0077, 0.2556, and 0.1626. The utility measure and regret measure are calculated using Eqns. (10) and (11), the VIKOR index of each alternative is calculated using Eqn. (12). The utility measure, regret measure, and VIKOR index are tabulated in Table 11.5.

TABLE 11.4 Normalized Assessment Values of RP Processes

SL. No.	RP Process	Normalized Assessment Value
1.	SLA 3500	0.0128
2.	SLS 2500	0.0088
3.	FDM 8000	–0.0275
4.	LOM 1015	0.0443
5.	Quadra T	–0.0370
6.	Z40 2	0.1882

TABLE 11.5 Utility and Regret Measure

SL. No.	RP Process	S_i	R_i	VIKOR Index
1.	SLA 3500	2.2346	0.1777	0.5493
2.	SLS 2500	2.3626	0.2220	0.7161
3.	FDM 8000	2.6863	0.1852	0.7974
4.	LOM 1015	2.3671	0.2740	0.8366
5.	Quadra T	1.7095	0.0974	0.0973
6.	Z40 2	1.7268	0.0547	0.0088

In DEA method, the weights of each output are calculated. The weights for each output are calculated using Eqn. (13) and is tabulated in Table 11.6. The weighted data is evaluated by multiplying weight fraction to output data to obtain weighted data using Eqn. (14) and is tabulated in Table 11.7. The multi-performance ranking index is calculated by dividing the larger the better performance characteristics by smaller-the-better performance characteristics using Eqn. (15) and is tabulated in Table 11.8.

11.4 RESULTS AND DISCUSSIONS

RP process selection raking is shown in Table 11.9. It is clear from table that the Z40 2 RP process technology is ranked first in all the three methods resulting in accuracy of 600, elongation of 1.0, tensile strength of 5.0, surface roughness of 15.5, cost of 0.135 and build time of 0.255. Both PCA-MOORA and PCA-VIKOR are based on the approach of distance from the ideal solution. Therefore, a similarity in the ranking structure is seen. Moreover, the deviation in the ranks by DEA method can be reasoned as it is based on a ratio approach.

TABLE 11.6 Table Showing Weights of Each Output Data

RP Process	A	S	E	R	C	B
SLA 3500	0.0941	0.3333	0.1235	0.0823	0.2055	0.1773
SLS 2500	0.1176	0.2051	0.2099	0.1582	0.2055	0.1773
FDM 8000	0.0980	0.1538	0.2469	0.2658	0.1834	0.2642
LOM 1015	0.1451	0.1282	0.2469	0.2532	0.1628	0.1454
Quadra T	0.0745	0.1538	0.1481	0.0443	0.2055	0.1454
Z40 2	0.4706	0.0256	0.0247	0.1962	0.0372	0.0904

TABLE 11.7 Weighted Data

RP Process	A	S	E	R	C	B
SLA 3500	11.2941	21.667	0.6173	0.5348	0.1531	0.0887
SLS 2500	17.6471	8.2051	1.7840	1.9778	0.1531	0.0887
FDM 8000	12.2549	4.6154	2.4691	5.5823	0.1220	0.1968
LOM 1015	26.8431	3.2051	2.4691	5.0633	0.0960	0.0596
Quadra T	7.07840	4.6154	0.8889	0.1551	0.1531	0.0596
Z40 2	282.3529	0.1282	0.0247	3.0411	0.0050	0.0231

TABLE 11.8 MPRI Values Obtained for Different RP Processes

SL. No.	RP Process	MPRI
1.	SLA 3500	43.2388
2.	SLS 2500	12.4509
3.	FDM 8000	3.27730
4.	LOM 1015	6.23070
5.	Quadra T	34.2123
6.	Z40 2	92.0447

TABLE 11.9 Rankings of RP Processes based on PCA-MOORA, PCA-VIKOR, and DEA Method

SL. No.	RP Process	PCA-MOORA Method		PCA-VIKOR Method		DEA Method	
		Normalized Assessment Value	Rank	VIKOR Index	Rank	MPRI	Rank
1.	SLA 3500	0.0128	3rd	0.5493	3rd	43.2388	2nd
2.	SLS 2500	0.0088	4th	0.7161	4th	12.4509	4th
3.	FDM 8000	–0.0275	5th	0.7974	5th	3.27730	6th
4.	LOM 1015	0.0443	2nd	0.8366	6th	6.23070	5th
5.	Quadra T	–0.0370	6th	0.0973	2nd	34.2123	3rd
6.	Z40 2	0.1882	1st	0.0088	1st	92.0447	1st

11.5 CONCLUSION

In this chapter, an optimal RP process selection has been carried out taking into account different criteria such as accuracy, tensile strength, elongation, surface roughness, cost of the part, and build time. For the selection of the RP process, three different hybrid MCDM Techniques have been applied, and the results have been compared to each other. The result of comparison clearly shows, Z40 2 RP Process yields first in all three methods. Apart from this, PCA MOORA and PCA VIKOR result shows the similarity in the ranking of RP processes as they are based on the approach of variation from the ideal solution. So, one of these two methods along with DEA can be used for further RP processes and other manufacturing process selections.

KEYWORDS

- **computer-aided design**
- **electron beam melting**
- **laminated object manufacturing**
- **optimization**
- **rapid prototyping**
- **RP process**
- **standard tessellation language**

REFERENCES

1. Mahesh, M., Fuh, J. Y. H., Wong, Y. S., & Loh, H. T., (2005). Benchmarking for decision making in rapid prototyping systems. *Proceedings of the 2005 IEEE International Conference on Automation Science and Engineering Edmonton.* Canada.
2. Hornberger, L. E., & Rapid Prototyping Program, (1993). Santa Clara University Santa Clara.
3. Masood, S., & Soo, A., (2002). A rule based expert system for rapid prototyping system selection. *Robotics and Computer Integrated Manufacturing, 18*, 267–274.
4. Byun, H. S., & Lee, K. H., (2005). A decision support system for the selection of a rapid prototyping process using the modified TOPSIS method. *International Journal of Advance Manufacturing Technology, 26*, 1338–1347.
5. Lan, H., Ding, Y., & Hong, J., (2005). Decision support system for rapid prototyping process selection through integration of fuzzy synthetic evaluation and an expert system. *International Journal of Production Research, 43*(1), 169–194.
6. Lokesh, K., & Jian, P. I., (2010). Selection of rapid prototyping technology. *Advance Production Engineering &Management, 5*, 2 75–84.
7. Shahrabi, M., & Javadi, M., (2014). Selection of rapid prototyping process using combined AHP and TOPSIS methodology. *International Journal of Information Science and System, 3*(1), 15–22.
8. Kumar, V., Kumar, L., & Haleem, A., (2016). Selection of rapid prototyping technology using an ANP based approach. *IOSR Journal of Mechanical and Civil Engineering (IOSR-JMCE), 13*(4, Ver. VI), 71–78.
9. Mondal, S., Singh, A. K., Chatterjee, P., & Chakraborty, S., (2017). Decision making for rapid prototyping process selection using complex proportional assessment method. *Proc. of the International Conference on Manufacturing Excellence (ICMAX-2017).*
10. Jagadish, & Amitava, R., (2016). Optimization of process parameters of green electrical discharge machining using principal component analysis (PCA). *The International Journal of Advanced Manufacturing Technology, 87*(5–8), 1299–1311.
11. Azizi, H., (2015). A note on: "A new approach for the selection of advanced manufacturing technologies: Data envelopment analysis with double frontiers." *Pakistan Journal of Statistics and Operation Research, 11*. 259. 10.18187/pjsor.v11i2.851.
12. Mahapatra, S. S., & Panda, B. N., (2013). Benchmarking of rapid prototyping systems using grey relational analysis. *International Journal of Services and Operations Management, 16*(4).
13. Syed, H. M., & Al-Alawi, M., (2002). The IRIS rapid prototyping system selector for educational and manufacturing users. *Int. J. Engng Ed., 18*(1), 66–77.
14. Wang, Y. M., & Chin, K. S. (2009). A new approach for the selection of advanced manufacturing technologies: DEA with double frontiers. *International Journal of Production Research, 47*(23), 6663–6679.

CHAPTER 12

AN INVESTIGATION INTO THE EFFECT OF CRITERIA INTERACTION ON THE TOMADA DE DECISAO INTERATIVA MULTICRITERIO (TODIM) METHOD

DIVYA ZINDANI[1], SAIKAT RANJAN MAITY[2], and SUMIT BHOWMIK[2]

[1]*Assistant Professor, Department of Mechanical Engineering, Sri Sivasubramaniya Nadar (SSN) College of Engineering, Kalavakkam, Tamil Nadu, India*

[2]*Assistant Professor, Department of Mechanical Engineering, National Institute of Technology, Silchar, Assam, India*

ABSTRACT

The effect of criteria interaction has been analyzed in the present work for TOmada de Decisao Interativa Multicriterio (TODIM) method. The impact has been analyzed through the Choquet-integral method of criteria interaction. The chapter incepts with a brief introduction to prospect theory. The literature review section presents literature on the Choquet Integral and TODIM methods. The methodology for arriving at material rankings through the Choqeuet-extended TODIM method has been delineated. The Choquet-extended TODIM method has been applied to solve five real case material selection problems solved by past researchers. A comparison analysis for the ranks obtained with and without criteria interaction has been made to study the impact of criteria interaction on the TODIM method. To the best of my knowledge, this work is the first to undertake such a study. A total of

Decision-Making Models and Applications in Manufacturing Environments. Pushpdant Jain, Kumar Abhishek, & Prasenjit Chatterjee (Eds.)

five real-world cases of material selection have been undertaken to study the effect of criteria interaction on the TODIM method.

12.1 INTRODUCTION

There are a number of multicriteria methods that are available to researchers and practitioners for making viable decisions. Different mathematical assumptions form the basis of such decision-making methods [1]. A number of developments have been made in the domain of decision making in the late 1970s, and the accomplishments are growing exponentially [2].

Decision theory with a behavior aspect at its core came into existence with the article of Ward Edwards [3] and since then it went through some major advances in the 1970s and 1980s. However, the traditional prospect theory, since its inception in 1979, remained a concept of historical importance. It is the cumulative prospect theory by Tversky & Kahneman [4] that is being used prominently as a descriptive decision theory tool under uncertainty. The main difference between the cumulative prospect theory and the traditional prospect theory is that weighting is applied to the cumulative probability function. Therefore, cumulative prospect theory lays the inception base for further developments of prospect theory. As such the limitations associated with the original prospect theory have been overcome with the cumulative prospect theory. As cumulative prospect theory aids in making decision under the environment of risk, it is considered to have more accuracy.

There have been a number of attempts to apply prospect theory to address decision making problems, the earliest one being made by Korhonen et al. [5]. The application of prospect theory in a deterministic multicriteria decision-making environment was tested by Salminen & Wallenius [6]. An interactive model for addressing deterministic multicriteria decision problems was proposed by Salminen [6] through the employability of linear prospect theory. Then approximations for the S-shaped value functions were then taken into consideration through piecewise linear marginal value functions. As such Salminen's prospect theory was valid only for convex preferences.

The TODIM method then came into existence in the early 1990s and since its inception, the method has been used to address a number of decision-making problems [7–10]. The aim of the present chapter lies in investigating the impact of criteria interaction on the TODIM method. The investigation has been done for the material selection problems. A total of five material selection problems have been considered for this purpose. A comparative analysis for the ranks obtained using TODIM and that obtained using TODIM with criteria interaction effect has been carried out.

12.2 LITERATURE REVIEW

The present section is split into two subsections: the first is discussed the literature review on Criteria interaction, and the other section discusses on TODIM method.

12.2.1 LITERATURE REVIEW ON CRITERIA INTERACTION

The problem on interaction between the criteria for multi-attribute theory (MAUT) was analyzed by Keeney & Raiffa [11]. Three model variants were investigated. A key term known as the sum of the weighted utilities was taken into consideration. Multi-linear model encompasses this important term and is one of the most general models. Setting of all the constants to one constant in case of multi-linear model results into multiplicative model. Another model being the derivative of the multi-linear model is the additive aggregation model wherein all the coefficients of interactions are set to zero. In essence, all the aforementioned models are additive.

Various MCDM methods are combined in order to take into account the relationships and dependencies between criteria. Few combinations have been discussed in the ensuing discussion. A hybrid approach combining ANP and CI was proposed by Gürbüz & Albayrak [12] for the evaluation of human resources. The proposed method was unique to the case of human resources. Furthermore, two different kinds of interactions were managed at the same time. The reason for employing ANP was that the decision-making problem has several criteria and that these criteria often exhibit interdependencies. On the other hand, the conjunctive or disjunctive interactions are handled by CI. Also, for the ERP selection problems, the same combinatorial approach had been used [13].

Nguyen et al. [14] developed a fuzzy MADM approach that considers the interactions between different criteria. The developed tool was employed for the machine tool selection problem. Fuzzy ANP and complex proportional assessment of alternatives with grey relations (COPRAS-G) encompassed the developed tool. The imprecise information emanating from the decision maker was handled using fuzzy ANP. Modeling of the criteria interactions, relationships, and their interdependencies as well as the weights were determined through the fuzzy ANP approach. COPRAS-G on the other hand aided in representing the preference ratios and also to derive the weighted priorities for the machine alternatives.

Interaction between criteria associated with sustainability factor was taken into consideration for the supplier selection problem [15]. To accomplish the aforementioned goal, interpretive structural modeling (ISM) and fuzzy DEMATEL were used wherein interaction between the evaluation criteria was established using the ISM method and the intensity of the interdependencies was established through the fuzzy DEMATEL methodology. The importance of ISM-DEMATEL approach for efficiently managing the evaluation of the criteria interdependencies. The proposed methodology was employed to solve the Iranian gas engineering problem and in the development of the companies.

FDEMATEL-FANP-FVIKOR model was proposed by Tadic et al. [16] and employed it for the selection of the concepts associated with the city logistics. Fuzzy ANP was used to determine the interdependencies between the factors and for the determination of criteria amongst the groups of factors, fuzzy DEMATEL was employed. FVIKOR was then used to establish the ranking for the alternatives. FDEMATEL and FANP were combined to provide weights for each criterion, and the determined weights are used in the next step to the FVIKOR method to establish the alternative rankings.

Interpolative Boolean Algebra (IBA) is another efficient tool that has been identified to efficiently establish relationship between the criteria/sub-criteria. IBA was combined with the TOPSIS method by Mandic et al. [17] and have used the combinatorial tool for the supplier selection problem. Dependencies between the criteria were established by the IBA tool with the aid of Boolean algebra and under the realm of Boolean laws. IBA was used in the study to present interdependencies between the various elements of decision-making. The rankings were established using the TOPSIS method.

Chatterjee et al. [18] proposed a hybrid meta-model encompassing design of experiments (DoEs) and EDAS method to deal with the criteria interaction. The proposed model was employed to address the material selection problem. The applicability of the proposed DOE-EDAS method was demonstrated through the gear material selection problem. The method was reported to be robust and straightforward to solve complex MCDM problems.

One of the important mathematical model that has aided in modeling of the criteria interactions is Choquet Integral [19]. Choquet integral is a way in which the expected utility of any uncertain event can be measured

[20]. Generalization of the arithmetic mean is the conceptual shadow for the Choquet integral method. Since last decade, the Choquet integral method has been employed for modeling interactions between the criteria in the domain of Multicriteria Decision Making (MCDM) [21, 22]. Roy [23] presented a critical analysis for the usage of Choquet integral method for modeling of the criteria interaction. Through critical analysis, it was revealed that the bipolar model, being a generalization of the Choquet integral needs to be utilized in order to capture the important aspects for modeling of criteria interaction. Choquet integral based linguistic operators were used for modeling the interactions in the material selection domain by Peng et al. [24].

12.2.2 LITERATURE ON TODIM

Traditional TODIM method originated from the prospect theory and aided to deal with the risk behavior associated with the decision maker. The method was developed in the 1990s [25] and since its inception has been used widely to address selection problems in various fields. Traditional TODIM method makes use of square root gains/losses function for evaluating the performance of the alternatives. Greater dominance values are produced with the lower weights for the losses part and as such values for the dominance values gets skewed and often incorrect results are obtained. As such, several modifications have been proposed [26], but there exists still a lacuna in explaining as to why the modifications on gains/losses function is required at all.

Some of the applications include health care [10], evaluation of rental evaluation [27], evaluation of real estate properties [7], behavioral analysis [28], ERP software selection [29], oil spill response [30], selection of fuel busses [31], evaluation of site location for thermal power plants [32], material selection [33, 34] and green supplier selection (GSS) [35].

12.3 METHODOLOGY

In this section, first the basic concepts associated with the TODIM method have been depicted. Next is discussed the conceptual framework of the Choquet integral method. Choquet-extended TODIM method has been discussed after the basic concepts on TODIM and Choquet integral method have been discussed.

12.3.1 TRADITIONAL TODIM METHOD

The TODIM method is employed to calculate the dominance degree of each alternative over the other alternative. This is achieved by establishing a multicriteria value function on the basis of prospect theory. The rankings of alternatives are determined using the dominance degree values. The capability of the TODIM method to capture the behavior of decision maker is one of the major advantages of the TODIM method. However, the TODIM method can only be used to deal with MCDM problems that have their criteria values in terms of crisp numbers. Following steps are followed to obtain alternative rankings through the TODIM method:

Let the material alternatives be represented by $M = \{1, 2, \ldots, 3\}$ and the number of criteria be represented by $N = \{1, 2, \ldots, 3\}$.

- **Step 1:** Decision matrix is identified, i.e., $X = (x_{ij})_{m\times n}$ and the identified decision matrix is normalized.
- **Step 2:** The relative weights, w_p are calculated next for each of the criteria in accordance with the following expression:

$$w_{jr} = w_j/w_r \tag{1}$$

  where; w_{jr} is the relative weight; w_r is the reference weight and is often the maximum of all the criteria weights; and w_j is the criteria weight.
- **Step 3:** The dominance of each alternative M_i over the other alternative M_k is calculated using Eqn. (2):

$$\partial(M_i, M_j) = \sum_{j=1}^{n} \phi_j(M_i, M_k) \tag{2}$$

  where;

$$\phi_j(M_i, M_j) = \begin{cases} \dfrac{w_{jr}(g_{ij} - g_{kj})}{\sum_{j=1}^{n} w_{jk}} & if\ g_{ij} - g_{kj} > 0 \\ 0 & if\ g_{ij} = g_{kj} = 0 \\ \dfrac{-1}{\theta}\sqrt{\dfrac{\sum_{j=1}^{n} w_{jk}(g_{ij} - g_{kj})}{w_{jk}}} & if\ g_{ij} - g_{kj} < 0 \end{cases} \tag{3}$$

The term $\varphi_j(M_i,M_k)$ signifies the contribution of the criteria C_j to the dominance degree function when a comparison between the alternative

M_i and M_k. The attenuation factor is represented by θ and signifies the losses. There are three cases as represented in Eqn. (3): gain is represented by $\varphi_j(M_i,M_k)$ for the first condition, i.e., $g_{ij}-g_{kj} > 0$, no loss no gain is represented by $\varphi_j(M_i,M_k)$ when $g_{ij}-g_{kj} = 0$ and loss is represented when $g_{ij}-g_{kj} < 0$.

- **Step 4:** The overall prospect value is calculated using Eqn. (4):

$$\phi(M_i) = \frac{\sum_{k=1}^{m} \partial(M_i, M_k) - \min_i \left\{ \sum_{k=1}^{m} \partial(M_i, M_k) \right\}}{\max_i \left\{ \sum_{k=1}^{m} \partial(M_i, M_k) \right\} - \min_i \left\{ \sum_{k=1}^{m} \partial(M_i, M_k) \right\}} \tag{4}$$

- **Step 5:** The alternatives are then ranked in accordance with the decreasing value of the overall prospect value.

12.3.2 CHOQUET INTEGRAL METHOD

One of the classes of the MCDM tools is the multi-attribute decision making (MADM). In multi-attribute decision making, the alternatives are ranked on the basis of their global performance score, which is the score evaluated over each criteria [21]. The process of obtaining the global performance score is termed as aggregation and this is done using an aggregation operator. Several mathematical properties [22] and behavioral properties [23] are required to be satisfied by the aggregation operators. Some of the aggregation operators such as simple weighted average, ordered weighted average, etc., works on the presumption that criteria are independent of each other [24] which is not true in real world problems. This is because of the interactive characteristics of the criteria [25]. Therefore, to deal with the conflicting or the interactive nature of the criteria Sugeno integral or Choquet integral is often used [26].

However, the application of Choquet integral is widening and has turned out to be the versatile operator to construct the global performance score [27–28]. The calculation of the Choquet integral requires the prior knowledge of the fuzzy measures, which not only represents the importance of each attribute but also considers the importance of all possible subsets of all the attributes.

Let there be a finite non-empty set represented as $F = \{1, 2, \ldots, K\}$ and a subsets of F be represented as 2^f. Then a fuzzy measure may be defined as a function μ: $2^f \rightarrow [0,1]$ such that:

- μ (φ) = 0 and μ (F)=1; the fuzzy measure is considered to be normalized when μ (F) = 1.
- Choquet integral in relation to the fuzzy measure can be obtained using Eqn. (5) given μ (A) ≤ μ (B), for all A ⊂ B, A, B ϵ F and there is a function a: F → R.

$$I_{\mu}(a)=\sum_{f=1}^{K}\left[\mu\left(\{f,\ldots,K\}\right)-\mu\left(\{f+1,\ldots,K\}\right)\right]a_{f}+\mu(K)a_{k} \tag{5}$$

where; $a_f = a(f)$; and $a_1 < a_1 <\ldots< a_k$.

The Choquet integral is the expected value in case μ is additive, i.e., as in the following Eqns. (6) and (7):

$$\left[\mu\left(\{f,\ldots,K\}\right)-\mu\left(\{f+1,\ldots,K\}\right)\right]=\mu(f) \tag{6}$$

$$I_{\mu}(a)=\sum_{f=1}^{K}\left[\mu(f)\right]a_{f} \tag{7}$$

12.3.3 *CHOQUET EXTENDED TODIM METHOD*

The relative dominance of alternative M_i over alternative M_k is obtained from the TODIM formulation using Eqn. (2).

However, with consideration of fuzzy measure, μ, the overall dominance degree can be obtained using the Eqn. (11). However, for arriving at Eqn. (10), the value of degree of criteria interaction and the fuzzy measures values are calculated using Eqns. (8) and (9), respectively.

$$\lambda+1=\prod_{i=1}^{n}(1+\lambda g_{i}) \tag{8}$$

$$g_{\lambda}\left(\{x_{1,}x_{2}\ldots x_{n}\}\right)=\sum_{i=1}^{n}g_{i}+\sum_{i_1=1}^{n-1}\sum_{i_2=i_1+1}^{n}g_{i_2}\cdot g_{i_1}+\ldots+\lambda^{n-1}g_{1}\cdot g_{2}\ldots g_{n}$$

$$g_{\lambda}\left(\{x_{1,}x_{2}\ldots x_{n}\}\right)=\frac{1}{\lambda}\left[\prod_{i=1}^{n}(1+\lambda g_{i})-1\right] \tag{9}$$

$$g_{\lambda}\left(\{x_{1,}x_{2}\ldots x_{n}\}\right)\lambda=\prod_{i=1}^{n}(1+\lambda g_{i})-1$$

$$\partial(M_{i},M_{j})=I_{\mu}(a)\phi_{j}(M_{i},M_{k}) \tag{10}$$

where; g_i denotes the weight of criteria.

12.4 CASE STUDIES FOR COMPARATIVE ANALYSIS

In this section, five real case examples solved by the past researchers have been considered and are solved using TODIM and Choquet-extended TODIM method. As such the impact of criteria interaction on MCDM tool has been exploited.

- **Example 1: Material Selection for Cryogenic Tank:** The example has been taken from Dehghan-Manshadi et al. [32] and is about the selection of the optimal material for the cryogenic storage tank application. Seven material alternatives and seven criteria have been considered. The data for this example has been tabulated in Table 12.1.

TABLE 12.1 Decision Matrix for Cryogenic Tank Material Selection Problem

SL. No.	Materials	Criteria						
		Toughness Index	Yield Strength	Young's Modulus	Density	Thermal Expansion Index	Thermal Conductivity	Specific Heat
1.	Al 2024-26	75.5	420	74.2	2.8	21.4	0.37	0.16
2.	Al 5052-0	95	91	70	2.68	22.1	0.33	0.16
3.	SS 301-FH	770	1,365	189	7.9	16.9	0.04	0.08
4.	SS 310-3AH	187	1,120	210	7.9	14.4	0.03	0.08
5.	Ti-6Al-4V	179	875	112	4.43	9.4	0.016	0.09
6.	Inconel 718	239	1,190	217	8.51	11.5	0.31	0.07
7.	70 Cu-30Zn	273	200	112	8.53	19.9	0.29	0.06

Out of the seven considered criteria, Density, thermal expansion index, thermal conductivity, specific heat are the cost criteria whereas toughness indicator, yield strength, and young's modulus are the benefit criteria. The weights identified were: $w_1 = 0.214$, $w_2 = 0.1555$, $w_3 = 0.119$, $w_4 = 0.119$, $w_5 = 0.155$, $w_6 = 0.083$, $w_7 = 0.083$.

TODIM method is applied, following the steps as delineated in the "Methodology" section. The overall prospect value and hence the rankings obtained using the TODIM method are depicted in Table 12.2.

Next Choquet-extended TODIM method has been applied to obtain the material rankings. The degree of criteria interaction is obtained using Eqn. (8) and is calculated to be $\lambda = 0.189145$. Next is obtained the fuzzy measures using Eqn. (9) and the values obtained have been tabulated in Table 12.3. Values of fuzzy measures are then used to obtain Choquet-dominance degree for each alternative over the other using Eqn. (11). Next is obtained the Choquet-overall prospect value using Eqn. (4). The Results obtained for the overall Choquet prospect values and therefore the material rankings have been tabulated in Table 12.2.

TABLE 12.2 Fuzzy Measures for Criteria Interaction in Cryogenic Tank Material Selection Problem

Criteria Interaction	Fuzzy Measure	Criteria Interaction	Fuzzy Measure
{}	0	{C4,C5}	0.307853
{C1}	0.137466	{C6}	0.091644
{C2}	0.171833	{C1,C6}	0.231494
{C1,C2}	0.313767	{C2,C6}	0.266456
{C3}	0.131739	{C3,C6}	0.225667
{C1,C3}	0.27263	{C4,C6}	0.225667
{C2,C3}	0.307853	{C5,C6}	0.266456
{C4}	0.131739	{C7}	0.091644
{C1,C4}	0.27263	{C1,C7}	0.231494
{C2,C4}	0.307853	{C2,C7}	0.266456
{C3,C4}	0.26676	{C3,C7}	0.225667
{C5}	0.171833	{C4,C7}	0.225667
{C1,C5}	0.313767	{C5,C7}	0.266456
{C2,C5}	0.349251	{C6,C7}	0.184877
{C3,C5}	0.307853	–	–

TABLE 12.3 Overall Prospect Value, Overall Choquet-Prospect Value and Rankings Obtained for Cryogenic Tank Material Selection Problem

Material	Overall Prospect Value	TODIM Rankings	Overall Choquet-Prospect Value	Choquet-Extended TODIM Ranking
Al 2024-26	1	1	0.824285	2
Al 5052-0	0.971095	2	0.798656	3
SS 301-FH	0.426808	5	0.535558	5
SS 310-3AH	0.290978	6	0.385757	6
Ti-6Al-4V	0	7	0	7
Inconel 718	0.668354	4	0.776888	4
70 Cu-30Zn	0.798054	3	1	1

- **Example 2: Material Selection for Gear:** The selection of material for gear has been solved by Milani et al. [33]. The data has been tabulated in Table 12.4. The five attributes considered were: surface hardness, core hardness, surface fatigue limit, bending fatigue limit and ultimate tensile strength (UTS). The weights identified were: $w_1 = 0.172$, $w_2 = 0.005$, $w_3 = 0.426$, $w_4 = 0.292$, $w_5 = 0.102$.

TABLE 12.4 Decision Matrix for Gear Material Selection Problem

SL. No.	Material	Criteria				
		Surface Hardness	**Core Hardness**	**Surface Fatigue Limit**	**Bending Fatigue Limit**	**Ultimate Tensile Strength**
1.	Cast iron	200	200	330	100	380
2.	Ductile iron	2,700	220	460	360	880
3.	S.G. Iron	7,900	240	550	340	845
4.	Cast alloy steel	4,500	270	630	435	590
5.	Through hardened alloy steel	8,900	270	670	540	1,190
6.	Surface hardened alloy steel	6,000	240	1,160	680	1,580
7.	Carburized steel	8,930	315	1,500	920	2,300
8.	Nitrided steels	750	315	1,250	760	1,250
9.	Through hardened carbon steels	185	185	500	430	635

The degree of criteria interaction is obtained using Eqn. (8) and is calculated to be λ=0.00871966. Next is obtained the fuzzy measures using Eqn. (9) and the values obtained have been tabulated in Table 12.5. The results obtained using TODIM and Choquet-extended TODIM method have been depicted in Table 12.6.

- **Example 3: Material Selection for Tool-Holder:** The present example for selection of material for tool holder [34] considered eight alternatives and four criteria. The considered criteria were: Specific strength, specific modulus, corrosion resistance and cost. The data has been tabulated in Table 12.7. Out of the considered criteria, only the cost criteria were the non-beneficial criteria. $w_1 = 0.291$, $w_2 = 0.079$, $w_3 = 0.206$, $w_4 = 0.188$, $w_5 = 0.098$, and $w_6 = 0.139$ were the identified weights.

TABLE 12.5 Fuzzy Measures for Criteria Interaction in Gear Material Selection Problem

Criteria Interaction	Fuzzy Measure	Criteria Interaction	Fuzzy Measure
{}	0	{C5}	0.102394
{C1}	0.172453	{C1,C5}	0.275001
{C2}	0.005389	{C2,C5}	0.107788
{C1,C2}	0.17785	{C1,C2,C5}	0.280403
{C3}	0.425743	{C3,C5}	0.528517
{C1,C3}	0.598836	{C1,C3,C5}	0.701765
{C2,C3}	0.431152	{C2,C3,C5}	0.533931
{C1,C2,C3}	0.604253	{C1,C2,C3,C5}	0.707187
{C4}	0.291014	{C4,C5}	0.393668
{C1,C4}	0.463905	{C1,C4,C5}	0.566713
{C2,C4}	0.296417	{C2,C4,C5}	0.399076
{C1,C2,C4}	0.469316	{C1,C2,C4,C5}	0.572128
{C3,C4}	0.717837	{C3,C4,C5}	0.820872
{C1,C3,C4}	0.89137	{C1,C3,C4,C5}	0.994559
{C2,C3,C4}	0.72326	{C2,C3,C4,C5}	0.8263
{C1,C2,C3,C4}	–	{C1,C2,C3,C4,C5}	0.999995

TABLE 12.6 Overall Prospect Value, Overall Choquet-Prospect Value and Rankings Obtained for Gear Material Selection Problem

Material	Overall Prospect Value	TODIM Rankings	Overall Choquet-Prospect Value	Choquet-Extended TODIM Ranking
Cast iron	1	1	1	1
Ductile iron	0.71201	2	0.592783749	5
S.G. iron	0.590622	4	0.507916903	6
Cast alloy steel	0.528303	6	0.862889552	3
Through hardened alloy steel	0.557109	5	0.506036142	7
Surface hardened alloy steel	0	9	0	9
Carburized steel	0.315575	8	0.608149576	4
Nitrided steels	0.648367	3	0.999727465	2
Through hardened carbon steels	0.509341	7	0.453129541	8

TABLE 12.7 Decision Matrix for Tool Holder Material Selection Problem

SL. No.	Material	Criteria					
		Young's Modulus	Compressive Strength	Fracture Toughness	Mechanical Loss Coefficient	Hardness	Cost
1.	AISI 1020	210	330	54.5	0.00111	150	0.673
2.	AISI 1040	212	632.5	46	0.00117	355	0.704
3.	AISI 4140	212	655	87.5	0.000515	305	0.864
4.	AISI 6150	206.5	1,575	38	0.00026	483	1.175
5.	AISI 8620	206.5	360	111.5	0.00089	190	0.866
6.	Maraging steel	187.5	1,825	80	0.00071	535.5	6.97
7.	AISI S5	210	1,930	21	0.00002055	771	7.99
8.	Tungsten	593	4,405	14.05	0.00135	1,250	79.6
9.	Fe-5Cr-Mo-V	212.5	1,655	120	0.00133	448.5	1.73

The degree of criteria interaction is obtained using Eqn. (8) and is calculated to be $\lambda=0$. Next is obtained the fuzzy measures using Eqn. (9) and the values obtained have been tabulated in Table 12.8. The results obtained using TODIM and Choquet-extended TODIM method have been depicted in Table 12.9.

TABLE 12.8 Fuzzy Measures for Criteria Interaction in Tool Holder Material Selection Problem

Criteria Interaction	Fuzzy Measure	Criteria Interaction	Fuzzy Measure
{}	0	{C1,C2,C3,C4,C5}	0.860763
{C1}	0.291199	{C6}	0.139533
{C2}	0.078866	{C1,C6}	0.43063
{C1,C2}	0.370008	{C2,C6}	0.218372
{C3}	0.206266	{C1,C2,C6}	0.509412
{C1,C3}	0.497315	{C3,C6}	0.345727
{C2,C3}	0.285092	{C1,C3,C6}	0.636676
{C1,C2,C3}	0.576084	{C2,C3,C6}	0.424526
{C4}	0.188066	{C1,C2,C3,C6}	0.715417
{C1,C4}	0.479128	{C4,C6}	0.327533
{C2,C4}	0.266895	{C1,C4,C6}	0.618495
{C1,C2,C4}	0.557901	{C2,C4,C6}	0.406335
{C3,C4}	0.394235	{C1,C2,C4,C6}	0.69724
{C1,C3,C4}	0.685149	{C3,C4,C6}	0.533631

TABLE 12.8 *(Continued)*

Criteria Interaction	Fuzzy Measure	Criteria Interaction	Fuzzy Measure
{C2,C3,C4}	0.473024	{C1,C3,C4,C6}	0.824444
{C1,C2,C3,C4}	0.763881	{C2,C3,C4,C6}	0.612393
{C5}	0.097066	{C1,C2,C3,C4,C6}	0.903149
{C1,C5}	0.388195	{C5,C6}	0.236565
{C2,C5}	0.175914	{C1,C5,C6}	0.527593
{C1,C2,C5}	0.466985	{C2,C5,C6}	0.315385
{C3,C5}	0.303282	{C1,C2,C5,C6}	0.606356
{C1,C3,C5}	0.594262	{C3,C5,C6}	0.44271
{C2,C3,C5}	0.382089	{C1,C3,C5,C6}	0.733588
{C1,C2,C3,C5}	0.673011	{C2,C3,C5,C6}	0.521489
{C4,C5}	0.285087	{C1,C2,C3,C5,C6}	0.812311
{C1,C4,C5}	0.576079	{C4,C5,C6}	0.424521
{C2,C4,C5}	0.363897	{C1,C4,C5,C6}	0.715412
{C1,C2,C4,C5}	0.654833	{C2,C4,C5,C6}	0.503304
{C3,C4,C5}	0.491206	{C1,C2,C4,C5,C6}	0.794138
{C1,C3,C4,C5}	0.78205	{C3,C4,C5,C6}	0.630569
{C2,C3,C4,C5}	0.569976	{C1,C3,C4,C5,C6}	0.921311

TABLE 12.9 Overall Prospect Value, Overall Choquet-Prospect Value and Rankings Obtained for Tool Holder Material Selection Problem

Material	Overall Prospect Value	TODIM Rankings	Overall Choquet-Prospect Value	Choquet-Extended TODIM Ranking
AISI 1020	0	9	0	9
AISI 1040	0.020823	8	0.037535	8
AISI 4140	0.178294	7	0.20658	7
AISI 6150	0.412944	5	0.486571	5
AISI 8620	0.191359	6	0.219533	6
Maraging steel	0.764902	3	0.897037	3
AISI S5	0.93315	2	0.945543	2
Tungsten	1	1	1	1
Fe-5Cr-Mo-V	0.559348	4	0.667738	4

➢ **Example 4: Material Selection of Flywheel:** Jee & Kang [35] addressed the material selection problem for design of flywheel. Fatigue limit, fracture toughness, price per unit mass and Fragmentability were considered as the attributes. Price per unit mass was considered as the non-beneficial criteria. Ten material alternatives were considered for the flywheel material selection problem. The data has been tabulated in Table 12.10. $w_1 = 0.4$, $w_2 = 0.3$, $w_3 = 0.2$, $w_4 = 0.1$ were the weights identified.

TABLE 12.10 Decision Matrix for Flywheel Material Selection Problem

SL. No.	Material	Criteria			
		Fatigue Limit	Fracture Toughness	Fragment Ability	Cost
1.	300M	100	8.6125	3	4,200
2.	2024-T3	49.6454	13.475	3	2,100
3.	7050-T73651	78.01422	12.5532	3	2,100
4.	Ti-6Al-4V	108.8795	26.0042	3	10,500
5.	E-glass epoxy FRP	70	10	9	2,735
6.	S-glass epoxy FRP	165	25	9	4,095
7.	Carbon-epoxy FRP	440.2516	22.0126	7	35,470
8.	Kevlar 29-epoxy FRP	242.8571	28.5714	7	11,000
9.	Kevlar 49-epoxy FRP	616.4384	34.2466	7	25,000
10.	Boron-epoxy FRP	500	23	5	3,15,000

The degree of criteria interaction is obtained using Eqn. (8) and is calculated to be $\lambda = -0.00248768$. Next is obtained the fuzzy measures using Eqn. (9) and the values obtained have been tabulated in Table 12.11. The results obtained using TODIM and Choquet-extended TODIM method have been depicted in Table 12.12.

TABLE 12.11 Fuzzy Measures for Criteria Interaction in Flywheel Material Selection Problem

Criteria Interaction	Fuzzy Measure	Criteria Interaction	Fuzzy Measure
{}	0	{C1,C4}	0.500002
{C1}	0.400002	{C2,C4}	0.400002
{C2}	0.300001	{C1,C2,C4}	0.800003
{C1,C2}	0.700003	{C3,C4}	0.300001
{C3}	0.200001	{C1,C3,C4}	0.700003
{C1,C3}	0.600002	{C2,C3,C4}	0.600002
{C2,C3}	0.500002	{C1,C2,C3,C4}	1
{C1,C2,C3}	0.900003	–	–

TABLE 12.12 Overall Prospect Value, Overall Choquet-Prospect Value and Rankings Obtained for Flywheel Material Selection Problem

Material	Overall Prospect Value	TODIM Rankings	Overall Choquet-Prospect Value	Choquet-Extended TODIM Ranking
300M	0.392575	6	0.404063	7
2024-T3	0.001958	9	0.004041	9
7050-T73651	0	10	0	10
Ti-6Al-4V	0.665955	5	0.79318	5
E-glass epoxy FRP	0.170412	8	0.21258	8
S-glass epoxy FRP	0.349061	7	0.469229	6
Carbon-epoxy FRP	0.889555	2	0.92006	3
Kevlar 29-epoxy FRP	0.674402	4	0.804358	4
Kevlar 49-epoxy FRP	0.824902	3	0.952607	2
Boron-epoxy FRP	1	1	1	1

- **Example 5: Material Selection for High-Speed Naval Craft:** The example is associated with the material selection of high speed naval craft [36]. Five alternatives and nine criteria were considered. Data have been tabulated in Table 12.13. $w_1 = 0.132$, $w_2 = 0.118$, $w_3 = 0.0833$, $w_4 = 0.0763$, $w_5 = 0.0694$, $w_6 = 0.125$, $w_7 = 0.0763$, $w_8 = 0.153$, $w_9 = 0.167$ were the calculated weights for the different considered criteria.

  The degree of criteria interaction is obtained using Eqn. (8) and is calculated to be $\lambda = -0.000682766$. Next is obtained the fuzzy measures using Eqn. (9) and the values obtained have been tabulated in Table 12.14. Only a part of the calculations have been depicted since the calculations are lengthy and will occupy more space in the chapter. The results obtained using TODIM and Choquet-extended TODIM method have been depicted in Table 12.15.

12.5 CONCLUSION

The present chapter has investigated the impact of criteria interaction on the rankings obtained using the TODIM methodology. A total of five real case examples from the material selection domain were taken. The cases were solved using both the TODIM method as well as the Choquet-extended TODIM method. Variations in rankings were revealed for cases where the degree of interaction was less than or greater than 0. However, similar rankings were obtained in case the degree of interaction was equivalent to zero. For –1

TABLE 12.13 Decision Matrix for High-Speed Naval Craft Material Selection Problem

SL. No.	Materials	Criteria								
		Yield Strength	**Young's Modulus**	**Fire Resistance**	**Reparability**	**Resistance to Corrosion**	**Fabrication Cost**	**Risk**	**Density**	**Weight Savings**
1.	Grade A steel	234.4	204.1	0.6818	0.7727	0.3182	0.5	0.3182	7,800	0
2.	Single skin aluminum (A5086-H34)	137.9	67	0.3182	0.6818	0.6818	0.3182	0.5	2,700	0.6818
3.	Aluminum sandwich (honeycomb core)	268.9	67	0.5	0.5	0.6818	0.5	0.5	1,800	0.7727
4.	LASCOR Stee	379.2	204.1	0.6818	0.5	0.6818	0.7727	0.6818	5,200	0.6818
5.	Composite (CFRP) carbon w/vinyl ester resin	1496.2	227.5	0.3182	0.5	0.7727	0.7727	0.5	1,800	0.7727
6.	DUCTAL (UHP2C)	220.6	53.9	0.7727	0.7727	0.7727	0.3182	0.7727	2,500	0.6818

TABLE 12.14 Fuzzy Measures for Criteria Interaction in High-Speed Naval Craft Material Selection Problem

Criteria Interaction	Fuzzy Measure	Criteria Interaction	Fuzzy Measure	Criteria Interaction	Fuzzy Measure
{}	0	⋮	⋮	⋮	⋮
{C1}	0.131985	⋮	⋮	⋮	⋮
{C2}	0.118092	⋮	⋮	⋮	⋮
{C1,C2}	0.250066	⋮	⋮	{C5,C6,C7,C8,C9}	0.590366
{C3}	0.0833588	⋮	⋮	{C1,C5,C6,C7,C8,C9}	0.722297
{C1,C3}	0.215336	⋮	⋮	{C2,C5,C6,C7,C8,C9}	0.70841
{C2,C3}	0.201444	⋮	⋮	{C1,C2,C5,C6,C7,C8,C9}	0.840331
{C1,C2,C3}	0.33341	⋮	⋮	{C3,C5,C6,C7,C8,C9}	0.673691
		⋮	⋮		
		⋮	⋮		
		⋮	⋮		
{C4}	0.0764122	{C1,C2,C5,C6,C8}	0.597309	{C1,C3,C5,C6,C7,C8,C9}	0.805615
{C1,C4}	0.20839	{C3,C5,C6,C8}	0.430641	{C2,C3,C5,C6,C7,C8,C9}	0.791729
{C2,C4}	0.194498	{C1,C3,C5,C6,C8}	0.562587	{C1,C2,C3,C5,C6,C7,C8,C9}	0.923642
{C1,C2,C4}	0.326465	{C2,C3,C5,C6,C8}	0.548698	{C4,C5,C6,C7,C8,C9}	0.666747
{C3,C4}	0.159767	{C1,C2,C3,C5,C6,C8}	0.680634	{C1,C4,C5,C6,C7,C8,C9}	0.798672
{C1,C3,C4}	0.291737	{C4,C5,C6,C8}	0.423696	{C2,C4,C5,C6,C7,C8,C9}	0.784785
{C2,C3,C4}	0.277845	{C1,C4,C5,C6,C8}	0.555643	{C1,C2,C3,C4,C5,C6,C7,C8,C9}	1.00001
{C1,C2,C3,C4}	0.409805	{C2,C4,C5,C6,C8}	0.541754	{C1,C3,C4,C6,C7,C8,C9}	0.812558
{C5}	0.0694656	{C1,C2,C4,C5,C6,C8}	0.67369	{C1,C3,C4,C5,C6,C7,C8,C9}	0.881986
{C1,C5}	0.201444	{C3,C4,C5,C6,C8}	0.507031	{C2,C3,C4,C5,C6,C7,C8,C9}	0.8681
{C2,C5}	0.187552	{C1,C3,C4,C5,C6,C8}	0.63897	–	–
{C1,C2,C5}	0.31952	{C2,C3,C4,C5,C6,C8}	0.625082	–	–

$< \lambda < 0$, the fuzzy measure is referred to as sub-additive and super additive for $0 < \lambda$. For $\lambda = 0$, the case is known as additive fuzzy measure. Hence it can be concluded that criteria interaction impacts the material rankings for either the sub-additive or super-additive case. Although, the Choquet-extended TODIM method is relatively lengthier method in comparison to TODIM methodology. However, through criteria interaction it tries to take into account the realistic scenario for arriving at material rankings.

TABLE 12.15 Overall Prospect Value, Overall Choquet-Prospect Value and Rankings Obtained for High Speed Naval Craft Material Selection Problem

Material	Overall Prospect Value	TODIM Rankings	Overall Choquet-Prospect Value	Choquet-Extended TODIM Ranking
Grade A steel	0.264714	4	0.267954	4
Single skin aluminum (A5086-H34)	0.44375	3	0.658331	2
Aluminum sandwich (honeycomb core)	0	6	0	6
LASCOR Stee	1	1	1	1
Composite (CFRP) carbon w/vinyl ester resin	0.105917	5	0.104547	5
DUCTAL (UHP2C)	0.788361	2	0.407742	3

As a future scope of the work, the impact of criteria interaction can be explored on other well-known MCDM tools.

KEYWORDS

- **Choquet integral**
- **criteria interaction**
- **interpretive structural modelling**
- **material selection**
- **multi-attribute decision making**
- **multi-attribute theory**
- **multicriteria decision making**
- **prospect theory**

REFERENCES

1. Greco, S., Słowiński, R., Figueira, J. R., & Mousseau, V., (2010). Robust ordinal regression. In: *Trends in Multiple Criteria Decision Analysis* (pp. 241–283). Springer, Boston, MA.
2. Wallenius, J., Dyer, J. S., Fishburn, P. C., Steuer, R. E., Zionts, S., & Deb, K., (2008). Multiple criteria decision making, multiattribute utility theory: Recent accomplishments and what lies ahead. *Management Science, 54*(7), 1336–1349.
3. Edwards, W., (1954). The theory of decision making. *Psychological Bulletin, 51*(4), 380.
4. Tversky, A., & Kahneman, D., (1992). Advances in prospect theory: Cumulative representation of uncertainty. *Journal of Risk and Uncertainty, 5*(4), 297–323.
5. Korhonen, P., Moskowitz, H., & Wallenius, J., (1990). Choice behavior in interactive multiple-criteria decision making. *Annals of Operations Research, 23*(1), 161–179.
6. Salminen, P., & Wallenius, J., (1993). Testing prospect theory in a deterministic multiple criteria decision-making environment. *Decision Sciences, 24*(2), 279–294.
7. Moshkovich, H. M., Gomes, L. F. A. M., & Mechitov, A. I., (2011). An integrated multicriteria decision-making approach to reale state evaluation: Case of the TODIM method. *Pesquisa Operacional., 31*(1), 03–20.
8. Gomes, L. F. A. M., (2009). An application of the TODIM method to the multicriteria rental evaluation of residential properties. *European Journal of Operational Research, 193*(1), 204–211.
9. Gomes, L. F. A. M., & Lima, M. M. P. P., (1992). TODIM: Basics and application to multicriteria ranking of projects with environmental impacts. *Foundations of Computing and Decision Sciences, 16*(4), 113–127.
10. Nobre, F. F., Trotta, L. T. F., & Gomes, L. F. A. M., (1999). Multicriteria decision making– an approach to setting priorities in health care. *Statistics in Medicine, 18*(23), 3345–3354.
11. Nicholas K. Jong & Peter Stone (1976). In: *Proceedings of the ICML-06 Workshop on Kernel Methods in Reinforcement Learning.*
12. Gürbüz, T., & Albayrak, Y. E., (2014). An engineering approach to human resources performance evaluation: Hybrid MCDM application with interactions. *Applied Soft Computing, 21*, 365–375.
13. GüRbüZ, T., Alptekin, S. E., & Alptekin, G. I., (2012). A hybrid MCDM methodology for ERP selection problem with interacting criteria. *Decision Support Systems, 54*(1), 206–214.
14. Nguyen, H. T., Dawal, S. Z. M., Nukman, Y., & Aoyama, H., (2014). A hybrid approach for fuzzy multi-attribute decision making in machine tool selection with consideration of the interactions of attributes. *Expert Systems with Applications, 41*(6), 3078–3090.
15. Mehregan, M. R., Hashemi, S. H., Karimi, A., & Merikhi, B., (2014). Analysis of interactions among sustainability supplier selection criteria using ISM and fuzzy DEMATEL. *International Journal of Applied Decision Sciences, 7*(3), 270–294.
16. Tadić, S., Zečević, S., & Krstić, M., (2014). A novel hybrid MCDM model based on fuzzy DEMATEL, fuzzy ANP and fuzzy VIKOR for city logistics concept selection. *Expert Systems with Applications, 41*(18), 8112–8128.
17. Mandic, K., & Radojevic, B. D. D., (2014). In: *Proc. 14th Joint International Conference of the INFORMS GDN Section and the EURO Working Group on DSS* (p. 134).
18. Chatterjee, P., Banerjee, A., Mondal, S., Boral, S., & Chakraborty, S., (2018). Development of a hybrid meta-model for material selection using design of experiments and EDAS method. *Engineering Transactions., 66*(2), 187–207.

19. Choquet, G. (1954). *Theory of Capacities* (Vol. 5, pp. 131–295). Annal Institute Fourier.
20. Gilboa, I., & Schmeidler, D. (1994). Additive representations of non-additive measures and the Choquet integral. *Annals of Operations Research, 52,* 43–65.
21. Grabisch, M., & Labreuche, C., (2005). In: Figueira, J., Greco, S., & Ehrgott, M., (eds.), *Multiple Criteria Decision Analysis: State of the Art Surveys* (Vol. 78). Springer Science & Business Media.
22. Grabisch, M., & Labreuche, C., (2010). A decade of application of the Choquet and Sugeno integrals in multicriteria decision aid. *Annals of Operations Research, 175*(1), 247–286.
23. Roy, B., (2009). À propos de la signification des dépendances entre critères: Quelle place et quels modes de prise en compte pour l'aide à la décision?. *RAIRO-Operations Research, 43*(3), 255–275.
24. Peng, A., Wen, X., & Wu, K., (2015). Material selection in engineering design using Choquet integral-based linguistic operators under hybrid environment. *Journal of Control Science and Engineering,* 35.
25. Gomes, L. F. A. M., & Lima, M. M. P. P. (1991). TODIMI: Basics and application to multicriteria ranking. *Found. Comput. Decis. Sci, 16*(3–4), 1–16.
26. Lourenzutti, R., & Krohling, R. A. (2015). TODIM based method to process heterogeneous information. *Procedia Computer Science, 55*, 318–327.
27. Gomes, L. F. A. M., & Rangel, L.A.D. (2009) An Application of the TODIM Method to the Multicriteria Rental Evaluation of Residential Properties. *European Journal of Operational Research, 193,* 204–211.
28. Gomes, L. F. A. M., & González, X. I., (2012). Behavioral multicriteria decision analysis: Further elaborations on the TODIM method. *Foundations of Computing and Decision Sciences, 37*(1), 3–8.
29. Kazancoglu, Y., & Burmaoglu, S., (2013). ERP software selection with MCDM: Application of TODIM method. *International Journal of Business Information Systems, 13*(4), 435–452.
30. Passos, A. C., Teixeira, M. G., Garcia, K. C., Cardoso, A. M., & Gomes, L. F. A. M., (2014). Using the TODIM-FSE method as a decision-making support methodology for oil spill response. *Computers & Operations Research, 42*, 40–48.
31. Lee, Y. S., & Shih, H. S., (2016). Incremental analysis for generalized TODIM. *Central European Journal of Operations Research, 24*(4), 901–922.
32. Soni, N., Christian, R. A., & Jariwala, N., (2016). Pollution potential ranking of industries using classical TODIM method. *Journal of Environmental Protection, 7*(11), 1645–1656.
33. Zindani, D., Maity, S. R., Bhowmik, S., & Chakraborty, S., (2017). A material selection approach using the TODIM (TOmada de Decisao Interativa Multicriterio) method and its analysis. *International Journal of Materials Research, 108*(5), 345–354.
34. Zindani, D., Maity, S. R., & Bhowmik, S., (2018). A decision-making approach for material selection of polymeric composite bumper beam. In: Kumar, K., & Davim, J. P., (eds.), *Composites and Advanced Materials for Industrial Applications* (pp. 112–128). IGI Global: USA.
35. Qin, J., Liu, X., & Pedrycz, W., (2017). An extended TODIM multicriteria group decision making method for green supplier selection in interval type-2 fuzzy environment. *European Journal of Operational Research, 258*(2), 626–638.
36. Dehghan-Manshadi, B., Mahmudi, H., Abedian, A., & Mahmudi, R., (2007). A novel method for materials selection in mechanical design: Combination of non-linear normalization and a modified digital logic method. *Materials & Design, 28*(1), 8–15.

37. Milani, A. S., & Shanian, A., (2006). Gear material selection with uncertain and incomplete data. Material performance indices and decision aid model. *International Journal of Mechanics and Materials in Design, 3*(3), 209–222.
38. Çalışkan, H., Kurşuncu, B., Kurbanoğlu, C., & Güven, Ş. Y., (2013). Material selection for the tool holder working under hard milling conditions using different multi criteria decision making methods. *Materials & Design, 45*, 473–479.
39. Jee, D. H., & Kang, K. J., (2000). A method for optimal material selection aided with decision making theory. *Materials & Design, 21*(3), 199–206.
40. Fayazbakhsh, K., Abedian, A., Manshadi, B. D., & Khabbaz, R. S., (2009). Introducing a novel method for materials selection in mechanical design using Z-transformation in statistics for normalization of material properties. *Materials & Design, 30*(10), 4396–4404.
41. Gilboa, I., & Schmeidler, D., (1994). Additive representations of non-additive measures and the Choquet integral. *Annals of Operations Research, 52*(1), 43–65.

CHAPTER 13

SELECTION OF AIR-CONDITIONING SYSTEM VIA VALUE ENGINEERING MANAGEMENT TOOL

SUNIL KUMAR TIWARI and VIKAS PAGEY

Department of Mechanical Engineering, Sagar Institute of Research Technology and Science, Bhopal, Madhya Pradesh, India

ABSTRACT

In the present time, it is apparent that the demand for air-conditioning (AC) system applications is increasing. The application has a penetration in the areas like apartment buildings, banks, office buildings, industrial plants, schools, restaurants, and in multi-segment avenues. The increasing demand for air conditioning system has forced manufacturers to develop a system that can be compatible in the global market and suffice user requirements and environmental aspects. A refrigerant plays an important role in producing desired cooling effects in AC system as required for providing comfort conditions. Currently, in the market, several AC systems are offered by different manufacturers with their own merits and demerits; thus, appropriate AC system selection is essential. The AC system selection procedure with value engineering (VE) approach allows selection from an extensive number of options to a manageable shortlist of potentially suitable AC system through sorting by a few critical selection attributes. Thereafter, paired comparison among different attributes followed by ranking of available alternatives on the basis of their feasibility for final selection.

Decision-Making Models and Applications in Manufacturing Environments. Pushpdant Jain, Kumar Abhishek, & Prasenjit Chatterjee (Eds.)

13.1 INTRODUCTION

Nowadays, the development in every aspect of society and economy to extract optimum productivity becomes crucial to pay attention for treating indoor air for comfort of occupants in all seasons. Brar [1] considered air-conditioning (AC) as a combined process that performs various functions simultaneously (i.e., conditioning of the air, transporting it and introducing it to the conditioned space). However, refrigerant is used as a cooling substance in AC system which is capable of absorbing heat from other substance (ice-water-air-brine solution) [1, 2]. Moreover, Agrawal [3]; and Roberts & Chambers [4] expresses the properties required in refrigerants which makes it suitable to use in AC system are high latent heat of vaporization, high density of suction as, non-corrosive, non-toxic, non-inflammable, cheaper in cost, ease of leak detection, environment friendly, ozone friendly and easily available also. Tripathi et al. [5] stated that different types of AC systems available in the market for diverse application are Split AC, Window AC, Packaged AC with air-cooled and Packaged AC With water-cooled condenser, etc. Moreover, he uses multi-criteria decision making (MCDM) TOPSIS technique in the selection of AC system and refrigerant selection also. Tiwari & Pande [6] use MCDM-TOPSIS techniques for selection of gear manufacturing process and materials. Tiwari et al. [7] uses VE as a tool for material optimization toll for design and fabrication of medical components. From the previous studies, it has been concluded that no one has given attention for selection of AC system using value engineering (VE). In this regard for doing analysis central library room is taken under various assumptions.

13.2 VALUE ENGINEERING (VE) METHODOLOGY

The refrigeration and AC have made significant strides over the past years and lots of manufacturers come in existence that were manufacturing verities of AC system satisfying the user's requirements. Tripathi et al. [8] stated that challenges are still having with this which is primarily associated to AC system and its selection which is able to produce desired effect with less cost. VE approach is used as a tool to manage these challenges. Mukhopadhyaya [9] considered VE as a proven effective and sustainable tool for continuous improvement. In addition to that Mukhopadhyaya [9] expressed its aims which are associated to identify unnecessary cost allied with any product, material, parts, component, system, or service. Moreover, VE involves analysis of functions and thereby to provide optimality through efficiently eliminating unnecessary functions without impairing the quality and functional reliability of the product. Mukhopadhyaya [9] stated that VE realize the necessary function consistently and uses the lowest life cycle cost. Besides, VE cultivates the potential to reduce the cost as well as realize the ways to improve the value of the product. Therefore, the enterprise's technological innovation could have the explicit direction which will lead enterprise competence which will lead enterprise or company could make more profit. The relation between cost, value, and function is expressed as: Value (V) = (Function/Cost) = (F/C). The function (F) expresses the exact work that a design or a product or part must perform. The ways to improve the value of product are one by decreasing production cost and other by removing product's unnecessary functions. The customer evaluates the mechanical product wants to buy considering the usefulness of the product and worth to buy it. Moreover, to produce high-quality and low-price products which give more profit is attainable by combining the product function with its cost and technical indicator with economic indicator. Mukhopadhyaya [9]; and Kiran et al. [10] said that application of VE is limited, and it includes areas like management information system, architectural field, customer value analysis, government bid procurement, etc. Mainly, inadequate consideration has been given in the optimal selection of AC system. Thus, the specific objective of the reported analysis work is to adopt the VE method in optimal selection of AC system satisfying the application requirements. The analysis approaches to get the answer to questions related what is the AC system? What function does AC system perform? What does AC system cost? What else AC system will do the function? What will be the AC system cast? VE practice stages are composed of orientation phase, information phase, functional analysis phase, development phase, presentation phase, implementation phase and lastly follow-up phase. First phase of VE is orientation phase. The orientation phase involves identification of problem very clearly, selection of project, lying down objective and target and in-depth training of all the members related with AC system. Second phase of VE is information phase which start after clearly identifying what is to be accomplished in the AC system and thereby to gather all the relevant information about AC system (i.e., related with material, drawing, and technical specification, manufacturing processes, detail cost break, performance/failure report, quality, procurement, and production problem is gathered in short). The third phase of VE is functional analysis phase which involves analysis and identification of functions of AC system. The fourth phase of VE is a creative phase. Creative phase is the essence of VE and all the possible alternatives are generated related with AC system. The generation of possible alternatives related with AC system can be achieved by brainstorming and other creative techniques for providing the functional requirement. The fifth phase of VE is the evaluation phase. The evaluation phase involves analysis of possible developed alternatives related with AC system. Furthermore, the cost of each idea is estimated related with AC system. The sixth phase of VE is critical evaluation. In this phase critical evaluation of all points of solution related with AC system is carried out. The idea which assures maximum saving are screened and short listed using short listing techniques (i.e., feasibility ranking, etc.). The development phase is seventh phase of VE and involves investigation of shortlisted ideas in depth to arrive at optimum and practical solution related with AC system. However, presentation phase is eight phase of VE and involves presentation of selected alternatives to the decision-makers (DMs) for approval and implementation of suitable AC system fulfilling users' requirements. The last phase of VE is the implementation phase and Defense Department, USA [11]; and Jay & Danny [12] stated that last phase of VE involves the implementation of results that is definite, specific, and tangible solution accepted to all. To impart creative ideas in VE a test checklist is used and consists of subject related with basic questions analysis, and answers. In the subject of AC systems, the basic question is asked what is the AC system specification? and analysis is done by asking the following questions like as can less expensive AC system be used?, can a more expensive AC system having more COP and eco-friendly properties?, are the newly developed AC system that can be used advantageously? and can the quantity or weight of the AC system is reduced? This will lead to get answers. By continuously applying VE technique in AC system can result saving in cost and results increases in functionality, acceptability, sustainability reliability and durability of the AC system Tripathi et al. [5]. The VE Technique is used for the selection of AC system based upon attributes like noise level, equipment cost, energy efficiency ratio (EER), ease of servicing, and reliability.

13.3 SELECTION OF AIR-CONDITIONING (AC) SYSTEM

The selection of AC systems using VE includes the following steps:

➢ **Step I: Orientation Phase:** The work reported is carried out in the following ways:

- **Condition 1 (Office Room with No AC):** First, by analyzing various available AC units, selecting an AC system with low operating cost and satisfying office cooling load with adequate comfort.

- **Condition 2 (Office Room with Installed Split AC):** Replace the installed AC with suitable AC having low operating cost and satisfying office cooling load with adequate comfort.

➢ **Step II: Information Phase:** In this phase, all the necessary information related with AC system and place specification where it is to be carried out is collected. The information contained is as follows:

- Room size: 120 ft × 50 ft.
- Area of space which is to be cooled: 120 × 50 ft^2.
- Number of rooms: 16 of size 15 × 20 ft.
- No. of split AC installed: 12 units having 1.5 TR each capacity.
- Average temperature in summer outside the room: 40°C.
- Comfort temperature: 23°C.
- Average temperature in winter outside the room: 8°C.
- Comfort temperature: 25°C.
- Humidity: 50%.
- Sunlight: Falls on one side of the room.

Furthermore, few assumptions were undertaken during the analysis work. The assumptions considered are underlined below:

➢ **Assumptions:**

- Assume no heat transfer through door and glass;
- Assume no heat transfer through radiation.

Moreover, the list of entity through which heat gain takes place in office hour (8:00 AM to 6:00 PM) is shown in Table 13.1.

TABLE 13.1 List of Entity which Cause Heat Gain During Office Hour

Entity	Quantity	Rate of Heat Gain (KW)
Sunlight	–	34.066
Occupant	120	14.04
Lightening equipment	40	2
Desktop	40	5.2
Total	–	56
Split AC system of capacity 1.5TR)	12	–

Furthermore, estimated total cooling load calculated and it is reported as 56 KW. The selected office space and designed plan view for the arrangement of ACs are shown in Figure 13.1.

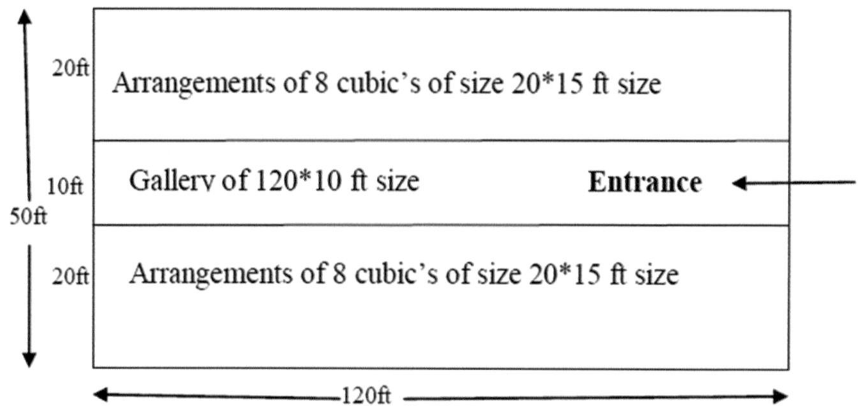

FIGURE 13.1 Plan view of office space designed for arrangement of ACs.

- **Step III: Function Analysis Phase:** The main function of AC system is to provide comfort to occupants by doing conditioning of air. The functional analysis of the AC system has been done by taking into account primary functions and secondary functions. Thereafter, the analysis is done either by changing the design of a product or by removing the undesired function or part of the product. The work reported may not consider any design changes in the AC system. Furthermore, removal of any function or part of the product has not been done. However, optimization is done through proper selection of AC system based on user's requirements with full satisfaction.
- **Step IV: Creative Phase:** The essence of create phase is brainstorming. During the brainstorming, the deciding factors are considered to select the best suitable AC, i.e., according to design consideration of office room, cooling load and human comfort.

  There are different types of AC available in the market and some of them were listed according to requirement. The required properties of AC unit are shown in Table 13.2. However, different types of AC system considered for analysis work is shown in Table 13.3.
- **Step V: Evaluation Phase:** The evaluation phase of the VE is crucial phase. Herein, it evaluates a variety of alternatives critically to select the best among all that meet the expectations of the laid down objectives as well as stated users' requirements. Thereafter it judges the ideas by considering parameters such as noise level, equipment cost, EER, ease of servicing, and reliability (*refer* to Table 13.4).

TABLE 13.2 Required Properties of AC Unit

Particular	Required Properties in AC System
Noise level	Minimum
Equipment cost	Minimum
EER (W/W)	Maximum
Ease of servicing	Quick
Reliability	Better

TABLE 13.3 Types of sAC System Considered for Study

Serial Number	AC System
1	Split AC
2	Window AC
3	Packaged AC with air-cooled
4	Packaged AC with water cooled condenser

TABLE 13.4 Idea Judging Parameters and Key Letters

Parameters	Key Letter
Noise level	A
Equipment cost	B
EER	C
Ease of servicing	D
Reliability	E

The description of different AC systems with parameters is shown in Table 13.5, wherein Table 13.6 shows weightage criteria for noise level and equipment cost. However, Table 13.7 shows weightage criteria for EER, ease of servicing, and reliability.

Each of these design criteria was given a weightage factor. The allocations of criteria values carried out by the way in which firstly, each of the above criteria compared with others, further depending on their relative importance, three categories are formed, viz. major, medium, and minor. A score of respectively 3, 2, and 1 are assigned to each of the levels (*refer* the Table 13.8). Table 13.9 shows paired comparison and Table 13.10 shows feasibility ranking of alternatives.

TABLE 13.5 Description of Different AC Systems with Parameters

Particular	AC System	Noise Level	Equipment Costs	EER (W/W)	Ease of Servicing	Reliability
Existing	Split AC 1.5 TR	3	3	3.51	4	4
Alternative-1	Window AC 1.5 TR	4	2	3.50	2	4
Alternative-2	Packed air-cooled AC	2	4	11	5	2
Alternative-3	Packed water-cooled AC	3	5	12	3	2

Source: Andrew et al. [13]; Brar [1]; Fei et al. [14]; Paul [15]; Pita [16]; Prashad [17]; Roberts & Chambers [4]; and Westra & Douglas [18].

TABLE 13.6 Weightage Criteria for Noise Level and Equipment Cost

Criteria	Weightage
High	5
Medium	4
Average	3
Below average	2
Low	1

TABLE 13.7 Weightage Criteria for EER, Ease of Servicing, and Reliability

Criteria	Weightage
Poor	1
Fair	2
Good	3
Very good	4
Excellent	5

TABLE 13.8 Weightage Analysis and Allocation of Points

Weightage Analysis	Points
Major difference	3
Medium difference	2
Minor difference	1
Equal	0

TABLE 13.9 Paired Comparison

	B	C	D	E	Score
A	A_2	C_2	D_1	A_2	4
	B	C_2	B_2	E_2	2
		C	C_2	E_1	6
			D	E_2	1
				E	5

TABLE 13.10 Feasibility Ranking

Parameters AC systems	A	B	C	D	E	Score	Rank
	4	2	6	1	5	–	–
Window AC	2/8	4/8	1/6	2/2	4/20	44	4
Split AC	3/12	3/6	2/12	4/4	4/20	54	2
Packaged air cooled AC	4/16	2/4	4/24	5/5	2/10	59	1
Packaged water-cooled AC	3/12	1/2	4/24	3/3	2/10	51	3

- **Step VI: Development Phase:** The development phase is very essence phase of the VE. The life cycle cost evaluation of best selected alternatives is shown in Table 13.11.

TABLE 13.11 Life Cycle Cost Evaluation

AC systems with quantity	Purchasing + Installation Cost in (Rs.)	Operating cost (1 yr.) in (Rs.)	Operating cost (15 yr.) in rupees	Maintenance cost (15 yr.)	Total cost (Rs.)
Window AC 12 nos.	4,10,000	7,30,080	10,951,200	30,000	12,121,280
Split AC 12 nos.	4,21,908	6,46,272	96,94,080	30,000	10,792,260
Packaged air cooled AC 4 nos.	11,24,776	2,03,904	30,58,560	40,000	44,27,240

13.4 RESULT AND DISCUSSION

This analysis focuses on selecting an appropriate AC system considering specific application considerations and comfort condition requirements. However, VE is used as a tool for selection of AC. The first stage of VE is a composition of the paired comparison matrix with the values of attributes/criteria like cost, energy efficiency ratio (EER), ease of service, availability, reliability, and noise level. On the basis of above matrix, feasible ranking matrix is constructed and weighted feasible ranking matrix is obtained when the paired comparison matrix and weights is assigned to the criteria. On the above parameters, ranking is done for different alternatives. Then the cost evaluation is done to calculate the value of different alternatives.

In the analysis, it is comprehended that among the specified alternatives, packaged air-cooled AC system is found to be the most preferable choice among four alternatives. Moreover, packaged air-cooled AC has the highest value as compared to the other available alternative, which is 13.32. However, it has a low operating cost over 15 years of its life cycle. The study would be beneficial for users who desire to optimize the resources.

KEYWORDS

- **air conditioning system**
- **air-conditioning**
- **energy efficiency ratio**
- **feasibility ranking**
- **global market**
- **multi-criteria decision making**
- **paired comparison**
- **value engineering**

REFERENCES

1. Brar, J. S., (2012). *Study, Modeling, Analysis, Evaluation, Selection and Performance Improvements of Air Conditioning System*. M. Tech Thesis, Thapar University, Patiala, India.
2. Piotr, A. D., (1998). Conference report: Refrigerants for the 21st century ASHRAE/NIST refrigerants conference Gaithersburg, MD. *Journal of Research of the National Institute of Standards and Technology, 103*(5), 529–533.
3. Agrawal, A. B., (2014). '*Eco-Friendly Refrigerants, Presentation on Slide Share*' [online] https://www.slideshare.net/pankajkumar123456/refrigeration-14185467?qid=b44edde9-d910-4bd9-b624 bf9ddaa90fbe&v=&b=&from_search=1 (accessed on 17 February 2022).
4. Roberts, N. A., & Chambers, O. R., (2004). Energy saving refrigerant blends comprising R125, R134a, R600 or R600a. *International Refrigeration and Air-Conditioning Conference.* Purdue, USA.

5. Tripathi, A. K., Dubey, S., Pandey, V. K., & Tiwari, S. K., (2015). Selection of refrigerant for air-conditioning system using MCDM-TOPSIS approach. *International Journal of Emerging Technology and Advanced Engineering, 5*(02), 384–388.
6. Tiwari, S. K., & Pande, S., (2017). Selection of gear materials using MCDM-TOPSIS approach. *International Journal of Manufacturing and Material Processing, 3*(2), 10–15.
7. Tiwari, S. K., Pande, & Bobade, S. M., (2016). Material optimization during medical components development in selective laser sintering process using value engineering. *International Journal of Research in Engineering and Technology, 5*(11), 186–194.
8. Tripathi, A. K., Tiwari, S. K., & Nema, A., (2015). Selection of air conditioning system using MCDM-TOPSIS approach. In: *Proc. of 3rd International Conference on Industrial Engineering (ICIE-2015)* (pp. 675–679). SVNIT, Surat, Gujrat, India.
9. Mukhopadhyaya, A. K., (2012). *Value Engineering–Concepts, Techniques and Applications*. Sage Publications, New Delhi, India.
10. Kiran, C. P., Clement, S., & Agrawal, V. P., (2011). Coding, evaluation and optimal selection of a mechatronic system. *Expert Systems with Applications, 38*, 9704–9712.
11. Defense Department, USA, (2011). *Value Engineering: A Guidebook of Best Practices and Tools.* USA.
12. Jay, M., & Danny, L. R., (2006). *Value Engineering Handbook*. Virginia, USA.
13. Andrew, D. A., Carl, H. T., Alfred, F. B., Daniel, C. B., & Gloria, M. B., (2017). *Modern Refrigeration and Air Conditioning* (20th edn.). G-W Publisher, Illinois, USA.
14. Fei, L., Hui, H., Yingjiang, M., & Rong, Z., (2008). *Research on the Air-Conditioning Water Heater System.* Paper presented at the international refrigeration and air-conditioning conference, Purdue, USA.
15. Paul, L. V., (2003). *Principles of Air Conditioning* (3rd edn.). CBS Publishers and distributors, New Delhi, India.
16. Pita, E. G., (2003). *Air Conditioning Principles and Systems* (4th edn.). Pearson Education, New York.
17. Prashad, M., (2011). *Refrigeration and Air Conditioning Data Book*. New Age International Publishers, New Delhi, India.
18. Westra, & Douglas, G., (2002). COP improvement of refrigerator /freezers, air-conditioners, and heat pump using nonazeotropic refrigerant mixtures. In: *NASA, Technology, The Third National Technology Transfer Conference and Exposition* (Vol. 2, pp. 164–173).
19. Pandey, V. K., Saxena, R., Tiwari, S. K., & Pande, S., (2013). Selection of refrigerant for air-conditioning system using MCDM-TOPSIS approach. In: *Proc. of 1st International Conference on Industrial Engineering (ICIE-2013)*. SVNIT, Surat, Gujrat, India.

CHAPTER 14

ON THE COMPARISON OF MCDM TECHNIQUES FOR THE SELECTION OF POLYMER-BASED ADDITIVE MANUFACTURING PROCESSES

JAIRAM RAIGAR[1], SHEKHAR SRIVASTAVA[1], VISHAL S. SHARMA[2], RAMESH CHAND[1], JAIVIR SINGH[1], and RAJIV K GARG[1]

[1]*Industrial and Production Engineering Department, Dr. B. R. Ambedkar National Institute of Technology, Jalandhar, Punjab, India*

[2]*School of Mechanical, Industrial and Aeronautical Engineering, University of Witwatersrand, Johannesburg, South Africa*

ABSTRACT

Additive manufacturing (AM) have been widely explored for the fabrication of end use components directly from a 3D CAD design in a single step. Recently, there are number of AM processes available for manufacturing a single component. Therefore, it is critical to select an appropriate AM process with reference to the component's quality, such as mechanical properties, dimensional accuracy, surface texture, fatigue life, production time and cost etc. In order to overcome these challenges Multi-Criteria Decision Making (MCDM) techniques are more suitable option. In this work a hybrid MCDM technique has been proposed to select an appropriate AM process according to the part's requirements. The hybrid approach includes best worst method (BWM) to determine optimal criteria weights, and the proximity indexed value (PIV) method to rank the selected AM processes.

Decision-Making Models and Applications in Manufacturing Environments. Pushpdant Jain, Kumar Abhishek, & Prasenjit Chatterjee (Eds.)

The rank 1, be the most appropriate AM process for manufacturing of the desired product attributes. Additionally, to benchmark the properties of an AM process, a conceptual model of spur gear was fabricated by four AM processes, vat photopolymerization (VP), material extrusion (ME), powder bed fusion (PBF), and material Jetting (MJ). Seven most significant criteria such as Surface roughness (R), dimensional accuracy (A), tensile strength (S), elongation at break (%E), heat deflection temperature (HDT), part cost (PC), and build time (BT) has been considered for database creation. Further, to validate the reliability of the results obtained from the hybrid MCDM approach, sensitivity analysis have been performed. The results obtained using the hybrid BWM-PIV MCDM approach is consistent and reliable.

14.1 INTRODUCTION

14.1.1 BACKGROUND

Additive manufacturing (AM) (3D printing) fabricates physical objects directly from a 3D CAD model data by successive addition of material layer upon layer as opposed to subtractive manufacturing technologies [1]. It can also be called rapid manufacturing (RM) when it is used to produce fully functional components [2]. Recently, AM technology has gained an impressive growth due to its capabilities of producing complex geometry parts with integrated assemblies to reduce development time and cost. AM has been applied in various applications of engineering industry, including aerospace, automotive, construction, jewelry, and biomedical, as well as other fields such as education, medicine, design, architecture, toys, and entertainment [3, 21]. According to ISO/ASTM 52900-2015 [3, 4], AM technologies are categorized into eight groups (vat photopolymerization (VatPP), material extrusion (ME), binder jetting (BJ), directed energy deposition, material jetting (MJ), powder bed fusion (PBF), direct-write, and sheet lamination (SL)) (Figure 14.1). There are more than 1,000 industrial AM machines, and 850 materials (e.g., 3D Hubs, Stratasys, 3D systems, EOS (electro-optical systems), Materialize, Shape ways, Sculpteo, 3Diligent, Think3D, Chizels) have been found in the global market [5].

Since various AM processes and materials are available, selecting a suitable one for a particular application is very problematic. Also, the selection of an appropriate AM process requires a thorough understanding of the interactions among process parameters and material properties [7]. Therefore, the selection of the most appropriate technology to fulfill the

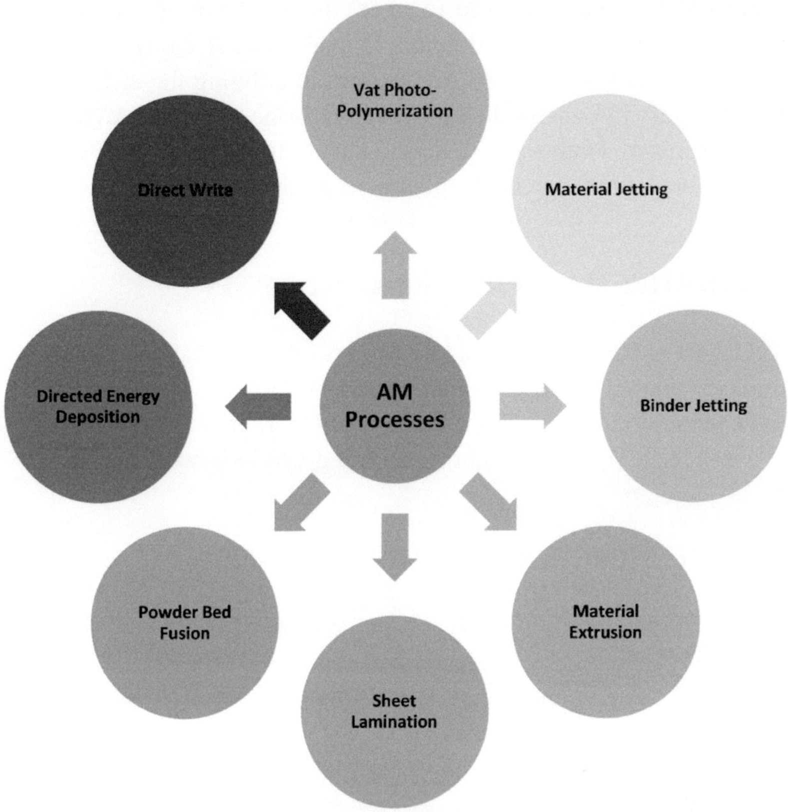

FIGURE 14.1 Classification of AM processes based on ISO/ASTM 52900:2015 standard [6].
Source: Reprinted from Ref. [6]. Copyright © 2017 Elsevier Ltd.

specific part's requirements is significantly essential, and thus the primary motivation for this study. Benchmarking plays a vital role in AM technology evaluation. The experimental results from tests are often more persuasive than those from other sources, and also data from benchmarking could be more reliable and believable [8, 9]. A systematic approach has to be espoused for ranking the candidate technologies. The AM process involves many criteria that could influence the selection of alternative AM technologies. Hence a methodology that takes into account all the most significant criteria and alternatives need to be employed. In that context, multiple criteria decision making (MCDM) approach could be more suitable. In the present study, newly developed MCDM methods, including best worst method (BWM) and Proximity Indexed Value (PIV) method, are used. A hybrid MCDM methodology has been proposed to select the most appropriate technology specifying parts requirements in terms of quality, cost, and time. BWM was employed to quantify the user's preferences in defining the criteria weights and subsequently PIV method employed to rank the candidate AM processes. Further, sensitivity analysis is done to validate the reliability of the results. The best-ranked one is the most suitable to produce the specified part.

14.2 ADDITIVE MANUFACTURING PROCESSES

The ISO/ASTM 52900 Standard had been formed in 2015 to standardize all terminology related to additive manufacturing (AM) as well as classify the different technologies into eight categories [3, 4]. The description of each process and the associated technologies are given in subsections.

14.2.1 VAT PHOTOPOLYMERIZATION (VATPP)

Vat Photopolymerization (VatPP) process includes a heat source selectively cures a liquid photopolymer resin contained in a vat (or tank). Layer by layer, a 3D physical object is built until completion. The popular Vat PP technologies include stereolithography (SLA), and digital light processing (DLP) [11].

14.2.2 POWDER BED FUSION (PBF)

This AM technique enables the manufacturing of a vast array of geometrically complex products. A heat source (laser or electron beam) is used to fuse powder particles layer-by-layer, forming a solid part. The PBF process includes selective laser sintering (SLS), direct metal laser sintering (DMLS) or selective laser melting (SLM) [10], electron beam melting (EBM), and multi-jet fusion (MJF) [14].

14.2.3 MATERIAL EXTRUSION (ME)

Material extrusion (ME) (fused deposition modeling (FDM)/fused filament fabrication (FFF)/liquefied filament fabrication (LFF)) was invented by S. Scott Crump in the late 1980s and was commercialized in 1990 by the Stratasys Inc., USA [15]. In this process, thermoplastic filament materials extrude through nozzle and deposits in layers on a platform. The head has two movable nozzles. The first nozzle deposits molten material and the second nozzle deposits the support material. The material is heated above its melting point so that it solidifies immediately after the extrusion [16, 52].

14.2.4 MATERIAL JETTING (MJ)

The printhead dispenses a photosensitive material like polymers, ceramics, metals, or wax that solidify when exposed to ultraviolet light or heat on the build platform. It allows different materials of different colors to be 3D printed within the same part [14, 15]. It requires support material which is dispensed along with the main material and removed during post-processing step. MJ process includes MultiJet printing (MJP) [17], drop on demand (DOD) [10, 15].

14.2.5 BINDER JETTING (BJ)

Binder jetting (BJ) was developed at Massachusetts Institute of Technology, USA in 1993 and commercialized by Z corporations in 1996 [17]. The powder is either ceramic-based (glass or gypsum) or metal-based (stainless steel). It is generally not suited for functional applications due to the brittle nature of the parts [15].

14.2.6 DIRECTED ENERGY DEPOSITION

The directed energy deposition (DED) AM technology, creates parts by directly melting materials and deposing them on the workpiece, layer by layer [1, 53]. This AM technique is mostly used with metal powders or wire source materials. It is also capable of fixing complex damaged parts, such as turbine blades or propellers. The DED AM technology includes laser engineering net shape (LENS) [19], and electron beam additive manufacturing (EBAM) [20].

14.2.7 SHEET LAMINATION (SL)

The sheet lamination (SL) process was developed in 1991 as laminated object manufacturing (LOM) [3] by Helisys, Torrance, CA. Later commercialized by M'C or and EnvisionTec Incorporation, USA.

14.2.8 DIRECT WRITE (DW)

Direct write (DW) AM technologies enable the selective deposition and patterning of liquid feedstock onto the substrate, which extends its application in the field of embedded electronics. Some DW processes are now used for the production of high-resolution 3D components. These are laser chemical vapor deposition (LCVD), focused ion beam DW, aerosol jet process, laser-induced forward transfer (LIFT), matrix-assisted pulsed-laser DW (MAPLE), and nozzle dispensing processes [3].

14.3 LITERATURE REVIEW

A substantial research input has been given for selection of an AM process over time. Previously, a few decision support systems using a multi-criteria decision making (MCDM) method have been established for AM process selection. Accordingly, Pham & Gault [22] offered an overview of the AM technologies with its strength and weakness by using a logical flowchart technique. It was suggested for the selection of AM process based on the part's requirement. Braglia & Petroni [23] proposed an analytical hierarchy process (AHP) based management support technique for the selection of the AM process considered with the subjective and objective criteria. Masood & Soo [25] developed a rule-based expert system considering 39 rapid prototyping (RP) systems commercially available from 21 manufacturers. The developed method for AM process selection was named as *Intelligent RP System Selector* based on the interactive question-answer session with the user. Byun & Lee [7] performed a benchmark study to develop a system for the selection of an appropriate AM process. The modified TOPSIS method and fuzzy approach had been used for ranking and assigning values for cost and build time, respectively. Rao & Padmanabhan [28] developed a methodology for the selection of the AM process using graph theory and matrix approach (GTMA). The relative importance of criteria had modeled in digraph and matrix-style and evaluated with permanent function. Munguia et al. [31] proposed an artificial intelligence (AI)-based information system integrating fuzzy logic, neural network and a relational database to execute decision-making. In this study, the process variables (part cost (PC) estimation, material selection and ranking of chosen AM systems) had been selected by the screening method and rule-based reasoning. Kumar & Jain [32] developed a systematic AHP approach with the help of potential industrial and academic users. Six different technologies had been selected in the study (SLS, SLA, FDM, 3DP, SGC, and LOM), by considering the various product, process, social, and environmental issues. Mahapatra & Panda [12] proposed a combined Grey Relational Analysis (GRA) and fuzzy TOPSIS approach for the selection of the AM process. Ghazy [2] developed an AM decision support system (AMDSS) to select appropriate AM process chains, materials, finishing methods, and machines. A simple multi-attribute rating technique (SMART) had been utilized for ranking the processes and compatible materials based on the desirability. Shende & Kulkarni [37] introduced an experimentation-based methodology for AM process selection using the GTMA technique. Additionally, a comparison had been made with the TOPSIS method considering part quality, cost, and time. Also, the important criteria, such as surface finish, accuracy, tensile strength, elongation, heat deflection temperature (HDT), PC, process time had been considered for ranking of four AM processes. Kek et al. [41] developed a decision support system using the fuzzy ANP-TOPSIS methodology to select an environment-friendly AM process with required mechanical properties for SLA, SLS, 3DP, and LENS process. Vinodh & Nagraj [39] applied the fuzzy VIKOR method for AM process selection in an agile environment by considering 20 criteria for selection of best AM process among SLA, SLS, and FDM processes. Peko et al. [47] applied AHP, fuzzy AHP and Preference ranking organization method for enrichment of evaluations (PROMETHEE) for the selection of four available AM processes. Further, the application of machine learning has also been extended to develop the decision support system for the determination of the candidacy of a part or assembly for an AM application. The authors [14] proposed a decision support system framework and implemented it as a web application with integrated cloud-based database and machine learning service. They have considered hip implants and a throttle pedal as a demonstrating examples.

MCDM methods (AHP, TOPSIS, fuzzy AHP-TOPSIS, and other methods) have been used widely to solve the problem of selection of AM processes [24–30, 33–36, 40, 42–48]. The AHP and TOPSIS methods have some drawbacks related to a large number of pairwise comparison for calculating weights. This drawback leads to inconsistency in results and rank reversal problems. Recently developed MCDM methods (best-worst method (BWM) and proximity index value (PIV) method) have been proposed by the researchers in manufacturing with simple calculation steps and minimum rank reversal problem [49–51]. Since BWM and PIV methods have certain advantages over previously used MCDM techniques and its application has not been explored in this field. Therefore, this article explored an application of a hybrid BWM-PIV MCDM technique to select the most appropriate AM process as per part requirements. In that context, BWM employed to enumerate the user's preferences in defining the criteria weights and subsequently PIV method employed to rank the AM processes. Subsequently, to compare the obtained results from the above method, a comparison has been made with conventional methods (hybrid AHP-TOPSIS method). Further, sensitivity analysis has been done to validate the reliability of the results—the best ranker is considered as the most suitable AM process to produce the specified part.

14.4 METHODOLOGY

The methodology adopted in this study for the selection of an appropriate AM process has shown in Figure 14.2.

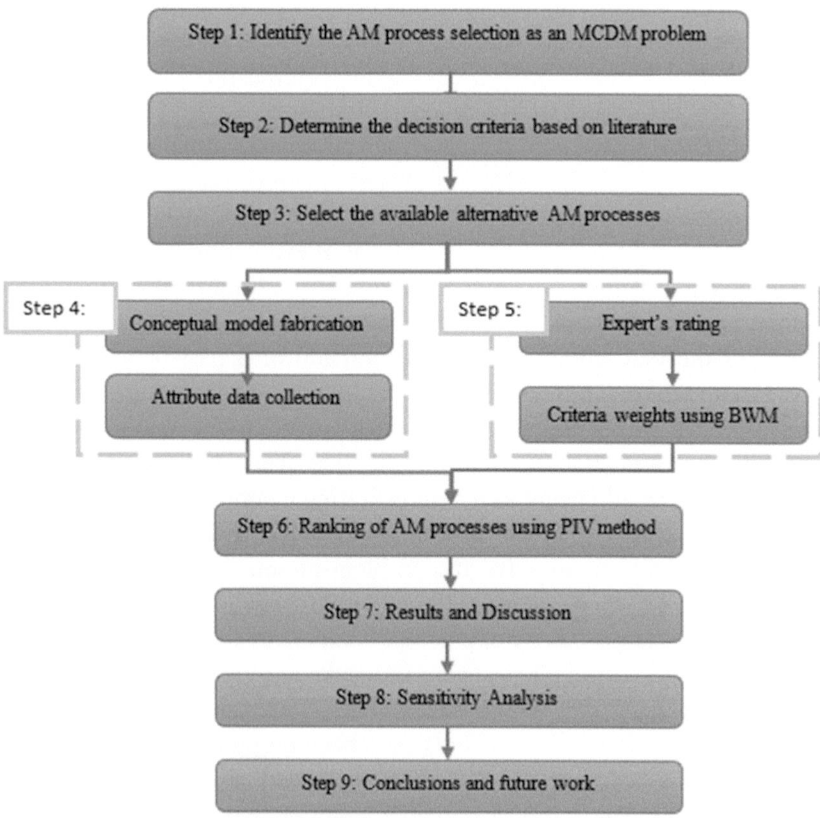

FIGURE 14.2 Methodology framework [62].

Source: Reprinted with permission from [62].

14.4.1 SELECTED DECISION CRITERIA

As per literature review and ISO 17296-3 (2014), the commonly used criteria in the selection of an AM process that fulfills specific part requirements related to quality, cost, and time are considered. These are dimensional accuracy (A) [7, 41], surface roughness (R) [41], tensile strength (S) [45], percentage elongation (E), HDT [38], part cost (PC), and build time (BT).

14.4.2 SELECTED AM PROCESSES

In this study, the considered AM processes are based on the AM technologies available in the market. The description of the considered AM process is as follow:

14.4.2.1 VAT PHOTOPOLYMERIZATION (VATPP): EKA DLP 3D PRINTER

EKA DLP 3D Printer is the fastest DLP 3D printer having printing speed 100 mm/hr. and manufactured by Make3D, India. This machine comes with a full HD UV LED projector engine, which gives 30,000 hours of working life.

14.4.2.2 MATERIAL EXTRUSION (ME): PRATHAM

Pratham desktop 3D printer is an accurate and reliable LFF technology-based 3D printer by Make3D, India. Aluminum heat bed, filament finish sensor, advanced smart cooling and multi-material support (ABS/PLA/metal fill/wood fill) are the key feature of this 3D printer.

14.4.2.3 POWDER BED FUSION (PBF): EOS P396

EOS P396 is a medium-sized, plastic industrial SLS 3D printer produced by EOS-GmbH, Germany. It uses a CO_2 laser to melt and solidify layers of powdered material into finished objects.

14.4.2.4 MATERIAL JETTING (MJ): PROJET MJP 2500 PLUS

ProJet MJP 2500 plus is a professional-grade 3D printer by 3D Systems, USA. It uses piezo printhead technology to deposit either photocurable plastic resin or casting wax material layer by layer.

14.4.3 CONCEPTUAL MODEL FABRICATION

A conceptual model of spur gear has been fabricated by four different AM process to benchmark its attribute abilities. The steps involved in the development of the physical 3D model from virtual CAD design are shown in Figure 14.3. The 3D model of spur gear was developed in CAD software PTC Creo4.0.

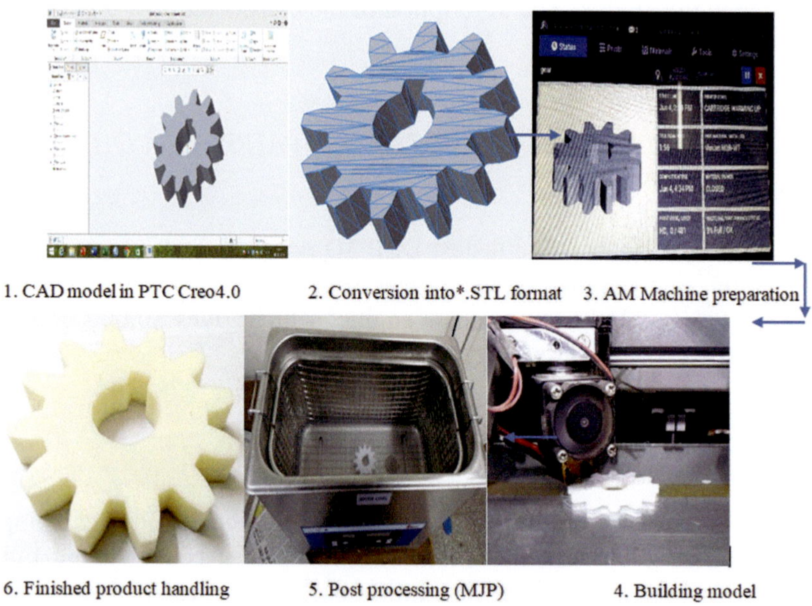

FIGURE 14.3 Model fabrication by AM process [62].

Source: Reprinted with permission from [62]. Copyright © 2020, Indian Academy of Sciences

14.4.4 AM PROCESS ATTRIBUTE DATA COLLECTION

The available AM processes like Vat Photo Polymerization, Material Extrusion, Powder Bed Fusion, and Material Jetting have been utilized for the fabrication of a test component (spur gear) as a geometric benchmark. The numerical values of dimensional accuracy (A) have been obtained by measuring the bore diameter with Mitutoyo Dial Vernier Caliper for each manufactured part. The surface roughness values were obtained through measuring the face part of the gear teeth (z-direction) by Roughness Tester (Taylor Hobson Surtronic S128). The measured values (three iterations) are shown in Table 14.1. The models fabricated by available four AM processes are shown in Figure 14.4. The numerical data values of all the attributes/criteria collected are shown in Table 14.2.

TABLE 14.1 Measured Values [62]

AM Process	Material	Dimensional Accuracy (±mm)				Surface Roughness-Ra (µm)			
Iterations		1	2	3	Average	1	2	3	Average
VP	ABS	0.81	0.88	0.54	0.743	3.87	2.80	2.50	3.06
ME	PLA	0.28	0.19	0.50	0.323	21.0	21.0	17.9	19.96
PBF	Nylon 12	0.15	0.20	0.06	0.136	15.7	10.6	9.07	11.79
MJ	Visijet M2R-WT	0.11	0.13	0.08	0.106	6.37	5.94	6.03	6.11

Source: Reprinted with permission from [62]. Copyright © 2020, Indian Academy of Sciences

PBF (SLS) PartVP (DLP) PartME (LFF) PartMJ (MJP) Part

FIGURE 14.4 Models fabricated by four different AM processes [62].

Source: Reprinted with permission from [62]. Copyright © 2020, Indian Academy of Sciences

14.4.5 DECISION MODELS

In this work, Analytical Hierarchy Process (AHP) and Best-Worst Method (BWM) has been used to determine the criteria weights whereas, Technique for Order Preference by Similarity to Ideal Solution (TOPSIS) and Proximity Indexed Value (PIV) methods have been used for ranking of considered AM processes. Later, a comparison has been made between the results of AHP coupled with TOPSIS and BWM coupled with PIV methods. However, following sub-section provides the details of the MCDM techniques used in this study.

TABLE 14.2 AM Process Benchmark Attributes Data [62]

AM Process	Machine	Material	Dimensional Accuracy* (A)	Surface Roughness (R)	Tensile Strength (S)	Elongation (%E)	Heat Deflection Temperature (HDT)	Process Cost (PC)	Build Time (BT)
Units	–	–	mm	μm	MPa	%	°C	Rs	Min
Material jetting (MJ)	ProJet MJP 2500 +	Visijet M2R-WT	0.106	6.11	45	30	51	1,000	140
Vat photopolymer-ization (VP)	EKA DLP 3D Printer	ABS TRU resin	0.743	3.06	49	10.2	50	700	30
Material extrusion (ME)	Pratham LFF 3D Printer	PLA Plus	0.323	19.96	60	29	52	300	72
Powder bed fusion (PBF)	EOS P 396	Nylon12 PA2200	0.136	11.79	48	18	163	1,180	160

**Deviation of bore diameter from the nominal value (in mm).*

Source: Reprinted with permission from [62].

14.4.5.1 ANALYTIC HIERARCHY PROCESS (AHP)

The AHP method was developed by Thomas Saaty in the 1970s [60]. The AHP method is used to calculate criteria weightage to prioritized alternative technologies through pairwise comparison. The steps involved in this method are as follows:

- **Step 1:** Develop the hierarchy structure of the decision-making problem considering the goal at the top, defined criteria at middle and available alternatives at the bottom level (Figure 14.5).
- **Step 2:** Determine the relative importance of criteria with respect to the goal through expert's judgments to avoid biasness in a pairwise comparison matrix (PCM) A, using fundamental scale of relative importance shown in Table 14.3.
- **Step 3:** Normalize the PCM A and transferred to matrix B, using Eqn. (1).

$$b_{ij} = \frac{a_{ij}}{\sum_{i=1}^{m} a_{ij}} \tag{1}$$

  where; a_{ij} are the elements of matrix A.

- **Step 4:** Determine the criteria weights (w_j) by taking the arithmetic average from the row of normalized comparison matrix using Eqn.(2).

$$w_j = \frac{\sum_{j=1}^{n} b_{ij}}{n} \tag{2}$$

  where; b_{ij} are the elements of matrix B.

- **Step 5:** Finally, to ensure the consistency of judgment matrix, consistency ratio (CR) is calculated as the ratio of consistency index (CI) and random index (RI).

$$CR = CI/RI \tag{3}$$

  where; $CI = (\lambda_{max} - n)/(n-1)$; where, 'n' is size of matrix and λ_{max} is maximum eigen value, and random index (RI) values suggested by T. Saaty are given as below:

n	2	3	4	5	6	7	8	9	10
RI	0	0.58	0.90	1.12	1.24	1.32	1.41	1.45	1.49

  If the CR<0.1, then judgment matrix is consistent and satisfactory. Otherwise, judgment matrix should be reviewed and repeated to obtained consistent matrix.

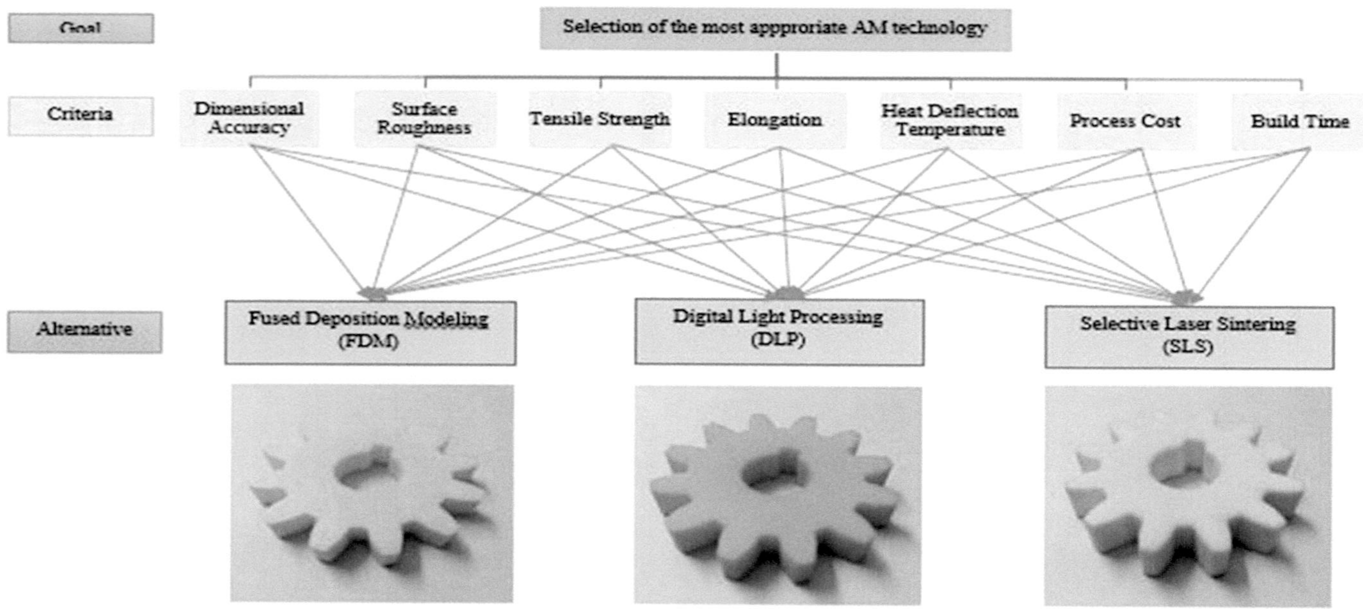

FIGURE 14.5 Hierarchy structure for AHP coupled with TOPSIS.

Source: Reprinted with permission from [62]. Copyright © 2020, Indian Academy of Sciences

TABLE 14.3 Scale of Relative Importance [62]

Relative Important Scale	Description
1	Equal importance
2	Somewhat between equal and moderate
3	Moderately more important
4	Somewhat between moderate and strong
5	Strongly more important
6	Somewhat between strong and very strong
7	Very strongly important
8	Somewhat between very strong and extreme
9	Extremely important

14.4.5.2 TECHNIQUE FOR ORDER PREFERENCE BY SIMILARITY TO IDEAL SOLUTION (TOPSIS) METHOD

The TOPSIS method was first developed by Hwang and Yoon in 1981 [61], for solving MCDM problems. It is based upon the concept that chosen alternatives should have minimum Euclidean distance from the positive ideal solution (PIS) and maximum from negative ideal solution (NIS). The PIS maximizes the benefit criteria and minimizes cost criteria and vice-versa for NIS. The ranking of the alternative AM technologies has obtained using closeness coefficient (CC_i). The steps involved in this method are as follows:

- **Step 1:** Identify the available alternatives Ai (i =1, 2, ..., m) and decision criteria Cj (j =1, 2, ..., n) involved in this decision-making problem.
- **Step 2:** Construct the decision matrix R by arranging alternatives in rows and criteria in columns as shown in Eqn. (4)

$$R = \left[r_{ij} \right] = \begin{bmatrix} r_{11} & r_{12} & \cdots & r_{1j} & \cdots & r_{1n} \\ r_{21} & r_{22} & \cdots & \cdots & \cdots & r_{2n} \\ \cdots & \cdots & \cdots & \cdots & \cdots & \cdots \\ r_{i1} & \cdots & \cdots & r_{ij} & \cdots & r_{in} \\ \cdots & \cdots & \cdots & \vdots & \ddots & \vdots \\ r_{m1} & r_{m2} & \cdots & r_{mj} & \cdots & r_{mn} \end{bmatrix} \tag{4}$$

  where; r_{ij} represents ith alternative actual value of jth criteria, where n is a number of criteria and m is a number of alternatives.
- **Step 3:** Normalize the decision matrix using vector normalization formulae given as:

$$r_{ij}^{*} = \frac{r_{ij}}{\sqrt{\sum_{1}^{m} r^{2}_{ij}}} \tag{5}$$

where; r_{ij}^{*} is the normalized value of ith alternative of jth criteria.

- **Step 4:** Calculate the weighted normalized decision matrix by using weights calculated by AHP method:

$$V_{ij} = r_{ij}^{*} \times w_{j} \tag{6}$$

where; V_{ij} is the weighted normalized value; w_j is the weight of criterion.

- **Step 5:** Determine the PIS (best) and NIS (worst) by using Eqn. (7):

$$V^{+} = [(\sum_{i}^{max} \text{Vij}, \forall \text{ j € J}), (\sum_{i}^{min} \text{Vij}, \forall \text{ j € J'}), \forall \text{ i= 1, 2, …, m}]$$

$$V^{-} = [(\sum_{i}^{min} \text{Vij}, \forall \text{ j € J}), (\sum_{i}^{max} \text{Vij}, \forall \text{ j € J'}), \forall \text{ i= 1, 2, …, m}] \tag{7}$$

where; J = 1, 2, …, n belongs to beneficial criteria; and J'=1, 2, …, n, belongs to cost criteria.

V_j^{+} Indicates the ideal value of the considered criteria among the values for different alternatives. In case of beneficial criteria, it is the higher value of the criteria and for cost criteria it is the lower value of the criteria and vice versa for V_j^{-}

- **Step 6:** Calculate the Euclidean distance of each alternative from PIS and NIS using Eqn. (8):

$$d_i^{+} = \sqrt{\{\sum_{j=1}^{n} (V_{ij} - V_j^{+})^{\wedge}2\}},\ d_i^{-} = \sqrt{\{\sum_{j=1}^{n} (V_{ij} - V_j^{-})^{\wedge}2\}},\ \text{i= 1, 2, …, m} \tag{8}$$

- **Step 7:** Determine the closeness coefficient (CC_i) by using Eqn. (9)

$$CC_i = \frac{d_i^{-}}{d_i^{+} + d_i^{-}} \tag{9}$$

- **Step 8:** Rank the alternative AM technologies based on decreasing values of CC_i. The higher value of CC_i has ranked first.

14.4.5.3 BEST WORST METHOD (BWM)

The best worst method (BWM) was developed by Jafar Rezaei in 2015 [49, 50], used to calculate the optimal weights of criteria. Selection of the best factor (most important) and worst factor (least important) are the major steps of this method. Subsequently, a pairwise comparison of best criteria and worst criteria with all other considered criteria is done using a 9-point scale. Thus, the comparison leads to the formation of two pairwise comparison vector, which are further utilized for the calculation of weights. A linear optimization model is used to determine optimal weights and consistency, which is used to check the reliability and optimality of criteria weights. These advantages of BWM method attracted the researchers to use it in other MCDM problems of different field of applications [54–56]. The necessary steps involved in this method are as follows:

- **Step 1:** Identify a set of decision criteria $\{c_1, c_2, \ldots, c_n\}$ for the selection of AM process. In this study, dimensional accuracy (A), surface roughness (R), tensile strength (T), elongation (E), heat deflection temperature (HDT), process cost (PC) and build time (BT) has been included.
- **Step 2:** Select the best and worst criteria. In this study, the best and worst criteria selected according to the expert's opinion.
- **Step 3:** Determine the preference of best criterion over other criteria through pairwise comparisons using a 9-point scale of relative importance.

  It leads to the formation of best to others (BO) vector as:

  $$X_B = [x_{B1}, x_{B2}, x_{B3}\ldots, x_{Bn}]$$

  where; x_{Bj} is the importance of best criterion over the jth criterion and hence $x_{BB} = 1$.
- **Step 4:** Similar to step 3, comparison of all the criteria with the worst criterion has done (Figure 14.6), which leads to the formation of others-to-worst (OW) vector as:

  $$Y_w = [y_{1w}, y_{2w}, y_{3w}, \ldots, y_{nw}]$$

  where; y_{jw} is the importance of jth criterion for the worst criterion; $y_{ww} = 1$.
- **Step 5:** Calculate the optimal weights $(w_1^*, w_2^*, w_3^*, \ldots, w_n^*)$. To determine the optimal weights, the absolute maximum differences $\{|w_B — x_{Bj}w_j|, |w_j — y_{jw}w_w|\}$ (for all j) should be minimum, which can be written as:

  $$\text{Min-max } \{| w_B — x_{Bj}w_j|, | w_j — y_{jw}w_w|\}$$

  Subject to:

  $$\Sigma_j w_j = 1$$

$$w_j \geq 0, \text{ for all } j \tag{10}$$

Eqn. (10), can be converted in terms of the linear programming (LP) model:

$$\text{Min } \xi^L$$

Subject to:

$$| w_B - x_{Bj} w_j | \leq \xi^L \text{for all } j$$

$$| w_j - y_{jw} w_w | \leq \xi^L \text{ for all } j$$

$$\Sigma_j w_j = 1$$

$$w_j \geq 0, \text{ for all } j \tag{11}$$

On solving Eqn. (11), unique solution for the optimal weights (w_1*, w_2*, w_3*, ..., w_n*) and the optimal value of the CR ξ^{L*} are obtained. The value of ξ^{L*} close to zero indicates a high degree of consistency and vice-versa. Ranking of alternative AM processes has been obtained by employing recently developed Proximity Indexed Value (PIV) method described in the following sub-section:

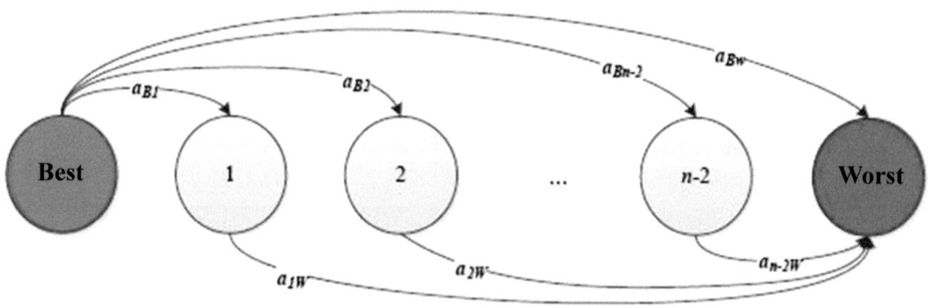

FIGURE 14.6 Preference comparisons in BWM [49].

Source: Adapted from Ref.[49]

14.4.5.4 PROXIMITY INDEXED VALUE (PIV) METHOD

The PIV method recently developed by Mufazzal and Muzakkir in 2018 [51]. The ranking of the alternative AM processes has been obtained by this method to minimize the rank reversal problem. Figure 14.7 shows the flowchart depicting rank reversal in PIV method. The steps involved in this method are as follows:

- **Step 1:** Identify the available alternatives Ai (i = 1, 2..., m) and decision criteria Cj (j = 1, 2, ..., n) involved in this decision-making problem.

➢ **Step 2:** Construct the decision matrix R by arranging alternatives in rows and criteria in columns, as shown in Eqn. (12)

$$R = \left[r_{ij} \right] = \begin{bmatrix} r_{11} & r_{12} & \cdots & r_{1j} & \cdots & r_{1n} \\ r_{21} & r_{22} & \cdots & \cdots & \cdots & r_{2n} \\ \cdots & \cdots & \cdots & \cdots & \cdots & \cdots \\ r_{i1} & \cdots & \cdots & r_{ij} & \cdots & r_{in} \\ \cdots & \cdots & \cdots & \vdots & \ddots & \vdots \\ r_{m1} & r_{m2} & \cdots & r_{mj} & \cdots & r_{mn} \end{bmatrix} \tag{12}$$

where; r_{ij} represents the ith alternative actual value of jth criteria, where n is the number of criteria, and m is the number of alternatives.

➢ **Step 3:** Normalize the decision matrix using vector normalization formulae given as:

$$r_{ij}^{*} = \frac{r_{ij}}{\sqrt{\sum_{1}^{m} r^{2}_{ij}}} \tag{13}$$

where; r_{ij}^{*} is the normalized value of ith alternative of jth criteria.

➢ **Step 4:** Calculate the weighted normalized decision matrix by using weights calculated by BWM method.

$$V_{ij} = r_{ij}^{*} \times w_{j} \tag{14}$$

where; V_{ij} is the weighted normalized value; W_j is the weight of criterion.

➢ **Step 5:** Determine the Weighted Proximity Index (WPI), u_i by using Eqn. (15)

$$u_i = \begin{cases} v_{max} - v_{i;} \text{ for beneficial criteria} \\ v_i - v_{\min;} \text{ for cost criteria} \end{cases} \tag{15}$$

➢ **Step 6:** Determine the overall proximity value, d_i which indicates the closeness of alternative w.r.t. the best alternative, which has been calculated by Eqn. (16)

$$d_i = \sum_{j=1}^{m} u_i \tag{16}$$

➢ **Step 7:** Rank the alternative AM processes based on increasing d_i values. The least value of d_i is ranked first as having a minimum deviation from the best.

14.5 RESULTS AND DISCUSSION

In this study, vat photo-polymerization (VP), material extrusion (ME), powder bed fusion (PBF), and material jetting (MJ) has been included for BWM coupled PIV method, whereas, vat photo-polymerization (VP), material extrusion (ME), powder bed fusion (PBF) has been included for AHP coupled TOPSIS method. The beneficial criteria whose higher values are preferred are indicated by (+) and cost criteria whose lower values are preferred are indicated by (–) in Table 14.4 along with other criteria obtained values. Table 14.4 has been considered as a decision matrix fora database in a decision support system for the selection of the AM process.

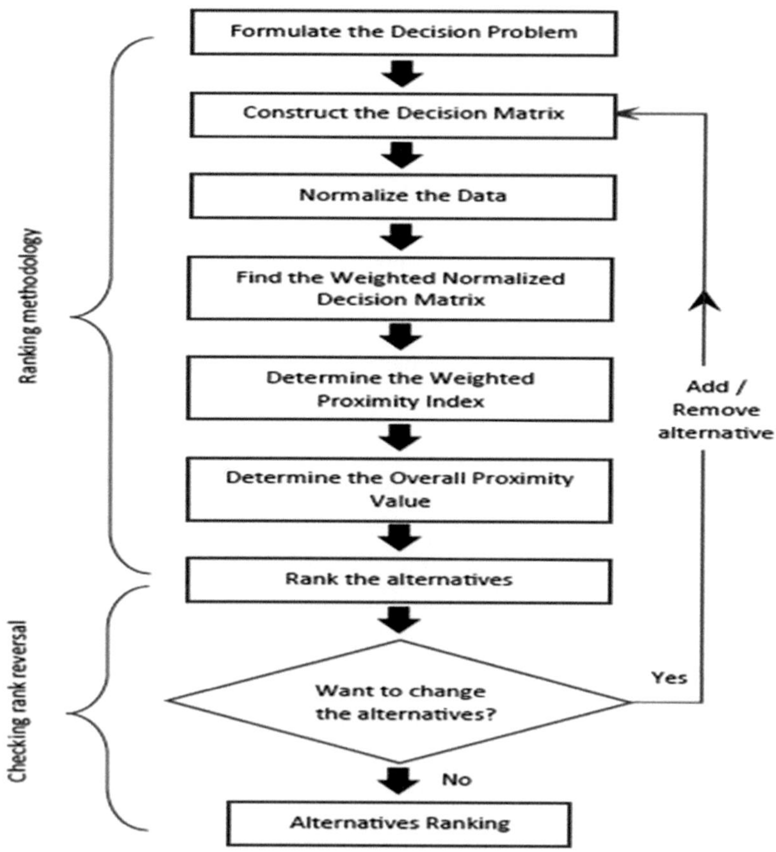

FIGURE 14.7 Flowchart for PIV method [51, 62].

Source: Adapted from Ref. [51, 62]

TABLE 14.4 Decision Matrix

AM Process	Machine	Material	Dimensional Accuracy (A)	Surface Roughness (R)	Tensile Strength (S)	Elongation (%E)	Heat Deflection Temperature (HDT)	Process Cost (PC)	Build Time (BT)
Units	–	–	mm	µm	MPa	%	°C	Rs	Min
Goal	–	–	(–)	(–)	(+)	(+)	(+)	(–)	(–)
Material jetting (MJ)	ProJet MJP 2500 +	Visijet M2R-WT	0.106	6.11	45	30	51	1,000	140
Vat photopoly-merization (VP)	EKA DLP 3D Printer	ABS TRU resin	0.743	3.06	49	10.2	50	700	30
Material extrusion (ME)	Pratham LFF 3D Printer	PLA Plus	0.323	19.96	60	29	52	300	72
Powder bed fusion (PBF)	EOS P 396	Nylon12 PA2200	0.136	11.79	48	18	163	1,180	160

Source: Reprinted with permission from [62]. Copyright © 2020, Indian Academy of Sciences

The data collected for the calculation of criteria weights is based on the interview conducted with five users (the first three were chosen as academic experts in AM processes, one as industry expert engineer, and one as manufacturer service bureau expert). The purpose has been explained to the experts during interaction for identification of the best and the worst criteria. The experts and users have been asked to provide their preferences of the best criterion over other criteria on a scale of 1 to 9, which forms the BO vector as given in Table 14.5. Similarly, they have also been asked to give importance to other criteria over the worst criterion on the same scale, which form the others to worst (OW) vector as given in Table 14.6. Whereas, using the same set of experts were asked to give their preference in PCM using the relative importance scale shown in Table 14.6. The PCM obtained was the one on which there is consensus shown in Table 14.7 and the normalized decision matrix is shown in Table 14.8.

TABLE 14.5 Best Criterion and Best to Others (BO) Vector – BWM [62]

Expert No.	Best Criterion	A	R	S	%E	HDT	PC	BT
1	A	1	1	3	5	7	6	6
2	R	1	1	4	6	6	7	8
3	A	1	2	1	7	8	9	7
4	S	1	3	1	2	5	7	7
5	R	1	1	2	5	6	9	8

Source: Reprinted with permission from [62]. Copyright © 2020, Indian Academy of Sciences

TABLE 14.6 Worst Criterion and Others to Worst (OW) Vector – BWM [62]

Expert No.	1	2	3	4	5
Worst Criterion	**HDT**	**BT**	**PC**	**BT**	**PC**
A	7	8	9	7	9
R	7	8	8	6	9
S	5	6	7	7	8
%E	4	5	5	4	7
HDT	1	7	6	5	5
PC	3	3	1	3	1
BT	3	1	2	1	3

Source: Reprinted with permission from [62]. Copyright © 2020, Indian Academy of Sciences

Further, the data collected in Tables 14.5 and 14.6 was used to calculate the optimal weights of each criterion for each expert using Eqn. (2). The

TABLE 14.7 Pairwise Comparison Matrix – AHP

	Dimensional Accuracy (AI)	Surface Roughness (R)	Tensile Strength (S)	Percentage Elongation (%E)	Heat Deflection Temperature (HDT)	Process Cost (PC)	Build Time (BT)
Dimensional accuracy (A)	1	1	3	5	7	9	7
Surface roughness (R)	1	1	3	5	7	9	7
Tensile strength (S)	1/3	1/3	1	3	4	8	7
Percentage elongation (%E)	1/5	1/5	1/3	1	2	8	4
Heat deflection temperature (HDT)	1/7	1/7	1/4	1/2	1	3	5
Process cost (PC)	1/9	1/9	1/8	1/8	1/3	1	2
Build time (BT)	1/7	1/7	1/7	1/4	1/5	1/2	1
Σ	2.930	2.930	7.851	14.768	21.533	38.500	36.000

values of the criteria weights and consistency ratio (CR) are listed in Table 14.8. For calculation, Microsoft Excel solver designed by BWM developer has been used, which is available on https://bestworstmethod.com/software/. The final optimal weight of each criterion has obtained by averaging the weights for all five experts, which is shown in the last row of Table 14.8.

TABLE 14.8 Normalized Decision Matrix – TOPSIS

AM Process	A	R	S	%E	HDT	PC	BT
Goal	(–)	(–)	(+)	(+)	(+)	(–)	(–)
DLP	0.904	0.131	0.537	0.286	0.556	0.557	0.168
FDM	0.393	0.854	0.658	0.814	0.292	0.238	0.404
SLS	0.165	0.504	0.526	0.505	0.914	0.795	0.899

It is observed from Table 14.10, that the average CR, ξ^{L*} (0.0876) is much smaller than 1. Hence, the weights of the criteria are optimal and reliable.

The normalized decision matrix obtained by using Eqn. (13), is presented in Table 14.10. For these calculations, a program for the PIV method was developed in MATLAB R2014b software (Tables 14.9–14.11).

TABLE 14.9 Weighted Normalized Decision Matrix ($V_{ij} = r^*_{ij} \times w_j$)–TOPSIS

AM Process	A	R	S	%E	HDT	PC	BT
Goal	(–)	(–)	(+)	(+)	(+)	(–)	(–)
DLP	0.278	0.040	0.088	0.030	0.017	0.016	0.004
FDM	0.121	0.262	0.107	0.086	0.018	0.006	0.010
SLS	0.051	0.155	0.086	0.054	0.056	0.022	0.022
V^+	0.051	0.040	0.107	0.086	0.056	0.006	0.004
V^-	0.278	0.262	0.086	0.030	0.017	0.022	0.022

TABLE 14.10 Optimal Weights of Criteria – PIV [62]

Expert No.	A	R	S	%E	HDT	PC	BT	Consistency-Ratio (ξ^{L*})
1	0.3163	0.3163	0.1275	0.0765	0.0357	0.0637	0.0637	0.0663
2	0.3268	0.3268	0.1100	0.0733	0.0733	0.0628	0.0266	0.1133
3	0.3216	0.2094	0.2718	0.0598	0.0523	0.0249	0.0598	0.0972
4	0.2780	0.1175	0.2780	0.1763	0.0705	0.0503	0.0290	0.0746
5	0.2976	0.2976	0.1922	0.0769	0.0640	0.0234	0.0480	0.0869
Average	**0.3080**	**0.2535**	**0.1959**	**0.0925**	**0.0591**	**0.0450**	**0.0454**	**0.0876**

Source: Reprinted with permission from [62]. Copyright © 2020, Indian Academy of Sciences

TABLE 14.11 Normalized Decision Matrix – PIV [62]

AM Process	A	R	S	%E	HDT	PC	BT
Goal	(–)	(–)	(+)	(+)	(+)	(–)	(–)
MJ	0.1279	0.2528	0.4427	0.6441	0.2725	0.5800	0.6182
VP	0.8970	0.1266	0.4821	0.2190	0.2993	0.4060	0.1324
ME	0.3899	0.8258	0.5903	0.6226	0.2779	0.1740	0.3179
PBF	0.1642	0.4878	0.4722	0.3865	0.8711	0.6844	0.7065

Source: Reprinted with permission from [62]. Copyright © 2020, Indian Academy of Sciences

Weights of the criteria calculated by the BWM which are shown in the last row of Table 14.10 are used in Eqn. (13) to determine the weighted normalized decision matrix, as shown in Table 14.12.

TABLE 14.12 Weighted Normalized Decision Matrix (V_{ij}) – PIV [62]

AM Process	A	R	S	%E	HDT	PC	BT
Goal	(–)	(–)	(+)	(+)	(+)	(–)	(–)
MJ	0.0394	0.0641	0.0867	0.0595	0.0161	0.0261	0.0280
VP	0.2763	0.0321	0.0944	0.0202	0.0176	0.0182	0.0060
ME	0.1201	0.2093	0.1156	0.0576	0.0164	0.0078	0.0144
PBF	0.0505	0.1236	0.0925	0.0357	0.0514	0.0307	0.0320

Source: Reprinted with permission from [62]. Copyright © 2020, Indian Academy of Sciences

The Weighted proximity index, u_i values obtained using Eqn. (14) has shown in Table 14.13.

TABLE 14.13 Weighted Proximity Index (u_i Values) – PIV [62]

AM Process	A	R	T	%E	HDT	PC	BT
Goal	(–)	(–)	(+)	(+)	(+)	(–)	(–)
MJ	0	0.0319	0.0289	0	0.0353	0.0182	0.0220
VP	0.2368	0	0.0212	0.0393	0.0338	0.0104	0
ME	0.0807	0.1772	0	0.0019	0.0350	0	0.0084
PBF	0.0111	0.0915	0.0231	0.0238	0	0.0229	0.0260

Source: Reprinted with permission from [62]. Copyright © 2020, Indian Academy of Sciences

Finally, the ranking of alternative AM processes has done based on the increasing overall proximity values (d_i), which were calculated by using Eqn. (15). The overall proximity values and ranking of the alternative processes presented in Table 14.14.

Table 14.14 and Figure 14.8 reveals the ranking of the alternative AM processes. Whereas Figure 14.9 show a ranking obtained ranking of alternate AM processes through TOPSIS method. The minimum value of d_i corresponds to material jetting (MJ) process (ranked 1) found to be the most suitable AM process. The ranking order obtained was MJ > PBF > ME > VP. However, in TOPSIS method, closeness coefficient has been calculated according to Eqn. (9) is presented in Table 14.15. The highest CC_i has been allotted rank 1 and vice versa.

TABLE 14.14 Overall Proximity (di) Value and Ranking – PIV [62]

AM Process	Overall Proximity Value (d_i)	Rank
MJ	0.1366	1
VP	0.3416	4
ME	0.3034	3
PBF	0.1987	2

Source: Reprinted with permission from [62]. Copyright © 2020, Indian Academy of Sciences

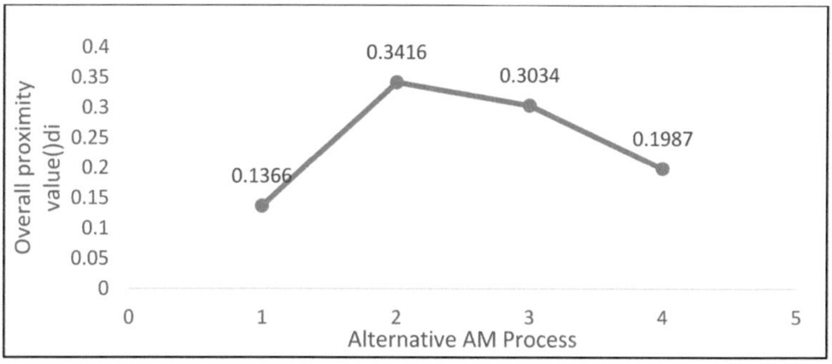

FIGURE 14.8 Ranking of alternative AM processes – PIV [62].

Source: Reprinted with permission from [62]. Copyright © 2020, Indian Academy of Sciences

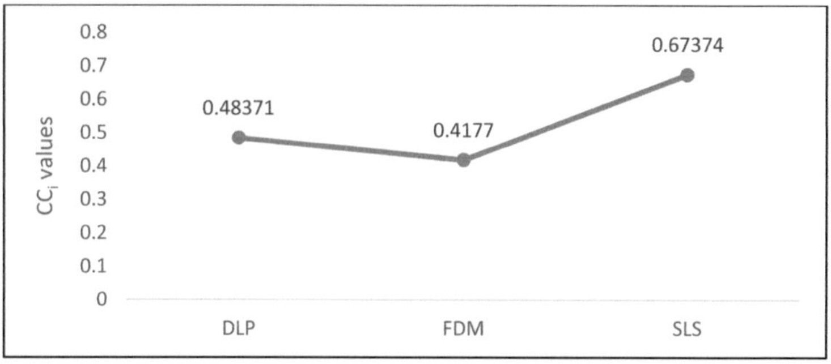

FIGURE 14.9 Ranking of alternative AM processes – TOPSIS [62].

Source: Reprinted with permission from [62]. Copyright © 2020, Indian Academy of Sciences

TABLE 14.15 Closeness Coefficient (CC_i) and Ranking – TOPSIS

AM Technology	d_i^+	d_i^-	CC_i	Rank
VP	0.238	0.223	0.484	2
ME	0.236	0.169	0.417	3
PBF	0.123	0.256	0.674	1

The ranking of the alternative AM processes has been obtained using the PIV method in Table 14.14. It clearly shows that the MJ process is the best choice among the four available AM processes. Vat photopolymerization (VP) is the worst choice as it gives lower dimensional accuracy (i.e., 0.743 mm) due to liquid material contraction. Although it has a high surface finish considering other all criteria, it is ranked 4. ME has the highest tensile strength though it is ranked 3. Considering all seven conflicting criteria the ranking obtained is first MJ process (MJP technology), second PBF process (SLS technology), third ME process (LFF/FDM technology), and fourth VP process (DLP technology). Depending upon the application, the ranking may be varied on providing different weights to the criteria.

14.5.1 SENSITIVITY ANALYSIS

A sensitivity analysis has been performed to ensure the authenticity of the results obtained in this study. The methodology available in the literature has been used to perform the sensitivity analysis [57–59]. It has been used to test the ranking obtained using varying the weights of all criteria in proportion to the highest weighted criterion. In this study, the highest weighted criterion is dimensional accuracy (A) as its weight is maximum, i.e., 0.3080 and therefore its weight has been varied from 0.1 to 0.9. Further weights of all other criteria have been calculated using the following equations [57]:

- if the weight of the P_{th} attribute changes by Δp, then the weight of other attributes changes by Δj, where:

$$\Delta j = \frac{\Delta p * w_j}{w_p - 1}; \mathrm{j} = 1, 2.\mathrm{k}; \ \mathrm{j} \neq p; \tag{17}$$

- The new weight of P_{th} attribute (Maximum weighted) changes as:

$$w'_p = w_p + \Delta p \tag{18}$$

- Weight of other attributes changes as:

$$w'_p = w_p + \Delta j \tag{19}$$

Table 14.16 shows the new weights during sensitivity analysis of all criteria calculated by varying the weight of dimensional accuracy from 0.1 to 0.9.

TABLE 14.16 Weights of the Criteria During Sensitivity Analysis [62]

Criteria	Normal	Modified Weights of All Criteria When the Weight of A is varied from 0.1 to 0.9								
A	0.3081	0.1000	0.2000	0.3000	0.4000	0.5000	0.6000	0.7000	0.8000	0.9000
R	0.2535	0.3298	0.2931	0.2565	0.2198	0.1832	0.1466	0.1099	0.0733	0.0366
S	0.1959	0.2548	0.2265	0.1982	0.1699	0.1416	0.1132	0.0849	0.0566	0.0283
%E	0.0926	0.1204	0.1070	0.0936	0.0803	0.0669	0.0535	0.0401	0.0268	0.0134
HDT	0.0592	0.0769	0.0684	0.0598	0.0513	0.0427	0.0342	0.0256	0.0171	0.0085
PC	0.0450	0.0586	0.0521	0.0455	0.0390	0.0325	0.0260	0.0195	0.0130	0.0065
BT	0.0454	0.0591	0.0525	0.0459	0.0394	0.0328	0.0263	0.0197	0.0131	0.0066
Sum	1	1	1	1	1	1	1	1	1	1

Source: Reprinted with permission from [62]. Copyright © 2020, Indian Academy of Sciences

Due to change in the weights of the criteria, the ranking of the alternative AM processes also changes. The ranking of the alternative AM processes is shown in Table 14.17.

TABLE 14.17 Ranking of AM Processes During Sensitivity Analysis [62]

Alternative AM Processes	Normal (0.3080)	Run 0.1	Run 0.2	Run 0.3	Run 0.4	Run 0.5	Run 0.6	Run 0.7	Run 0.8	Run 0.9
MJ	1	1	1	1	1	1	1	1	1	1
VP	4	2	4	4	4	4	4	4	4	4
ME	3	4	2	3	3	3	3	3	3	3
PBF	2	3	3	2	2	2	2	2	2	2

Source: Reprinted with permission from [62]. Copyright © 2020, Indian Academy of Sciences

It is evident from Table 14.17, that when the weight of the dimensional accuracy (A) is varied from 0.1 to 0.9, then alternative AM process MJ occupies rank 1. But during 0.1 and 0.2, the rank of the alternative VP and ME process get interchanged from 4 to 2, and vice versa and PBF process occupy rank 3. Table 14.17 also reveals that for all the weights' variation from 0.2 to 0.9, the alternative VP occupies the last rank, i.e., 4.

The weights of criteria and ranking of alternative AM processes derived from the sensitivity analysis are also shown in Figures 14.10 and 14.11, respectively.

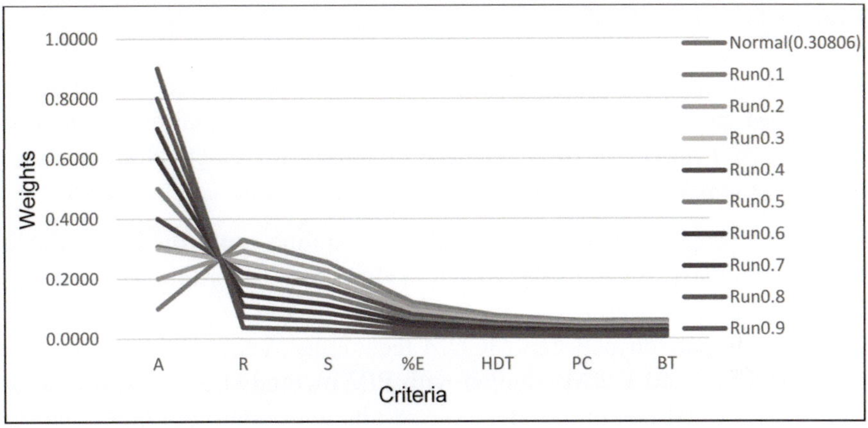

FIGURE 14.10 Weights of all the criteria during sensitivity analysis [62].

Source: Reprinted with permission from [62]. Copyright © 2020, Indian Academy of Sciences

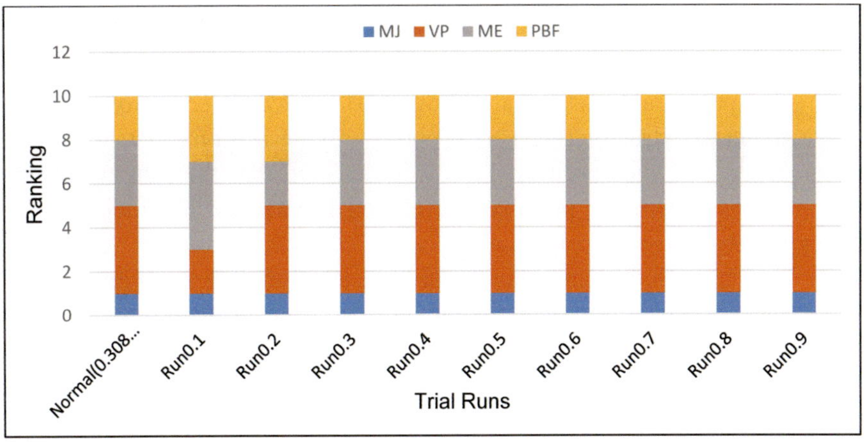

FIGURE 14.11 Ranking of alternative AM processes during sensitivity analysis [62].

Source: Reprinted with permission from [62]. Copyright © 2020, Indian Academy of Sciences

Sensitivity analysis yields dimensional accuracy (A) and surface roughness (R) is the most critical criteria in the selection of a suitable AM process for design verification purpose. Further, as material jetting (MJ) process possesses rank 1, is the most suitable AM process out of four considered

AM processes. Hence the results obtained using PIV are highly consistent and reliable.

14.6 CONCLUSIONS

This study presents a methodology for a decision support system in the selection of AM processes. A conceptual model (spur gear) was fabricated on four available AM processes to benchmark its ability for practical application. Based on the results obtained, the following conclusions can be drawn:

- The proposed new hybrid MCDM technique can provide a suitable framework for the selection of AM process according to the desirability of end-user.
- A comparison between MCDM techniques, i.e., AHP coupled with TOPSIS and BWM coupled with PIV method shows that the latter is more efficient to produce the results with minimum rank reversal problem.
- The data obtained through benchmarking can form a basis for database creation for a decision support system for the selection of AM processes.
- Experimentally obtained data are more precise and persuasive.
- The ranking order obtained to achieve desired dimensional accuracy and part quality: material jetting (MJ)> powder bed fusion (PBF) > material extrusion (ME) > vat photo-polymerization (VP).
- Results of the sensitivity analysis revealed that the ranking results derived from the proposed combined BWM-PIV methodology are highly consistent and reliable.
- The proposed methodology can be used in the selection problem of other applications area manufacturing, such as material selection, machine tool selection.

14.7 SCOPE FOR FUTURE WORK

In future work, for the selection of most suitable AM processes based on conflicting criteria, the following suggestions could be made.

- More number of decision criteria and alternative AM processes may include more options.
- More number of experts, customers, and users can be involved to obtain the rating of criteria weights to obtain more precise and unbiased results.

- Other MCDM methods may be used to compare the result obtained in this study.
- Based on the result obtained and proposed hybrid MCDM methodology, a web-based computer program software can be developed for the larger group of users.

KEYWORDS

- **additive manufacturing**
- **best worst method**
- **multiple criteria decision making**
- **photo-polymerization**
- **proximity indexed value**
- **stereolithography**

REFERENCES

1. *ISO: ISO/ASTM 52900:2015 (en)–Additive Manufacturing General–Principles–Terminology*. https://www.iso.org/obp/ui/#iso:std:iso-astm:52900:ed-1:v1:en (accessed on 17 February 2022).
2. Ghazy, M. M., (2012). *Development of an Additive Manufacturing Decision Support System (AMDSS).* https://theses.ncl.ac.uk/jspui/bitstream/10443/1692/1/Ghazy%2012.pdf (accessed on 17 May 2022).
3. Gibson, I., Rosen, D. W., & Stucker, B., (2010). Additive manufacturing technologies. *CIRP Encycl. Prod. Eng.* doi: 10.1007/978-1-4419-1120-9.
4. Srivastava, S., Garg, R. K., Sharma, V. S., Alba-Baena, N. G., Sachdeva, A., Chand, R., & Singh, S., (2020). Multi-physics continuum modelling approaches for metal powder additive manufacturing: A review. *Rapid Prototyping Journal, 26*(4), 737–764. doi: https://doi.org/10.1108/RPJ-07-2019-0189.
5. Senvol, L. L. C., (2015). *Senvol Database*. http://senvol.com/database/ (accessed on 17 February 2022).
6. Lee, J., An, J., & Chua, C. K., (2017). Fundamentals and applications of 3D printing for novel materials. *Appl. Mater. Today, 7*, 120–133. doi: 10.1016/j.apmt.2017.02.004.
7. Byun, H. S., & Lee, K. H., (2005). A decision support system for the selection of a rapid prototyping process using the modified TOPSIS method. *Int. J. Adv. Manuf. Technol., 26*, 1338–1347. doi: 10.1007/s00170-004-2099-2.
8. Wang, Y., Blache, R., & Xu, X., (2017). Selection of additive manufacturing processes. *Rapid Prototype. J., 23*, 434–447. doi: 10.1108/RPJ-09-2015-0123.

9. Gibson, I., Rosen, D., & Stucker, B., (2015). *Additive Manufacturing Technologies: 3D Printing, Rapid Prototyping, and Direct Digital Manufacturing* (2nd edn.). Springer.
10. *Additive Fabrication (Rapid Prototyping, Tooling)*. https://www.custompartnet.com/wu/additive-fabrication (accessed on 17 February 2022).
11. *The Free Beginner's Guide–3D Printing Industry*. https://3dprintingindustry.com/3d-printing-basics-free-beginners-guide#04-processes (accessed on 17 February 2022).
12. Lu, W., Lin, F., Han, J., Qi, H., & Yan, N., (2009). Scan strategy in electron beam selective melting. *Tsinghua Sci. Technol., 14*, 120–126. doi: 10.1016/S1007-0214(09)70078-1.
13. *Electron Beam Melting.* Additively. https://www.additively.com/en/learn-about/electron-beam-melting (accessed on 17 February 2022).
14. *MJF 3D Printing, Multi Jet Fusion 3D Printing Service in India.* think3D. https://www.think3d.in/multi-jet-fusion-mjf-3d-printing-service-india/ (accessed on 17 February 2022).
15. *Introduction to 3D Printing–Additive Processes* | Make, https://make.3dexperience.3ds.com/processes/introduction-to-additive-processes (accessed on 17 February 2022).
16. Pham, D. T., & Dimov, S. S., (2003). Rapid prototyping and rapid tooling—The key enablers for rapid manufacturing. *Proc. I Mech E Part C J. Mech. Eng. Sci., 217*, 1–23. doi: 10.1243/095440603762554578.
17. Molitch-Hou, M., (2018). Overview of additive manufacturing process. In: *Additive Manufacturing: Materials, Processes, Quantifications and Applications* (pp. 1–38). Elsevier Inc.
18. Yang, S., Page, T., Zhang, Y., & Zhao, Y. F., (2020). Towards an automated decision support system for the identification of additive manufacturing part candidates. *Journal of Intelligent Manufacturing.* In Press. doi: https://doi.org/10.1007/s10845-019-01490-z.
19. Hanemann, T., (2007). Rapid prototyping and rapid tooling techniques for the manufacturing of silicon, polymer, metal and ceramic microdevices. In: *MEMS/NEMS Handbook – Techniques and Applications* (pp. 187–255). Springer, Leondes.
20. *EBAM® Technology for Metal Additive Manufacturing* | Sciaky, http://www.sciaky.com/additive-manufacturing/electron-beam-additive-manufacturing-technology (accessed on 17 February 2022).
21. Peko, I., Bajić, D., & Veza, I., (2015). Selection of additive manufacturing process using the AHP method. In: *International Conference Mechanical Technologies and Structural Materials* (pp. 119–129).
22. Pham, D. T., & Gault, R. S., (1998). A comparison of rapid prototyping technologies. *Int. J. Mach. Tools Manuf., 38*, 1257–1287.
23. Braglia, M., & Petroni, A., (1999). A management-support technique for the selection of rapid prototyping technologies. *J. Ind. Technol., 15*, 1–6.
24. Bibb, R., (1999). *The Development of a Rapid Prototyping Selection System for Small Companies*. http://hdl.handle.net/10369/6518 (accessed on 17 February 2022).
25. Masood, S. H., & Soo, A., (2002). A rule based expert system for rapid prototyping system selection. In: *Robotics and Computer-Integrated Manufacturing* (pp. 267–274).
26. Mahesh, M., Loh, H. T., Wong, Y. S., & Fuh, J. Y. H., (2003). *Integrated Decision Support System for Selection of RP Processes*. http://utw10945.utweb.utexas.edu/Manuscripts/2003/2003-16-Mahesh.pdf (accessed on 10 February 2022).
27. Lan, H., Ding, Y., & Hong, J., (2005). Decision support system for rapid prototyping process selection through integration of fuzzy synthetic evaluation and an expert system. *Int. J. Prod. Res., 43,* 169–194. doi: 10.1080/00207540410001733922.

28. Rao, R. V., & Padmanabhan, K. K., (2007). Rapid prototyping process selection using graph theory and matrix approach. *J. Mater. Process. Technol., 194*, 81–88. doi: 10.1016/j.jmatprotec.2007.04.003.
29. Smith, P. C., & Rennie, A. E. W., (2008). Development of an additive layer manufacturing (ALM) selection tool for direct manufacturing of products. *Ind. Res.*, 507–518.
30. Armillotta, A., (2008). Selection of layered manufacturing techniques by an adaptive AHP decision model. *Robot. Comput. Integr. Manuf., 24*, 450–461. doi: 10.1016/j.rcim.2007.06.001.
31. Munguia, J., Bernard, A., Erdal, M., & Munguia, J., (2011). Proposal and evaluation of a KBE-RM selection system. *Rapid Prototyp. Journal, 17*(4), 236–246. doi: 10.1108/13552541111138351.
32. Kumar, L., & Jain, P. K., (2010). Selection of rapid prototyping technology. *Adv. Prod. Eng. Manag., 5*, 75–84.
33. Shames, H., (2010). *Development of a Selection Program for Additive Manufacturing Systems*. https://core.ac.uk/download/pdf/37324147.pdf (accessed on 17 May 2022).
34. Borille, A., Gomes, J., Meyer, R., & Grote, K., (2010). Applying decision methods to select rapid prototyping technologies. *Rapid Prototyp. J., 16*, 50–62. doi: 10.1108/13552541011011712.
35. Khrais, S., Al-Hawari, T., & Al-Araidah, O., (2011). A fuzzy logic application for selecting layered manufacturing techniques. *Expert Syst. Appl., 38*, 10286–10291. doi: 10.1016/j.eswa.2011.02.061.
36. Zhang, Y., Xu, Y., & Bernard, A., (2014). A new decision support method for the selection of RP process: Knowledge value measuring. *Int. J. Comput. Integr. Manuf., 27*, 747–758. doi: 10.1080/0951192X.2013.834474.
37. Mahapatra, S., & Panda, B., (2013). Benchmarking of rapid prototyping systems using grey relational analysis. *Int. J. Serv. Oper. Manag., 16*, 460–477. doi: 10.1504/IJSOM.2013.057509.
38. Shende, V., & Kulkarni, P., (2014). Decision support system for rapid prototyping process selection. *Int. J. Sci. Res. Publ., 4*, 2250–3153.
39. Vinodh, S., Nagaraj, S., & Girubha, J., (2014). Application of fuzzy VIKOR for selection of rapid prototyping technologies in an agile environment. *Rapid Prototyp. J., 20*, 523–532. doi: 10.1108/RPJ-07-2012-0060.
40. Mançanares, C. G., De S. Zancul, E., Cavalcante Da, S. J., & Cauchick, M. P. A., (2015). Additive manufacturing process selection based on parts' selection criteria. *Int. J. Adv. Manuf. Technol., 80*, 1007–1014. doi: 10.1007/s00170-015-7092-4.
41. Vimal, K. E. K., Vinodh, S., Brajesh, P., & Muralidharan, R. (2016). Rapid prototyping process selection using multi criteria decision making considering environmental criteria and its decision support system. *Rapid Prototyp. J., 22*, 225–250.
42. Çetİnkaya, C., Kabak, M., & Özceylan, E., (2017). 3D printer selection by using fuzzy analytic hierarchy process and PROMETHEE. *Int. J. Informatics Technol., 10*, 371–380. doi: 10.17671/gazibtd.347610.
43. Mondal, S., Singh, A. K., Chatterjee, P., & Chakraborty, S., (2017). Decision making for rapid prototyping process selection using complex proportional assessment method. In: *Proc. of the International Conference on Manufacturing Excellence*.
44. Park, H. S., & Tran, N. H., (2017). A decision support system for selecting additive manufacturing technologies. In: *Proc. of the International Conference on Information System and Data Mining* (pp. 151–155). Charleston, SC, USA.

45. Anand, M. B., & Vinodh, S., (2016). Application of fuzzy AHP – TOPSIS for ranking additive manufacturing process for microfabrication. *Rapid Prototyp. J., 24*, 424–435. doi: 10.1108/rpj-10-2016-0160.
46. Kretzschmar, N., Ituarte, I. F., & Partanen, J., (2018). A decision support system for the validation of metal powder bed-based additive manufacturing applications. *Int. J. Adv. Manuf. Technol., 96*, 3679–3690. doi: 10.1007/s00170-018-1676-8.
47. Peko, I., Gjeldum, N., & Bilić, B., (2018). Application of AHP, fuzzy AHP and PROMETHEE method in solving additive manufacturing process selection problem. *Tech. Gaz., 25*, 453–461. doi: 10.17559/TV-20170124092906.
48. Wang, Y., Zhong, R. Y., & Xu, X., (2018). A decision support system for additive manufacturing process selection using a hybrid multiple criteria decision-making method. *Rapid Prototyp. J., 24*, 1544–1553. doi: 10.1108/RPJ-01-2018-0002.
49. Rezaei, J., (2015). Best-worst multi-criteria decision-making method. *Omega (United Kingdom), 53,* 49–57. doi: 10.1016/j.omega.2014.11.009.
50. Rezaei, J., (2016). Best-worst multi-criteria decision-making method: Some properties and a linear model $. *Omega, 64*, 126–130. doi: 10.1016/j.omega.2015.12.001.
51. Mufazzal, S., & Muzakkir, S. M., (2018). A new multi-criterion decision making (MCDM) method based on proximity indexed value for minimizing rank reversals. *Comput. Ind. Eng., 119*, 427–438. doi: 10.1016/j.cie.2018.03.045.
52. *Make3D–3D Printer Manufacturer in India*. https://www.make3d.in/ (accessed on 17 February 2022).
53. *Laser-Sinter-System EOS P 396 for Additive Manufacturing of Serial Components and Functional Prototypes–EOS*, https://www.eos.info/systems_solutions/plastic/systems_equipment/eos_p_396 (accessed on 17 February 2022).
54. Sadjadi, S. J., & Karimi, M., (2018). Best-worst multi-criteria decision-making method: A robust approach. *Decis. Sci. Lett., 7*, 323–340. doi: 10.5267/j.dsl.2018.3.003.
55. Abouhashem, A. F., Ghasemian, S. I., Arab, A., Alavi, A., & Karachi, H., (2017). Application of the best-worst method in evaluation of medical tourism development strategy. *Decis. Sci. Lett., 7*, 77–86. doi: 10.5267/j.dsl.2017.4.002.
56. Badri, A. H., Kusi-Sarpong, S., & Rezaei, J., (2017). Assessing the social sustainability of supply chains using best worst method. *Resour. Conserv. Recycle, 126*, 99–106. doi: 10.1016/j.resconrec.2017.07.020.
57. V, P., & M, K., (2014). Sensitivity analysis of TOPSIS technique: An approach to archives websites' performance evaluation in our country with interval intuitionistic fuzzy information. *IOSR J. Dent. Med. Sci., 13*, 89–94. doi: 10.9790/0853-13828994.
58. Prakash, C., & Barua, M. K., (2015). Integration of AHP-TOPSIS method for prioritizing the solutions of reverse logistics adoption to overcome its barriers under fuzzy environment. *J. Manuf. Syst., 37,* 599–615. doi: 10.1016/j.jmsy.2015.03.001.
59. Kumar, S., Kumar, P., & Kumar, M., (2015). Risk analysis in green supply chain using fuzzy AHP approach: A case study. *Resour. Conserv. Recycl., 104*, 375–390. doi: 10.1016/j.resconrec.2015.01.001.
60. Saaty, R. W., (1987). The analytic hierarchy process-what it is and how it is used. *Math. Model., 9*, 161–176. doi: 10.1016/0270-0255(87)90473-8.
61. Srivastava, S., Kumar, S., & Garg, R K. (2021). A multi-objective optimization of TIG welding parameters using response surface methodology. *International Journal of Computer Aided Engineering and Technology, 14*(4), 575–590. DOI: https://doi.org/10.1504/IJCAET.2021.115363.
62. Raigar, J., Sharma, V. S., Srivastava, S., Chand, R., & Singh, J., (2020). A decision support system for the selection of an additive manufacturing process using a new hybrid MCDM technique. *Sadhana Journal, 45*, 1–14. doi: https://doi.org/10.1007/s12046-020-01338-w.

APPENDIX

Questionnaire Sample for the Selection of Additive Manufacturing Process Using the Best Worst Method

---

Name: ....

Designation: ....

Organization: ....

Experience: ....

Criteria/Attributes: (Size: ϕ50 mm, height: 10 mm, Material: Polymer)

1. Dimensional accuracy (mm)
2. Surface roughness (μm)
3. Tensile strength (MPa)
4. Percentage elongation (%)
5. Heat deflection temperature (°C)
6. Part cost (Rs)
7. Build time (Min)

Considering the goal (selecting the most appropriate AM process to fabricate a Gear part as a test prototype for design verification purpose), select the most important and worst criteria and rate other criteria according to their relative importance as given below:

Relative Important Scale	Description
1	Equal importance
2	Somewhat between equal and moderate
3	Moderately more important
4	Somewhat between moderate and strong
5	Strongly more important
6	Somewhat between strong and very strong
7	Very strongly important
8	Somewhat between very strong and extreme
9	Extremely important

Select the Best Criterion: ....

Comparison of Best criterion to other criteria: (Give rating using 1 to 9)

Best to Other (BO)	**Rating**
Dimensional accuracy	
Surface roughness	
Tensile strength	
% Elongation	
Heat deflection temperature	
Part cost	
Build time	

Ex: How much important is best criteria than Tensile Strength?

Ans: If "3," It means the best criteria is "Moderately more Important" than Tensile Strength.

Select the Worst Criterion: ....

Comparison of others to Worst: (Give rating using 1 to 9)

Others to Worst (OW)	**Rating**
Dimensional accuracy	
Surface roughness	
Tensile strength	
% Elongation	
Heat deflection temperature	
Part cost	
Build time	

Ex: How much important is Dimensional Accuracy than worst criteria?

Ans: If "7," It means Dimensional Accuracy is "Very strongly Important" than Worst Criteria.

Comments if any:

********************Thank You for your Feedback*******************

MATLAB Code for PIV method

```
% Decision Matrix
% A(mm) R(micm) S(MPa) E(%) HDT(C) PC(RS) BT(min)
X=[0.106 06.11 45 30.0 051 1000 140; %MJ
0.743 03.06 49 10.2 056 0700 30; %DLP
0.323 19.96 60 29.0 052 0300 72; %LFF
0.136 11.79 48 18.0 163 1180 160]; %SLS
% Criteria Weights
%W=[0.3080 0.2535 0.1959 0.0925 0.0591 0.0450 0.0454];
%W=[0.1000 0.3297 0.2548 0.1203 0.0769 0.0585 0.0590];
```

```
%W=[0.2000 0.2931 0.2265 0.1070 0.0684 0.0521 0.0525];
%W=[0.3000 0.2565 0.1982 0.0936 0.0598 0.0455 0.0459];
%W=[0.4000 0.2198 0.1699 0.0803 0.0513 0.0390 0.0394];
%W=[0.5000 0.1832 0.1416 0.0669 0.0427 0.0325 0.0328];
%W=[0.6000 0.1466 0.1132 0.0535 0.0342 0.0260 0.0263];
%W=[0.7000 0.1099 0.0849 0.0401 0.0256 0.0195 0.0197];
%W=[0.8000 0.0733 0.0566 0.0268 0.0171 0.0130 0.0131];
W=[0.9000 0.0366 0.0283 0.0134 0.0085 0.0065 0.0066];
Wcriteria=[0 0 1 1 1 0 0];% 1-Beneficial 0-Non beneficial criteria
Xval=length(X(:,1));
Y = zeros([Xval,length(W)]);
%% calculating the normalized matrix
for j=1:length(W)
for i=1:Xval
Y(i,j)=X(i,j)/sqrt(sum((X(:,j).^2)));
end
end
Normalized_Matrix = num2str([Y])
%% calculating the weighted normalized matrix
for j=1:length(W)
for i=1:Xval
Yw(i,j)=Y(i,j).*W(j);
end
end
Weighted_Normalized_Matrix = num2str([Yw])
%% calculating the weighted proximity index, ui
for i=1:Xval
for j=1:length(W)
ifWcriteria(1,j)== 0
U(i,j)=Yw(i,j)-min(Yw(:,j));
else
U(i,j)=max(Yw(:,j))-Yw(i,j);
end
end
```

```
end
weighted_proximity_index_U = num2str([U])
%% Overall proximity value
D=zeros(Xval,1);
for i=1:Xval
D(i)=sum(U(i,:));
end
Overall_proximity_value = num2str([D])
[sortedranks Ranks] = sort(D,'ascend');
Ranks
```

CHAPTER 15

STRENGTH ANALYSIS OF VIBRATORY FRICTION STIR WELDED AA 2024 AND AA 7075 DISSIMILAR METALS

GUNJI BALA MURALI[1], YOGESH DHAKAL[2], and
AMEY AKSHAY ACHARYA[2]

[1]*Assistant Professor (Sr.), School of Mechanical Engineering, VIT University, Vellore, Tamil Nadu, India*

[2]*School of Mechanical Engineering, VIT University, Vellore, Tamil Nadu, India*

ABSTRACT

Friction stirs welding (FSW) gained its importance due to its non-consumable electrode, which is used to join the two workpieces and to attain superior mechanical properties to the arc welding process. In the current research work, AA2024 and AA7075 dissimilar materials are being considered for welding using the FS welding process to improve the characteristics of the weld bead by mechanically vibrating the fixture normal to the weld bead using DC motors. To generate the vibrations for the fixture during welding, DC motors are mounted at the legs of the fixture. FSW and Vibratory Friction Stir welding (VFSW) are carried out on the same dissimilar materials, and strength is analyzed to know the weld bead strength of the two welding processes. For analyzing the strength, a tensile test has been carried out using the digital universal testing machine (DUTM). From the results, the strength of the VFSW weld bead is high compared to the FSW weld bead.

Decision-Making Models and Applications in Manufacturing Environments. Pushpdant Jain, Kumar Abhishek, & Prasenjit Chatterjee (Eds.)

15.1 INTRODUCTION

Welding is the process that can be performed in a solid or melted state on similar or dissimilar materials with filler or without filler by application of pressure or without pressure. In the modern era of manufacturing, different types of welding processes have been developed to join the metal plates. Out of those, FS welding is one of the solid-state welding processes that is widely applied these days for the metal joining process. The friction stir welding method is energy-effective, adaptable, and harmless to the ecosystem. In specific, FS welding is used to join high-strength alloys that are difficult to weld by the ordinary welding process. Aluminum alloys, AA2024, and AA7075 are mostly utilized in developing expanses in the aero sector and marine. The maximum of the research work on FSW is processed parameters optimization, microstructural analysis of similar and dissimilar metals, optimization of tool angle, and fixture design for holding the workpiece. In this research work, the authors attempted to perform FSW on dissimilar metals AA2024 and AA7075 under controlled vibration conditions to improve the strength of the weld bead.

Research has been carried out to know the strength of the weld bead by creating the vibrations to the tool normal to the welding direction using ultrasonic vibration setup during the joining of AA6061-T6 specimens. The authors analyzed the tool force, the temperature at the weld bead, and the strength of the weld bead for vibratory friction stir welded specimens. From the observations, UV to the tool during FS welding, strength, and ductility at the weld region is increased [1, 2].

Sajad et al. conducted a study on Al5052 alloy specimens during the FS welding process by generating the vibrations perpendicular to the weld line to improve the microstructural properties at the weld bead region. In the study, characterization and strength analysis have been carried out for the workpieces of the Al50502 alloy with FS and VFS weld processes. From the studies, it is found that strength and ductility at weld bead increase with the vibrations [3].

Rahim et al. carried out the experimentation by creating the vibrations perpendicular to the weld direction during friction stir welding on Al5052 alloy specimens. From the experimentation, the grain size of the Al5052 alloy specimen at the weld region decreases by 45%. In addition to the grain size decrement, the strength, and ductility of the weld specimens increase compared to the friction stir weld specimens [4].

Sadeesh et al. conducted the experimentation using FSW on 5 mm thick plates for analyzing the consequences of rotating speed, traverse speed, and tensile strength at the weld bead. To analyze the consequences, researchers used five different tool profiles to analyze the properties of the specimens. From the studies by varying the tool profiles the process parameters rotation speed, microstructure of the weld bead, and tensile strength are greatly affected [5].

Park et al. conducted the experimentation using FSW by providing the UV in the path of the tool feed. The studies carrying the FSW with vibrations will improve the weld bead quality, increases the tool life, and reduces the welding force [6].

Kumar et al. [7] study is to provide a quick review related to applications of UV in FS welding processes that can significantly contribute to improving mechanical as well as microstructural properties. In the study, ultrasonic integration to the FS welding process improves robustness, harmless, and having good weld characteristics compared to FS welding. With the studies, it has been proved that ultrasonic assistance in FSW significantly improves mechanical working as well as proper flow of weld material at the weld region by eliminating the several defects which occur during welding as well as enhancing weld mechanical and microstructure properties.

Cavaliere et al. [8] conducted a study during friction stir welding for 20204 and 7075 aluminum alloys to know the properties and their effect on the weld bead. Specimens are welded according to the different tool traverse directions and a strength test is carried out. From the observations, tensile strength at weld bead increases in the traverse direction. Saeid Amini et al. [9] highlighted the mechanism of friction stir welding and how it proves to be cheap, efficient, versatile, and environment friendly. Friction stir welding also overcomes the problem of evaporation encountered during welding.

Liu et al. use sonotrode to generate the ultrasonic vibrations near the 2024Al-T4 workpieces during the FS welding process to improve the properties at the weld bead. From the investigation, the authors found that properties at the weld bead increase compared to the FS welding weld bead workpieces [10].

The authors carried out the FSW experiments by changing the parameters like welding speed and shoulder diameter of the pin on the tensile strength of the weld bead by conducting the experimentation on Al alloy 2219 using FS welding. From the studies, the authors came to know that tensile strength was effecting by weld speed and tooltip diameter [11].

Yi et al. developed an empirical relation for estimating the heat input using welding parameters. To develop the empirical relation, the authors use the multiple regression analysis results. The authors also investigated the grain size effect by changing the heat input and developed a numerical relation between the heat input, grain size, and input variables [12].

Suhuddin et al. studied about the arrangement of grains and its appearance during the FS welding process of 6016 aluminum sheets using the stop-action technique. From the microstructural analysis, the authors found that there is a continuous development of recrystallization at the bead. The authors also found that at the circumference of the pin during welding undergoes simple shear deformation by which recrystallization grain structure will have good characteristics [13].

Singh et al. [14] have applied response surface methodology for process parameter optimization during vibratory welding. The authors optimized the process parameters like current required for welding operation, speed of the tool, and vibrations frequency induced during welding by developing two separate regression equations for the tensile strength and hardness. From the results, the authors found that vibration parameter is the most effecting factor for strength and hardness.

The authors developed a novel setup called the UV enhancement FS welding process to improve the joint quality by lowering the welding loads. From the developed setup the authors came to know that traverse force and tool torque can be reduced [15].

By the discussed literature, very few researchers are concentrated on providing the vibrations to the workpiece to test the mechanical properties of the weld specimens. Keeping this limitation in view, in this research article vibration friction stir welding is carried out by providing the vibrations to the fixture. Mechanical properties are tested for the friction stir welded specimen and vibratory friction welded specimen, from the results 50% efficiency has been observed in mechanical properties compared to the friction stir weld specimen.

15.2 MATERIALS AND METHODS

Aluminum plates AA2024 and AA7075 with a thickness of 6 mm and size as 150 mm × 50 mm were used as workpieces for welding. To carry the vibratory friction stir welding process, a 30–300 NC type FS welding machine is considered and the same is reflected in Figure 15.1.

Figure 15.2 shows the design considerations of the FS welding tool for welding the workpieces. High-Speed Stainless steel (HSS) material is considered as a tool material for welding the workpieces. Generally, in most of the vibratory friction stir welding processes, vibrations are provided to the tooltip but in this research work, vibrations are provided to the workpiece base. The setup shown in Figure 15.3 is considered for providing the vibrations to the workpieces table. Four DC motors are provided to the setup for generating the vibrations in a controlled manner. Before setting up the vibration table setup to the FS welding machine bed, frequency calculation has to be done to ensure that the generated vibrations are within the range and to avoid the resonance.

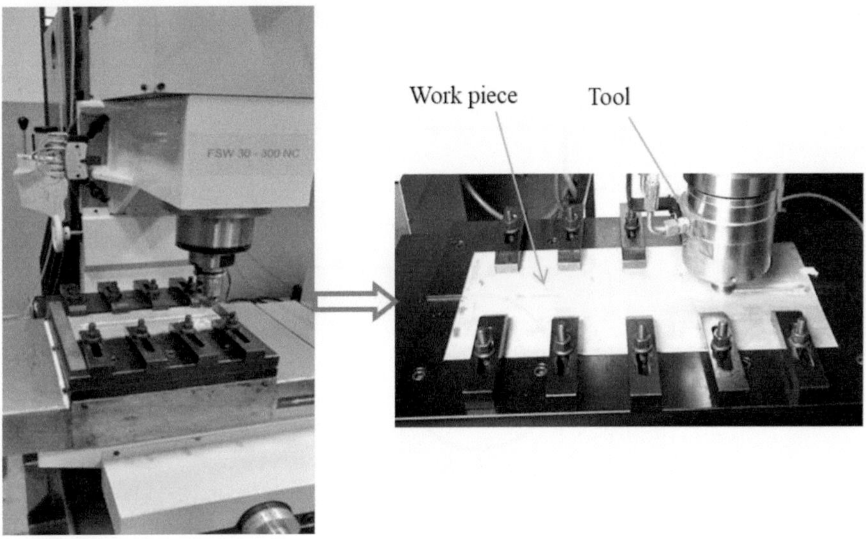

FIGURE 15.1 FS welding setup with the workpiece.

Table 15.1 represents the material and parts used in the vibration setup with the dimensions. A schematic design of the vibration table fixing to the FS welding bed is shown in Figure 15.4.

15.3 EXPERIMENTAL PROCEDURE

15.3.1 FREQUENCY CALCULATION

In this research work, before conducting the experimentation of FS welding and vibratory FS welding a frequency calculation has been carried out to ensure the developed vibration setup does not fall under resonance during experimentation. To calculate the frequency of the vibration table, the accelerometer is used to measure the vibration on the specimen region while the setup was running which is illustrated in Figure 15.5. The motor was designed in a way that vibrations are created to the upper fixture around the specimen region with a frequency of 1.2 kHz.

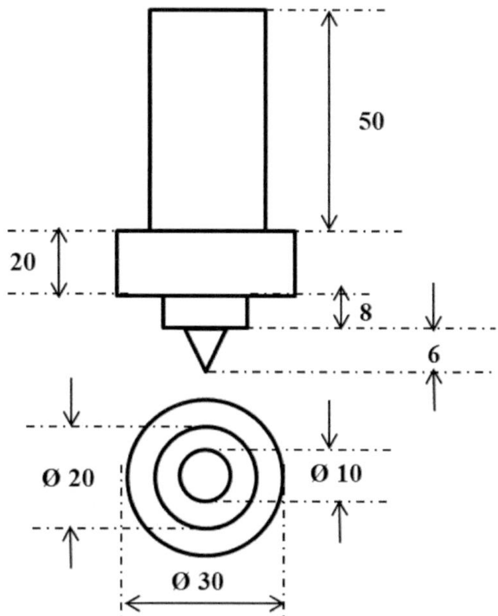

FIGURE 15.2 Tooltip design considerations (all dimensions are in mm).

FIGURE 15.3 Vibration table setup.

TABLE 15.1 Bill of the Material of Vibrational Setup

SL. No.	Part Name	Material	Quantity	Dimensions (mm)
1.	Base plate	Mild steel	1	180 × 350 × 10
2.	Top plate	Mild steel	1	200 × 300 × 10
3.	Rods	Mild steel	4	30 diameters
4.	Top clamps	Mild steel	6	70 × 20 × 10
5.	Bolts	Mild steel	12	M6
6.	Nuts	Mild steel	12	M6
7.	Motors	–	4	70 × 50 diameter

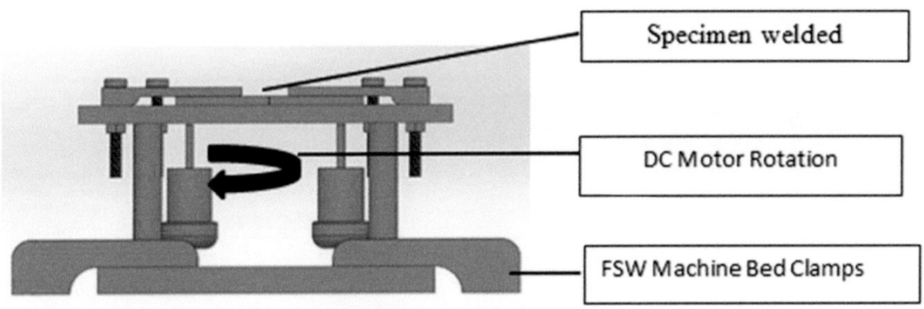

FIGURE 15.4 Fixture setup design for FS welding bed.

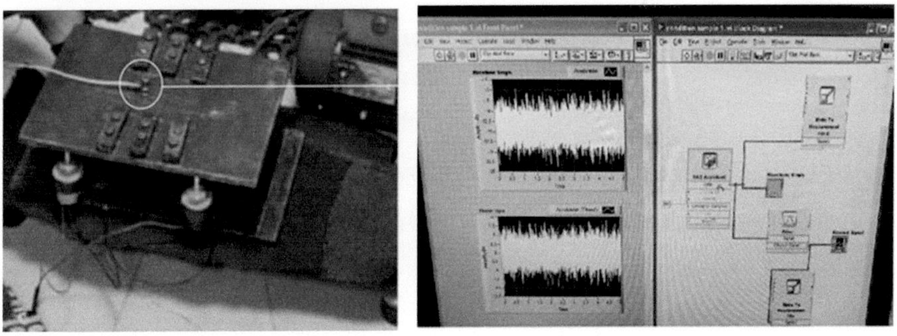

FIGURE 15.5 Experimentation of frequency calculation for the vibration table.

Frequency graphs are plotted as shown in Figures 15.6 and 15.7 for the natural frequency of the table and operating frequency of the table while vibrating at 1.2 kHz to know the resonance condition. From the figures, the amplitude of operating frequency and the natural frequency is different, and it is very much safe to operate the table at the considered frequency.

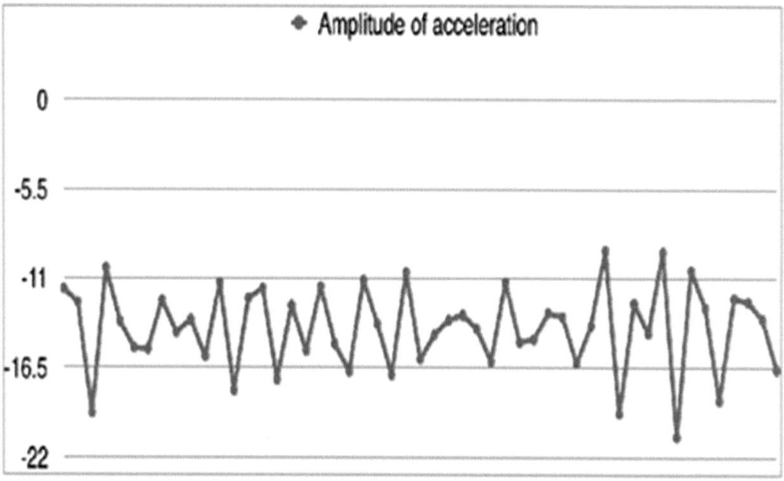

FIGURE 15.6 Operating frequency graph.

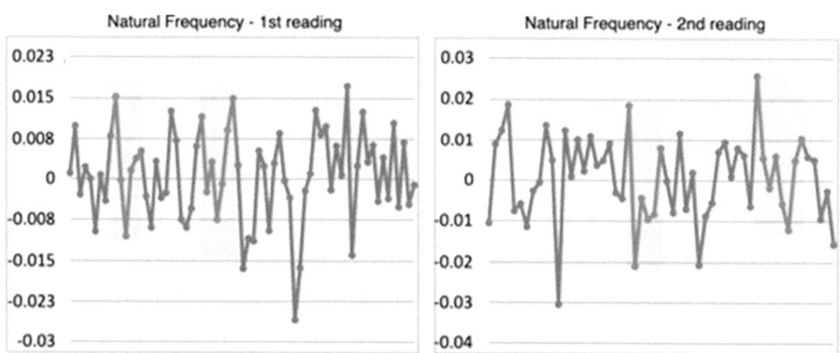

Frequency = 1/(0.00025*6) = **666.7 Hz** Frequency = 1/(0.00025*6) = **666.7 Hz**

Frequency = 1/(0.00025*5) = **800 Hz** Frequency = 1/(0.00025*7) = **571 Hz**

FIGURE 15.7 Natural frequency graph.

15.3.2 WELDING PROCESS

A schematic experimental setup of VFS welding is shown in Figure 15.8. Initially, the edges of the specimen were cleaned to remove oil and debris properly to place the fixture for VFSW, which is illustrated in Figure 15.8.

The vibration was applied through the vibrational stand setup as shown in Figure 15.3. The motors present to the vibrating table allow to rotate at a speed of 1,000 rmp to produce a frequency of 1.2 KHz, which transforms the vibration to the specimen attached to the upper fixture. The power for fixture vibration was supplied by a battery of a total voltage of 72 V. The batteries were connected in series while motors we connected in parallel with the combination of 2 DC motors with 36 V battery and next 2 DC motors with 36 V battery.

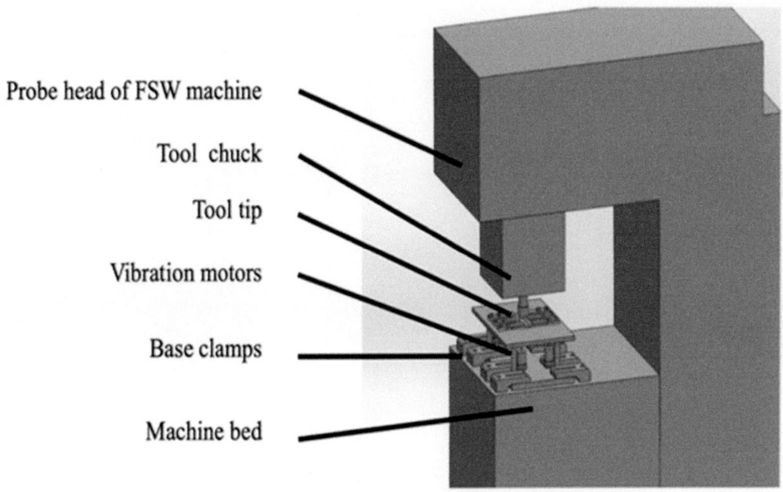

FIGURE 15.8 Schematic design for VFSW welding.

Figure 15.9 shows the vibratory FS welding process carried out by providing the vibrations to the worktable. At first, the aluminum plates were placed in the FS welding machine in the butt-welding position for carrying out normal FS welding and after completion of FS welding, the vibrational fixture setup was placed in the same FS welding machine to carry out Vibrational friction stir welding. The welded specimen with FS welding and VFS welding is illustrated in Figure 15.10. Welding conditions for FS welding and VFS welding are illustrated in Table 15.2. Except for the vibration other conditions for FS welding and VFS welding are the same.

15.4 RESULTS AND COMPARISON

15.4.1 EXPERIMENTATION

Experimentation is performed using the Instron testing machine with a travel speed of 5 mm/min as shown in Figure 15.11 to know the tensile strength at the weld bead of the joints obtained by FS welding and VFS welding. For the tensile studies, the samples are made using a wire cut electric discharge machine according to ASTM E8 standards normal to the weld line which is shown in Figure 15.12 with dimensions. To eliminate the surface irregulates during the welding process, all the specimens are polished thoroughly using a grinding machine.

FIGURE 15.9 Welding process.

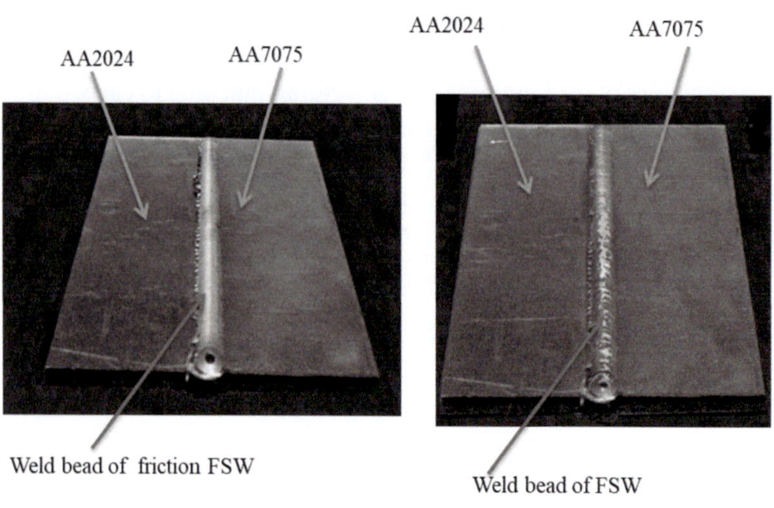

FIGURE 15.10 (a) FS welding welded specimen; and (b) VFS welding welded specimen.

TABLE 15.2 Welding Conditions During Experimentation

Type of Test	Mode	Spin Speed (rpm)	Tool Moving Speed (mm/min)
FS welding	No vibration	1,200	30
VFS welding	Vibration is provided	1,200	30

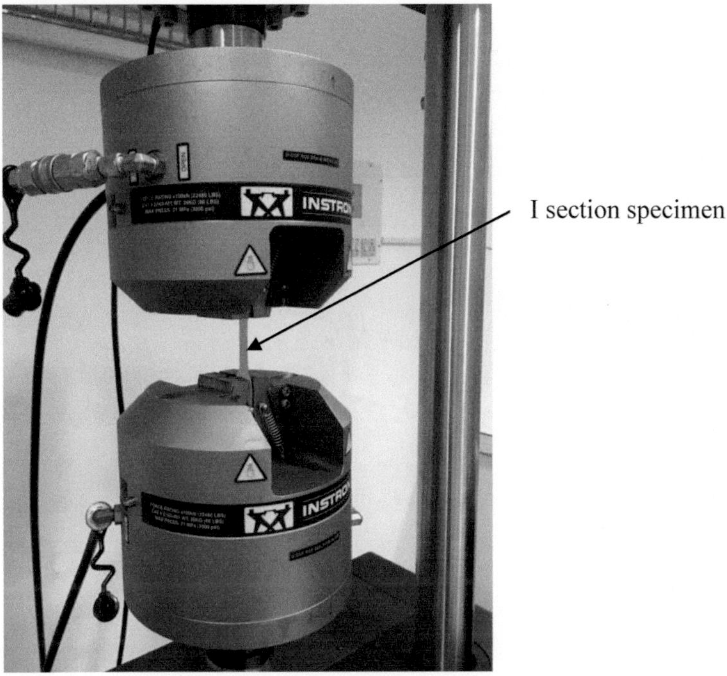

FIGURE 15.11 Tensile strength testing machine.

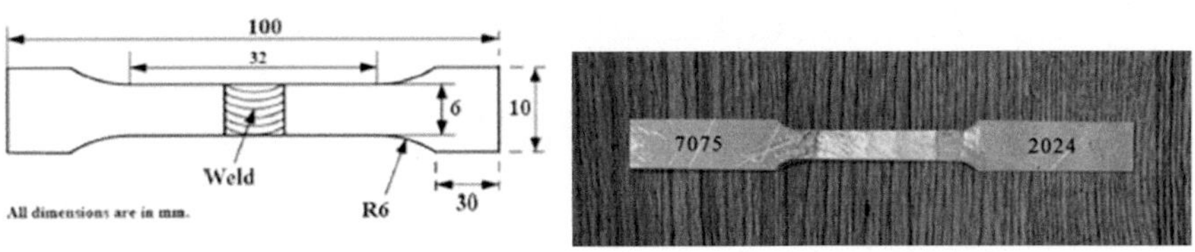

FIGURE 15.12 (a) Test sample design with dimensions; and (b) test sample.

From the experimentation yield strength, UTS, and percentage of elongation at fracture have been calculated and listed in Table 15.3. Based on the results from the table the strength and ductility values of VFS welding specimens are greater than those of FS welding, which means friction-stir welded specimens with vibration have better strength and mechanical properties.

A comparative study of the stress-strain graph between the friction stirs weld and vibration friction stir weld specimens and the corresponding experimentation values are listed in Table 15.3 and are shown in Figure 15.13. The UTS and yield strength values of tensile-tested specimens for both FS welding and VFS welding specimens lie between the induced stresses of the Al alloys (AA2024 and AA7075). From the results, yield strength and UTS for VFS welding are greater than those of the FS welding specimen.

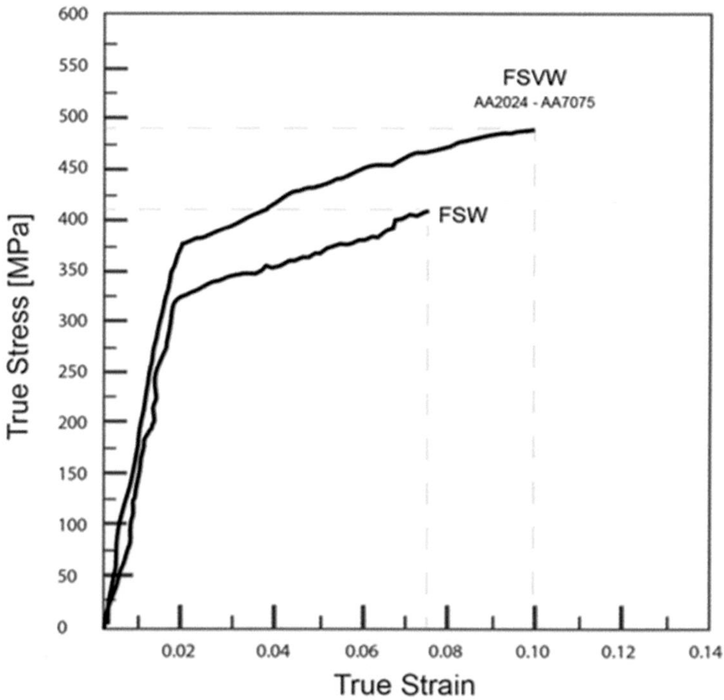

FIGURE 15.13 Stress-strain curve.

TABLE 15.3 Mechanical Properties of 2024-7075 Joints Compared with FS Welding and VFS Welding

	Tensile Strength (MPa)	Ultimate Tensile Strength (MPa)	Elongation (%)
2024-7075 FS welding	325	420	7
2024-7075 VFS welding	350	480	10

15.5 CONCLUSION

The present work discussed the benefits of vibrating the workpiece during the FS welding procedure for improving weld bead strength. In the study, vibrations are developed using DC motors for the fixture by fixing at the legs of the fixture. AA2024 and AA7075 specimens were welded using FS welding and VFS welding processes and mechanical properties were compared. It is seen that having vibration during welding enhances the mechanical properties at the weld zone. Additionally, from Table 15.3, the UTS value and ductility of the VFS welded specimen are greater than the friction stir welded specimen.

KEYWORDS

- **DC motors**
- **dissimilar metals**
- **fixture**
- **natural frequency**
- **strength**
- **vibratory friction stirs welding**

REFERENCES

1. Amini, S., & Amiri, M. R., (2014). Study of ultrasonic vibrations' effect on friction stir welding. *Int. J. Adv. Manuf. Technol., 73*, 127–135.
2. Amini, S., Nazari, M., & Rezaei, A., (2015). Bending vibrational tool for friction stir welding process. *The International Journal of Advanced Manufacturing Technology, 84*. 10.1007/s00170-015-7834-3.
3. Fouladi, S., Ghasemi, A. H., Abbasi, M., Abedini, M., Khorasani, A. M., & Gibson, I., (2017). The effect of vibration during friction stir welding on corrosion behavior, mechanical properties, and machining characteristics of stir zone. *Metals, 7*(10), 421.
4. Rahmi, M., & Abbasi, M., (2017). Friction stir vibration welding process: Modified version of friction stir welding process. *Int. J. Adv. Manuf. Technol., 90*, 141–151.
5. Sadeesh, P., Kannan, M. V., Rajkumar, V., Avinash, P., Arivazhagan, N., Ramkumar, K. D., & Narayanan, S. (2014). Studies on friction stir welding of AA 2024 and AA 6061 dissimilar metals. *Procedia Engineering, 75,* 145–149.

6. Park, K., (2009). *Development and Analysis of Ultrasonic Assisted Friction Stir Welding Process*. PhD Diss.
7. Kumar, S., Wu, C. S., & Padhy, G. K., (2017). "Ultrasonic vibrations in friction stir welding: State of the art. In: *7th International Conference on Welding Science and Engineering (WSE 2017) in Conjunction with 3rd International Symposium on Computer-Aided Welding Engineering (CAWE 2017)* (pp. 272–276). Shandong University, Jinan, China.
8. Cavaliere, P., Cerri, E., & Squillace, A., (2005). Mechanical response of 2024–7075 aluminum alloys joined by friction stir welding. *Journal of Materials Science, 40*(14), 3669–3676.
9. Amini, S., Mohammad, N. M., & Rezaei, A., (2016). Bending vibrational tool for friction stir welding process. *The International Journal of Advanced Manufacturing Technology, 84*(9), 1889–1896.
10. Liu, X. C., Wu, C. S., & Padhy, G. K., (2015). Improved weld macrosection, microstructure and mechanical properties of 2024Al-T4 butt joints in ultrasonic vibration enhanced friction stir welding. *Science and Technology of Welding and Joining, 20*(4), 345–352.
11. Arora, K. S., et al., (2010). Effect of process parameters on friction stir welding of aluminum alloy 2219-T87. *The International Journal of Advanced Manufacturing Technology, 50*(9), 941–952.
12. Yi, D., Onuma, T., Mironov, S., Sato, Y. S., & Kokawa, H., (2017). Evaluation of heat input during friction stir welding of aluminum alloys. *Science and Technology of Welding and Joining, 22*(1), 41–46.
13. Suhuddin, U. F. H. R., Mironov, S., Sato, Y. S., & Kokawa, H., (2010). Grain structure and texture evolution during friction stir welding of thin 6016 aluminum alloy sheets. *Materials Science and Engineering: A, 527*(7, 8), 1962–1969.
14. Singh, P. K., Deepak, K. S., Patel, D., & Prasad, S. B., (2017). Optimization of vibratory welding process parameters using response surface methodology. *Journal of Mechanical Science and Technology, 31*(5), 2487–2495.
15. Shi, L., Wu, C. S., & Gao, S., (2018). Analysis of welding load reduction in ultrasonic vibration-enhanced friction stir welding. *The International Journal of Advanced Manufacturing Technology, 99*(1), 373–385.

CHAPTER 16

PERFORMANCE EVALUATION OF CONVEYORS THROUGH FUZZY AHP-BASED FUZZY GREY THEORY ANALYSIS UNDER MCDM ENVIRONMENT

BIPRADAS BAIRAGI

Department of Mechanical Engineering, Haldia Institute of Technology, West Bengal, India

ABSTRACT

This chapter aims to combine the fuzzy analytic hierarchy process (AHP) with fuzzy grey theory (FGT) for the performance evaluation of conveyors under multi-criteria decision-making (MCDM) environment. Fuzzy AHP is applied to find fuzzy weights of criteria under consideration. FGT is employed to assess and choose suitable alternatives. It scientifically and rationally integrates the fuzzy MCDM tools in order to achieve an appropriate decision. An empirical problem on conveyor evaluation and selection is solved and its result is demonstrated that shows the effectiveness of the method. The present investigation finds that the methodology applied in the current endeavor is effective and useful to uncover more perfect decision regarding selection of the best conveyors under fuzzy MCDM environment.

16.1 INTRODUCTION

Selection of appropriate conveyor for handling material be capable of diminishing production lead time, improving product quality, enhancing

Decision-Making Models and Applications in Manufacturing Environments. Pushpdant Jain, Kumar Abhishek, & Prasenjit Chatterjee (Eds.)

material flow efficiency, increasing utilization facility, shortening product life cycles and enhance productivity. Materials handling with conveyors involve considerable costs and have a substantial consequence on the tactical position of the company in the industry. Hence conveyor selection for material handling in manufacturing system design is an imperative function of the decision-makers (DMs). This research aims at assisting DMs towards appropriate decision making. A good number of previous researchers have focused on the broad area under consideration. A review on the literature has been carried out and furnished below.

Soufi et al. [1] introduced an analytic hierarchy process (AHP)-based multi-criteria decision-making (MCDM) methodology for the evaluation and selection of material handling equipment to be utilized in manufacturing systems. Satyam et al. [2] applied a multi-attribute decision making approach for evaluation and selection of conveyors as material handling equipment. Nguyen et al. [3] advocated a combined multi-criteria decision making (MCDM) model for the evaluation and selection of conveyor as the material handling equipment based on fuzzy analytical hierarchy process and fuzzy additive ratio assessment (ARAS) with vague and imprecision information. Mathewa & Sahua [4] made a comparison among the novel multi-criteria decision-making approaches by solving a problem on material handling equipment selection. Karsak et al. [5] suggested a FMCDM model based on regression for robot evaluation and selection. Velocity, cost, load capacity, and repeatability were considered as criteria. Tansel et al. [6] suggested a decision support system for selection of robot namely ROBSEL. Using FAHP, robots are ranked in the next phase. This model cannot be applied for group decision making. Parkan et al. [7] advocated an MCDM process for robot selection problem. Though this paper makes an attempt to remove the inconstancy of ranking order, but it is also not free from the drawback of handling fuzzy environments. Chu and Lin proposed a fuzzy-based Technique for Order of Preference by Similarity to Ideal Solution (TOPSIS) procedure. Performance rating of diverse alternative robots along with various subjective criteria and weight of are evaluated by linguistic variables (LVs). Though this paper can handle both objective and subjective criteria, it cannot reduce the rank reversal phenomenon [8]. Kahraman et al. [9] developed a TOPSIS method based on hierarchical fuzziness employed for the selection of robotic system for industrial purpose. Sujono & Lashkari [10] suggested a method for selection of material handling system and for operation allocation simultaneously. Shih [11] suggested a group TOPSIS to select robots based on incremental benefit cost ratio. Chatterjee et al. [12] found the solution of the problem on robot selection using two different MCDM procedures to compare the comparative performance. Tuzkaya [13] proposed an integrated FMCDM approach MHE selection combining fuzzy set theory, Preference Ranking Organization Method for Enrichment Evaluations (PROMETHEE) and Analytic network process (ANP) methods. Genetic algorithm has been used for selection and allocation of MHE [14].

None of the models has made any attempt of removal inconsistency of selection procedures in the conventional models. Bairagi et al. [15] suggested a novel multiplicative technique to lessen and solve the robot selection problem. Liu [16] developed a TOPSIS process using LV and an interval 2-tuple to solve problem on robot selection problem in fuzzy environment. Bairagi et al. [17] applied three tools, viz. FTOPSIS, FVIseKriterijumska Optimizacija I Kompromisno Resenje (FVIKOR) and COmplex PRoportional ASsessment (COPRAS)-G integrated with FAHP in fuzzy environment for selection of industrial robots for foundry. Bellman & Zadeh [18] introduced fuzzy theory for decision making. While fuzzy theory is integrated with MCDM the concept is termed as FMCDM. FMCDM technique can aid DMs in making decision of selecting the right alternative [19]. This study uses LV to extract performance of alternatives and weights of criteria. Best non-fuzzy performance (BNP) is the defuzzification process [20]. Though there are different methods of defuzzification, center of area (COA) is an easy method for defuzzification [21].

None of the above models have made any attempts to remove inaccuracy in ranking orders and decision making. In the current approach, triangular fuzzy number (TFN) is utilized for its capability of handing LVs, simplicity, and ease of use. A linguistic term/variable are words or sentences in artificial or in a natural language [22, 23]. Researchers in general apply LVs in the difficult circumstances that are hard to characterize and enumerate with usual processes [24].

The above literature survey reveals that very few number of past researchers have applied MCDM techniques for solving conveyor selection problems. However, this effort is inadequate for exhaustive and extensive decision making regarding proper selection of conveyors from thousands of available alternatives under multiple conflicting criteria of diverse categories. Therefore, an effort in the chapter is made to suggest a combined multi-criteria fuzzy AHP based fuzzy grey theory (FGT) approach for performance evaluation of conveyors under MCDM environment. The real life situation of conveyor evaluation is associated with vague, imprecise, and ambiguous information. Hence imprecise data in a volatile environment deserves theory of fuzzy set in MCDM approaches. In the evaluation and ranking of conveyors, the present chapter uses fuzzy MCDM approaches which are thus more logical and appropriate.

The objective of this chapter is to investigate and explore ways to aid the DMs in the selection of conveyors for industrial purpose. The remaining part of the chapter is classified into some sections. Section 2 presents the fuzzy multiple criteria decision-making approaches to be integrated, that describes FAHP for estimation of weight and FGT for performance evaluation and the selection of conveyors. An empirical selection problem on conveyor evaluation is described in section 3. Section 4 explains the result and discusses some most important points of the chapter. Lastly, section 5 states some necessary conclusions on the overall investigation and indicates some idea for the possible directions of potential research.

16.2 FAHP AND FGT

16.2.1 FAHP

Saaty [25] developed Analytical Hierarchy Process (AHP) to exact ratio in 1–9 scale by comparison between every pair. The idea of integration of fuzzy set theory with AHP is termed as FAHP. In practical situation pair wise comparison is easy using linguistic terms. Buckley [26] expanded Saaty's AHP to FAHP for fuzzy weight estimation. The theory of FAHP is described below:

- **Step 1:** Build the weight matrix by pair-wise comparison every two criteria with the help of each decision maker.

$$\tilde{Q}_{(h)} = \begin{bmatrix} \tilde{1} & \tilde{q}_{12(h)} & \cdots & \tilde{q}_{1n(h)} \\ \tilde{q}_{21(h)} & \tilde{1} & \cdots & \tilde{q}_{2n(h)} \\ \vdots & \vdots & \ddots & \vdots \\ \tilde{q}_{m1(h)} & \tilde{q}_{m2(h)} & \cdots & \tilde{1} \end{bmatrix} = \begin{bmatrix} \tilde{1} & \tilde{q}_{12(h)} & \cdots & \tilde{q}_{1n(h)} \\ \tilde{q}^{-1}_{12(h)} & \tilde{1} & \cdots & \tilde{q}_{2n(h)} \\ \vdots & \vdots & \ddots & \vdots \\ \tilde{q}^{-1}_{1n(h)} & \tilde{q}^{-1}_{2n(h)} & \cdots & \tilde{1} \end{bmatrix} \quad (1)$$

$\tilde{q}_{ij(h)}$ be a fuzzy number assessed by h^th^ decision maker for i^th^ alternative and j^th^ criteria.

- **Step 2:** Build average weight matrix as follows by using the formula:

$$\tilde{Q} = \begin{bmatrix} \tilde{1} & \tilde{q}_{12} & \cdots & \tilde{q}_{1n} \\ \tilde{q}_{21} & \tilde{1} & \cdots & \tilde{q}_{2n} \\ \vdots & \vdots & \ddots & \vdots \\ \tilde{q}_{m1} & \tilde{q}_{m2} & \cdots & \tilde{1} \end{bmatrix} = \begin{bmatrix} \tilde{1} & \tilde{q}_{12} & \cdots & \tilde{q}_{1n} \\ \tilde{q}^{-1}_{12} & \tilde{1} & \cdots & \tilde{q}_{2n} \\ \vdots & \vdots & \ddots & \vdots \\ \tilde{q}^{-1}_{1n} & \tilde{q}^{-1}_{2n} & \cdots & \tilde{1} \end{bmatrix} \quad (2)$$

$\tilde{q}_{ij} = (q_1, q_2, q_3)$ is average of all fuzzy numbers assessed by different DMs. $\tilde{q}_{ij}$ can be computed using $\tilde{q}_{ij} = \left\{\tilde{q}_{ij(1)} \oplus \tilde{q}_{ij(2)} \oplus \ldots \tilde{q}_{ij(n)}\right\} / h$.

- **Step 3:** Compute Fuzzy Geometric Mean of the weights of every criterion using the formula.

$$\tilde{r}_i = (\tilde{q}_{i1} \otimes \tilde{q}_{i2} \otimes ... \tilde{q}_{in})^{1/n} = \left\{ \left(\prod_{j=1}^{n} Lq_{ij} \right)^{1/n}, \left(\prod_{j=1}^{n} Mq_{ij} \right)^{1/n}, \left(\prod_{j=}^{n} Uq_{ij} \right)^{1/n} \right\} \quad (3)$$

- **Step 4:** Evaluate fuzzy weight of the criteria with the normalization formula:

$$\tilde{w}_i = \tilde{r}_i \otimes (\tilde{r}_1 \oplus \tilde{r}_2 \oplus ... \oplus \tilde{r}_n)^{-1} \quad (4)$$

$\tilde{w}_i$ is fuzzy weight of criterion *i,* can be represented as $\tilde{w}_i = (Lw_i, Mw_i, Uw_i)$ i a TFN.

16.2.1 FUZZY GREY THEORY (FGT)

Fuzzy grey theory (FGT) is a well-organized method for analysis of mathematical systems specified by vague information and inadequate facts in the fuzzy environment [27]. The procedural steps with various normalization techniques used for different sense of the criteria are furnished below [28]:

- **Step 1:** Set the reference data:

$$x_0 = (x_{01}, x_{02}, \dots, x_{0n}) \quad (5)$$

Reference data consists of the most favorable value or target value for each criterion.

- **Step 2:** Determine performance values:

$$x_i = (x_{i1}, x_{i2}, \dots, x_{in}) \quad (6)$$

For every alternative regarding every criterion, where $j = 1,2\dots m.$, $j = 1,2\dots n$. *m* denotes alternative number and n criteria number.

- **Step 3:** Accomplish normalization of performance values by using the equations:

$$r_{ij} = \frac{x_{ij} - (x_j)_{\min}}{(x_j)_{\max} - (x_j)_{\min}}, j \in B \quad (7)$$

$$r_{ij} = \frac{(x_j)_{\max} - x_{ij}}{(x_j)_{\max} - (x_j)_{\min}}, j \in C \quad (8)$$

$$r_{ij} = 1 - \frac{\left|x_{ij} - x_j^*\right|}{\max\left|x_{ij} - x_j^*\right|}, \ \mathrm{j} \in O \tag{9}$$

x_j^* is optimum value associated with the jth criterion.

- **Step 4:** Compute the differences between data sets by the equation.

$$\Delta_i = \left\{\left|x_{01} - x_{i1}\right|, \left|x_{02} - x_{i2}\right|, \ldots, \left|x_{0n} - x_{in}\right|\right\} \tag{10}$$

Also, determine global maximum $(\Delta_j)_{max}$ and global minimum $(\Delta_j)_{min}$ for each criterion.

- **Step 5:** Evaluate gray relation coefficient by the equation:

$$\lambda_i(j) = \frac{(\Delta_j)_{\min} + \mu(\Delta_i)_{\max}}{\Delta_i(j) + \mu(\Delta_j)_{\max}} \tag{11}$$

Value of coefficient μ is in the interval $0 \le \mu \le 1$ but usually 0.5 is taken as the value of μ. The coefficient μ is used to mitigate the effect of global maximum value $(\Delta_j)_{max}$.

- **Step 6:** Compute the fuzzy gray relational grade (GRG_i) by means of the subsequent formula.

$$GRG_i = \sum_{j=1}^{n} w_j \otimes \lambda_i(j) \tag{12}$$

w_j is the weight of jth the criterion in fuzzy number. The alternative with the highest value of objective gray relational grade is the best solution and the alternative with the least value of objective gray relational grade is the worst solution.

- **Step 7:** Determine the modified gray relational grade (GRG_i) by using the equation.

$$MGRG_i = \frac{GRG_i - GRG_{\min}}{GRG_{\max} - GRG_{\min}} \tag{13}$$

where; $MGRG_i$ refers to the modified gray relational grade of the ith conveyor, GRG_{min} refers to the minimum Grey Relational Grade, where GRG_{max} refers to the Maximum Grey Relational Grade. GRG_{min} and GRG_{max} are crisp value defuzzified by taking an average of the three points of a TFN.

The appliance of the suggested methodology mentioned above is exemplified with a real life problem on conveyor selection presented in section 3.

16.3 PROBLEM DEFINITION

An Indian manufacturing organization desires to select the best conveyor before commissioning its own manufacturing units. For this purpose, the management of the organization preliminary screens nine different conveyors and considers ten selection criteria. The list of conveyors is furnished below:

- **E1: Flat Belt Conveyor:** These are extensively used as the principle conveying system in delivering sand in foundry, fuel in power plants, for ground level and underground haulage of coal and ores in coal mining, between processing step in the light and food industries.
- **E2: Gravity Discharge Bucket Conveyor:** It is used to move bulk material in a horizontal, inclined or vertical path over suitable drive, corner, and take-up terminals. Unload of buckets are mechanically done the last part of the run. Gravity discharged buckets are attached to an endless cable, belt or chain. These are simple and relatively low cost. Buckets are automatically discharged under the gravity at each run of conveyors.
- **E3: Apron Chain Conveyor:** It is the best for conveying bulk material such as chips formed by machine tools. Specifications can be changed according to applications. The carrying surface of the apron chain conveyor consists of overlapping metal apron. Aprons are partitioned with plates. Stop plates are used for sharp inclined conveyor. The stop plates can be fitted at any preferred pitch. Small holed aprons are available for freezing and drying parts or material.
- **E4: Gravity Chute Conveyor:** These are useful for many kind of material and objects, interflow moves, interlevel moves, ware houses, stores, loading, and unloading carriers A slide usually made of metal or other material and shaped to guide material during moving. These may be used for horizontal, inclined or spiral path between extreme levels. Chute conveyor preferably is powered by gravity. The key advantages of chute conveyors are inexpensiveness, no power requirement, low maintenance and economic use of space.
- **E5: Pipeline Pneumatic Conveyor:** A pneumatic conveying system is used to move granules, powders, and other dry bulk materials. The motive force in pneumatic conveying system is due to the combination of differential pressure and the stream of air supplied by an air mover like blower or fan. The system is entirely enclosed and normally has a small number of moving parts.

- **E6: Gravity Roller Conveyor:** The two main types of gravity conveyor are roller or skate wheel. Gravity roller conveyor is cost effective equipment for moving materials either by using a gravity fall or manual pushing. Gravity roller type is a superior choice for movement of unstable/heavier loads. Gravity conveyor with skate wheel employs wheels of 48 mm diameter. These are made of either plastic or aluminum. The advantage is their low inertia and orientation characteristics. The disadvantage is their dependency on the manufactured goods/items having a good quality flat base.
- **E7: Screw Conveyor:** It is simple in design, easy to maintain. This conveyor allows intermediate discharge of materials at a number of desired points. The screw conveyor has small width and dust-tight jacket covering the trough. It consumes high power due to the friction of material with the screw and trough, wear of conveyor components, crushing of material. It has low and medium capacity up to 100 cubic meters per hour. They are mostly 30 to 40 meter long and exceptionally 50 to 60 meters.
- **E8: Vibrating Conveyor:** These are largely utilized in food grade that essentially requires wash down, sanitation, and little maintenance. Vibrating conveyors are also useful for very hot, dirty, harsh, and corrosive atmosphere.
- **E9: Wheel Conveyor:** Gravity wheel conveyor is non-powered light weight and economical conveyor suitable for moving light weight unit products and packages. The free spinning wheel permits products and packages to convey with minimal manual efforts or gravity takes over. The conveyor is suitable for portable applications due to its light construction. Aluminum option ensures higher portability.

Though there are nine types of conveyors in the problem, their basic function is same-conveyance of material with diverse capacity and flexibility. Therefore, the conveyors can be considered as alternatives. Since the problem is related to the finding of the best alternative conveyor among a set of feasible alternative conveyors hence it is a performance evaluation and selection problem.

DMs' make a selection committee involves experts, from diverse sectors of the organization, having complete knowledge of material handling devices. A cut-off is set at 10 percentile above which a criterion is considered. Using this method the following ten criteria are selected for conveyor evaluation. For selection of criteria:

- **C1:** Speed (ft/min);
- **C2:** Load carrying capacity (quintal/min);
- **C3:** Path flexibility;
- **C4:** Material flexibility;
- **C5:** Life period (years);
- **C6:** Elevated angle (degree);
- **C7:** Maximum distance covered (m);
- **C8:** Maintenance required;
- **C9:** Space required (m^2);
- **C10:** Cost ($).

First seven (C1–C7) are benefit criteria and their higher values are desirable, whereas the last three (C8–C10) are cost/non-benefit criteria whose lower values are desirable for the optimized performance of the conveyors. Here cost refers procurement cost for the equipment. For solving the conveyor selection problem, a decision matrix with ratings and a weights matrix with weights. Ratings and weights are represented by two different sets of five grades of LVs evaluated by one or more authentic persons like manufacturers, operators, DMs who are associated with conveyors. Though selection criteria C1, C2, C5, C6, C7, C9, and C10 are subjective in nature, due to imprecise, incomplete, and ambiguous nature of available data, these are evaluated in terms of linguistic terms. Now the problem of conveyor selection is solved using the methodologies described in the chapter. The results and discussions on the given FMCDM problem of conveyor selection are summarized in the following section.

16.4 CALCULATION AND DISCUSSIONS

In this method, ratings are normalized with particular equations different for benefit criteria, cost criteria and criteria with optimum or target value. The sense of all normalized performance ratings is positive. Differences between data sets are computed by the variation of normalized performance rating from the reference data set by DMs. Grey relation coefficient of every conveyor regarding every criterion is calculated. Fuzzy numbers for performance rating, associated LVs, corresponding acronyms, and TFNs are presented in Table 16.1. Fuzzy numbers for weights of criteria, associated LVs, scales of corresponding TFN and $(\text{TFN})^{-1}$ is shown in Table 16.2.

TABLE 16.1 Fuzzy Numbers, Linguistic Variables, Acronyms, TFNs

Fuzzy Numbers	Linguistic Variables	Acronyms	Scale of TFNs
$\tilde{1}$	Extremely poor	EP	(0, 1, 3)
$\tilde{3}$	Poor	P	(1, 3, 5)
$\tilde{5}$	M	M	(3, 5, 7)
$\tilde{7}$	Good	G	(5, 7, 9)
$\tilde{9}$	Extremely good	EG	(7, 9, 10)

TABLE 16.2 Fuzzy Numbers, Linguistic Variables, Scales of TFN and $(\text{TFN})^{-1}$

Fuzzy Numbers	Linguistic Variables	Scales of TFN	$(\text{TFN})^{-1}$
$\tilde{1}$	Extremely low	(1, 1, 2)	(1/2, 1, 1)
$\tilde{3}$	Low	(2, 3, 4)	(1/4, 1/3, 1/2)
$\tilde{5}$	Medium	(4, 5, 6)	(1/6, 1/5, 1/4)
$\tilde{7}$	High	(6, 7, 8)	(1/8, 1/7, 1/6)
$\tilde{9}$	Extremely high	(8, 9, 9)	(1/9, 1/9, 1/8)

Fuzzy decision matrix in terms of linguistic terms is in Table 16.3. Fuzzy weight matrix containing fuzzy numbers in terms of TFNs are given in Table 16.4. FGRG, Defuzzified DGRG, Modified DGRG and Rank of the Conveyor are furnished in Table 16.5.

TABLE 16.3 Fuzzy Decision Matrix in Terms of Linguistic Terms

Ei	C1 (+)	C2 (+)	C3 (+)	C4 (+)	C5 (+)	C6 (+)	C7 (+)	C8 (–)	C9 (–)	C10 (–)
E1	EG	EG	P	G	M	P	G	P	G	M
E2	G	M	E P	P	M	EG	M	P	P	P
E3	P	M	P	EG	G	M	M	E P	M	G
E4	M	M	M	G	M	M	P	P	P	EG
E5	M	P	G	P	G	EG	P	E P	P	P
E6	P	M	P	M	G	E P	M	E P	M	M
E7	E P	G	E P	P	M	EG	E P	P	M	EG
E8	E P	M	M	G	M	P	P	P	M	EG
E9	M	P	M	G	M	M	M	P	G	EG

TABLE 16.4 Fuzzy Weight Matrix in Terms of TFNs

	C1 (+)*	C2 (+)	C3 (+)	C4 (+)	C5 (+)	C6 (+)	C7 (+)	C8 (–)**	C9 (–)	C10 (–)	Fuzzy Weights
C1	$\tilde{1}$	$\tilde{5}$	$\tilde{7}^{-1}$	$\tilde{5}$	$\tilde{3}$	$\tilde{3}^{-1}$	$\tilde{5}$	$\tilde{1}$	$\tilde{3}^{-1}$	$\tilde{3}$	(.030.037.052)
C2	$\tilde{5}^{-1}$	$\tilde{1}$	$\tilde{9}$	$\tilde{1}$	$\tilde{3}$	$\tilde{9}$	$\tilde{1}$	$\tilde{5}$	$\tilde{9}$	$\tilde{3}$	(.057.067.091)
C3	$\tilde{7}$	$\tilde{9}^{-1}$	$\tilde{1}$	$\tilde{9}^{-1}$	$\tilde{9}^{-1}$	$\tilde{1}$	$\tilde{9}^{-1}$	$\tilde{7}^{-1}$	$\tilde{1}^{-1}$	$\tilde{9}^{-1}$	(.008.009.012)
C4	$\tilde{5}^{-1}$	$\tilde{1}^{-1}$	$\tilde{9}$	$\tilde{1}$	$\tilde{3}$	$\tilde{9}$	$\tilde{1}$	$\tilde{5}$	$\tilde{9}$	$\tilde{3}$	(.056.067.082)
C5	$\tilde{3}^{-1}$	$\tilde{3}^{-1}$	$\tilde{9}$	$\tilde{3}^{-1}$	$\tilde{1}$	$\tilde{9}$	$\tilde{1}$	$\tilde{3}$	$\tilde{3}$	$\tilde{1}$	.026.031.043)
C6	$\tilde{3}$	$\tilde{9}^{-1}$	$\tilde{1}^{-1}$	$\tilde{9}^{-1}$	$\tilde{9}^{-1}$	$\tilde{1}$	$\tilde{9}^{-1}$	$\tilde{3}^{-1}$	$\tilde{1}$	$\tilde{9}^{-1}$	(.008.009.121)
C7	$\tilde{5}^{-1}$	$\tilde{1}^{-1}$	$\tilde{9}^{-1}$	$\tilde{1}^{-1}$	$\tilde{1}^{-1}$	$\tilde{9}$	$\tilde{1}$	$\tilde{5}$	$\tilde{9}$	$\tilde{9}$	(.032.043.049)
C8	$\tilde{1}^{-1}$	$\tilde{5}^{-1}$	$\tilde{7}^{-1}$	$\tilde{5}^{-1}$	$\tilde{3}^{-1}$	$\tilde{3}$	$\tilde{5}^{-1}$	$\tilde{1}$	$\tilde{3}$	$\tilde{3}^{-1}$	(.011.014.019)
C9	$\tilde{3}$	$\tilde{9}^{-1}$	$\tilde{1}$	$\tilde{9}^{-1}$	$\tilde{3}^{-1}$	$\tilde{1}^{-1}$	$\tilde{9}^{-1}$	$\tilde{3}^{-1}$	$\tilde{1}$	$\tilde{9}^{-1}$	(.009.010.014)
C10	$\tilde{3}^{-1}$	$\tilde{3}^{-1}$	$\tilde{9}$	$\tilde{3}^{-1}$	$\tilde{1}^{-1}$	$\tilde{9}$	$\tilde{9}^{-1}$	$\tilde{3}$	$\tilde{9}$	$\tilde{1}$	(.027.034.044)

**(+) denotes benefit criterion; and*

***(–) cost criterion.*

TABLE 16.5 FGRG, Defuzzified DGRG, Modified DGRG and Rank of the Conveyor

Conveyors	FGRG*	DGRG**	MGRG***	Rank
E 1	(0.1902, 0.2334, 0.3104)	0.7340	1	1
E 2	(0.1339, 0.1663, 0.2208)	0.5210	0.3064	5
E 3	(0.1691, 0.2092, 0.2717)	0.6501	0.7268	2
E 4	(0.1228, 0.1521, 0.2119)	0.4869	0.1952	7
E 5	(0.1530, 0.1886, 0.2507)	0.5922	0.5384	3
E 6	(0.1487, 0.1852, 0.2511)	0.5850	0.5147	4
E 7	(0.1100, 0.1325, 0.1843)	0.4269	0	9
E 8	(0.1342, 0.1644, 0.2130)	0.5116	0.2757	6
E 9	(0.1169, 0.1464, 0.2000)	0.4633	0.1184	8

**FGRG*: Fuzzy grey relational grade;

***DGRG*: Defuzzified grey relational grade;

****MGRG*: Modified grey relational grade.

Data set are fixed value with positive sense, i.e., higher value is better. Grey relation coefficients have a negative sense, i.e., the lower value is better. Figure 16.1 is the graphical representation of fuzzy gray relation coefficients (FGRC). It is clearly noted from the graph that the variations of the FGRC are quite wide. In another word, FGRC frequently fluctuates to a large extent. FGRC is utilized to compute fuzzy gray relational grade (FGRG) which is defuzzified by taking BNP. Again, FGRG is modified/normalized by setting the lowest MGRG at 0 (zero) and the highest MGRG at 1 (unity) as summarized in Table 16.5. FGRG of each conveyor is graphically represented in Figure 16.2 that gives a preface of the best performance of conveyor E1 (Flat belt). A comparison of DGRG with MGRG is depicted in Figure 16.3. MGRGs of the conveyors are 1, 0.3064, 0.7268, 0.1952, 0.5384, 0.5147, 0, 0.2757, and 0.1184 respectively. The ranking of the conveyors is E1>E5>E2>E7>E3>E4>E9>E6>E8. Figure 16.4 represents the ranking order of the conveyors. This ranking indicates that E1 (Flat belt) is the best conveyors and E8 is the worst one.

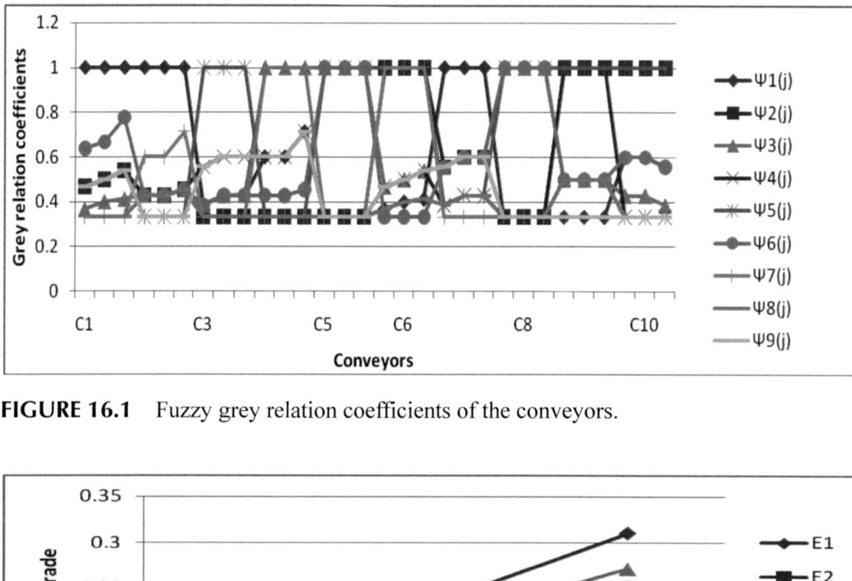

FIGURE 16.1 Fuzzy grey relation coefficients of the conveyors.

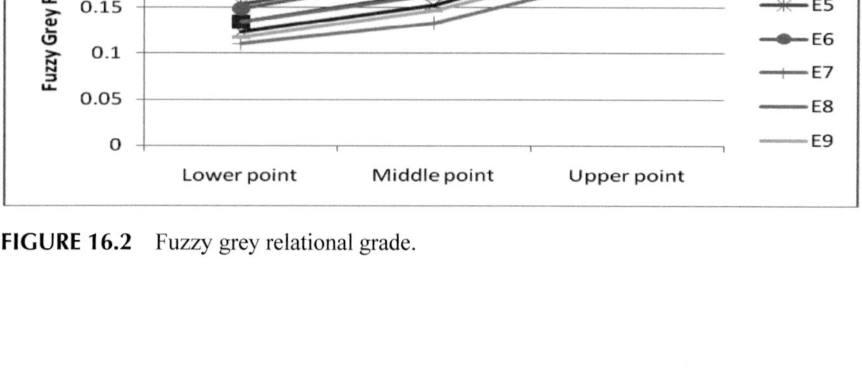

FIGURE 16.2 Fuzzy grey relational grade.

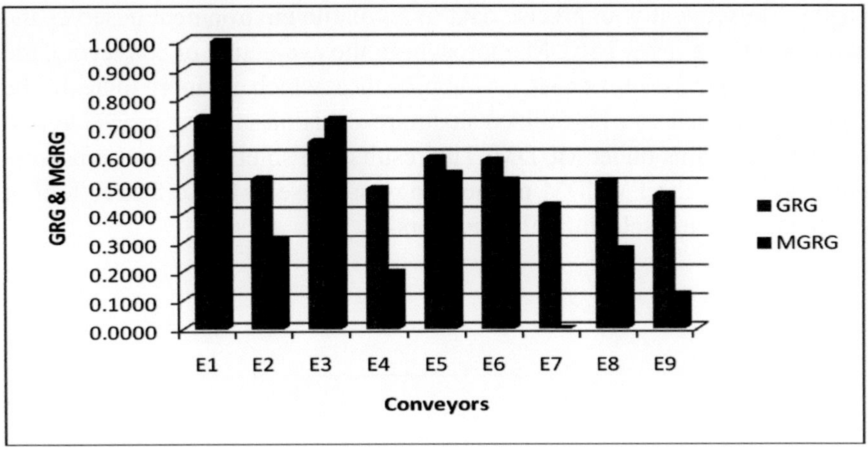

FIGURE 16.3 Comparison of grey relational grade with modified grey relational grade.

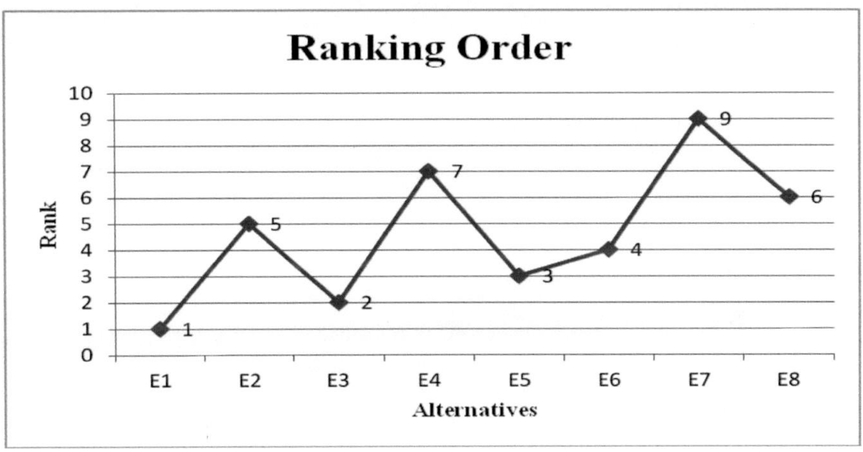

FIGURE 16.4 Ranking order of the conveyors.

16.5 CONCLUSIONS

This chapter integrates FGT with fuzzy analytical hierarchical process (FAHP) for the evaluation and selection of conveyors. The present chapter is different from the past literature on conveyor evaluation and selection in four ways. It takes a larger number of selection criteria into account than any past literature. Thus the current method is more suitable and satisfactory. The practical situation of conveyor evaluation and selection is uncertain and vague. Consequently, imprecise data in a volatile environment deserves the application of a fuzzy MCDM approach. In the evaluation of conveyors, the present chapter uses fuzzy MCDM approaches, which are more logical. The applied approach can be utilized in future decision-making procedures in real-life situations under MCDM. The results and solution of the conveyors problem prove the FMCDM methodology as a useful and effective tool in the evaluation and selection of conveyors.

KEYWORDS

- **analytic hierarchy process**
- **decision-makers**
- **FMCDM**
- **fuzzy AHP**
- **fuzzy grey theory**
- **multi-criteria decision-making**
- **selection of conveyor**

REFERENCES

1. Soufi, Z., David, P., & Yahouni, Z., (2021). A methodology for the selection of material handling equipment in manufacturing systems. *IFAC-Papers OnLine, 54*, 122–127.
2. Satyam, F., Satywan, K., & Avinash, K., (2021). *Application of Multi-Attribute Decision-Making Methods for the Selection of Conveyor*. Research Square. doi: 10.21203/rs.3.rs-1033410/v1.
3. Huu-Tho Nguyen, Siti Zawiah Md Dawal, Yusoff Nukman, Achmad P. Rifai, & Hideki Aoyama (2016). An integrated MCDM model for conveyor equipment evaluation and selection in an FMC based on a fuzzy AHP and fuzzy ARAS in the presence of vagueness. *PLOS One, 11*. doi: 10.1371/journal.pone.0153222.
4. Mathewa, M., & Sahua, S., (2018). Comparison of new multi-criteria decision-making methods for material handling equipment selection. *Management Science Letters, 8*, 139–150.
5. Karsak, E. E., Sener, Z., & Dursun, M., (2012). Robot selection using a fuzzy regression-based decision-making approach. *International Journal of Production Research, 50*(23), 6826–6834.
6. Tansel, İ. Y., & Dengiz, B., (2013). Development of a decision support system for robot selection. *Robotics and Computer-Integrated Manufacturing, 29*, 142–157.

7. Parkan, C., & Wu, M. L., (1999). Decision-making and performance measurement models with application to robot selection. *Computers & Industrial Engineering, 36*, 503–523.
8. Chu, T. C., & Lin, Y. C., (2003). A fuzzy TOPSIS method for robot selection. *International Journal of Advanced Manufacturing Technology, 21*, 284–290.
9. Kahraman, C., Cevik, S., Ates, N. Y., & Gulbay, M., (2007). Fuzzy multi-criteria evaluation of industrial robotic systems. *Computers and Industrial Engineering, 52*, 414–433.
10. Sujono, S., & Lashkari, R. S., (2007). A multi-objective model of operation allocation and material handling system selection in FMS design. *International Journal of Production Economics, 105*, 116–133.
11. Shih, H. S., (2008). Incremental analysis for MCDM with an application to group TOPSIS. *European Journal of Operational Research, 186*, 720–734.
12. Chatterjee, P., Athawale, V. M., & Chakraborty, S., (2010). Selection of industrial robots using compromise ranking and outranking methods. *Robotics and Computer-Integrated Manufacturing, 26*, 483–489.
13. Tuzkaya, G., Gülsün, B., Kahraman, C., & Özgen, D., (2010). An integrated fuzzy multi-criteria decision-making methodology for material handling equipment selection problem and an application. *Expert Systems with Applications, 37*, 2853–2863.
14. Poon, T. C., Choy, K. L., Cheng, C. K., Lao, S. I., & Lam, H. Y., (2011). Effective selection and allocation of material handling equipment for stochastic production material demand problems using genetic algorithm. *Expert Systems with Applications, 38*, 12497–12505.
15. Bairagi, B., Dey, B., Sarkar, B., & Sanyal, S., (2012). A novel multiplicative model of multi criteria analysis for robot selection. *International Journal on Soft Computing, Artificial Intelligence and Applications, 1*, 1–9.
16. Liu, H. C., Ren, M. L, Wu, J., & Lin, Q. L., (2013). An interval 2-tuple linguistic MCDM method for robot evaluation and selection. *International Journal of Production Research, 51*, 1–14.
17. Bairagi, B., Dey, B., Sarkar, B., & Sanyal, S., (2014). Selection of robot for automated foundry operations using fuzzy multi-criteria decision-making approaches. *International Journal of Management Science and Engineering Management, 9*(3), 221–232.
18. Bellman, R. E., & Zadeh, L. A., (1970). Decision-making in a fuzzy environment. *Management Science, 17*, B141–B164.
19. Abass, S. A., (2012). A solution procedure for estimating the scaling constants of the additive utility function under a fuzzy environment. *International Journal of Management Science and Engineering Management, 7*(4), 280–283.
20. Opricovic, S., & Tzeng, G. H., (2007). Extended VIKOR method in comparison with outranking methods. *European Journal of Operational Research, 178*, 514–529.
21. Wu, H. Y., Tzeng, G. H., & Chen, Y. H., (2009). A fuzzy MCDM approach for evaluating banking performance based on the balanced scorecard. *Expert Systems with Applications, 36*, 10135–10147.
22. Chen, Y. C., Lien, H. P., Tzeng, G. H., & Yang, L. S., (2008). Fuzzy MCDM approach for selecting the best environment-watershed plan. *Applied Soft Computing, 11*(1) 265–275.
23. Haldar, A., Ray, A., Banerjee, D., & Ghosh, S., (2014). Resilient supplier selection under a fuzzy environment, *International Journal of Management Science and Engineering Management, 9*(2), 147–156.

24. Zadeh, L. A., (1975). The concept of a linguistic variable and its application to approximate reasoning. *Information Science, 8*, 199–249(I), 301–357 (II).
25. Saaty, T. L., (1980). *The Analytic Hierarchy Process*. New York: McGraw-Hill.
26. Buckley, J. J., (1985). Fuzzy hierarchical analysis. *Fuzzy Sets and Systems, 17*, 233–247.
27. Ozcan, T., Celebi, N., & Esnaf, S., (2011). Comparative analysis of multi-criteria decision-making methodologies and implementation of a warehouse location selection problem, *Expert Systems with Applications, 38,* 9773–9779.
28. Yuan, X., (2007). Grey relational evaluation of final situation of listed company. *Journal of Modern Accounting and Auditing, 3*, 41–44.

CHAPTER 17

AN INNOVATIVE COMPUTATIONAL APPROACH FOR PROCESSING AND CHARACTERIZATION OF POLYMER WELDING

JINESH KUMAR JAIN, RAJEEV AGRAWAL, and PANKAJ SONIA

Department of Mechanical Engineering, Malaviya National Institute of Technology, Jaipur, Rajasthan, India

ABSTRACT

The chemical chain of monomers after polymerization becomes a polymer. The polymer is a high-performance low weight material. The co-polymerization, reinforcement of various alloying element facilitates polymer as a gradient structure material. The gradient structure facilitates polymer to manipulate its properties with an optimum value in different regions. The gradient polymer material found its application in plastic gasoline tanks, biomedical implants and damping materials for a high-frequency range. The high-performance polymers can withstand at high temperatures like polyamide, polyimide, poly-ether-ether-ketone, poly-phthal-amide, and poly-phenylene sulfide have 250°C to 280°C glass transition temperatures. These thermoplastic polymers found application in bearings, bushing, retainers, thrust plate, thrust washers, rotary seal ring, and gears, etc. The industrial application of polymeric components is revolutionary growing in the R&D area because the processing techniques of these components are more environmentally friendly.

This chapter covers the performance analysis and process optimization related to processing and characterization of various methods of polymer

Decision-Making Models and Applications in Manufacturing Environments. Pushpdant Jain, Kumar Abhishek, & Prasenjit Chatterjee (Eds.)

welding, i.e., LASER welding, infrared welding, spin welding, stir welding, and vibration welding with the help of different optimization methods, i.e., ANN, Taguchi's, ANOVA, and GRA. The chapter also describes the conventional methods of welding like hot gas welding, hot wedge welding, extrusion welding, hot plate welding, resistant implant welding, induction welding, dielectric welding (DW), and microwave welding (MWW). The chapter covers the modeling and simulation aspects related to the polymer to polymer and polymer to composite welding. The chapter highlights the complication related to fusion, heat transfer, issues related to the strength of joint and their solutions with futuristic prospects of polymer welding for empowering the polymer to become an absolute substitute of metals.

17.1 INTRODUCTION

The polymer used for industrial applications is categorized as thermoplastic and thermosetting plastic polymers. The joining of thermosetting polymers is not possible mainly because thermosets have a high cross-linked molecular chain. In most of the thermosets, the polymerization is permanent and irreversible. The joining of thermoplastic polymer only facilitates a fusion welding process. The process of welding of polymer categorizes according to their source of heat generation. Polymer/plastic are becoming a necessary part for mankind. There is no possible field where plastic is not present. The dependence of humans on plastic is increasing day by day. After a long journey of polymer science, polymers are capable of replacing metals with a very good aspect ratio related to weight to strength ratio and good corrosion resistance. The properties of the polymer can be manipulated by adding reinforcing elements such as glass fiber, carbon fiber, banana fiber, Basalt fiber, etc., known as fiber polymer composite [1–4]. If the reinforcing agent is in the form of particles (nano- or micro- in size), such a composite is well known as a particulate polymer composite [5–7]. The high strength polymer frequently uses in automobile, aerospace industries, naval applications, and medical/surgical equipment [8–10]. Advanced types of high-performance thermoplastics have been developed that have achieved temperature resistance of 200°C like Porous polyamide fibers in the framework have a high temperature-tolerance-characteristic up to 170°C [11–13]. In other word earth field to satellite field, plastic/polymer have made his own space. In general-purpose, the polymer can be classified into two basic categories, i.e., thermoplastic and thermosetting type polymers. The thermosetting polymer is processed under an irreversible cross-linked reaction; thus, such polymers are not capable to weld by softening. In this chapter, only the welding of thermoplastic has been described. The pattern toward more noteworthy utilization of these materials will most likely proceed for a few reasons, i.e., engineered polymer materials give numerous helpful properties, fabricating methods are moderately basic, also, a sufficient stockpile of oil feedstock's, presently increased by organic sources, is accessible. Continuous developments in plastic/polymers, polymers are frequently used by engineering as well as biomedical industries [14–16]. There are various methods of manufacturing plastic components such as casting, injection molding, hot pressing, extrusion, etc. The complex shape of polymeric/plastic components can be manufactured by the extrusion process. Manufactured polymeric components are required in joining/welding techniques for their extension/maintenance. The welding of polymers is required in various applications such as:

- Automobile industry, aircraft industry, ships, and submarines;
- Buildings, bridges, pressure vessels, girders, pipelines, machine tools, offshore structures, nuclear power plants, etc.;
- Household products, farm, mining, oil industry, jigs, and fixtures, boilers, furnaces, railways, etc.;
- Biomedical application in permanent and biodegradable implants [17].

There are three basic ways to joining and welding of polymeric components, i.e., mechanical fastening, adhesive bonding, and welding. Adhesive bonding and mechanical fastening of polymers are similar to the other materials. The welding methods have been classified according to the heat applied. If the heat is applied on the surface, the process is known as an external method, and if the heat concentrates inside the surface, the process is known as an internal method. The detail classification of polymer welding is shown in Figure 17.1. Formation of a decent quality weld doesn't just rely upon the welding techniques, yet in addition weldability of base materials also. Consequently, the assessment of weldability is of higher significance than the welding method (see rheological weldability) for plastics. The welding of plastic includes the following meritable prospects:

- High productivity;
- Strong joints;
- Reduced requirement for surface preparation;
- Weldability;
- Recyclability.

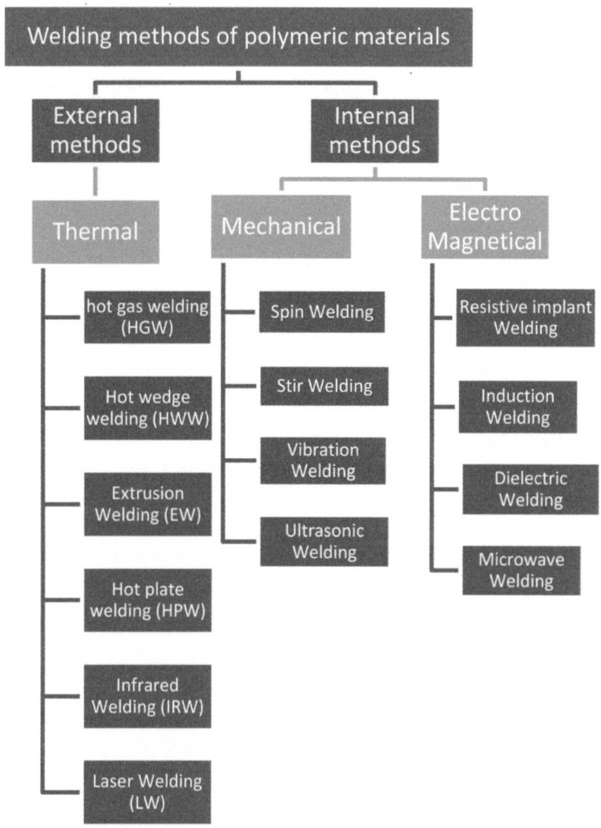

FIGURE 17.1 Classification of welding methods of polymers.

17.2 POLYMERS

The polymer is a Greek word, i.e., composed of 'poly + mer,' means many parts. The polymer is a large molecular chain of repeating units. The chain structure of the polymer has classified the polymer in two distingue categories, i.e., thermosetting and thermoplastic. The thermoplastic polymers are subdivided into two categories: Amorphous and semi-crystalline the polymer molecules chains, in amorphous polymers, are arbitrarily looped and interlaced with different chains [18]. On warming, they go through a temperature called the Glass Transition temperature (T_g), at which mellowing starts. They don't liquefy; rather, as the temperature is raised over the T_g, the consistency keeps on diminishing until it arrives at a level helpful for welding or molding. Since the crystal structures framed in semicrystalline polymers dissolve pointedly at temperatures higher than the Tg, they seem to have a clear liquefying temperature. In any event, when softened, be that as it may, the structure extremely thick materials. Unadulterated semicrystalline polymers are normally opaque; undefined polymers are mostly transparent. Examples of common amorphous polymers include acrylic, polycarbonate, polyvinyl chloride, acrylonitrile-butadiene-styrene (ABS), polystyrene (PS), polysulfone, and polyetherimide. Semicrystalline polymers include polyethylene, polypropylene, polyphenylene sulfide (PPS), polyetheretherketone (PEEK), and polyamides [19].

In other words, a polymer is a macromolecule composed of substance. These molecules have a relatively high molecular mass, the structure of such molecules have repetitive multiple units which are derived from the molecules of relatively low molecular mass. The polymer is categorized mainly in two types, i.e., natural polymer and the other one is a synthetic polymer. The detail description of the polymers is shown in Figure 17.2.

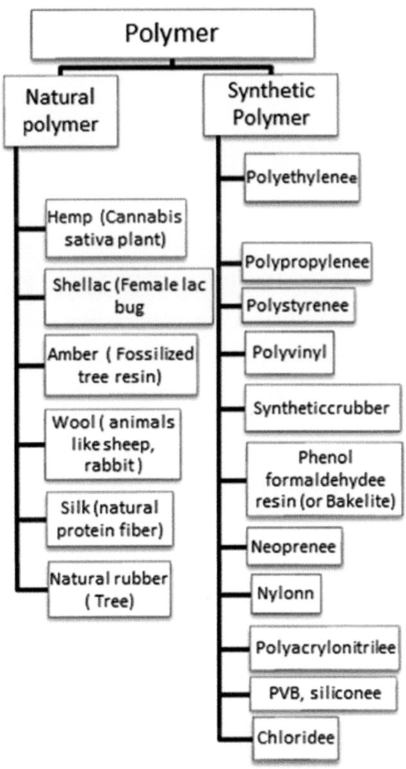

FIGURE 17.2 Types of polymers under natural and synthesis category.

Every thermoplastic has a “preparing window,” a time of peak processing efficiency characterized as the temperature, pressure, and welding time run in which fruitful flow is accomplished before polymer deterioration can happen. Welding forms must be controlled to prevent decomposition. All polymer welding processes involve applying heat and pressure to cause flow and joining.

In distinction to metals, polymers are having an insulating characteristic. The amount of heat required in polymers is less as compared to metals. The low conductivity of polymers confines the heat to the weld zone during welding, so the total heat concentrates on the surface, because of that more attention required for dwell time, fixturing, of parts.

The small molecular chain of the monomer is combined with some chemical reaction with the help of a curing agent. The process of cross-linking of monomers with the curing agent is known as polymerization [20] (Figure 17.3).

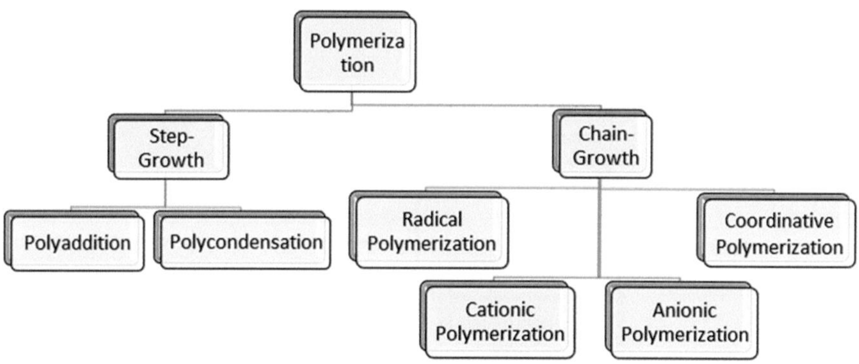

FIGURE 17.3 Polymerization process.

17.3 A COMPUTATIONAL TECHNIQUE FOR PROCESS OPTIMIZATION AND PERFORMANCE PREDICTION

Process parameter optimization is the way of adjusting a process to optimize a few specific sets of parameters without contravening any constraint. The most general objectives are reducing costs and enhancing productivity. The optimization technique is an essential tool for industries to decide the operating parameters. Process optimization can lead the industry to decrease time and money and the best utilization of resources in a process. There are various optimization techniques available for a selection of parameters and performance prediction.

17.3.1 ARTIFICIAL NEURAL NETWORKS (ANN)

ANNs are biologically motivated computer programs proposed to simulate according to the human brain processes information (Figure 17.4). ANN is trained by gathering knowledge through a pattern in the data set and by experience. An ANN is formed from hundreds of single units, processing elements (PE), and artificial neurons or connected with weights (coefficients), which constitute the neural structure and are organized in layers. The computational capability of NNs comes from the interconnecting neurons. Each PE has a transfer function, weighted inputs, and one output. The nature of a NN is evaluated by the transfer functions (TF) of its neurons (Figure 17.4), by the learning rule, and by the structure itself. The adaptable parameters are the weights, in that means; a NN is a parameterized system. The weighted addition of the inputs constitutes the excitation of the neuron [21].

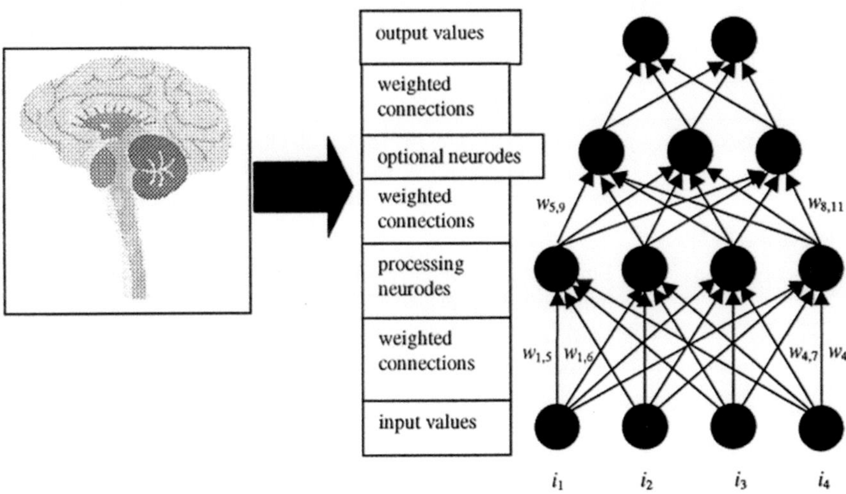

FIGURE 17.4 Analogies between a biological neuron and an artificial neuron [22].

Source: Reprinted/Adapted with permission from Ref. [22]. Copyright 2019. Elsevier.

The Eqns. (1) and (2) formulate the weighted sum of net input. The TF is used to pass the activation signal to produce a unit output of neurons (Eqn. (3)), for the optimization by NN the series of iteration during network training is continuously performed, until and unless the error of prediction is reduced, and network achieved a specific level of accuracy. Once the network becomes trained and validated then it can be used for the performance prediction of output with a new value of the input. The step-by-step processing and range of parameters are noted in Table 17.1. There are various types of NNs have been modeled priorly and researchers trying to develop a new model of NN. The main consideration is that all can be described by a TF of their neurons, by the learning rule and by connection formulas. ANNs is very suitable modeling and prediction technique, suitably when the data set having a non-linear relationship [23].

$$y_{in} = x_1.w_1 + x_2.w_2 + x_3.w_3 \dots\dots\dots x_m.w_m \tag{1}$$

The net input:

$$y_{in} = \sum_i^m x_i.w_i \tag{2}$$

The output can be calculated by applying activation function over the net input:

$$Y = F\,(y_{in}) \tag{3}$$

TABLE 17.1 Parameter Setting for ANN Modeling [24]

Parameter	Parameter Setting
Types of ANN	Feedforward neural network, radial basis function neural network, multilayer perceptron, convolutional neural network, recurrent neural network, etc.
Training algorithm	Backpropagation Levenberg-Marquardt algorithm
Transfer function	Linear, binary step, piecewise linear, sigmoid, gaussian, hyperbolic tangent
Data usage	Most commonly 70% for training, 15% for validation, 15% for verification
Learning rate	$0 \leq$ learning rate ≤ 1
Momentum	Selected according to problem like '$m_c = 0.9$'

Source: Reprinted/Adapted with permission from Ref. [24]. Copyright 2011. Elsevier.

There are many types of neural network available. The most commonly used networks are mentioned below:

- Feedforward neural network – artificial neuron;
- Radial basis function (RBF) neural network;
- Multilayer perceptron;
- Convolutional neural network;
- Recurrent neural network (RNN) – long short-term memory;

- Modular neural network (MNN);
- Sequence-to-sequence models.

17.3.1.1 BACK-PROPAGATION ARTIFICIAL NEURAL NETWORK (BP-ANN)

BP-ANN is the most broadly castoff supervised/trained learning algorithm, it consists of a forward and backward surpass. The Bp-ANN constructed by three layers, i.e., input layer (input parameters), hidden layer (weight from input layer) and output layer (output or objectives). By using the back-propagation algorithm, the weights and biases, related to the neurons, are tailored to reduce the predicting error (i.e., the classification error of the training set). After repetitive iteration of the training data patterns to the BP-ANN, the ANN is supposed to train when the weights and biases of architecture become stable [25, 26].

17.3.1.2 GENERALIZED FEED-FORWARD (GFF-MLP) NEURAL NETWORK

This is one of the most usual ANN that applied for the connection between units. In this network, no feedback is used since the information travels only forward direction from the input layer to the hidden layer and passes to the output layer (no loops) as shown in Figure 17.5. Overall, GFF-MLP's neural networks are initially used for trained learning in cases where the data to be learned is neither chronological nor time-dependent. The weight (w) indicates the strength of connection between neurons and 'h' denote the hidden layer and 'L' connection of neurons with output (Figure 17.5) [26].

17.3.1.3 RADIAL BASIS FUNCTION (RBF) NEURAL NETWORK

RBF network is consisting of three layers, i.e., input, hidden, and output layer. The neurons in the hidden layer are actuated by RBF. RBF networks are used to resolve the problems related to organization in science and engineering. Every RBF neuron has a specific set of values known as "reference vector" for comparison with an input set of the same cardinality. The governing equation of individual neurons is based on the multivariate Gaussian function (Eqn. (4)) [27, 28]

$$\varphi(r) = e^{-\frac{1}{2\sigma_j^2} x - t_j^2} \quad (4)$$

where; x is the input vector for the neuron; tj is the set of reference values.

> *σj = standard deviation (σ2 is the variance) of the function for each of the centers (j)*
> *r (||x – tj||) = Euclidean distance between a center vector and the set of data points*

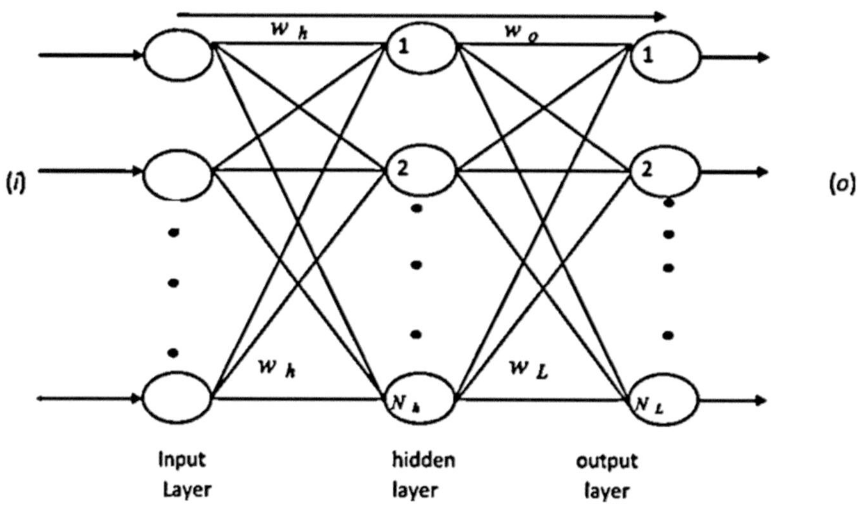

FIGURE 17.5 Neural network.

17.3.1.4 MULTILAYER PERCEPTRON

A multilayer perceptron has at least three layers. It is utilized to order information that can't be isolated directly. It is a sort of fake neural system that is completely associated. This is on the grounds that each and every hub in a layer is associated with every hub in the accompanying layer. A multilayer perceptron utilizes a nonlinear enactment work (most commonly used transfer function – log-sigmoid transfer function, hyperbolic tangent or logistic function) [29].

17.3.1.5 CONVOLUTIONAL NEURAL NETWORK

A convolutional NN (CNN) utilizes a variety of the multilayer perceptron. A CNN contains at least one than one convolutional layer. These layers can either be totally interconnected or pooled. Prior to passing the outcome to the following layer, the convolutional layer utilizes a convolutional procedure on the info. Because of this convolutional activity, the system can be a lot further yet with many less boundaries. Because of this capacity, convolutional neural systems show exceptionally successful outcomes in picture and video acknowledgment, common language handling, and recommended frameworks. Convolutional neural systems additionally show extraordinary outcomes in semantic parsing and reword location. They are additionally applied in signal handling and picture characterization. CNNs are additionally being utilized in picture examination and acknowledgment in horticulture where climate highlights are extricated from satellites like LSAT to foresee the development and yield of a land parcel [30].

17.3.1.6 MODULAR NEURAL NETWORK (MNN)

A modular neural network (MNN) is a combination of different network, those function individually and perform subtasks. All individual networks are not interacting or transmit a signal to each other during the processing of computation. Each network works independently to attain the output. By the entire processing, a huge and complex computational process is accomplished significantly faster by performing sub-task independently. The high computation speed can be achieved because the networks are not interacting/signaling or connected to each other. The visual representation of an MNN (Figure 17.6) [31].

17.3.2 PARTICLE SWARM OPTIMIZATION

PSO is inspired by the social behavior of species that live in an organized group or colonies like an ant, bee swarm birds et. The design concept of PSO has opted from the information transfer mechanism of a group of swarm birds. These birds fly in a group with a specific pattern at a very high (VH) speed. Every person in the swarm has a position and speed characterized, the calculation takes a gander at each case to build up the best result utilizing the present swarm, and afterward, the entire swarm moves to the new relative area. The PSO can be applied in several areas in optimization. This technique can also be applied with the combination of another existing algorithm. This technique plays out the hunt of the ideal arrangement through operators, alluded to as particles, whose directions are balanced by a stochastic and a deterministic segment. Every molecule is affected by its 'best' accomplished position and the gathering 'best' position, yet it will in general move haphazardly. The sum of all three vectors from the resultant vector is considered as the velocity vector for the next iteration/trial and added this velocity vector to the current position vector to find the next position of individual particles as shown in Eqn. (5). The position of the particle in a swarm forgiven number of particles after a number of iterations can be calculated by Eqn. (6) [32].

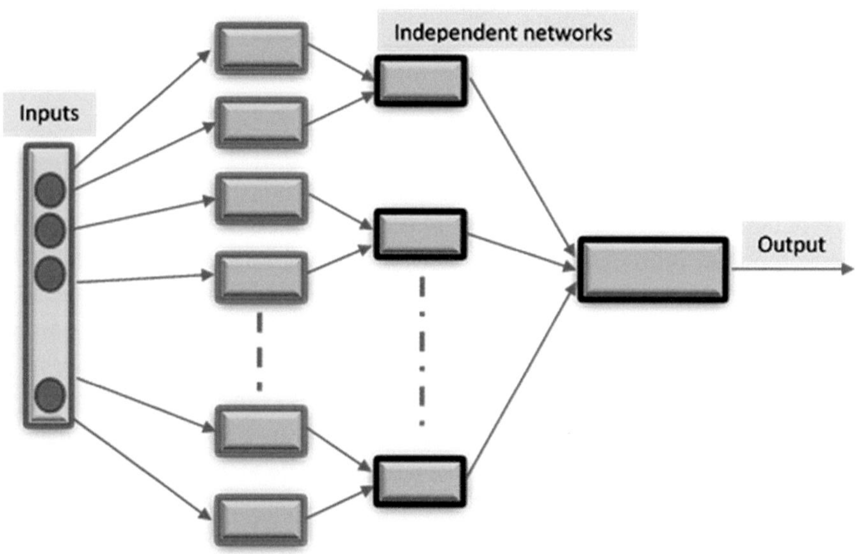

FIGURE 17.6 Modular neural network (MNN).

$$vi\ (t+1)=w-vi(t)+c1-r1-\ [Li(t)-xi(t)]+c2-r2-[G(t)-xi(t)] \tag{5}$$

$$xi\ (t+1)=xi(t)+vi(t+1) \tag{6}$$

where; $vi(t)=Velocity\,of\ i_{th}\ particle\,in\,t_{th}\ iteration$ (The velocity initialized of every particle with 0)

$$\text{w} = weightage\ given\ to\ velocity\ vector$$

$$i = 1\ to\ 200$$
$$t = 1\ to\ 100$$
$$0 - r1, r2 - 1(random\ numbers)$$

$c1, c2 = 2$ (*c1, c2 are the weightages given to individual best and global best directions*)

$Li(t)$ = *individual best for* i_{th} *particle in* t_{th} *iteration*

$G(t)$ = *Global best for individual best for* i_{th} *iteration*

$xi(t)$ = *Position of* i_{th} *particle in* t_{th} *iteration*

- **PSO Algorithm:**
  1. Initialization:
     A. For each particle I in a swarm population size P
        i. Initialize X_i randomly;
        ii. Initialize V_i randomly;
        iii. Evaluate fitness $f(X_i)$;
        iv. Initialize pbest$_i$ with a copy of X_i.
     B. Initialize pbest$_i$ with a copy of X_i with the best fitness
  2. Repeat until a stopping criterion is satisfied:
     A. For each particle i
        i. Update V_i^t and X_i^t
        ii. Evaluate the fitness $f(X_i^t)$
           c) $pbest_i \leftarrow X_i^t\ iff(pbest_i < f(X_i^t)$
           d) $gbest_i \leftarrow X_i^t\ iff(gbest_i < f(X_i^t)$

17.3.3 GREY RELATION ANALYSIS

The Grey relational analysis is multiple response characterization techniques. GRA is best suited for the gray system model. The gray system model is those systems in any organization where partial information of the system is known and partial information is unknown. GRA established an inter-relationship between multiple responses. GRA provides a solution by converting several responses into an equivalent one gray relational grade function. The GRA performs in three steps, first start with pre-processing of data second transmit a reference sequence and at last comparable/equivalent sequence. In the initial step, the responses from the experiment are normalized from 0 to 1. There are mainly three performance characteristics, i.e., smaller is better, nominal is better and larger is better. In the processing step start by maximization of response, then normalization is completed with the sequence "larger the better" performance attribute normalized as by Eqn. (7) and smaller the better is normalized by Eqn. (8) [33, 34]:

$$x_i^*(k) = \frac{x_i^0(k) - \min x_i^0(k)}{\max x_i^0(k) - \min x_i^0(k)} \tag{7}$$

Grey relational generation, the normalized by smaller-the-better (SB) criterion.

$$X_t(k) = \frac{\max y_t(k) - y_t(k)}{\max y_t(k) - \min y_t(k)} \tag{8}$$

where; $x_i^*(k)$ *is the sequence after the data pre – processing*

$x_i^0(k)$ *is the reference sequence;*

$\min x_i^0(k)$ *is the minimum value in the refecence sequence;*

$\max x_i^0(k)$ *is the maximum value in the refecence sequence;*

$i = 1, 2, \ldots, n; m = 1, 2, \ldots$

'm' is the number of an experiment; 'n' is the number of experimental data.

The significance of the grey relational coefficient (GRC) is to explicate the relationship between the reference sequence and comparability sequence. The comparability sequences for every experiment were evaluated by Eqn. (9)

GRC can be calculated by using the formula:

$$\xi_t(k) = \frac{\Delta\min + \Psi\Delta max}{\Delta o_t(k) + \Psi\Delta max} \tag{9}$$

After averaging the GRC, the gray relational grade (GRG) can be calculated by Eqn. (10).

$$G_t = \frac{1}{n}\sum_{k=1}^{n} \xi_t(k) \tag{10}$$

17.3.4 TAGUCHI'S METHOD FOR PROCESS OPTIMIZATION

The Taguchi method is a very efficient process parameter optimization tool. The statistical approach is used to optimize the process parameters and improve the quality of components that are manufactured. The full factorial design needs a large number of experiments to be performed and it's very costly, laborious, time-consuming, and complex, Taguchi suggested that if the number of factors increases then full factorial design is not feasible.

Taguchi design a partial factorial design to reduce the number of experiments by a special design method. The orthogonal array is generated by the entire parametric study with a lesser number of experiments. Taguchi proposes the loss function to find the performance characteristics that are deviating the desired objective value. The value of this loss function is converted into S/N ratio (Signal to Noise). The Taguchi is a very useful tool to identify the individual and combined effect of parameters on the output. To identify the significant parameter which most or least affects the output, an ANOVA is used.

There are three categories of the performance characteristics to calculate the S/N ratio for nominal the better, larger the better and smaller the better by given Eqns. (11)–(13) [33, 35].

I. Nominal the better $\left[\frac{S}{N} ratio = -10 log_{10}(\frac{\bar{y}^2}{s^2})\right]$ (11)

II. Larger the better $\left[\frac{S}{N} ratio = -10 log_{10}(\frac{1}{n}\sum_{i=1}^{n}\frac{1}{y_i^2})\right]$ (12)

III. Smaller the better $\left[\frac{S}{N} ratio = -10 log_{10}(\frac{1}{n}\sum_{i=1}^{n} y_i^2)\right]$ (13)

17.3.5 ANALYSIS OF VARIANCE (ANOVA)

As per the discussion that the Taguchi design method is used to find the optimum parameter for single and multiple objectives, but the Taguchi method is not capable to find the significant parameter or in other words to which parameter most or least affect the output and also in the correlations. Analysis of Variance (ANOVA) is a very useful tool to find a significant factor in the data set [35].

The ANOVA strategy depends on separating the change between the trials to inside the trial. In the event that 'between variety' is a lot bigger than the 'inside variety,' implies various trial won't be equivalent. In the event that 'among varieties' and 'inside varieties' is around equivalent, at that point there will be no significant distinction between tests implies. There are two ways to deal with populace change, one depends on inside trial difference and the other depends on between tests fluctuation. At that point said two assessments of populace change are contrasted and F-test, wherein we work out the *F*-Ratio by Eqn. (14)

$$F = \frac{Estimate\ of\ population\ variance\ based\ on\ between\ samples\ variance}{\text{An estimate of population variance based on within samples } variance} \quad (14)$$

According to the number of variables ANOVA technique categorize in two different ways, i.e., one-way ANOVA and another one is two-way ANOVA. If the problem is having one independent variable with two levels, such a problem is known as one-way ANOVA, and if the problem is having two independent variables with multiple levels. The two-way ANOVA subcategories as two-way ANOVA with replication and other one is without replication [36, 37]. The algorithm to solve the problem in one way and two ways tabulated in Tables 17.2 and 17.3 simultaneously.

TABLE 17.2 One Way ANOVA

Source of Variation	**Sum of Squares (SS)**	**Degree of Freedom (DOF)**	**Mean Square (MS)**	**F-Ratio**
Between sample or categories variation	$= n_1(\bar{X}_1 - \bar{\bar{X}})^2 + n_2(\bar{X}_2 - \bar{\bar{X}})^2 + n_k(\bar{X}_k - \bar{\bar{X}})^2$	$(k - 1)$	$\frac{SS\ between}{(k-1)}$	$\frac{MS\ between}{MS\ within}$
Within sample or categories variation	$=\sum(X_{1i} - \bar{X}_1)^2 + \ldots \sum(X_{ki} - \bar{X}_k)^2$	$(n - k)$	$\frac{SS\ within}{(n-k)}$	
Total	$=\sum(\boldsymbol{X}_{ij} - \bar{\bar{\boldsymbol{X}}})^2$			

TABLE 17.3 Two Ways ANOVA

Source of Variation	**Sum of Squares (SS)**	**Degree of Freedom (DOF)**	**Mean Square (MS)**	**F-Ratio**
Between column treatment	$=\sum\frac{(T_j)^2}{n_j} - \frac{(T)^2}{n}$	$(c - 1)$	$\frac{SS\ between\ column}{(c-1)}$	$\frac{MS\ between\ column}{MS\ residual}$
Between rows treatment	$\sum\frac{(T_1)^2}{n_j} - \frac{(T)^2}{n}$	$(r - 1)$	$\frac{SS\ between\ row}{(r-1)}$	$\frac{MS\ between\ row}{MS\ residual}$
Residual or error	$= Total\ SS - (SS\ between\ column + SS\ between\ rows)$	$(c.r - 1)$	$\frac{SS\ residual}{(-1).(c-1)}$	
Total	$=\sum \boldsymbol{x}_{ij}^2 - \frac{\boldsymbol{T}}{\boldsymbol{n}}$			

17.4 WELDING OF POLYMERS

17.4.1 EXTERNAL METHODS OF WELDING

17.4.1.1 HOT GAS WELDING

This process is very much similar to flame torch welding in gas welding. The only difference is that hot gas is used instead of flame heating. In the HGW process, it is optional to use filler or not. The composition of filler must be identical to the base polymer. The joint preparation in the HGW process is similar to joint preparation in metal welding [38]. It includes butt joint lap joint and filet weld, etc., as shown in Figure 17.7.

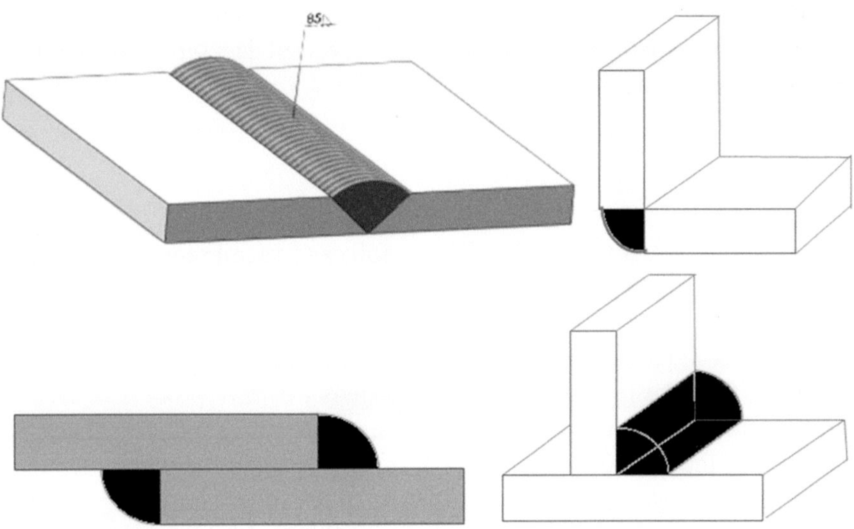

FIGURE 17.7 Commonly used weld joint in HGW.

The process of HGW is that the stream heated gas will focus on the joint surface or beveled surface. The work surface and filler start softening, now filler rod is continuously fed in the workpiece for the required fusion. But it is not necessary to melt the filler as this will result in slower welding duration and enhance the possibility of polymer decomposition by excess heating [39].

To begin a weld, the filler rod is embedded into the tip, and the welding torch is quickly set in contact with the joint and held at 90 (degrees) to the substrate. The filler bar is continuously fed until it touches the substrate at the beginning of the weld. If required, then the light is lifted to permit the bar to go under the pressure foot. The torch is pulled along the weld zone to begin the weld. When the weld is begun, the angle of the torch can be decreased to around 45° to facilitate the preheating of the weld zone. The torch angle most provably depends on base material and welding speed. The welder must find some kind of relativity between the pace of filler-bar heating and the pace of base-polymer preheating. In the case of preheating is unnecessary, frequently showed by staining of the polymer in the weld zone, the angle of the torch can be raised to control the excessive heating preheating substrate.

The process parameters which affect the weld quality by HGW are the type of gas, the flow rate of the gas and temperature, filler rod incident angle relative to the workpiece, speed of welding or traverse speed, and applied pressure after weld. The most significant factor out of these parameters is gas flow rate and gas temperature and traverse speed to decompose of polymer with an adequate rate [40].

17.4.1.2 HOT WEDGE WELDING

This process is used to join thin sheet, foil, and membrane of polymers/plastics. The HWW process is quite similar to the sheet rolling process. The heated wedge shape tool is used to melt the faying surface sheet to be welded and the sheet is passed between to roller. Rollers pressed the sheet and as a result of that, the faying surfaces fuse together. HWW process is used to join the long plastic or polymeric sheets in a continuous manner. There are two types of the joint can be produced: single seam and the other one is the double seam. In a single seam, only one set of the roller is used to weld and in the double seam, two sets of rollers along with dual split wedge are used. In the double seam joint, a blank space is left between two seams, this gap can be utilized to check the weld defect by filling this gap with high-pressure air or water. The cross section of dual and single seam type joint is shown in Figure 17.8 [41].

17.4.1.3 EXTRUSION WELDING

Extrusion welding is a variation of hot gas welding in which molten thermoplastic material is forced against preheated workpieces to produce the weld.

When pressure is applied, the molten polymer is extruded into the weld joint (Figure 17.9), creating the weld. The different size and geometry of the weld tip are used according to the type of joint and thickness of the sheet. This method is particularly useful when welding thick polymer sections. Good quality joints can be produced in one pass in the material that is 25 mm [39].

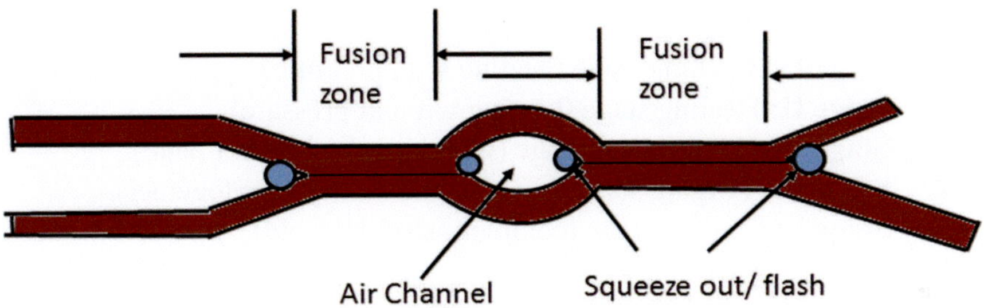

FIGURE 17.8 Cross-section of dual seam weld joint.

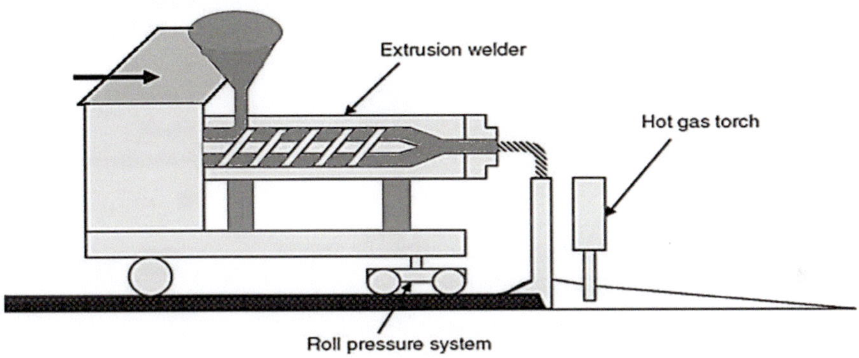

FIGURE 17.9 Extrusion welding machine.

17.4.1.4 HOT TOOL WELDING (HTW)

This is a very common plastic welding method and frequently performed in domestic applications. In the engineering terminology, these processes can also be known as a hot plate, mirror, butt, platen, and fusion welding. This technique is extensively used to weld of extruded and injection molding components. In this process a heated plate of metal used for welding. The interface of parts to be welded is in contact with the hot plate until the interface becomes melt or softens. When the interface becomes melt or softens then hot metal plate removes in between the polymer interface. Now the interface of parts is made in contact under slight pressure for a required time of fused surface solidification. Heating time depends on the geometry of the workpieces. The correct heating time can be evaluated mathematically or by experiments. Hot plate welding cycle times can range from seconds to minutes. The complete welding process can be described in the following steps (Figure 17.10) [39, 42]:

1. **Stage I:** Matching stage (heating with pressure)
2. **Stage II:** Heating stage (heating without pressure).
3. **Stage III:** Changeover stage (withdrawal of the hot plate).
4. **Stage IV:** Forging stage (pressing the workpieces together).
5. **Stage V:** Cooling stage (completion of the weld).

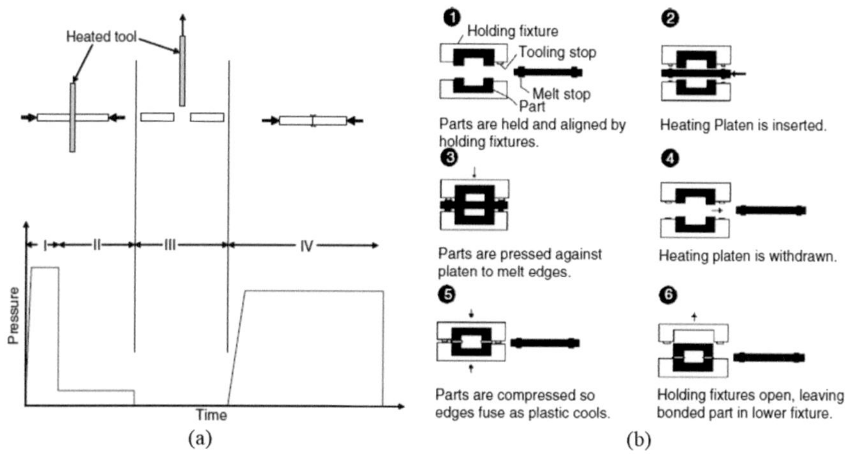

FIGURE 17.10 (a) Four phases of heated tool welding; and (b) schematic of welding process.
Source: Reprinted with permission from Ref. [42]. Copyright Elsevier

Process parameter of HTW: The parameters that affect the quality of weld in the HTW process are: (i) temperature of the hot plate (ii) dwell time (iii) applied pressure; and (iv) geometry of interface. The controlling of given parameters can produce flash free sound joints. The adequate value of parameters varies with the type of polymer/plastic to be welded. The advantage, disadvantages, and applications of the HTW process are tabulated in a Table 17.4.

Hot tool welding (HTW) parameters for different polymeric materials:

TABLE 17.4 Tabulation of Advantages, Disadvantages, and Applications

Advantages	Disadvantages	Applications
Simple economical technique high strength, hermetic welds can be achieved.	More number of steps involve in a complete cycle, so the cycle time is more as compared to other common techniques such as vibration or ultrasonic welding.	It is capable of joining a very small size of Parts can be join in a range of a few centimeters. Also, a part of large as (1,600 rpm) 63 inches in dia., such plastic coupling, plastics pipes, cylinders.
Suitable for both large and small parts.	High temperatures required for melting.	Welding of vehicle taillights and indicators.
Flat, curved, or complex geometries can be welded.	Some cases can cause plastic degradation or sticking to the hot plate.	Housing, made of acrylonitrile-butadiene-styrene (ABS).
Surface irregularities can be smoothed out during the heating phase.	Weld flash is produced when the parts have pressed each other.	Different polymers of similar melting point can be welded by the single hot plate.
Dissimilar materials that are compatible.	Part dimensions cannot always be controlled reliably due to variations in the molten film thickness.	Two polymers of different melting points can be welded by dual hot plate.

Source: Reprinted with permission from Ref. [42]. Copyright Elsevier

1. **Triax PC/ABS:**
   - Hot plate temperature: 250–410°C (482–770°F)
   - Joining pressure: 0.2–0.9 N/mm^2 (30–130 psi)
2. **Bayblend PC/ABS:**
   - Hot plate temperature: 250–410°C (482–770°F)
   - Joining pressure: 0.3–0.9 N/mm^2 (40–130 psi)
3. **Desmopan:**
   - Hot plate temperature: 270–320°C (518–608°F).
   - Joining pressure: 0.3–1.0 N/mm^2 (44–145 psi)

17.4.1.4.1 Taguchi Optimization of Hot Plate Welding Parameters

The parameter selected to optimize by is plate temperature, welding displacement and heating time with three levels. The problem is multiple objectives, and the selected objects are mean joint strength and joint efficiency The DOE is designed for analysis (Table 17.5). The L9 orthogonal array is used for DOE and the given number of experiments performed according to the set parameter in Table 17.6. The S/N ratio calculated for the larger is a better criterion as mentioned in Eqn. (12). By analyzing the S/N ratio value for the given parameter by minimum and maximum value to calculate the rank for Taguchi as shown in Table 17.7 [45].

TABLE 17.5 Welding Parameter and Levels for DOE

Welding Parameter	Symbol	Unit	Level I	Level 2	Level 3
Plate temperature	A	°C	270.00	280.00	290.00
Welding displacement	B	mm	0.70	1.25	2.00
Heating time	C	s	15.00	20.00	25.00

Source: Reprinted/Adapted with permission from Ref. [32]. Copyright 2020. Elsevier.

17.4.1.5 INFRARED WELDING

Infrared welding is very similar to the hot plate welding process, except that the workpiece is heated by infrared radiation that will produce by electromagnetic radiation. In the IR process, the contact surface exposed to the infrared lamp for adequate heating, radiation absorbed by the thermoplastic material and convert into heat, after that withdraw the lamp. The workpiece rapidly aligned and pressed together until it solidifies and cools. The working step are shown in Figure 17.11 [46].

TABLE 17.6 Experimental Result According to L9 (3^4) Orthogonal Array [45]

Experiment No: #	Welding Parameters			Mean Joint Strength (Mpa)	Joint Efficiency (%)	Calculated S/N Ratios (dB)
	A	B	C			
	Temperature (°C)	Displacement (mm)	Time (s)			
1	270.00	0.7000	15.00	8.33.	15.900.	17.79.
2	270.00	1.2500	20.00	14.50.	27.600.	23.16.
3	270.00	2.0000	25.00	11.66.	212.000.	21.30.
4	280.00	0.7000	20.00	9.74.	18.600.	19.47.
5	280.00	1.2500	25.00	14.01.	26.700.	22.92.
6	280.00	2.0000	15.00	10.47.	19.900.	20.12.
7	290.00	0.7000	25.00	14.64.	27.900.	23.21.
8	290.00	1.2500	15.00	15.93.	30.300.	23.87.
9	290.00	2.0000	20.00	15.75.	30.000.	23.93.

Source: Adapted from Ref. [45].

TABLE 17.7 Response Table

Symbol	Welding Parameters	Means						
		Level I	Level 2	Level 3	Max.	Min.	Delta	Rank
A	Plate temperature	11.500	11.410	15.440	15.440	11.410	04.030	1
B	Welding displacement	10.900	14.810	12.630	14.810	10.900	03.910	2
C	Heating time	1158.000	13.330	13.440	13.440	11.580	01.860	3

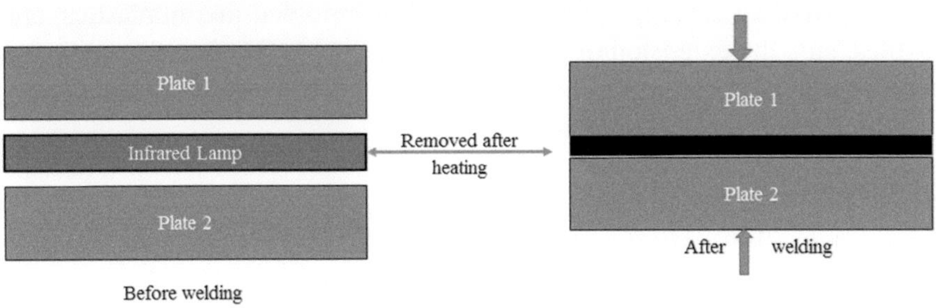

FIGURE 17.11 Steps involved in the infrared welding process.

17.4.1.6 *LASER WELDING*

A high-intensity beam of laser light strikes on the work surface. The beam melts the joint at the faying surface by absorption of the beam. According to the laser contact to the surface, the welding process is divided into two categories: direct laser welding and transmission laser welding. When the laser comes in the contact of surface, a part of energy absorbed, part of energy reflects and a part of energy transmitted. If the sufficient amount of energy for melting is absorbed by the surface, then the welding is known as direct laser welding. When the laser beam strike to the upper joining plate is irradiated by the laser beam without affecting the components (Figure 17.12). The bottom part to be joined absorbs the laser energy and once the laser energy translates into thermal energy, parts start to soften and melt. The upper plate is receiving heat by conducted from the lower plate as a result of that upper plate also in the melting stage. The adequate clamping force applied to allow fusion without deformation. With this process, a weld between the two parts is accomplished. Sometime carbon black is used to increase the absorbing capability of the lower plate.

The correct amount of energy interaction at the correct location signifies the quality of the weld joint. The interaction of the beam depends on the following parameters:

- Type of material and surface coating;
- The intensity of a laser beam in term of power, spot area;
- Traverse speed of beam over the surface.

From the above parameters, the material and surface coating can't be altered. The surface energy and chain length of the polymer decide the required beam intensity. The spot area of the laser beam can be control from micro-size to mm so this process is quite suitable for micro-welding as well as large welding structure such as sheet, film, molded thermoplastics, and textiles, etc., the extension of laser plastic welding is facilitated to join very sensitive electronic components, delicate pars packaging, biomedical applications, human care, automobile, etc. [47].

17.4.1.6.1 ANN Model to Predict the Weld Performance for LASER Welding

Model development: The ANN model is developed to predict the weld performance. There are four input variables, selected, i.e., Welding speed (mm/min), power (W), clamp pressure (MPa) and stand-off distance (mm), with two output/objective variables namely: weld-seam width (mm) and lap-shear strength (N/mm) are used in a single network to found the logical relationship between inputs and outputs data. A multilayer feed-forward NN with the back-propagation system is used to model the laser transmission welding process. In this learning, the backpropagation is used according to the network architecture. Thus, several measures are tried out to select a best grouping as presented in Figure 17.13. To train the network, experimental data are used. The twenty data set of input and output are selected for training the network from the twenty-six experimental results as mentioned in Table 17.8. The remaining six results are utilized for trial data authentication. All the input and output variables are stabilized from 0 to 1 [25].

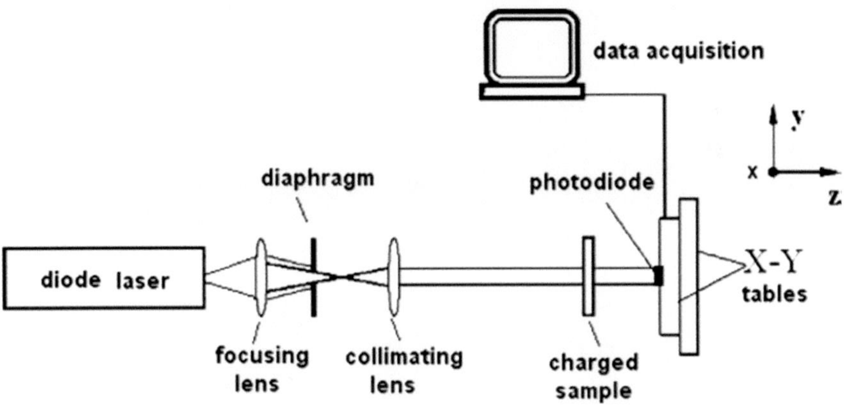

FIGURE 17.12 Schematic of transmission type LASER plastic welding [48].

Source: Reprinted with permission from Ref. [48]. Copyright 2007. Elsevier

Weights and biases of the network are applied to reduce the performance function, i.e., *'mean squared error – the average squared error'* between the network outputs and the real outputs.

The network is qualified by the Initial learning rate = 0.05. The limit set for network training by a maximum of 2,00,000 iterations. The training will be stopped after several iterations, or when the error is reduced by the set presentation goal. To check the probability of the generated model of NN, mean forecast errors (%) are considered for the lap shear strength and weld seam width by Eqn. (15) the performed value and predicted value are comprised in Tables 17.9 and 17.10 according to network architecture.

The mean prediction error:

$$= \frac{1}{p}\sum_{1}^{p}\frac{|actual\ value - predicted\ value|}{actual\ value}\times 100 \tag{15}$$

TABLE 17.8 Experimental Data of Input and Output [25]

SL. No.	Network Architecture	Mean Prediction Error (%)	Maximum Prediction Error (%)	Minimum Prediction Error (%)
1.	4-4-2	9.450	36.240	0.670
2.	4-5-2	8.700	30.070	0.170
3.	4-6-2	7.660	26.640	0.260
4.	4-7-2	5.370	17.580	0.450
5.	4-8-2	6.520	24.010	0.510
6.	4-9-2	19.20	47.240	0.760
7.	4-10-2	6.040	25.620	0.430
8.	4-11-2	9.260	39.360	0.730
9.	4-12-2	7.710	19.040	0.430
10.	4-13-2	10.850	25.390	0.640
11.	4-14-2	13.440	38.470	0.770
12.	4-15-2	13.170	40.850	0.680
13.	4-16-2	13.27	49.48	0.670
14.	4-17-2	9.38	27.33	0.770
15.	4-18-2	11.470	35.570	0.290
16.	4-19-2	12.22	28.24	0.770
17.	4-20-2	11.26	30.500	0.710
18.	4-4-3-2	9.470	29.970	0.690
19.	4-4-4-2	6.950	16.300	0.710
20.	4-4-5-2	4.12	15.590	0.110
21.	4-5-4-2	11.990	49.130	0.570
22.	4-5-7-2	10.080	29.280	0.780
23.	4-6-3-2	7.940	26.830	0.420
24.	4-6-4-2	11.480	35.020	0.070
25.	4-6-5-2	14.320	63.680	0.670
26.	4-7-5-2	6.510	23.600	0.200
27.	4-7-7-2	10.29	18.020	0.650
28.	4-7-9-2	9.800	36.140	0.710
29.	4-7-10-2	13.690	61.2	0.800
30.	4-8-4-2	10.550	36.800	0.190

Source: Reprinted/Adapted with permission from Ref. [25]. Copyright 2011. Elsevier.

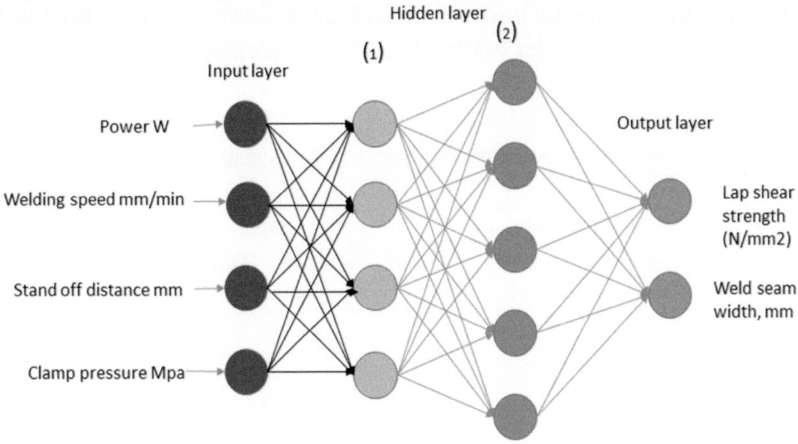

FIGURE 17.13 Performance of BP neural networks with different architectures.

TABLE 17.9 Comparable Description of Outcomes for Actual and Predicted Test Data of Lap-Shear Strength [25]

SL. No.	Welding Parameters				Lap-Shear Strength (N/mm)			
	Power (W)	Welding Speed (mm/min)	Stand-Off Distance (mm)	Clamp Pressure (MPa)	Actual	ANN Output	Error	% Error
1.	19.0	420.0	28.0	63.0	36.2	36.9	0.7	205.0
2.	25.0	420.0	36.0	6.3	28.6	29.1	51.0	1.8
3.	19.0	300.0	28.0	6.3	3523.0	34.9	32.0	0.9
4.	19.0	300.0	36.0	123.0	4603.0	43.2	226.0	621.0
5.	22.0	360.0	32.0	9.3	4743.0	4738.0	0.1	0.1
6.	25.0	420.0	28.0	123.0	38.1	4405.0	594.0	15.6

Source: Reprinted/Adapted with permission from Ref. [25]. Copyright 2011. Elsevier.

The above analysis shows the correlation and effectiveness between performance and prediction; the data shows in Tables 17.8 and 17.9. The bar graph indicates the percent error is very less and the predicted value is very close to the actual value (Figure 17.14(a) and (b)) [25].

17.4.1.6.2 ANN Model for Joining of Biodegradable Polyethylene Terephthalate/Polyethylene Vinyl Acetate by LASER Transmission Welding

Laser transmission welding is an abrupt, simple, and suitable technique to join plastic materials for a few mechanical domains. The fundamental test for producers is still on the best way to pick the information cycle boundaries to accomplish the best joint exhibition. The network architecture of NN for joining is shown in Figure 17.15 [28].

TABLE 17.10 Comparable Description of Outcomes for Actual and Predicted Test Data of Weld-Seam Width [25]

SL. No.	Welding Parameters				Weld-Seam Width (mm)			
	Power (W)	Welding Speed (mm/min)	Stand-OFF distance (mm)	Clamp Pressure (MPa)	Actual Outcomes	Output of ANN	Error	% Error
1.	19.0	420.0	28.0	6.30	1.92	1.94.	2.00	1.15
2.	25.0	420.0	36.0	6.30	2.43	2.58	0.15	6.27
3.	19.0	300.0	28.0	6.30	2.250	2.04	0.21	9.12
4.	19.0	300.0	36.0	12.30	2.88	3.030	0.15	5.30
5.	22.0	360.0	32.0	9.30	2.40	2.310	0.11	4.54
6.	25.0	420.0	28.0	12.30	2.220	2.110	0.11	4.79

Source: Reprinted/Adapted with permission from Ref. [25]. Copyright 2011. Elsevier.

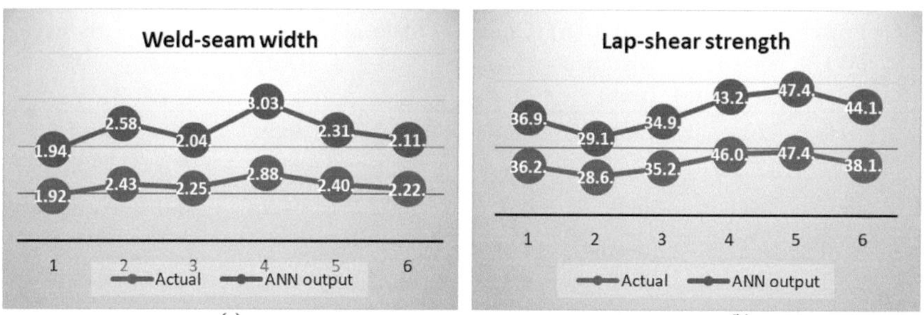

(a) (b)

FIGURE 17.14 Actual and ANN predictions of (a) weld seam width; and (b) lap-shear strength.

17.4.1.6.3 Process Optimization by Using GRA-Based Taguchi Analysis

If the problem having multiple output parameters, then GRA based Taguchi method is most suited for the analysis. In this optimization technique, initially experimental data is normalized by assigning the values from 0 to 1 (Table 17.11). Later, the gray relation coefficient was calculated by Eqn. (9) (Table 17.12), which shows the interconnection between the desired and found experimental data. Additionally, the problem was transacted to a single objective function by calculating the gray relational grade by Eqn. (10) The combination of parameters with the uppermost value of gray relational grade will be the optimal solution [33, 49, 50].

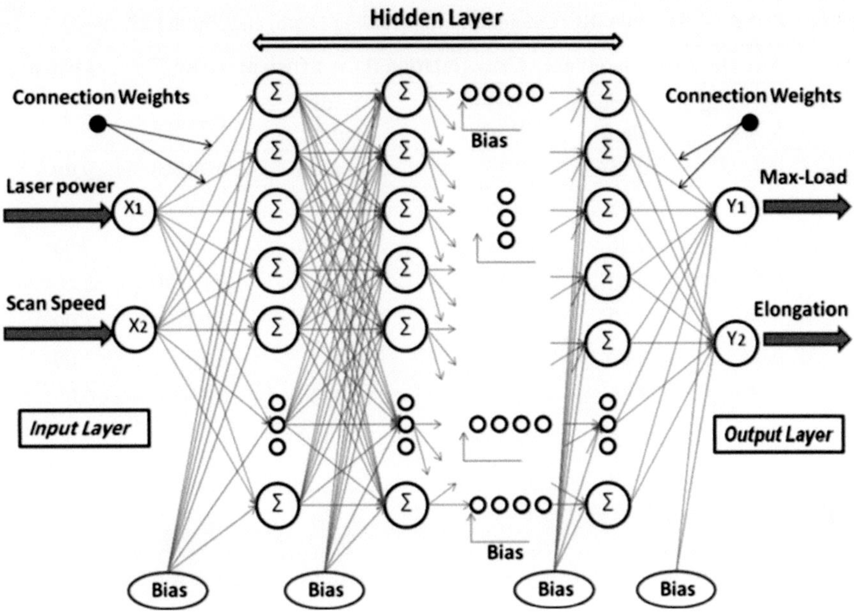

FIGURE 17.15 The applied neural network with five hidden layers [28].

Source: Adapted from Ref. [28]

TABLE 17.11 Normalized Value [33]

No. Exp.	Normalized Value	
	Welding Width	**Welding Depth**
1	0.0611	0.0437
2	0.09410	0.5243
3	0.0	0.641
4	0.03145	0.399
5	1.0	0.813
6	0.53320	1.0
7	0.6088	0.0
8	0.5801	0.5243
9	0.4961	0.839

Source: Adapted from Ref. [33]

Grey relational generation, the normalized top bead width, and HAZ width correspond to the smaller-the-better (SB) criterion according to Eqn. (8)

$$X_t(k) = \frac{\max y_t(k) - y_t(k)}{\max y_t(k) - \min y_t(k)}$$

TABLE 17.12 GRC and GRG Calculation for Welding Depth and Width [33]

No. Exp.	The Deviation Sequence		Cray Relational Co-Efficient		Gray Relational Grade	Order
	Welding Width	Welding Depth	Welding Width	Welding Depth		
1	0.9388	0.9563	0.35	0.34330	0.34540	9.00
2	0.905	0.48	0.35560	0.31250	0.43	7.00
3	1.00	0.36	0.33330	0.38210	0.46	5.00
4	0.96855	0.60	0.34047	0.45410	0.39730	8.00
5	0.00	0.19	1.00	0.73	0.96390	1.00
6	0.47	0.00	0.52	1.00	0.76	2.00
7	0.39110	1.00	0.56	0.33330	0.45	6.00
8	0.42	0.48	0.54358	0.51250	0.5280	4.00
9	0.5039	0.16	0.50	0.76	0.63	3.00

Source: Adapted from Ref. [33]

Penetration should follow the larger-the-better (LB) criterion as mentioned priorly in Eqn. (7),

$$X_i(k) = \frac{y_i(k) - min\, y_i(k)}{\max y_i(k) - \min y_i(k)}$$

where;

$$xi\ (k) is\, the\, value\, after\, the\, grey\, relational\, generation,$$
$$min\, yi\ (k) is\, the\, smallest\, value\, of\ yi\ (k)\, for\, the\, kth\, response, and$$
$$max\, yi\ (k) is\, the\, largest\, value\, of\ yi\ (k)\, for\, the\, kth\, response$$

GRC can be calculated by using the formula as discussed by Eqn. (9)

$$\xi_i(k) = \frac{\Delta \min + \Psi \Delta max}{\Delta o_i(k) + \Psi \Delta max}$$

where; Δo_i is the difference of the absolute value $x_0(k)$ and $x_i(k)$, is the normalized value for a given set of input parameters.

Ψ is a distinguishing coefficient 01 ≤≤ for all quality characteristics

$$\Delta \min = \text{smallest value } \nabla o_i$$
$$\Delta \max = \text{largest value } \nabla o_i$$

After averaging the gray relational coefficients can be calculated by Eqn. (10)

$$G_i = \frac{1}{n}\sum_{k=1}^{n} \xi_i(k) \qquad \text{(n is the number of process response)}$$

17.4.2 INTERNAL METHOD

17.4.2.1 SPIN WELDING

Spin welding and stir welding are the variances of the friction welding process. The spin welding process divided into four-stage, i.e., both the part to be weld should be coaxial. One part will be in stationary condition and others will rotate at required RPM. The frictional heat is generated at the interface; the moving part smoothly fed into the stationary part and in the stage stops the rotation of the moving part. After solidification and cooling a sound, joint has accomplished. Spin welding is a pollution-free welding process. The time required for welding is very less. Material or polymers of different melting points can be weld. With all merits, the spin welding process has few critical issues like a very accurate fixturing is required to hold the parts co-axially, collar formation (upsetting) due to deformation of the material. Sometime fiber in the material is misaligned or pullout from the spinning zone. These critical issues can be overcome by controlling the parameters, i.e., spinning time, feeding force and RPM of the rotating part [51].

17.4.2.2 FRICTION STIRS WELDING

FSW is a solid-state welding process, most widely used to join metal to metal and metal to polymers. Frictional heat is used to forge parts together under an axial force. In this method, the generated heat does not exceed beyond the melting point of the material, the material is heated to its forging temperature and is stirred together by a rotating tool. This process is capable of joining metals having a different melting point. In the FSW process, a stir tool is used to produce frictional heat. The plates to be weld are in closed contact at the interface (butt joint) with proper fixturing. The stir tool of metal will rotate on the interface line of the surface after a few seconds polymer start softens due to frictional heat. Now the tool is started transverse motion along the line of the interface (Figure 17.16). The plate at the interface is mixed together by stirring [52, 53]. After cooling a sound joint is formed. This process of welding is categorized in green manufacturing because FSW is an eco-friendly welding process with high productivity at low production cost. The weld joint can be accomplished by optimizing a few parameters, such as rotational speed, transverse speed, plunge depth of shoulder, and tool pin geometry. The many researchers reported that the welding by a tapered threaded pin provides an excellent weld quality [54].

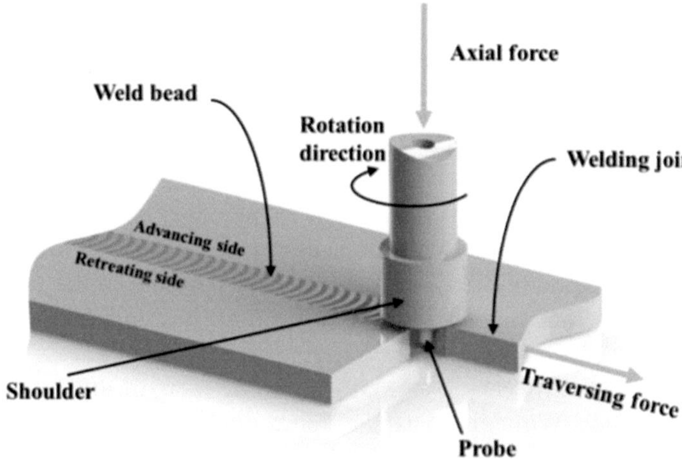

FIGURE 17.16 Processing of friction stir welding [55].

Source: Reprinted/Adapted with permission from Ref. [55]. Copyright 2015. Elsevier

The FSW is most widely used for the joining of dissimilar materials like metal to the polymer. The melting point between metals and polymers has a great difference. This huge temperature difference creates a complicacy in solid-state fusion. Once the metal start heating, the polymer reaches a melting point. Welding of fiber polymer composite also encountered by various defects such as blowholes, tunnel defects, flash, root thinning [52, 56], etc., as shown in Figure 17.17.

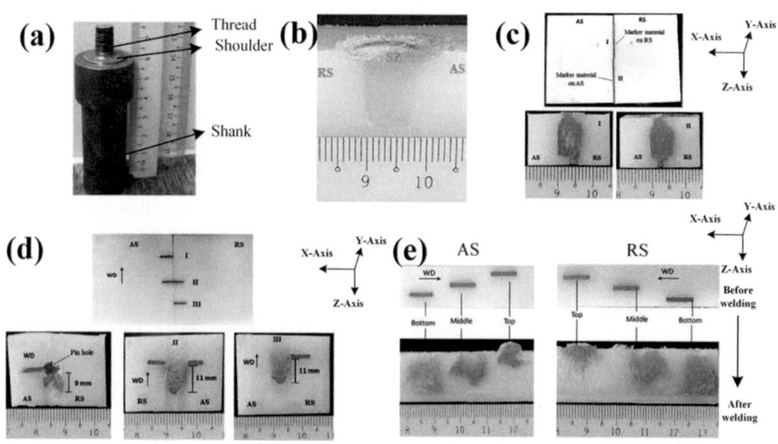

FIGURE 17.17 16 mm thick nylon 6 (polyamide 6) welded (butt joint) [52].

Source: Adapted from Ref. [52]

The new innovative method of friction welding is 'threaded hole friction spot welding' (THFSW). In this joining process, a pre-threaded hole in the metal sheet is filled by the melted and re-solidified polymer during the process. In the THFSW process, a rotating tool gets coaxial to the pre-threaded hole in the Al plate. The tool gets plunge in the Al plate by less than 1 mm. The heat is generated between the tool and Al plate, and heat is conducted at the interface. The polymer starts melting, due to the effect of plunging pressure the molten metal come out with filling a threaded hole. A film of molten polymer is spread over the surface of the Al plate and after removing the tool, the polymer solidifies in the thread and on the surface (Figure 17.18) [57].

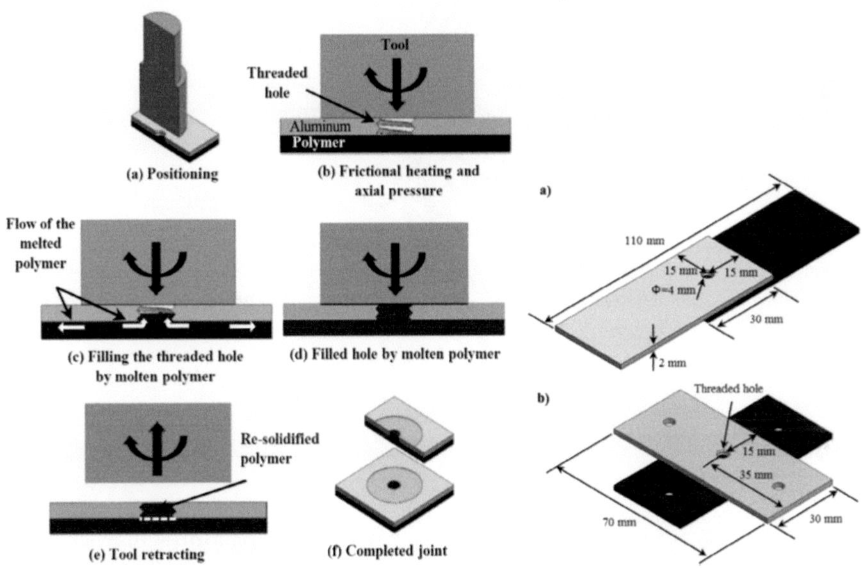

FIGURE 17.18 Process step of THFSW and joint preparation [57].

Source: Reprinted with permission from Ref. [57]. Copyright 2017. Elsevier

17.4.2.2.1 ANN Model for Process Optimization of FSW

The information layer comprises of four hubs which represent tool-pin profile, tool rotational speed, transverse speed and tilt angle which used to anticipate the reaction and the yield layer comprise one hub which represent to reaction GRG by training a few networks, the quantity of hidden layers and neurons in the hidden layer has been resolved. The quantity of hidden layers and neurons in the hidden layer has been controlled via training a few networks, 4 hidden layers and 10 neurons in each hidden layer were picked as optimum (Figure 17.19). So as to prepare the ANN model, 70% readings for preparing, 15% for approval and 15% for testing were taken. The correlation coefficients for GRG at the training, validation, and testing stages were 0.9989, 0.9999, and 0.9994, respectively [58].

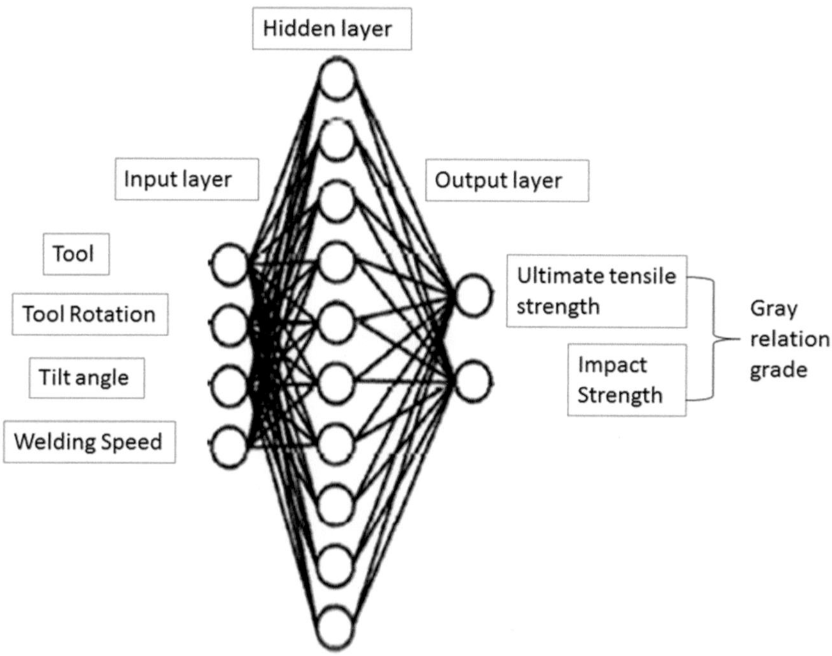

FIGURE 17.19 The architecture of ANN [58].

Source: Reprinted/Adapted with permission from Ref. [58]. Copyright 2018. Elsevier

17.4.2.2.2 WPCA-ANN-PSO Approach

WPCA-ANN-PSO approach to perform multi-response optimization (MRO) of Friction Stir Welding (FSW) processes is completed in three phase process as shown in Figure 17.20.

- **First Stage:** The multiple responses are converted into a single multi-performance index (MPI) using Taguchi "weighted principal component analysis" (WPCA) approach.
- **Second Stage:** ANN model is generated to predict the MPI for assign sets of control factors.

- **Third Stage:** PSO algorithm is utilized to find the overall optimal solution by using the developed ANN model.

Further, to demonstrate the efficiency of the projected model, a case study on MRO of the FSW process has performed to join two plates [24].

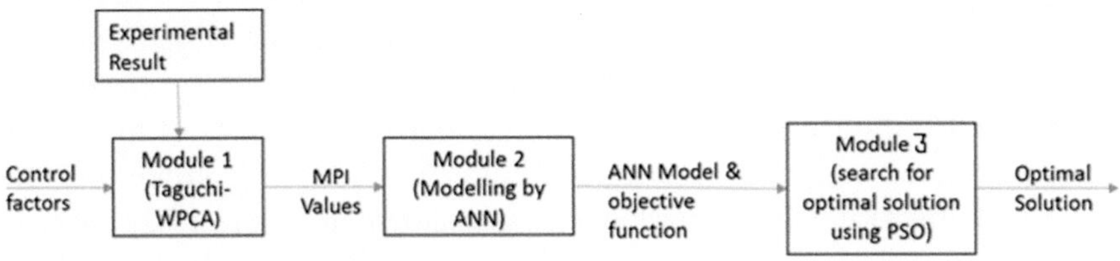

FIGURE 17.20 Flow diagram for the representation of the WPCA-ANN-PSO approach [24].
Source: Adapted from Ref. [24]

17.4.2.3 VIBRATION WELDING

Vibration welding is nothing but a linear friction welding. In this process, parts to be weld is linearly sliding over stationary part and after adequate frictional heating for melting of polymer, moving part immediately comes in steady state on the stationary part. The pressure is applied for fusion. This process required an efficient fixture. The sound joint can be produced by optimum process parameters such as frequency, amplitude, and pressure. This is a very efficient polymer welding process with a narrow heat-affected zone [59].

17.4.2.4 ULTRASONIC WELDING (UW)

Ultrasonic welding (UW) is a solid-state welding process. The domestic power supply of 220–250 V, 50 Hz frequency is converted into ultrasonic frequency, with the help of frequency generator. The high-frequency electrical signal is converted by transducers into a high frequency (10–60 kHz) mechanical vibration. The Magnetostrictive (for low voltage high current) and Piezoelectric (high voltage low current) transducer are used for this application. High-frequency vibration energy is utilized to produce heat. The process involves three fundamental steps: (i) the column is lowered, and the ultrasonic power activated; (ii) the faying surfaces are melted, final pressure is applied, then welding occurs; and (iii) the power is turned off and the components are held together as the weld solidifies. These steps (cycle time) completed in few seconds. UW is mainly used for processing amorphous polymers like polystyrene (PS), Acrylonitrile-Butadiene-Styrene (ABS). The spot type weld is produced by UW. General practice, the part to be weld by UW need to be adequately rigid or the material required to have a sufficiently high storage modulus to pass on the vibrations to the weld interface. Materials with low modulus, for example, thermoplastic elastomers, easily deform and repel without transmitting the vibrations to the weld interface. Therefore, only foils, thin films of thermoplastic elastomers can be welded ultrasonically. The ultrasonic machine consists of four basic units: frequency generator, transducer, ultrasonic horn, sonotrode tool the assembled unit of the above components has shown in Figure 17.21 [60].

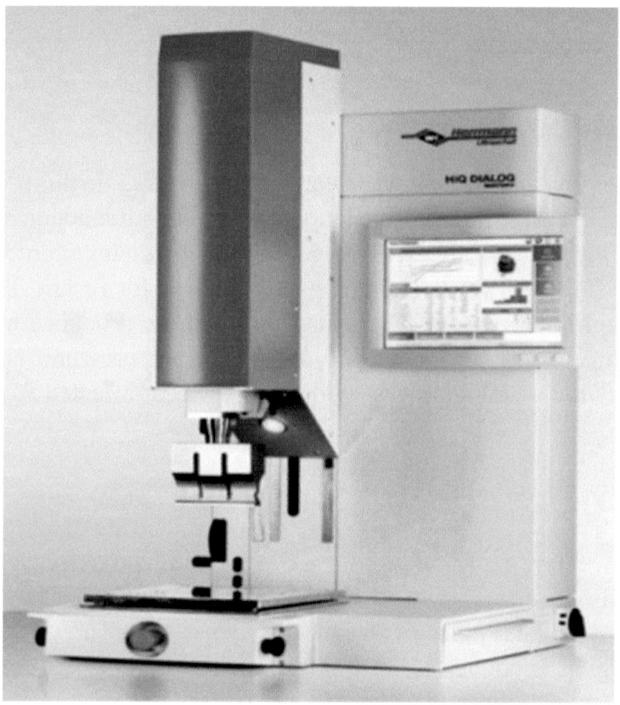

FIGURE 17.21 Ultrasonic welding machine [61].

Source: Adapted from Ref. [61]

17.4.2.4.1 ANN Model for Welding Strength in USW of Polymer

The trained network is used for finding a considerable input and output parameter as well as predicting the performance for the other input, which is not experimentally performed but fit in the range of training. The hidden layer neurons are linked with the output neurons by a purlin transfer function. The output is evaluated as the addition of entering signals as expressed in Eqns. (16) and (17) [62].

$$net_k^{(3)} = \sum_j w_j b_j^{(2)}(x) \tag{16}$$

$$N_k^{(3)}(x) = f_k^{(3)}\left\{net_k^{(3)}(x)\right\} = net_k^{(3)}(x) \tag{17}$$

26 the triple layer neurons are used to specify the data sets for inputs (Table 17.13), an output neuron is applied to identify the weld strength of the joint, and these functions are found from absolutely randomized models. The result of parameters optimization for forecast the weld strength value verse specimen for the network is shown in Figure 17.22.

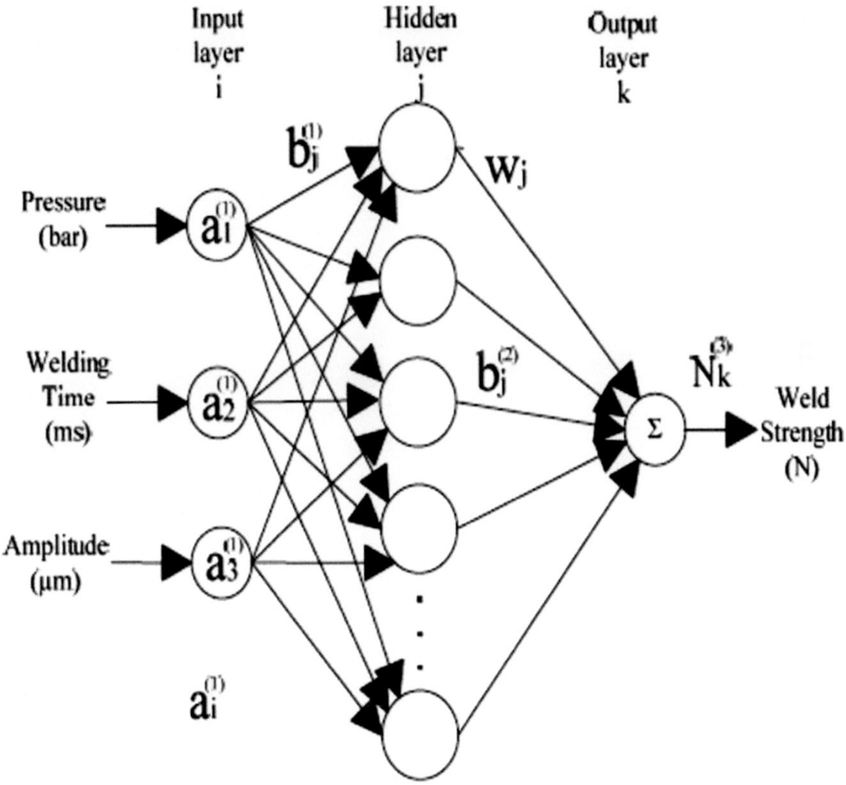

FIGURE 17.22 Network architecture.

Source: Reprinted/Adapted with permission from Ref. [23]. Copyright 2020. Elsevier.

TABLE 17.13 ANN Model to Predict the Weld Strength of Polymer [62]

SL. No.	Welding Pressure (Bar)	Welding Time (ms)	Amplitude (μm)	Strength (N)
1.	3.0	400	100	570.3
2.	3.5	500	100	791.7
3.	3.5	500	100	803.8
4.	3.5	500	100	819.3
5.	3.5	500	100	826.6
6.	3.5	500	100	799.0
7.	4.0	500	100	798.3
8.	4.0	500	100	785.8
9.	4.0	550	100	792.5
10.	4.0	550	100	813.7
11.	4.0	550	100	858.5
12.	4.0	550	100	833.9
13.	6.0	600	100	916.8
14.	5	600	100	811.8
15.	5	600	100	808.0
16.	5	600	100	797.3
17.	5.0	650	100	835.3
18.	5.0	650	100	789.7
19.	5.0	650	100	839.2
20.	5.0	650	100	878.7
21.	5.0	650	100	862.1
22.	5	700	100	827.8
23.	5.5	700	100	815.9
24.	6.0	750	100	847.2
25.	6.0	750	100	840.9
26.	6.0	750	100	826.4

Source: Adapted from Ref. [62]

The best-fit performance curve (Figure 17.23) shows the correlation between ANN trained data, validation data, and test data. The fit performance marked on the curve for the consideration. The regression analysis is applied to obtain the correlation coefficient. The correlation coefficient indicates that the closeness between the predicted value and obtained value. If the value of "$R \approx 1$"it means proximity between the predicted value and experimental value, and "0" signify a mismatch. In the above result, the various regression coefficient (R) is obtained:

$R = 0.9692$ *is obtained for the set of training data,*

$R = 0.94552$ *for the set of testing data,*

$R = 0.98841$ *for set of validation data and finally*

$R = 0.95851$ *was obtained for all the data presented as a whole.*

Hence, we can consider that the forecast of weld strength can be found using the ANN model [62].

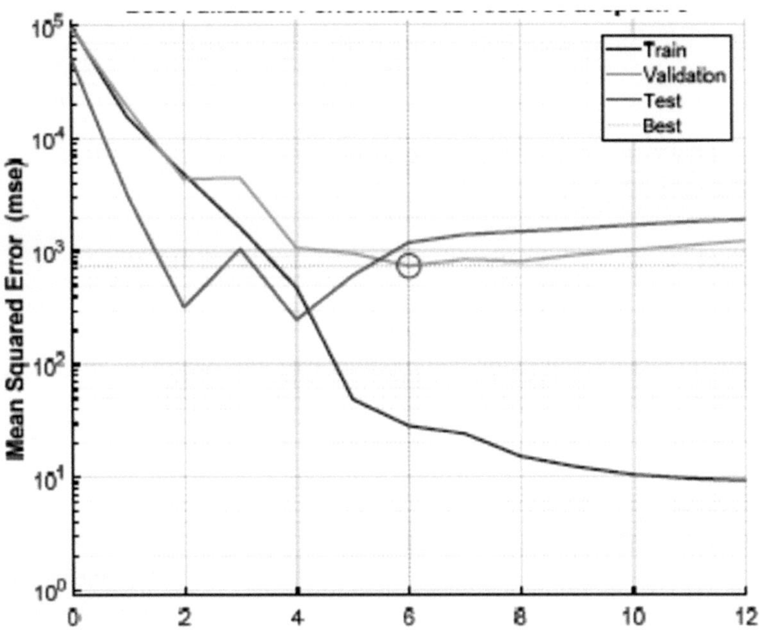

FIGURE 17.23 Best fit validation performance [62].

Source: Adapted from Ref. [62]

17.4.2.4.2 TAGUCHI METHOD FOR PROCESS OPTIMIZATION IN USW

In Taguchi method, the deviation between the experimental value and the desired value is known as a loss function. Generally, the three categories of the quality characteristic in the signal-to-noise (S/N) ratio analysis, i.e., the lower-the-better, the larger the better, and nominal the better. Larger-is better quality characteristic is considered to obtain best welding parameters. The value of S/N ratio calculated by use of Eqn. (12).

$$\text{Larger the better}\left[\frac{S}{N}ratio = -10log_{10}(\frac{1}{n}\sum_{i=1}^{n}\frac{1}{y_i^2})\right]$$

For the development of the design of experiments (DOE) according to the given parameter and their levels as shown in Table 17.14, L27 orthogonal array selected to find the multiple responses. The experiments were performed according to the DOE. The result is shown in Table 17.15.

TABLE 17.14 Factors and Levels [35]

SL. No.	Factors	Designation	Level 1	Level 2	Level 3
1.	Pressure (bar)	A	3	3.5	4
2.	Time (sec.)	B	2.5	3	3.5
3.	Amplitude (μm)	C	28	42.5	57

Source: Adapted from Ref. [35]

TABLE 17.15 Signal to Noise Response for Average Strength [35]

Factors	Level 1	Level 2	Level 3	Delta	Rank
Pressure in bar	23.542	24.461	23.636	0.919	2
Time in sec.	23.977	23.573	24.088	0.515	3
Amplitude in μm	23.5313	23.445	24.663	1.218	1

Source: Adapted from Ref. [35]

The optimum parameter combinations for maximizing the weld strength were identified. Parameters like weld pressure, weld time and amplitude have a significant effect on the weld strength. Also, the interaction between weld pressure and weld time, weld pressure and amplitude, and weld time and amplitude have a significant effect on the weld strength [35].

17.4.3 ELECTROMAGNETIC WELDING OF PLASTIC

Most of the thermoplastic polymers do not occurrence a temperature rise when irradiated under the electromagnetic wave. Though, the addition of a microwave vulnerable implant at the interface allows localizing heating. If the interface is exposed to microwaves and an exerted pressure, the melting of the adjacent polymer/plastic results and a weld is accomplished. Suitable implants contain metals, carbon or conducting polymers. Electromagnetic wave heating-based welding methods are resistive implants welding, induction welding (IW), dielectric welding (DW), and microwave welding (MWW) [63].

17.4.3.1 RESISTANCE IMPLANT WELDING (RIW)

Polymers usually don't have a potential for inductance heating, because most of the polymers are not electrically and electromagnetically conductive. The implants of conductive material are imposed at the interface of the weld. The implants are heated with the influence of electromagnetic waves and melt the interface; pressure is applied for welding and left of the solidification. The implants used in this process should be consumable. The polymer composite reinforced with metal particles, fibers are not required implants additionally because metal powder, fibers also act as consumable implants. The implants induction heating also has demonstrated in Figure 17.24 [64].

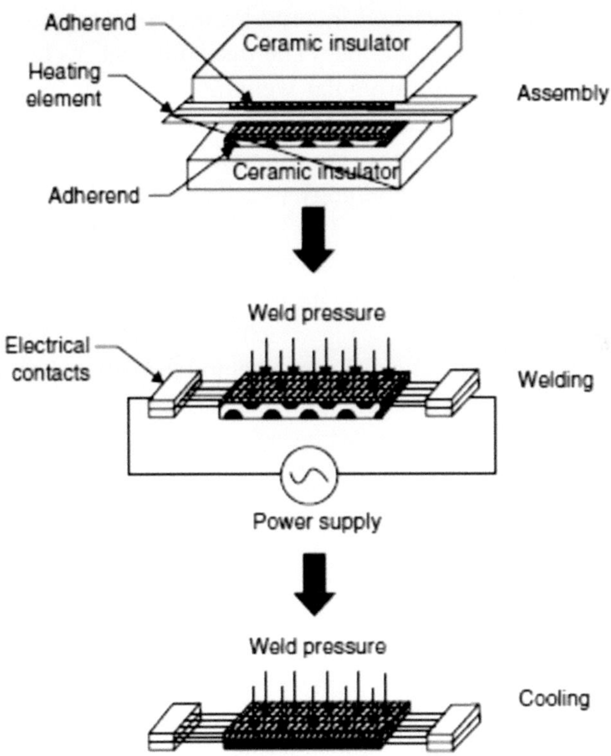

FIGURE 17.24 Resistive implant welding [64].

Source: Reprinted with permission from Ref. [64]. Copyright © 2009 William Andrew Inc. Published by Elsevier Inc.

17.4.3.2 INDUCTION WELDING (IW)

Induction welding (IW) is fundamentally the same as the resistance implant welding (RIW) measure. The heat is produced in IW by eddy current or hysteresis losses. For this situation, the heat is created by an induction field for example set up by eddy flows or to hysteresis losses. A work curl is put close to the interface, associated with a force gracefully. As the electric flow of high recurrence flexibly through the work loop, a powerful attractive field is created, whose motion connects to the susceptible/embed. As the embed warms up, the contiguous thermoplastic mellow and melts, the weight is applied to the joint and a weld structure as the joint cools. IW is utilized for connecting metalize tops to plastic bottles. It is likewise reasonable for a wide scope of composite materials [65].

The IW process can be performed in two ways: continuous manner and the other one is a discontinuous manner.

In persistent sort, IW measures incorporate that the workpiece or the loop is moved comparative with another. The discontinuous welding utilizes a fixed arrangement. Although irregular welding is a simple cycle, persistent welding is more critical because of the lesser process duration and the capacity to weld critical shapes, for example, airplane structures and car undercarriage. A general methodology is called semi-consistent enlistment welding (SCIW), which is portrayed by a static arrangement where warming and union happen at a similar spot and same time and afterward followed by a transportation step between the necessary welding areas. The continues type of IW process (Figure 17.25) in which a coil raises the temperature of the interface and the follower roller move behind the coil and apply pressure on the surface for fusion. The simulation image (Figure 17.25) of the process indicates the temperature profile with cooling. The continuous IW process mostly designs for large applications [63].

17.4.3.3 DIELECTRIC WELDING (DW)

Dielectric welding (DW) welding is also known as radio frequency (RF) welding. In this method, thin sheets of the polar thermoplastic polymer are joined together. It uses extremely high frequency (13 to 100 MHz) of electromagnetic energy utilized to produce heat for fusion. An alternating electric field is established between two metallic rods. The polar molecules found in some thermoplastics when this molecule influenced by an electromagnetic field resulting in molecules to oscillate and orient themselves with respect to the field. The energy generated by this process causes a temperature increase resulting in the melting of the materials. The uniform clamping pressure is maintained until the interface solidifies. A high-frequency radio signal is used to impart molecular vibration in two similar or dissimilar polymers. RF welding is one of the methods for manufacturing that allows the designer to weld different types of plastics. This welding method can be applied to various commercially significant medical-grade polymers and PVC of varying hardness [46].

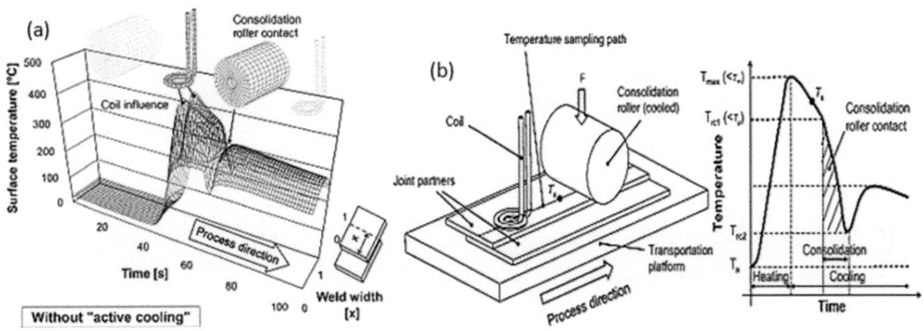

FIGURE 17.25 Simulation of continuous type induction welding with temperature profile [63].
Source: Reprinted/Adapted with permission from Ref. [63]. Copyright © 2013 Elsevier Ltd.

17.4.3.4 MICROWAVE WELDING (MWW)

MWW is a kind of electromagnetic welding process. MWW process is similar to IRW, IW, resistive implants welding. The Susceptible implants are required to apply for absorbing microwave energy as similar to RIW. A type of polymer, which has a polar group in their molecular structure are capable of absorbing heat in the microwave field, can be weld directly. The absorption of energy from the microwave in a polar conducting medium depends on the various parameters such as conductivity of polymer, dielectric constant (in the atmosphere and in a vacuum), frequency of radiation. The process of Microwave implant welding is performed in the following steps.

- Mounting of implants at the weld interface;
- Clamping of parts to be weld;
- Heat absorption by implants from microwave;
- Heat transfer to the polymer by conduction for melting of the polymer;
- Apply pressure for fusion and cooling.

MWW is a very advanced welding process; it has numerous advantages like a very fast welding process, volumetric heating, capable to weld complex and three-dimensional (3D) joint geometries, power consumption is very less, and low setup costs. In this process, implants are required to be applied on the interface, so this adds one additional activity in the welding process. Sometimes implants in the polymer after welding create an issue in recycling. Those polymers are not required implants; they melt very rapidly by the influence of microwaves. Therefore, it's necessary to control the parameters of microwave energy [42, 43].

KEYWORDS

- **acrylonitrile-butadiene-styrene**
- **artificial neural networks**
- **polyetheretherketone**
- **polypropylene, polyphenylene sulfide**
- **processing elements**
- **transfer functions**

REFERENCES

1. Ku, H., Wang, H., Pattarachaiyakoop, N., & Trada, M., (2011). A review on the tensile properties of natural fiber reinforced polymer composites. *Compos. Part B Eng.* doi: 10.1016/j.compositesb.2011.01.010.
2. Pothan, L. A., Oommen, Z., & Thomas, S., (2003). Dynamic mechanical analysis of banana fiber reinforced polyester composites. *Compos. Sci. Technol.* doi: 10.1016/S0266-3538 (02)00254-3.
3. Goh, G. D., et al., (2018). Characterization of mechanical properties and fracture mode of additively manufactured carbon fiber and glass fiber reinforced thermoplastics. *Mater. Des.* doi: 10.1016/j.matdes.2017.10.021.
4. Czigány, T., (2005). *Basalt Fiber Reinforced Hybrid Polymer Composites*. doi: 10.4028/www.scientific.net/msf.473-474.59.
5. Saba, N., Tahir, P. M., & Jawaid, M., (2014). A review on potentiality of nano filler/natural fiber filled polymer hybrid composites. *Polymers*. doi: 10.3390/polym6082247.
6. Hewitt, C. A., Kaiser, A. B., Roth, S., Craps, M., Czerw, R., & Carroll, D. L., (2012). Multilayered carbon nanotube/polymer composite based thermoelectric fabrics. *Nano Lett.* doi: 10.1021/nl203806q.
7. Kumar, S., Doshi, H., Srinivasarao, M., Park, J. O., & Schiraldi, D. A., (2002). Fibers from polypropylene/nano carbon fiber composites. *Polymer (Guildf).* doi: 10.1016/S0032-3861(01)00744-3.

8. Vaidyanathan, H. P., Murty, P., & Eswara, S. P., (2011). *Hybrid Natural Fiber Composites Molded Auto-Body Panels/ Skins (Hybrid NFPC): Processing, Characterization & Modeling*. doi: 10.4271/2011-01-0219.
9. Kumar, R., Giriraj, B., Kumar, S., & Suresh, A. R., (2018). Study of mechanical properties of jute/carbon fiber reinforced polymer hybrid composites for automotive applications. *J. Balk. Tribol. Assoc.*
10. Schumacher, K. H., Kirsch, H. P., & Rupaner, R., (1996). Safe and environmentally friendly nonwovens for medical uses. *Nonwovens Rep. Int.*
11. Allred, R. E., et al., (2003). The influence of sizings on the durability of high-temperature polymer composites. *High Perform. Polym.* doi: 10.1177/09540083030154002.
12. Li, Y. H., et al., (2013). A novel polymer electrolyte with improved high-temperature tolerance up to 170°C for high-temperature lithium-ion batteries. *J. Power Sources*. doi: 10.1016/j.jpowsour.2013.01.148.
13. Cao, S., Wu, Z., & Wang, X., (2009). Tensile properties of CFRP and hybrid FRP composites at elevated temperatures. *J. Compos. Mater.* doi: 10.1177/0021998308099224.
14. Venkatesan, J., Bhatnagar, I., Manivasagan, P., Kang, K. H., & Kim, S. K., (2015). Alginate composites for bone tissue engineering: A review. *International Journal of Biological Macromolecules*. doi: 10.1016/j.ijbiomac.2014.07.008.
15. Sridhar, R., Sundarrajan, S., Venugopal, J. R., Ravichandran, R., & Ramakrishna, S., (2013). Electrospun inorganic and polymer composite nanofibers for biomedical applications. *Journal of Biomaterials Science, Polymer Edition*. doi: 10.1080/09205063.2012.690711.
16. Aminabhavi, T. M., & Deshmukh, A. S., (2016). Polymeric hydrogels as smart biomaterials. In: *Polymer and Composite Materials*.
17. Dziadek, M., Stodolak-Zych, E., & Cholewa-Kowalska, K., (2017). Biodegradable ceramic-polymer composites for biomedical applications: A review. *Materials Science and Engineering C*. doi: 10.1016/j.msec.2016.10.014.
18. De Gennes, P. G., (1971). Reptation of a polymer chain in the presence of fixed obstacles. *J. Chem. Phys.* doi: 10.1063/1.1675789.
19. Khan, I., Saeed, K., & Khan, I. (2019). Nanoparticles: Properties, applications and toxicities. *Arabian Journal of Chemistry, 12*(7), 908–931.
20. Ryan, A. J., & Frank, F. C., (2007). Polymer science. In: *100 Years of Physical Chemistry*.
21. Agatonovic-Kustrin, S., & Beresford, R., (2000). Basic concepts of artificial neural network (ANN) modeling and its application in pharmaceutical research. *Journal of Pharmaceutical and Biomedical Analysis*. doi: 10.1016/S0731-7085(99)00272-1.
22. Yang, Z. R., & Yang, Z., (2014). Artificial neural networks. *Compr. Biomed. Phys., 6*, 1–17. doi: 10.1016/B978-0-444-53632-7.01101-1.
23. Salah Alaloul, W., & Hannan Qureshi, A. (2020). Data Processing Using Artificial Neural Networks. Dynamic Data Assimilation – Beating the Uncertainties. doi: 10.5772/intechopen.91935.
24. Ambekar, M., & Kittur, J., (2019). Multiresponse optimization of friction stir welding process parameters by an integrated WPCA-ANN-PSO approach. *Mater. Today Proc.*, (xxxx). doi: 10.1016/j.matpr.2019.11.152.
25. Acherjee, B., Mondal, S., Tudu, B., & Misra, D., (2011). Application of artificial neural network for predicting weld quality in laser transmission welding of thermoplastics. *Appl. Soft Comput. J., 11*(2), 2548–2555. doi: 10.1016/j.asoc.2010.10.005.

26. Mohamed, A., et al., (2020). Nature-inspired algorithms for feed-forward neural network classifiers: A survey of one decade of research. *Ain Shams Eng. J.*, (xxxx). doi: 10.1016/j.asej.2020.01.007.
27. Liu, J. (2013). Radial Basis Function (RBF) neural network control for mechanical systems: design, analysis and Matlab simulation. Springer Science & Business Media.
28. Mehrpouya, M., Gisario, A., Rahimzadeh, A., & Barletta, M., (2019). An artificial neural network model for laser transmission welding of biodegradable polyethylene terephthalate/polyethylene vinyl acetate (PET/PEVA) blends. *Int. J. Adv. Manuf. Technol., 102*(5–8), 1497–1507. doi: 10.1007/s00170-018-03259-9.
29. Hagan, M. T., Demuth, H. B., & De Jesús, O., (2002). An introduction to the use of neural networks in control systems. *Int. J. Robust Nonlinear Control, 12*(11), 959–985. doi: 10.1002/rnc.727.
30. Albawi, S., Mohammed, T. A., & Al-Zawi, S., (2018). *Understanding of a Convolutional Neural Network.* doi: 10.1109/ICEngTechnol.2017.8308186.
31. Happel, B. L. M., & Murre, J., M. J., (1994). Design and evolution of modular neural network architectures. *Neural Networks*. doi: 10.1016/S0893-6080(05)80155-8.
32. Kumar, D., Ghosh, S., Kuar, A. S., & Paitandi, S., (2020). Laser transmission welding of thermoplastic with beam wobbling technique using particle swarm optimization. *Mater. Today Proc.*, (xxxx). doi: 10.1016/j.matpr.2019.12.422.
33. Hubeatir, K. A., (2020). *Laser Transmission Welding of PMMA Using IR Semiconductor Laser Complemented by the Taguchi Method and Grey Relational Analysis*. doi: 10.1016/j.matpr.2019.09.167.
34. Shanmugarajan, B., Shrivastava, R., Sathiya, P., & Buvanashekaran, G., (2016). Optimization of laser welding parameters for welding of P92 material using Taguchi based grey relational analysis. *Def. Technol., 12*(4), 343–350. doi: 10.1016/j.dt.2016.04.001.
35. Srinivasan, V. P., & Elangovan, S. (2013). Optimization of ultrasonic metal welding process parameters for copper wire with aluminium sheet using taguchi method and finite element modeling of stress distribution. *CiiT Int. J. Programmable Device Circuits Syst, 5,* 31–37.
36. Kim, T. K., (2017). Understanding one-way ANOVA using conceptual figures. *Korean J. Anesthesiol.* doi: 10.4097/kjae.2017.70.1.22.
37. Holcomb, Z. C., Cox, K. S., Holcomb, Z. C., & Cox, K. S., (2018). Two-way ANOVA II. In: *Interpreting Basic Statistics*.
38. Ageorges, C., Ye, L., & Hou, M., (2001). Advances in fusion bonding techniques for joining thermoplastic matrix composites: A review. *Compos. – Part A Appl. Sci. Manuf.* doi: 10.1016/S1359-835X(00)00166-4.
39. Grewell, D., & Benatar, A., (2007). *Welding of Plastics: Fundamentals and New Developments.* doi: 10.3139/217.0051.
40. Hot-gas welding. In: *Encyclopedic Dictionary of Polymers*.
41. Haberstroh, E., Wolters, L., & Stegmaier, M., (2000). *Plastics Welding – Part 1: Skilled Manual Processes.* Weld. Res. Abroad.
42. Troughton, M. J., (2009). Chapter 7 – extrusion welding. In: *Handbook of Plastics Joining* (2nd edn., pp. 73–79). Boston: William Andrew Publishing.
43. Alojaly, H., & Benyounis, K. Y., (2020). Joining of plastics and polymeric materials. In: *Reference Module in Materials Science and Materials Engineering*.
44. Benatar, A., (2017). Plastics joining. In: *Applied Plastics Engineering Handbook: Processing, Materials, and Applications* (2nd edn.).

45. Ülker, A., Öztoprak, N., Sayer, S., & Yeni, C., (2018). Optimization of welding parameters of hot plate welded PC/ABS blends by using the Taguchi experimental design method. *J. Elastomers Plast., 50*(2), 162–181. doi: 10.1177/0095244317740735.
46. Infrared Welding, (2009). *Handb. Plast. Join.*, 97–103. doi: 10.1016/B978-0-8155-1581-4.50011-1.
47. Jansson, A., Kouvo, S., Salminen, A., & Kujanpää, V., (2003). *The Effect of Parameters on Laser Transmission Welding of Polymers.* doi: 10.2351/1.5060071.
48. Ilie, M., Kneip, J. C., Matteï, S., Nichici, A., Roze, C., & Girasole, T., (2007). Through-transmission laser welding of polymers–temperature field modeling and infrared investigation. *Infrared Phys. Technol., 51*(1), 73–79. doi: 10.1016/j.infrared.2007.02.003.
49. Kumar, N., Rudrapati, R., & Pal, P. K., (2014). Multi-objective optimization in through laser transmission welding of thermoplastics using grey-based Taguchi method. *Procedia Mater. Sci., 5*, 2178–2187. doi: 10.1016/j.mspro.2014.07.423.
50. Shanmugarajan, B., Shrivastava, R., Sathiya, P., & Buvanashekaran, G., (2016). *Optimization of Laser Welding Parameters for Welding of P92 Material Using Taguchi Based Grey Relational Analysis.* doi: 10.1016/j.dt.2016.04.001.
51. Troughton, M. J., (2009). Chapter 4–spin welding. In: *Handbook of Plastics Joining* (2nd edn., pp. 49–55). Boston: William Andrew Publishing.
52. Huang, Y., et al., (2018). Friction stir welding/processing of polymers and polymer matrix composites. *Composites Part A: Applied Science and Manufacturing.* doi: 10.1016/j.compositesa.2017.12.005.
53. Barmouz, M., Shahi, P., & Asadi, P., (2014). *Friction Stir Welding/Processing of Polymeric Materials*. Woodhead Publishing Limited.
54. Bilici, M. K., (2012). Effect of tool geometry on friction stir spot welding of polypropylene sheets. *Express Polym. Lett.* doi: 10.3144/expresspolymlett.2012.86.
55. Eslami, S., Ramos, T., Tavares, P. J., & Moreira, M., G. P. P., (2015). Effect of friction stir welding parameters with newly developed tool for lap joint of dissimilar polymers. *Procedia Eng., 114*, 199–207. doi: 10.1016/j.proeng.2015.08.059.
56. Zafar, A., Awang, M., Khan, S. R., & Emamian, S., (2016). Investigating friction stir welding on thick nylon 6 plates. *Weld. J., 95*(6), 210s–218s.
57. Karami, P. H., Movahedi, M., & Kokabi, A. H., (2017). A new refill friction spot welding process for aluminum/polymer composite hybrid structures. *Compos. Struct., 174*, 59–69. doi: 10.1016/J.COMPSTRUCT.2017.04.053.
58. Wakchaure, K. N., Thakur, A. G., Gadakh, V., & Kumar, A., (2018). Multi-objective optimization of friction stir welding of aluminum alloy 6082-T6 using hybrid Taguchi-grey relation analysis–ANN method. *Mater. Today Proc., 5*(2), 7150–7159. doi: 10.1016/j.matpr.2017.11.380.
59. Troughton, M. J., (2009). Chapter 3 – vibration welding. In: *Handbook of Plastics Joining* (2nd edn., pp. 37–47). Boston: William Andrew Publishing.
60. Ultrasonic Welding, (2009). *Handb. Plast. Join.*, 15–35. doi: 10.1016/B978-0-8155-1581-4.50004-4.
61. Society, A. W., (1987). *Welding Handbook, 5.*
62. Chinnadurai, T., Saravanan, S., Prabaharan, N., Pandian, M. K., & Deebika, S., (2018). *Analyzing the Weld Strength of Ultrasonic Polymer Welding Using Artificial Neural Networks*. doi: 10.1016/j.matpr.2018.10.116.

63. Bayerl, T., Duhovic, M., Mitschang, P., & Bhattacharyya, D., (2014). The heating of polymer composites by electromagnetic induction: A review. *Composites Part A: Applied Science and Manufacturing*. doi: 10.1016/j.compositesa.2013.10.024.
64. Troughton, M. J., (2009). Chapter 10 – resistive implant welding. In: *Handbook of Plastics Joining* (2nd edn., pp. 105–111). Boston: William Andrew Publishing.
65. Banik, N., (2018). A review on the use of thermoplastic composites and their effects in induction welding method. *Mater. Today Proc., 5*(9), 20239–20249. doi: 10.1016/J.MATPR.2018.06.395.
66. Yussuf, A. A., Sbarski, I., Hayes, J. P., Solomon, M., & Tran, N., (2005). Microwave welding of polymeric-microfluidic devices. *J. Micromechanics Microengineering*. doi: 10.1088/0960-1317/15/9/011.
67. Microwave Welding. (2009). *Handb. Plast. Join.*, 135–138. doi: 10.1016/B978-0-8155-1581-4.50017-2.

CHAPTER 18

PERFORMANCE ANALYSIS OF MATERIAL HANDLING DEVICE USING INTEGRATED COPRAS-G AND FUZZY AHP

BIPRADAS BAIRAGI

Department of Mechanical Engineering, Haldia Institute of Technology, West Bengal, India

ABSTRACT

This chapter employs Grey COmplex PRoportional ASsessment (COPRAS-G) method combined with the fuzzy analytic hierarchy process (FAHP) for performance assessment of material handling devices (MHDs). Due to the verified compatibility in the treatment of fuzzy data, COPRAS-G is exploited to assess, rank, and select the most suitable alternative MHD. FAHP is utilized to compute the weights of criteria in a fuzzy environment. The fuzzy set theory with five degrees of diverse linguistic variables (LVs) is applied to extract ratings of alternatives. FAHP and COPRAS-G are combined to compute the utility index of each individual device used for material handling purposes. Arrangement of alternatives is done in the decreasing order of concerned utility indices. The alternative having the maximum utility index is regarded as the paramount alternative. An empirical example on evaluation and selection of device for the function of material handling is demonstrated. The analysis of the result clearly shows that this approach is effective and useful for the evaluation and selection of devices associated with material handling to aid managerial decision-makers (DMs) regarding this aspect.

Decision-Making Models and Applications in Manufacturing Environments. Pushpdant Jain, Kumar Abhishek, & Prasenjit Chatterjee (Eds.)

18.1 INTRODUCTION

Material handling device (MHD) selection is a complex and time-consuming multi-criteria decision-making dilemma owing to the existence of numerous realistic alternatives and contradictory criteria. Appropriate selection of MHD for industrial application is extremely essential. Though, past researchers have given attention in selection of MHD. Yet there is still absolute requirement of further research to finding simple and effective way to find the best solution of decision-making problem. Some research works on broad area of evaluation and selection of devices for the material handling by past researchers are cited as follows.

This chapter first classifies the literature on evaluation and selection of MHD that has received substantial consideration. Past literature on MHD selection can be classified based on the models used: models developed by using computer, numerical, optimization, and Multi-criteria decision-making (MCDM) models. Some researchers have developed numerical models. Layek and Lars [1] advocated a decision framework using analytical method for selection of machining centers and robots. Geometry is determined and the mating piece is designed using the framework. As per the framework, it is capable of selecting the robot as well as machining center. It employs the performance of the cell to diverse constraints. This research work is unable to assess the performance of robots in fuzzy MCDM environment. Khouja et al. presented a statistical model for performance measuring, ranking, and selection of best robot as an MHD. Models developed by computers assistance have been suggested by different researchers to undertake a lot of robot characteristics as well as accessible robots [2]. Goswami & Behera [3] made an investigation for the capability and applicability of two well-known MCDM approaches Additive Ratio ASsessment (ARAS) and COmplex PRoportional ASsessment (COPRAS) for the evaluation and selection of conveyors, automated guided vehicle (AGV) and robots as material handling equipment. Soufi et al. [4] introduced an AHP-based MCDM methodology for the evaluation and selection of material handling equipment to be utilized in manufacturing systems. Satyam et al. [5] applied a multi-attribute decision making approach for evaluation and selection of conveyors as material handling equipment. Nguyen et al. [6] advocated a combined multi-criteria decision making (MCDM) model for the evaluation and selection of conveyor as the material handling equipment based on fuzzy AHP and fuzzy ARAS with vague and imprecision information. Mathewa & Sahua [7] made a comparison among the novel multi-criteria decision-making approaches by solving a problem on material handling equipment selection. Fonseca et al. [8] discussed the development of an expert system to select conveyor for industrial purpose. Kulak [9] developed a decision support system for MHD selection under multiple criteria analysis considering fuzziness.

The above literature survey reveals that past researchers have applied MCDM techniques for selection of material handling equipment. However, this effort is inadequate for exhaustive and extensive decision making regarding proper selection of MHD from thousands of available alternatives under multiple conflicting criteria of diverse categories. Therefore, an effort is made in the chapter to suggest a combined multi-criteria fuzzy AHP-based COPRAS-G approach for performance evaluation of MHD under MCDM environment. The real life situation of MHD evaluation is associated with vague, imprecise, and ambiguous information. Hence imprecise data in a volatile environment deserves the implementation of the fuzzy set theory in MCDM approaches. The objective of this investigation is to guide and aid the management in MHD selection by providing a useful but simple way.

The rest of part of the chapter has been planned by dividing into some sections. In Section 18.2 theory of fuzzy set along with linguistic variables (LVs) is presented in brief. Section 18.3 presents FAHP as a weight evaluation technique and COPRAS-G method as a way of measurement of subjective performance ratings of alternatives. A case study on selection of MHD is presented in Section 18.4. The discussion on the result is summarized in Section 18.5. Overall conclusions with few suggestions for further research are furnished in Section 18.6.

18.2 FUZZY SET THEORY

Decision-makers (DMs) usually have a preference for subjective to objective assessment of fuzzy information. Theory of fuzzy set is used to convert these subjective data into numerical (objective) values [10]. A number of important definitions of fuzzy set are presented in the following subsection 2.1.

18.2.1 SOME IMPORTANT FUZZY DEFINITIONS

- **Definition 1:** A fuzzy set $\tilde{A}$ is defined in a universe of discourse denoted by X specified by $u_{\tilde{A}}(x)$, called membership function, which connects every member x (a real number) in X in an interval where x belong to [0,1] [11].

- **Definition 2:** A Triangular fuzzy number (TFN) $\tilde{q}$ is defined as a triplet (q_1,q_2,q_3), where membership function is characterized as [12]:

$$\mu_{\tilde{q}}(x)=\begin{cases} 0, & x<q_1 \\ \dfrac{x-q_1}{q_2-q_1}, & q_1\le x\le q_2 \\ \dfrac{x-q_3}{q_2-q_3}, & q_2\le x\le q_3 \\ 0, & x>q_3. \end{cases}$$

The membership function of a TFN $\tilde{q}=(q_1,q_2,q_3)$ is depicted in Figure 18.1.

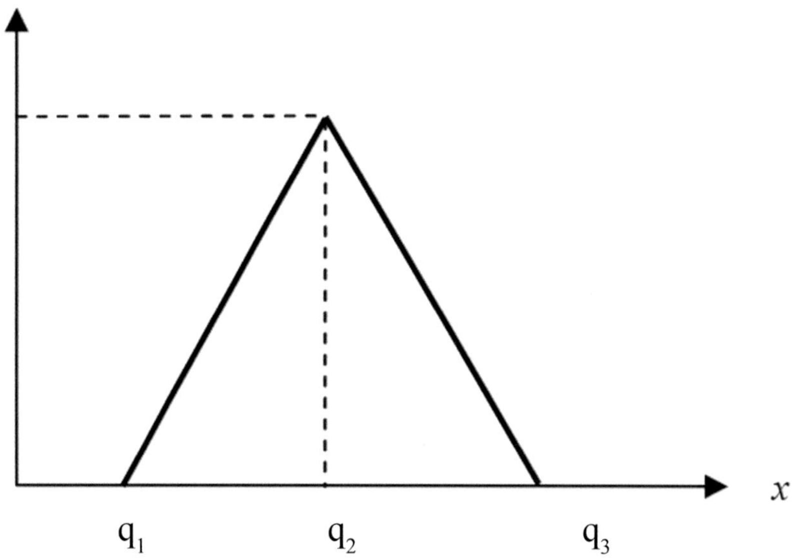

FIGURE 18.1 Membership function of a TFN $\tilde{q}\,(q_1,q_2,q_3)$.

- **Definition 3:** Let $\tilde{q}=(q_1,q_2,q_3)$ and $\tilde{r}=(r_1,r_2,r_3)$ be two TFNs, then the distance between the two fuzzy numbers can be calculated as:

$$d(\tilde{q},\tilde{r})=\sqrt{\frac{1}{3}\left[(q_1-r_1)^2+(q_2-r_2)^2+(q_3-r_3)^2\right]}$$

This method of calculating distance between two fuzzy numbers is termed as vertex method.

18.2.2 FUZZY OPERATIONS

Let $\tilde{q} = (q_1,q_2,q_3)$ and $\tilde{r} = (r_1,r_2,r_3)$ are two TFNs.

a. Addition: $\tilde{q} \oplus \tilde{r} = (q_1,r_1,q_2 + r_2,q_3,q_3)$;
b. Multiplication of a fuzzy number $\tilde{q} = (q_1,q_2,q_3)$ with a real number k;

$$k \otimes \tilde{q} = (kq_1,kq_2,kq_3) \text{ where } k \geq 0 \ \& \ k \in R$$

c. Multiplication commutative property $k \otimes \tilde{q} = \tilde{q} \otimes k$ where $k \geq 0$ & $k \in R$;
d. Division of a fuzzy number $\tilde{q} = (q_1,q_2,q_3)$ with a real number k;

$$\tilde{q}(\div)k = \left(\frac{q_1}{k},\frac{q_2}{k},\frac{q_3}{k}\right)$$

where; $k \geq 0$ & $k \in R$.

e. A TFN $\tilde{q} = (q_1,q_2,q_3)$ can be defuzzified by the equation:

$$BNP_i = \frac{\left[(q_3 - q_1)+(q_2 - q_1)\right]}{3} + q_1$$

BNP_i be the best non-fuzzy performance. When a fuzzy number is defuzzified then this value can be called best non-fuzzy performance value (BNP) [13]. Alpha-cut, center of area (COA), mean of maximal are defuzzification method [14]. Each fuzzy data expressed by DMs is transformed into equivalent TFN. In the current method, TFN is chosen for its capability of handing LVs, simplicity, and ease of use.

18.2.3 CONCEPT OF LINGUISTIC VARIABLE (LV)

A Linguistic variable (LV) is represented by a word or a sentence in a language. To evaluate performance rating LVs have been graded into five degrees, namely "very low (VL)," "Low (L)," "Medium (M)," "High (H)" and "very high (V)." These LVs are changed into respective TFNs (0, 1, 3), (1, 3, 5), (3, 5, 7), (5, 7, 9), and (7, 9, 10).

Weight of a criterion is estimated by a set of five-degree different LVs namely "equally important," "weakly important," "essentially important," "very strongly important" and "absolutely important." The variables for criteria weight are transformed into TFNs with (1,1,2), (2,3,4), (4,5,6), (6,7,8), and (8,9,9), respectively.

18.3 FAHP AND COPRAS-G

18.3.1 FAHP

Analytical Hierarchy Process (AHP) is a MCDM technique. It was by developed by T. L. Saaty [15]. FAHP is the integration of fuzzy concept and AHP. Buckley added fuzzy theory to Analytical Hierarchy Process which is called FAHP. The FAHP concept is reviewed below:

- **Step 1:** Build a weight matrix consisting of the weights applying pair-wise comparison by every decision maker.

$$\tilde{P}_{(h)} = \begin{bmatrix} \tilde{1} & \tilde{p}_{12(h)} & \cdots & \tilde{p}_{1n(h)} \\ \tilde{p}_{21(h)} & \tilde{1} & \cdots & \tilde{p}_{2n(h)} \\ \vdots & \vdots & \ddots & \vdots \\ \tilde{p}_{m1(h)} & \tilde{p}_{m2(h)} & \cdots & \tilde{1} \end{bmatrix} = \begin{bmatrix} \tilde{1} & \tilde{p}_{12(h)} & \cdots & \tilde{p}_{1n(h)} \\ \tilde{p}_{12(h)}^{-1} & \tilde{1} & \cdots & \tilde{p}_{2n(h)} \\ \vdots & \vdots & \ddots & \vdots \\ \tilde{p}_{1n(h)}^{-1} & \tilde{p}_{2n(h)}^{-1} & \cdots & \tilde{1} \end{bmatrix}$$

  $\tilde{p}_{ij(h)}$ = be a fuzzy number. It is obtained by transforming LV assessed by decision maker *h*, for criterion *i* over criterion *j*.

- **Step 2:** Build average weight matrix as follows by using the formula:

$$\tilde{P} = \begin{bmatrix} \tilde{1} & \tilde{p}_{12} & \cdots & \tilde{p}_{1n} \\ \tilde{p}_{21} & \tilde{1} & \cdots & \tilde{p}_{2n} \\ \vdots & \vdots & \ddots & \vdots \\ \tilde{p}_{m1} & \tilde{p}_{m2} & \cdots & \tilde{1} \end{bmatrix} = \begin{bmatrix} \tilde{1} & \tilde{p}_{12} & \cdots & \tilde{p}_{1n} \\ \tilde{p}_{12}^{-1} & \tilde{1} & \cdots & \tilde{p}_{2n} \\ \vdots & \vdots & \ddots & \vdots \\ \tilde{p}_{1n}^{-1} & \tilde{p}_{2n}^{-1} & \cdots & \tilde{1} \end{bmatrix}$$

  $\tilde{p}_{ij}$ is average of fuzzy numbers computed by using the equation $\tilde{p}_{ij} = \{\tilde{p}_{ij(1)} \oplus \tilde{p}_{ij(2)} \oplus \ldots \oplus \tilde{p}_{ij(n)}\}/k$. $\tilde{p}_{ij}$ be TFN and expressed as $\tilde{p}_{ij} = (p_1, p_2, p_3)$.

- **Step 3:** Geometric Mean of fuzzy weight is determined by using the equation.

$$\tilde{q}_i = (\tilde{p}_{i1} \otimes \tilde{p}_{i2} \otimes \ldots \tilde{p}_{in})^{1/n} = \left\{ \left(\prod_{j=1}^{n} Lp_{ij} \right)^{1/n}, \left(\prod_{j=1}^{n} Mp_{ij} \right)^{1/n}, \left(\prod_{j=}^{n} Up_{ij} \right)^{1/n} \right\}$$

- **Step 4:** Computation of criteria weight in fuzzy form is carried out by the normalization equation:

$$\tilde{w}_i = \tilde{q}_i \otimes (\tilde{q}_1 \oplus \tilde{q}_2 \oplus \ldots \oplus \tilde{q}_n)^{-1}$$

  $\tilde{w}_i = (t_1, t_2, t_3)$ be weight of ith criterion in TFN.

18.3.2 COPRAS-G

The concept of integration of gray system theory with COPRAS (complex proportional assessment) is termed as COPRAS-G [16, 17]. COPRAS-G is a relatively is a fresh technique for the alternative appraisal in interval considering multiple criteria. COPRAS-G method has the following simple steps:

- **Step 1:** Form a decision matrix consisting of performance rating of alternatives in interval.

  $y_{ij}(c_{ij}; d_{ij})$ =. Here, c_{ij} is a lower boundary and d_{ij} is the upper boundary of the interval.

- **Step 2:** Carryout the normalization of performance rating by applying the equation.

$$\tilde{y}_{ij} = \left(\tilde{c}_{ij}\tilde{d}_{ij}\right) = \left[\frac{c_{ij}}{\frac{1}{2}\left(\sum_{i=1}^{m} c_{ij} + \sum_{i=1}^{m} d_{ij}\right)}; \frac{d_{ij}}{\frac{1}{2}\left(\sum_{i=1}^{m} c_{ij} + \sum_{i=1}^{m} d_{ij}\right)}\right]$$

- **Step 3:** Construct the weight matrix:

$$W = [\tilde{w}_1, \ldots \tilde{w}_j \ldots \tilde{w}_n], \tilde{w}_j \text{ is weight of jth criterion}$$

- **Step 4:** Compute weighted normalized decision matrix applying the formula.

$$\hat{y}_{ij} = \left(\hat{c}_{ij}; \hat{d}_{ij}\right) = \left(\tilde{w}_j \otimes \overline{c}_{ij}; w_j \otimes \overline{d}_{ij}\right), \text{ is fuzzy weight of j}^{\text{th}} \text{ criterion.}$$

- **Step 5:** Calculate the average value of weighted normalized rating for every alternative with respect to each benefit criterion.

$$D_i^+ = \frac{1}{2}\sum_{j=1}^{n}\left(\hat{c}_{ij} + \hat{d}_{ij}\right), \quad j \in B$$

- **Step 6:** Compute the average value of weighted normalized rating for every alternative with respect to each cost criterion.

$$D_i^- = \frac{1}{2}\sum_{j=1}^{n}\left(\hat{c}_{ij} + \hat{d}_{ij}\right), \quad j \in C$$

- **Step 7:** Evaluate the relative weight of every alternative by utilizing the equation.

$$Z_i = Z_i^+ + Z_i^- = D_i^+ + \frac{\sum_{i=1}^{m} D_i^-}{D_i^- \sum_{i=1}^{m} \frac{1}{D_i^-}}$$

Z_i denotes relative weight for the alternative i.

- **Step 8:** Determine the degree of utility (U_i) for every alternative by using the equation.

$$U_i = \frac{Z_i}{Z_{max}} \times 100\% \text{ Here } Z_{max} = \text{Maximum } (Z_i).$$

- **Step 9:** Determine modified utility index (MUI) by using the normalizing equation.

$$MUI_i = \frac{MUI_i - MUI_{min}}{MUI_{max} - MUI_{min}}$$

where; MUI_i refers to the modified utility index of the ith alternative, MUI_{min} refers to the minimum modified utility index and MUI_{max} refers to the maximum modified utility index.

- **Step 10:** Select the alternative having the highest value of utility index or modified utility index.

18.4 CASE STUDY

An East Indian manufacturing company decided to select the best MHD for its day-to-day functioning in plants. Preliminary screen test allows a set of nine conveyors capable of doing the expected functions. The set of nine MHD is denoted by{Ei: i = 1, 2, 3, 9}. A set of ten selection criteria are considered and listed as: C1: Speed (ft/min), C2: Load carrying capacity (quintal/min), C3: Path flexibility, C4: Material flexibility, C5: Life period (years) C6: Elevated angle (degree) C7: Maximum distance covered (m) C8: Maintenance required C9: Space required (m^2) C10: Cost ($). First seven criteria (C1–C7) are benefit criteria and last three criteria (C8–C10) are cost criteria. Performance rating of every MHD with every criterion is estimated in LVs which are presented in Table 18.1.

TABLE 18.1 Conveyor Performance Ratings in Terms of Linguistic Variables

	C1 (+)	C2 (+)	C3 (+)	C4 (+)	C5 (+)	C6 (+)	C7 (+)	C8 (–)	C9 (–)	C10 (–)
E1	Very high	Very high	Low	High	Medium	Low	High	Low	High	Medium
E2	High	Medium	Very low	Low	Medium	Very high	Medium	Low	Low	Low
E3	low	Medium	Low	Very high	High	Medium	Medium	Very low	Medium	High
E4	Medium	Medium	Medium	High	Medium	Medium	Low	Low	Low	Very high
E5	Medium	Low	High	Low	High	Very high	Low	Very low	Low	Low
E6	Low	Medium	Low	Medium	High	Very low	Medium	Very low	Medium	Medium
E7	Very low	High	Very low	Low	Medium	Very high	Very low	Low	Medium	Very high
E8	Very low	Medium	Medium	High	Medium	Low	Low	Low	Medium	Very high
E9	Medium	Low	Medium	High	Medium	Medium	Medium	Low	high	Very high

The weight matrix is determined in LV. It is then transformed from LV to TFNs which is presented in Table 18.2. The calculation, results, and discussions on the given problem of MHD selection are summarized in section 5.

TABLE 18.2 Fuzzy weight matrix in terms of TFNs

C_i	C1	C2	C3	C4	C5	C6	C7	C8	C9	C10	Fuzzy Weights
C1	$\tilde{1}^{-1}$	$\tilde{5}^{-1}$	$\tilde{7}^{-1}$	$\tilde{5}^{-1}$	$\tilde{3}$	$\tilde{3}^{-1}$	$\tilde{5}^{-1}$	$\tilde{1}^{-1}$	$\tilde{3}^{-1}$	$\tilde{3}^{-1}$	(0.030, 0.037, 0.052)
C2	$\tilde{5}^{-1}$	$\tilde{1}$	$\tilde{9}^{-1}$	$\tilde{1}^{-1}$	$\tilde{3}$	$\tilde{9}^{-1}$	$\tilde{1}^{-1}$	$\tilde{5}^{-1}$	$\tilde{9}^{-1}$	$\tilde{3}^{-1}$	(0.057, 0.067, 0.091)
C3	$\tilde{7}$	$\tilde{9}^{-1}$	$\tilde{1}^{-1}$	$\tilde{9}^{-1}$	$\tilde{9}^{-1}$	$\tilde{1}^{-1}$	$\tilde{9}^{-1}$	$\tilde{7}^{-1}$	$\tilde{1}^{-1}$	$\tilde{9}^{-1}$	(0.008, 0.009, 0.012)
C4	$\tilde{5}^{-1}$	$\tilde{1}^{-1}$	$\tilde{9}^{-1}$	$\tilde{1}^{-1}$	$\tilde{3}$	$\tilde{9}^{-1}$	$\tilde{1}^{-1}$	$\tilde{5}^{-1}$	$\tilde{9}^{-1}$	$\tilde{3}^{-1}$	(0.056, 0.067, 0.082)
C5	$\tilde{3}^{-1}$	$\tilde{3}^{-1}$	$\tilde{9}^{-1}$	$\tilde{3}^{-1}$	$\tilde{1}$	$\tilde{9}^{-1}$	$\tilde{1}^{-1}$	$\tilde{3}^{-1}$	$\tilde{3}^{-1}$	$\tilde{1}^{-1}$	(0.026, 0.031, 0.043)
C6	$\tilde{3}$	$\tilde{9}^{-1}$	$\tilde{1}^{-1}$	$\tilde{9}^{-1}$	$\tilde{9}^{-1}$	$\tilde{1}^{-1}$	$\tilde{9}^{-1}$	$\tilde{3}^{-1}$	$\tilde{1}^{-1}$	$\tilde{9}^{-1}$	(0.008, 0.009, 0.121)
C7	$\tilde{5}^{-1}$	$\tilde{1}^{-1}$	$\tilde{9}^{-1}$	$\tilde{1}^{-1}$	$\tilde{1}^{-1}$	$\tilde{9}^{-1}$	$\tilde{1}^{-1}$	$\tilde{5}^{-1}$	$\tilde{9}^{-1}$	$\tilde{9}^{-1}$	(0.032, 0.043, 0.049)
C8	$\tilde{1}^{-1}$	$\tilde{5}^{-1}$	$\tilde{7}^{-1}$	$\tilde{5}^{-1}$	$\tilde{3}^{-1}$	$\tilde{3}$	$\tilde{5}^{-1}$	$\tilde{1}^{-1}$	$\tilde{3}^{-1}$	$\tilde{3}^{-1}$	(0.011, 0.014, 0.019)
C9	$\tilde{3}$	$\tilde{9}^{-1}$	$\tilde{1}$	$\tilde{9}^{-1}$	$\tilde{3}^{-1}$	$\tilde{1}^{-1}$	$\tilde{9}^{-1}$	$\tilde{3}^{-1}$	$\tilde{1}^{-1}$	$\tilde{9}^{-1}$	(0.009, 0.010, 0.014)
C10	$\tilde{3}^{-1}$	$\tilde{3}^{-1}$	$\tilde{9}^{-1}$	$\tilde{3}^{-1}$	$\tilde{1}^{-1}$	$\tilde{9}^{-1}$	$\tilde{9}^{-1}$	$\tilde{3}$	$\tilde{9}^{-1}$	$\tilde{1}^{-1}$	(0.027, 0.034, 0.044)

18.5 CALCULATION AND DISCUSSIONS

In COPRAS-G technique, each performance ratings are expressed with an interval having lower limit and upper limit that correspond to lower and upper points of a TFN respectively. These performance ratings are normalized and multiplied by their respective weights to determine weighted normalized performance ratings (WNPR). Each WNPR under benefit criteria retains positive sense and each WNPR under cost criteria negative sense. WNPR are summarized in the form of the weighted normalized decision matrix. WNPRs graphically represented in Figure 18.2.

The curve area under the benefit criteria represents positive performance, i.e., greater area is desirable. On the contrary, the curve area under the cost criteria represents negative performance, hence lower area is desirable. The average weighted normalized performance rating (AWNPR) is shown in Table 18.3. Relative weight as well as utility index of each conveyor is computed and shown in Table 18.4. The utility indices are 100, 73.14, 79.05, 72.77, 73.08, 79.37, 58.51, 66.35, and 69.72 which are graphically represented in Figure 18.3.

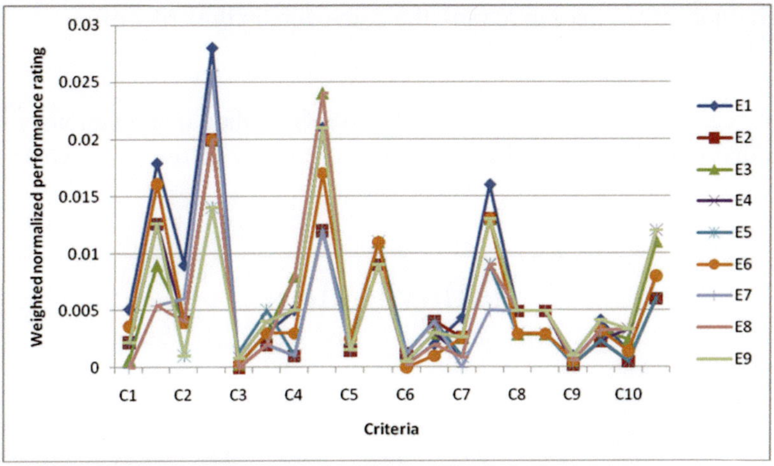

FIGURE 18.2 Weighted normalized performance rating by COPRAS-G.

TABLE 18.3 Average Weighted Normalized Matrix (COPRAS-G)

MHD	C1	C2	C3	C4	C5	C6	C7	C8	C9	C10
E1	0.0115	0.0185	0.0014	0.0134	0.0051	0.0010	0.0051	0.0027	0.0026	0.0049
E2	0.0074	0.0118	0.0008	0.0065	0.0051	0.0025	0.0051	0.0027	0.0013	0.003
E3	0.0048	0.0118	0.0014	0.0157	0.0069	0.0016	0.0069	0.0015	0.0019	0.0066
E4	0.0074	0.0118	0.0022	0.0134	0.0051	0.0016	0.0051	0.0027	0.0013	0.0077
E5	0.0074	0.0077	0.0030	0.0065	0.0069	0.0025	0.0069	0.0015	0.0013	0.0033
E6	0.0099	0.0118	0.0014	0.0099	0.0069	0.0006	0.0069	0.0015	0.0019	0.0049
E7	0.0027	0.0159	0.0008	0.0065	0.0051	0.0025	0.0051	0.0027	0.0019	0.0077
E8	0.0027	0.0118	0.0008	0.0157	0.0051	0.0010	0.0051	0.0027	0.0019	0.0077
E9	0.0074	0.0077	0.0022	0.0134	0.0051	0.0016	0.0051	0.0027	0.0026	0.0077

TABLE 18.4 Utility Index and Ranking of MHD Using COPRAS-G

E_i	$Z^{+}(D)$	D–	∑D–	(1/D–)	∑(1/D–)	Z–	Zi	Ui	Rank
E1	0.0655	0.0102	0.091	98.05	951.4	0.0094	0.0749	100	1
E2	0.0415	0.0072	0.091	139.22	951.4	0.0133	0.0548	73.14	4
E3	0.0497	0.0100	0.091	100.09	951.4	0.0096	0.0592	79.05	3
E4	0.0463	0.0116	0.091	86.08	951.4	0.0082	0.0545	72.77	6
E5	0.0387	0.0060	0.091	167.67	951.4	0.0160	0.0547	73.08	5
E6	0.0480	0.0083	0.091	120.31	951.4	0.0115	0.0595	79.37	2
E7	0.0361	0.0123	0.091	81.39	951.4	0.0078	0.0438	58.51	9
E8	0.0419	0.0123	0.091	81.39	951.4	0.0078	0.0497	66.35	8
E9	0.0449	0.0130	0.091	77.19	951.4	0.0074	0.0522	69.72	7

These utility indices give rank the conveyors in the order of:

E1>E4>E3>E6>E5>E2>E9>E8>E7.

Hence, the selection based on COPRAS-G method unambiguously indicates that E1 (Flat belt) is the best conveyor and E7 (screw conveyor) is the worst conveyor.

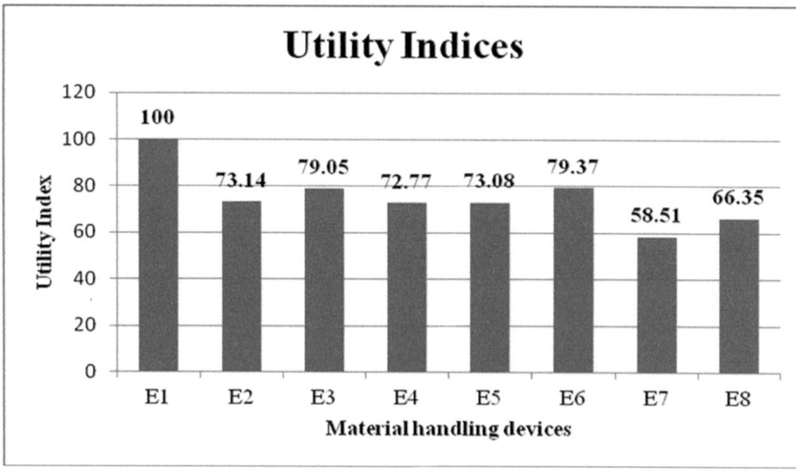

FIGURE 18.3 Utility index of the material handling device.

18.6 CONCLUSIONS

This chapter aims to find out an integrated and simple fuzzy MCDM approach having the capability of selecting the best MHD considering several subjective criteria. This method integrates Fuzzy AHP with COPRAS-G. Fuzzy AHP estimates the weights of different criteria and COPRAS-G assess the performance rating. The amalgamation of the two well-known tools gives rise to a suitable technique which solves the purpose and aids the DMs for appropriate decision regarding appropriate selection of MHDs.

The technique can also be applied in the future decision making process in real life problem under FMCDM. The results of the MHD selection problem prove that the current FMCDM technique is an effective and helpful tool for precise performance evaluation. The incorporation of tangible, intangible, and interdependent factors in the selection process might be some probable direction for further research.

KEYWORDS

- **automated guided vehicle**
- **complex proportional assessment**
- **fuzzy analytic hierarchy process**
- **material handling device**
- **multi-criteria decision-making**
- **COPRAS-G**
- **fuzzy AHP**
- **fuzzy set theory**
- **MHD selection**

REFERENCES

1. Layek, A. M., & Lars, J. R., (2000). Algorithm-based decision support system for the concerted selection of equipment in machining/assembly cells. *International Journal of Production Research, 38*, 323–339.
2. Khouja, M., Booth, D. E., Suh, M., & Mahaney, Jr. J. K., (2000). Statistical procedures for task assignment and robot selection in assembly cells. *International Journal of Computer Integrated Manufacturing, 13,* 95–106.
3. Goswami, S. S., & Behera, D. K., (2021). Solving material handling equipment selection problems in an industry with the help of entropy integrated COPRAS and ARAS MCDM techniques. *Process Integration Optimization and Sustainability, 5*, 947–973.
4. Soufi, Z., David, P., & Yahouni, Z., (2021). A methodology for the selection of material handling equipment in manufacturing systems. *IFAC-Papers OnLine, 54*, 122–127.
5. Satyam, F., Satywan, K., & Avinash, K., (2021). *Application of Multi-Attribute Decision-Making Methods for the Selection of Conveyor*. Research Square. doi: 10.21203/rs.3.rs-1033410/v1.
6. Huu-Tho Nguyen, Siti Zawiah Md Dawal, Yusoff Nukman, Achmad P. Rifai, & Hideki Aoyama (2016). An integrated MCDM model for conveyor equipment evaluation and selection in an FMC based on a fuzzy AHP and fuzzy ARAS in the presence of vagueness. *PLOS One, 11*. doi: 10.1371/journal.pone.0153222.
7. Mathewa, M., & Sahua, S., (2018). Comparison of new multi-criteria decision-making methods for material handling equipment selection. *Management Science Letters, 8*, 139–150.
8. Fonseca, D. J., Uppal, G., & Greene, T. J., (2004). A knowledge-based system for conveyor device selection. *Expert Systems with Applications, 26*, 615–623.
9. Kulak, O., (2005). A decision support system for fuzzy multi-attribute selection of material handling equipment's. *Expert Systems with Applications, 29*, 310–319.

10. Klir, J. G., & Yuan, B., (2005). *Fuzzy Sets and Fuzzy Logic: Theory and Applications.* New Delhi: Prentice Hall of India Private Limited.
11. Zadeh, L. A., (1965). Fuzzy sets. *Information and Control, 8*, 338–353.
12. Keufmann, A., & Gupta, M. M., (1991). *Introduction to Fuzzy Arithmetic: Theory and Application.* New York: Van Nostrand Reinhold.
13. Opricovic, S., & Tzeng, G. H., (2007). Extended VIKOR method in comparison with outranking methods. *European Journal of Operational Research, 17*, 514–529.
14. Wu, H. Y., Tzeng, G. H., & Chen, Y. H., (2009). A fuzzy MCDM approach for evaluating banking performance based on the balanced scorecard. *Expert Systems with Applications, 36*, 10135–10147.
15. Saaty, T. L., (1980). *The Analytic Hierarchy Process*. New York: McGraw-Hill.
16. Chen, Y. C., Lien, H. P., Tzeng, G. H., & Yang, L. S., (2008). Fuzzy MCDM approach for selecting the best environment-watershed plan. *Applied Soft Computing, 11*, 265–275.
17. Zavadskas, E. K., Kaklauskas, A., Turskis, Z., & Tamosaitiene, J., (2008). International Conference, Contractor selection multi-attribute model applying COPRAS method with grey interval numbers of 20th Euro mini conference "Continuous Optimization and Knowledge-Based Technologies" (EurOPT-2008). Neringa, Lthuania.

CHAPTER 19

ROLE OF IOT, BIG DATA, AND AI IN MANUFACTURING AND INDUSTRIAL SECTORS DURING COVID-19 PANDEMIC: AN ISM APPROACH

HRISHIKESH DEKA, SUSHANTA TRIPATHY, and DEEPAK SINGHAL

School of Mechanical Engineering, KIIT University, Bhubaneswar, Odisha, India

ABSTRACT

COVID-19 has led to a devastating effect on the manufacturing sector. The lockdown due to Novel Corona virus has forced the manufacturing sector into a standstill which resulted in lower production due to lower off-take and availability of raw materials for production, derailing of the entire world supply chain. Digital innovation such as artificial intelligence (AI), big data and the internet of things (IoT) are the most advanced technologies that are revamping the manufacturing industries at an impressive course of action. In the 21st-century manufacturing industries are transforming into smart and intelligent manufacturing industries. AI, IoT are remodeling the ecosystem of the manufacturing sectors. This chapter scrutinizes and reviews the role of big data, IoT, and AI in assisting the manufacturing industries in the COVID-19 pandemic situation. This chapter also critically dissects and reviews the existing literature review and research papers related to the advancement of technologies like AI, IoT, and big data in upgrading the manufacturing sector and tackling any critical situation in the manufacturing industry. This chapter has ventured to pinpoint the accomplishment factors for the role of AI, IoT, and Big Data in manufacturing industries and dispense an interpretive

Decision-Making Models and Applications in Manufacturing Environments. Pushpdant Jain, Kumar Abhishek, & Prasenjit Chatterjee (Eds.)

structural model (ISM) to thrive a map of complicated interconnection and enormity among the pin-pointed ten challenges and Cross-Impact Matrix Multiplication Applied to Classification (MICMAC) analysis has classified them according to one's driving and dependence power along with an exhibition of SSIM and reachability matrix of the identified factors. The chapter also illustrates a case study on digital transformation of an Indian-prominent tire manufacturing industry.

19.1 INTRODUCTION

COVID-19 has fabricated a pandemic situation all over the globe affecting the lives of mankind and all the sectors of industries as well. The advancement in technology since the recent years has created an opportunity to transform the manufacturing sector to more efficient and computerized ecosystem [1]. Big data, artificial intelligence (AI) and internet of things (IoT) are the main three supporting pillars in the modernization and intelligent manufacturing in the 21st century. To enhance automation in manufacturing and services these technologies provide wireless connectivity [2]. They are dependent on each other and have made manufacturing intelligent, flexible, and more adaptive [3]. Smart manufacturing is important for rapid production and upgradation of manufacturing industries. It is a combination of industrialization and information technology. Intelligent manufacturing processes through all aspects of production, service, product design and management [4].

Since the last few decades, the field of AI has grown towards computerizing human thinking. AI is the brain of the modern manufacturing industries. AI is one of the cutting edge technology that makes the computer to mimic and provide reasons like the human brain do or beyond [5]. AI can generate new knowledge on its own or solve new tasks which have not been seen earlier by deriving conclusions and analyzing the data available from all the previous task or services. AI uses algorithms to analyze the previous data which was stored in the devices by IoT to plan and generate a more advance reasoning capability brain to tackle any new task. In AI, we don't need to command the computer what to do, it learns and decides itself what has to be done from the experience gained [6]. AI has been dispensing Industrial digital twin (DT) and cyber physical system (CPS) to the conventional manufacturing industries [7]. AI can be implanted with existent services and products for making them durable, safe, and effectual [8].

IoTs are used in manufacturing industries where hordes of data are generated. The interconnection of a large number of physical objects with the internet to track and trace the status of physical objects and collect data creates a smart environment which has transformed the traditional manufacturing industries into digitalized manufacturing sector. IoT monitors the statuses of physical objects and captures enormous amount of data generated from IoT connected devices for actionable results by AI algorithms and machine learning. Adoption of IoT by manufacturing industries increases the volume of data which transforms the industrial data into industrial big data [9]. A simple IoT system should consist of software defined networks and self-optimizing network. The real value of IoT is about delivering smarter insights and creating smarter products. The increased adoption of AI providing advanced data analytical options improves the utilization of IoT [3]. The goal of IoT is uniting everything in this world under a common platform. The core concept behind IoT is the cognizance that multiple devices are communicating among themselves immaculately. Industrial IoT equips industrial communication technology with automation and lay out communication between machine to machine [10]. One of the pre requirement for IoT is radio frequency (RF) identification (RFID) which can apprehend quantity change and manufacturing environment fluctuations [11]. Big data refers to huge complicated structured and unstructured data sets that are of huge importance to the industry, which are collected by various IoT devices and are stored at cloud storage in the form of industrial Big data. Big data involves extremely large stockpiles of data that are analyzed computationally in order to divulge the trends, patterns, associations, and decision support correlating to human behavior and interaction. It revolves out new technology paradigm for data which comes at high velocity and are fabricated at high volume and received from a variety of sources [12, 13]. Big data has four dimensions: Volume denoting that amount of data is very huge, veracity pointing trustworthiness and quality of the data, Variety denoting that data is gathered in different kinds of formats like image, video, etc., and Velocity connoting that data processing speed is very high (VH) [13]. Using data analytics along with Big data can lay out new animation of finer business intelligence and understandings [14].

19.2 LITERATURE REVIEW

Literature review is the nucleus in the field of research and development and provides a medium to carry out further investigation and conclusion from the existing research. Substantial consignment of research has been divulged in the field of accomplishment of AI, IoT, and Big Data in contemporary manufacturing system. In this report, a critical elongated review of research executed on contrasting implementation of the cutting edge technologies has been highlighted in Tables 19.1 and 19.2. Table 19.1 illustrates the objective of AI, BD, and IoT and the application areas along with findings of different research journals which will help to get a clear picture of the practical implementations of these technologies from a time span of 2002 to 2020. Table 19.2 interprets the classification of challenges faced by AI, BD, and IoT technologies from a period of 2002 to 2020, highlighting the need of further research and development in the particular fields.

In Table 19.1, most of the researchers have intensely focused on transforming the obsolete manufacturing sector with modern cutting-edge technology. In the case of big data, researchers have highlighted a strategic innovation by processing big data into vital information along with challenges and applications in intelligent and predictive manufacturing [11, 27, 33]. Researchers in the field of AI have focused on predictive manufacturing by designing predictive models based on the data available in the industry [4, 5, 16, 38]. Some researchers have prioritized in AI collaboration with robotics in the manufacturing sector [15, 40] along with human centered intelligent manufacturing [7, 19]. Inventory management and supply chaisn management (SCM) have been a special focus for the researchers to apply AI for smooth functioning and transparency of the whole manufacturing system [21, 29, 31, 35]. In the field of IoT researchers have highlighted the use of IoT sensors and data collectors for various methodology like DT, Bayesian method, Pi mind, mechanism, RFID, WSNs, etc., for smoothening of the manufacturing designs and models in the industry [10, 13, 15, 36].

In the above-mentioned Table 19.2, researchers have focused on the challenges faced by the world trending technologies in order to carry out further research and development in their respective fields and overcome the current obstacles. Researchers in the field of AI has mentioned that even if manmade intelligent system is made more intelligent than human brain but it still can't mimic sympathy and other human characteristics and thus complete fluidity cannot be achieved in the manufacturing [32]. Furthermore, data division among different fields, assorted data types and different sorts of information cannot be complied with 100% accuracy and colossal dataset become unnecessarily tangled and hard to decrypt [21, 39]. Security, data protection and range have been a major concern for the researchers [18, 26] along with the complexity of the data and its pre-processing [3]. One of the most significant issues is the information securing and absence of confidence regarding the data collected. The most challenging part is regarding the visualization of the data and its results [17]. Unlike IoT, Big Data also faces the concern regarding the data security and privacy issues [30]. Researchers have focused on the

TABLE 19.1 Objective of AI, BD, and IoT and Application Areas

Application Area	Technology Used	Findings	References
Clinical research and development	Artificial intelligence (AI)	This paper highlights the use of AI in the field of clinical research and development during the crisis of COVID-19. AI helps to improve treatment consistency by creating valuable calculations. It is showing an accuracy of 90% advancement in medical clinic.	[1]
Robotic system in industry		AI with robotics collaboration for upgrading the manufacturing sector, problems in realization of real world mobile industrial robot and establishing RA-MIPM to find optimal solution. Key novelties include: mathematical modeling of robot, pre-assumptions, and time window using various algorithms and experimenting production ratios and safety performance by robotic system.	[15]
Smart manufacturing and designing predictive models		This paper focuses on Predictive manufacturing in smart industries and concept drift in smart manufacturing. Designing predictive MODELS from time to time such as hybrid models and data driven models from the data gathered and using it in the factory is necessary in order to boost the efficiency of all the department roles in the industry.	[4]
Human-centered intelligent manufacturing and real-time assembly operation recognition		Framework design taking into consideration of data sensing collection and preparation, assembly task and deep learning approach in operational management A developed model is evaluated which achieves 95% accuracy. Fog computing application utilizing image frames for operation recognition in human centered intelligent manufacturing.	[16]
Inventory management		This paper highlights a vital activity that increases the profit and efficiency of firm by decreasing uncertainty and lead time in MTS (making to stock) production systems. This paper exploits TPM issues under some generalized input parameters and offering iSMART for solving the issue.	[17]
Efficiency in manufacturing		This paper presents the utilization of many machine learning algorithms for predicting key performance indicators and applying ai for explicit predictions.	[18]

TABLE 19.1 *(Continued)*

Application Area	Technology Used	Findings	References
Manufacturing sector		AI has become a pathway for the machines to think and act like human beings and is a medium for further growth of manufacturing sector. This paper highlights the application of AI in manufacturing and the benefits and challenges along with it.	[19]
Supply chain		AI has become a priority for many organizations around the world. The use of AI in increasing with time. This paper exploits the use of AI in eliminating manual workforce, supply chains, e-commerce, warehouse optimization and logistics, smart manufacturing, etc.	[20]
Supply chain		The paper mentions different applications of AI in the manufacturing industries such as image processing, predictive maintenance, etc., and forecast the future value of AI by 2025. The paper also highlights some basic uses of AI in traveling, shopping, healthcare, robotics, entertainment, etc., and the challenges faced by it.	[21]
Manufacturing		The researcher has stated the importance of AI in the manufacturing industry for checking quality insurance, welfare of the working place, defect detection, etc. The paper also informs us about precise data collection, processing, and storage.	[22]
Manufacturing		Successful application of AI in manufacturing industry upgrading the quality of production, smart maintenance, human robot collaboration and designing better predictive models from data gathered. AI has successfully changed the design process, supply chain, administration thus making it easy, reliable, and more efficient for production and supply by boosting the performance.	[8]
Intelligent manufacturing		Use of three levels of intelligent manufacturing has been highlighted here: intelligent manufacturing equipment, Intelligent manufacturing system, intelligent manufacturing service. AI powered industrial robots, 3D printers, intelligent sensors, etc., are used to upgrade the traditional manufacturing sector; man-machine intelligence are will make the future manufacturing sector more efficient.	[7]

TABLE 19.1 *(Continued)*

Application Area	Technology Used	Findings	References
Smart Manufacturing		Upgrading of AI 1.0 to AI 2.0 including exploring data driven potentiality for deep learning ad human-machine hybrid intelligence. AI2.0 developing new intelligent manufacturing wit new models and intelligent system architecture. From the outlook of industrial manufacturing, the current development of AI is discussed.	[23]
Inventory management		The experimental results concluded that TRAINGDX training, TANSIG transfer function and 8 hidden neurons displayed foremost presentation for prediction and planning the ordering structure for a company.	[24]
Inventory management		Efficient management of inventory by replacing conventional ABC analysis with modern ABC criteria. Back propagation networks and genetic algorithms, artificial intelligence-based classification have verified to be methodical mechanism for categorizing inventory items. Experimental results have described AI classification superior then conventional MDA procedure.	[25]
Supply chain		This paper reviews detailed information about artificial neural networks, machine learning, agent-based system, etc. The researcher evaluates the sustainability and theoretical advances of AI in the field of supply chain management and also exploring AI uses in SCM that have not been explored before.	[26]
Manufacturing with IMS		This paper focuses on learning processes models, automatic recognition of different process changes, degradation, and intervening process, complexity of problems makes it necessary for machine learning to work with IMSs Intelligent Manufacturing Systems. Hybrid Artificial Intelligence and machine learning are given special importance along with agent-based systems in managing complex production.	[5]
Manufacturing system	Internet of things (IoT)	In manufacturing sectors IoT sensors are used to collect real time data and in cyber side digital twin (DT) is used as an apparatus for collecting data; less than 1% of data collected by IoT sensors are utilized whereas 99% IoT data is used for predictive maintenance and smart optimization.	[27]

TABLE 19.1 *(Continued)*

Application Area	Technology Used	Findings	References
Model in manufacturing		This paper highlights exploration and analysis of the systemic innovation problems by proposing a rough Bayesian framework with the help of BR-DNP and modified Bayesian VIKOR methods. The research resulted in a hybrid systemic evaluation model that addresses practical issues and can be used for future innovation.	[28]
Manufacturing and data collection		This paper highlights IoT in terms of worldwide movement and key technology, main motive of IoT is to capture the real time data from the IoT sensors in the plant to turn into actionable insights to guide the future technology to be made more efficient using the available data.	[9]
Intelligent manufacturing		The paper introduces Pi-mind as a mechanism for cloning and apprehending necessary framework of models that lay hold from premium human decision creator and manufacturing models that are capable and executable of automated decision author in the industry.	[12]
Intelligent manufacturing		Analysis to describe evolutionary impact on manufacturing industry. Significant steps: modular, radical, incremental, and architectural innovation by integrating Henderson and Clark model, impact of IoT in manufacturing industry, framework added to case of additive manufacturing, 3D modeling.	[29]
Manufacturing and supply chain		Core technology playing a principal role in IoT: RFID, WSNs wireless sensor network, big data and cloud computing; applications of IoT: automation, proactive maintenance, supply chain management, energy management has been pinpointed.	[30]
Smart Manufacturing	Big data (BD)	This paper highlights big data analytics from an IoT authorized manufacturing shop to examine RFID data for visualization and data processing. Studies conducted for validation of the approach predicted 50% of validation.	[11]

TABLE 19.1 *(Continued)*

Application Area	Technology Used	Findings	References
Manufacturing and innovation		Special focus on strategic innovation of an organization's capabilities has been discussed. Processing Big data into vital information is the path to sustainable innovation in the manufacturing industry.	[31]
Intelligent manufacturing		This paper mainly focuses on big data analytics and BDA in cloud computing. Evolution of manufacturing information system. Data for information systems; challenges and application are discussed. A thorough case study on BDA in cloud manufacturing has been sited which helps to understand BDA to roots.	[32]
Predictive manufacturing		Conceptual framework of predictive manufacturing system by with acquisition of data from monitoring assets. Predictive manufacturing means making machine have self-aware capability with the help of CPS, AI, IoT, and big data.	[33]

TABLE 19.2 Literature Regarding Technology Used

Technology Used	Challenges	Findings	Author and Year
Artificial intelligence (AI)	Man-made intelligence can figure, yet can't pass judgment, as it definitely needs significant human characteristics, for example, sympathy.	This paper highlights the effect of intelligent from the point of view of challenges and problems of the manufacturing sector, and also major opportunity and core technology from current to future trends; human ingenuity is focused over AI, new generation AI technology is focused, AI under sustainability analysis perspective is highlighted.	[34]
	Data separation among various fields, assorted data types and various types of information, absence of a cycle security situated dynamic system.	Green manufacturing importance, current status, problems are clearly highlighted like information isolation on multiple fields, lack of safety-oriented settlement making systems, lack of early warning, etc., some problems are solved by using AI, Bayesian network and deep learning; technical challenges like error data, accurate fusion of data, etc., are also discussed.	[35]
	Enormous dataset, network geography become excessively convoluted and hard to decipher, time compelling.	The paper describes the use of artificial neural algorithm (ANN) for refining the gross the staging and regulation for the management of supply chain which minimizes operational cost. Performance resulted 2–3% hike using AI based ANN algorithm for supply chain.	[6]
Internet of things (IoT)	Versatility, restricted range and transmission capacity, security, and protection issues.	Adoption of technology like IoT, AI, blockchain, big data, cloud computing in order to tackle COVID-19 challenges. Pinpointing application and challenges of IoT in health care and manufacturing has been highlighted.	[10]
	Information securing, information pre-processing and capacity.	Survey of technologies of Big Data analytics in manufacturing. Life circuit ion of big data Analytics in MIoT: Input accession, data pre-processing and storage, input analytics along with Necessity of Big Data analytics: upgrading factory production and operation, enhancing supply chain, etc. In contrasts of big data analytics challenges are laid in manufacturing with the internet of things (MIoT).	[36]

TABLE 19.2 *(Continued)*

Technology Used	Challenges	Findings	Author and Year
	Security, similarity, and complexity, fake stupidity, absence of confidence	IoT and AI are the same sides of a coin which work by depending on each other; almost all the factory applications are completed by AI, IoT, and Big data by depending on each other. This paper highlights challenges of AI with IoT and application of AI in IoT	[3]
	visualization of the results, cost-proficient attachment and-play perspective	Acquisition of IoT paradigm in manufacturing sector that generates industrial BD. Industries are upgrading the concept of IoT to a more futuristic way using new monitoring services and IoT tools. Demonstration of converting industrial real time data from IoT sensors into industrial big data. IoT paradigm transformed industry to more flexible and efficient production condition.	[37]
Big data (BD)	Privacy, security, ownership.	Discussion on big data benefiting the fabricating industry to discussion of moral issues. Ethical issues of BD in manufacturing from the point of view of Individuals and organization.	[38]
	Big data controls, big data trust, big data monetization.	This paper highlights leveraging of Fog computing, AI, IoT, blockchain; using big data in Health care, urban planning, agriculture, and industry; challenges faced by big data, integration of AI, IoT, Big data and blockchain technology, security necessity of Big data and BD skills.	[14]
	Heterogenous and complex data, control and monitoring, storage, and transmission	Possible future research of big data in the field of BD analysis, cloud-based storage relation to BD, use of big data by respondents include: 49% customer centric outcomes, 18% operational optimization, 15% risk management and 4% employee collaboration.	[36]

complex issue regarding big data controls, complexity, trust issue, storage, monitoring, and transmission issue [20, 31].

19.3.1 PROPOSED STRATEGY AND MODEL DEVELOPMENT

19.3.1.1 ISM MODELING

Interpretive structural modeling (ISM) is the evaluated and manifested expansion to get the insight and blueprint the anatomy of critically aggregate problem. It is an appliance that is utilized to develop the interconnection betwixt the diversity of inconsistency influencing a concern. The partial opinion of experts and potency of association between the factors are the lead fragility of this technique. Nevertheless, a fine number of researchers have exploited the ISM in countless areas. Pinpointing the constituents which act as impediment via literature review along with expert opinion, enlargement of complex relationships betwixt the variables, development of structural self-interaction matrix (SSIM) then emergence of reachability matrix, division of reachability matrix into various levels, ISM modeling and driver dependence diagram (MICMAC analysis) have been explained in the following.

19.3.1.2 SSIM MATRIX

Four symbols represent the contextual interrelationship between the factors I and j. The following are the contextual symbols: (a) V, if factor I has an effect on factor j; (b) A if factor j has an impact on factor I; (c) X indicates that factors I and j will interact; and (d) O indicates that factors I and j are unrelated. The SSIM matrix for the challenges is shown in Table 19.3.

19.3.1.3 REACHABILITY MATRIX

The SSIM modeling has advanced the reachability matrix. The SSIM matrix is converted to a binary matrix by replacing the 1s and 0s in A,O,X,V. The following are the rules that govern the transformation: (a) if the SSIM's (i,j) position is V, the reachability matrix's (i,j) position is 1 and the (j,i) position is 0; (b) if the SSIM's (i,j) position is A, the reachability matrix's (i,j) position is 0 and the (j,1) position is 1; (c) if the SSIM's (i,j) position is X, the reachability matrix's (i,j) position becomes 1 as well as the position

TABLE 19.3 SSIM Matrix

Factor	Notation	LQD	LSW	TI	PDM	SCM	LU	IU	CP	LS	CA
Lack of high-quality data	LQD	–	A	X	A	V	A	A	A	A	A
Lack of skilled workers	LSW	–	–	V	V	V	O	V	V	O	V
Trust issues	TI	–	–	–	A	V	A	A	A	A	A
Poor data management	PDM	–	–	–	–	V	A	V	X	A	V
Security challenges	SC	–	–	–	–	–	A	A	A	A	A
Lack of understanding	LU	–	–	–	–	–	–	V	V	O	V
Insufficient updates	IU	–	–	–	–	–	–	–	A	A	X
Complexity	CP	–	–	–	–	–	–	–	–	A	V
Lack of speed	LS	–	–	–	–	–	–	–	–	–	V
Cloud attacks	CA	–	–	–	–	–	–	–	–	–	–

(j,i); and (d) the reachability matrix's (j,i) position becomes 0 if the SSIM's (i,j) position is 0. The reachability matrix depicts each factor's driving force and dependence. Factors 2, 6, and 9 have the largest driving power, while Factor 5 has the most reliance in our study. The reachability matrix obtained from SSIM is shown in Table 19.4.

TABLE 19.4 Reachability Matrix

Factor	LQD	LSW	TI	PDM	SC	LU	IU	CP	LS	CA	Driving Power
1	1	0	1	0	1	0	0	0	0	0	3
2	1	1	1	1	1	0	1	1	0	1	8
3	1	0	1	0	1	0	0	0	0	0	3
4	1	0	1	1	1	0	1	1	0	1	7
5	0	0	0	0	1	0	0	0	0	0	1
6	1	0	1	1	1	1	1	1	0	1	8
7	1	0	1	0	1	0	1	0	0	1	5
8	1	0	1	1	1	0	1	1	0	1	7
9	1	0	1	1	1	0	1	1	1	1	8
10	1	0	1	0	1	0	1	0	0	1	5
Dependence	9	1	9	5	10	1	7	5	1	7	1

19.3.1.4 LEVEL PARTITION

With the use of the reachability matrix, the reachability set and antecedent sets for each factor are created. The reachability set includes the same factors as well as factors that impact a factor, whereas the antecedent set includes the same factor as well as additional elements that influence achieving that factor. Intersection sets are the common factors that exist between the reachability set and the antecedent set. When the values of the reachability set and the intersection set are similar, level one or top level is determined. The reachability set, intersection set, antecedent set, and level of all factors are highlighted in Table 19.5.

19.3.1.5 ISM MODEL

The level partition table and digraph help in the formation of ISM model. The interpretive structural model is developed by the means of vertices or nodes and lines of edges from the final reachability matrix. The arrow pointing from i to j highlights a relationship between the factors i and j. The structural relationship among the key factors is displayed by the ISM model. Figure 19.1 displays the ISM model.

TABLE 19.5 Level Partition

Factor	Notation	Reachability Set	Antecedent Set	Intersection Set	Level
1	LQD	1,3,5	1,2,3,4,6,7,8,9,10	1,3	2
2	LSW	1,2,3,4,5,7,8,10	2	2	5
3	TI	1,3,5	1,2,3,4,6,7,8,9,10	1,3	2
4	PDM	1,3,4,5,7,8,10	2,4,6,8,9	4,8	4
5	SC	5	1,2,3,4,5,6,7,8,9,10	5	1
6	LU	1,3,4,5,6,7,8,10	6	6	5
7	IU	1,3,5,7,10	2,4,6,7,8,9,10	7,10	3
8	CP	1,3,4,5,7,8,10	2,4,6,8,9	4,8	4
9	LS	1,3,4,5,7,8,9,10	9	9	5
10	CA	1,3,5,7,10	2,4,6,7,8,9,10	7,10	3

19.3.1.6 MICMAC ANALYSIS

Driving powers and dependence diagrams are also known as Matriced Impacts croises-multiplication applique and classification (MICMAC) analysis. The obtained driving powers with dependence are used as inputs for the MICMAC analysis.

The key factors are divided into three categories: (a) autonomous factors, which include factors with low driving powers and low dependence; (b) dependent factors, which include factors with low driving powers but high dependence; and (c) the third quadrant elements are referred to as linkage factors because they have high dependency with high driving powers, (d) whereas the fourth quadrant factors are referred to as the driver because they have low dependence with high driving powers.

19.4 RESULTS AND DISCUSSION

The chapter has successfully evaluated the Interrelationship of the cutting edge technologies and the struggle to integrate these technologies with human effort. Much has been discussed and analyzed on the roles IoT, AI, and Big Data in manufacturing Industries during the Pandemic of COVID-19. Research conducted by various researchers in this field of IoT, AI, and Big Data has helped us to draw the conclusions about their respective applications in the intelligent manufacturing sector during COVID-19 pandemic which is helping mankind. The research pursues to scrutinize and develop an organizational relationship amid the vital components influencing the production industrial sector during the COVID-19 with the support of the ISM model skeleton frame. It pinpoints the pyramid of factors having the least concern and placed at lower extent of the ISM model affecting the factors at the higher extent. The MICMAC analysis (Figure 19.2) provides some precious insight on the respective importance and interconnection among the key factors. The main findings of the study are as following:

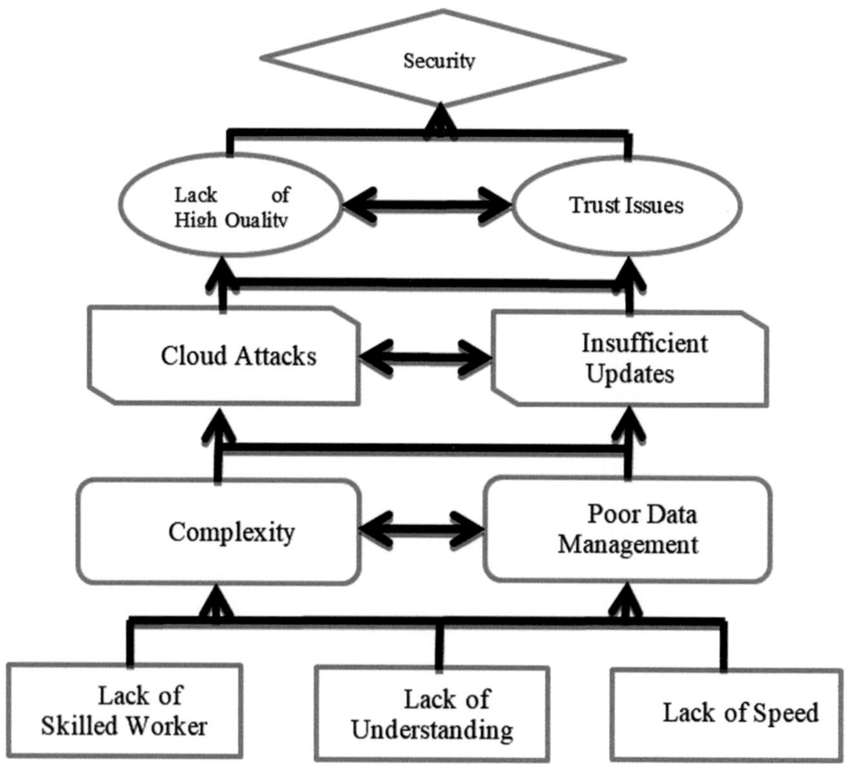

FIGURE 19.1 Interpretive structural modeling.

- There is the absence of any autonomous factors. Autonomous factors are usually considered to be weak driver and weak dependent. The autonomous factors don't have ample impact on the supplementary factors of the system. It is concluded that there are no unfocused factors which subsist among the key factors. Thus all the key factors are treated with equal force and importance.

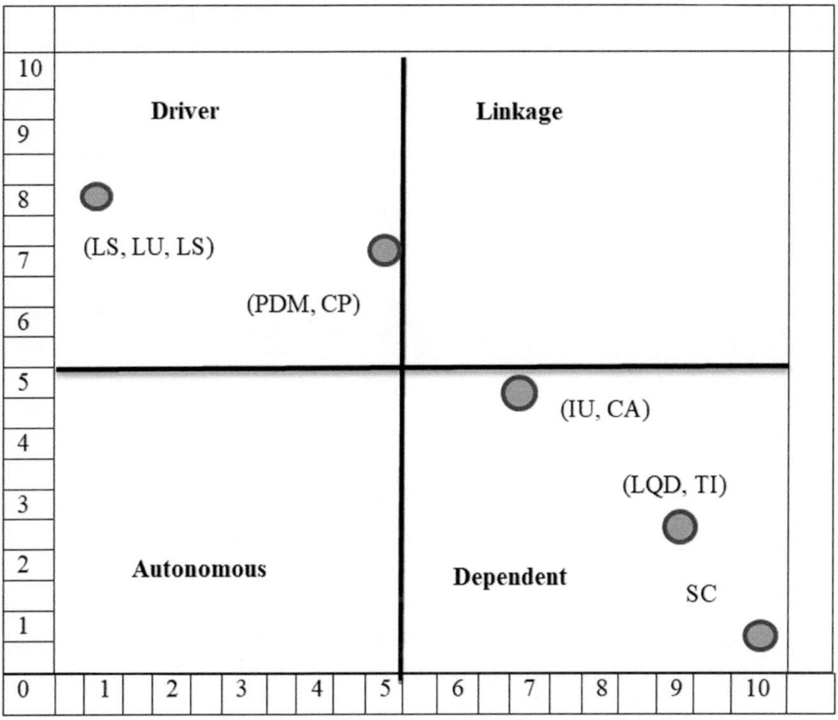

FIGURE 19.2 MICMAC analysis.

- Five factors are categorized under the dependent category such as Insufficient Updates, Cloud attacks, Lack if high quality data, Trust Issues and Security Concerns. These are the elements with a high dependency power but low driving power. As seen in the ISM model, these factors are positioned at the apex. The results displays Security concern is a long term issue that can be sorted out by tackling the issues of Insufficiency of elevated standard data and Trust issues. Further Insufficiency of elevated standard data and Trust Issues can be resolved by handling the issues of Cloud Attacks and Insufficient updates. These are the solutions for undertaking issues at the Peak of ISM model.

- Further five factors such as insufficiency of speed, insufficiency of understanding, lack of skilled workers, poor data management and complexity are designated as driving factors. These factors have more driving powers with less dependent powers. They are positioned at the extremity of the ISM model. Lack of speed, lack of understanding and insufficient skilled workers are the most dominant defiance during the COVID-19 situation. These bottom factors must be enhanced properly in order for the manufacturing industries to achieve their Objectives with 100% success. They play an important role in influencing the factors at the top. Thus these problems must be solved properly to increase efficiency.

19.5 CONCLUSION

A detailed study has been cited on some of the applications of these technologies and the challenges faced by them. The manufacturing sector is one of the best places for the application of IoT, AI, and Big Data in order to study them and develop more research works and predictive hybrid models using them. We are still not able to reap 100% benefits from these technologies. Much work has to be achieved in the field of these sophisticated technologies in order to attain 100% successful Industry 4.0 by achieving full automation. More research work has to be carried out in procuring new advanced benefits from the trio. The Trio IoT, AI, and Big Data will bring a more futuristic revolutionary change in the field of manufacturing in recent upcoming years.

AI may be more intelligent than the human brain, but there is no substitution for inventiveness and insight for the human brain in dealing with unexpected alteration in taste of product or demand. The basic overview of the present-day working of manufacturing industries can be harnessed from this chapter and utilized it for further future deep study and innovation. Further deep study can be done in the field of machine learning, artificial neural networks (ANNs), radio frequency identification system (RFID), cloud computing, CPS, etc., which is a vast fragment of the same realm. The capacious implementation of these world trending technologies will grasp the manufacturing techniques and efficiency to a more refined homogenous, safe, and reliable human-machine intelligence development. Further research and development can solve the key problems laid down by these technologies for smooth and effective experience. The hierarchy-based ISM model displays the critical factors which are affecting the manufacturing industries and need immediate attention in order to increase their efficiency.

KEYWORDS

- **artificial intelligence**
- **big data**
- **COVID-19**
- **digital twin**
- **Internet of things**
- **interpretive structural modeling**
- **manufacturing industry**
- **MICMAC analysis**
- **radio frequency identification**

REFERENCES

1. Hussain, A. A., Bouachir, O., Al-Turjman, F., & Aloqaily, M., (2020). AI techniques for COVID-19. *IEEE Access: Practical Innovations, Open Solutions, 8*, 128776–128795. https://doi.org/ 10.1109/ACCESS.2020. 3007939.
2. Boru, A., Dosdogru, A. T., Göçken, M., & Erol, R., (2019). A novel hybrid artificial intelligence-based methodology for the inventory routing problem. *Symmetry, 11*, 717.
3. Mohamed, E., (2020). The relation of artificial intelligence with internet of things: A survey. *Journal of Cybersecurity and In-formation Management, 1*(1), 30–24.
4. Kronberger, G., Bachinger, F., & Affenzeller, M., (2020). Smart manufacturing and continuous improvement and adaptation of predictive models. *Procedia Manufacturing, 42*, 528–531.
5. Monostori, L., (2003). AI and machine learning techniques for managing complexity, changes and uncertainties in manufacturing. *Eng. Appl. Artif. Intelligence, 16*, 277–291.
6. Praveen, U., Farnaz, G., & Hatim, G., (2019). Inventory management and cost reduction of supply chain processes using AI based time-series forecasting and ANN modeling. *Procedia Manufacturing, 38*, 256–263.
7. Yin, D., & Gengbao, H., (2018). Intelligent manufacturing: A way to upgrade manufacturing industry. *IOP Conf., Materials Science and Engineering, 439*, 032020.
8. Sreelekha, P., (2018). Impact of AI in manufacturing industries. *International Research Journal of Engineering and Technology (IRJET), 05*(11), 1765–1767.
9. Zhong, R. Y., & Ge, W., (2018). Internet of things enabled manufacturing: A review. *International Journal of Agile Systems Management, 11*(2), 126–154.

10. Kamal, M., Aljohani, A., & Alanazi, E., (2020). *IoT Meets COVID-19: Status, Challenges, and Opportunities*. https://arxiv.org/abs/2007.12268.
11. Kho, D. D., Lee, S., & Zhong, R. Y., (2018). Big data analytics for processing time analysis in an IoT-enabled manufacturing shop floor. *Procedia Manufacturing, 26*, 1411–1420.
12. Terziyan, V., Gryshko, S., & Golovianko, M., (2018). Patented intelligence: Cloning human decision models for industry 4.0. *Journal of Manufacturing Systems, 48*(Part C), 204–217.
13. Yin, S., & Kaynak, O., (2015). Big data for modern industry: Challenges and trends. In: *Proceedings of the IEEE* (Vol. 2, pp.143–146).
14. Rabah, K. V., (2018). Convergence of AI, IoT, big data and blockchain: A review. *The Lake Institute Journal, 1*(1), 1–18.
15. Xiao, H., Balaan, M., & Kadry, S., (2020). Artificial intelligence with robotics for advanced manufacturing industry using robot-assisted mixed-integer programming model. *Intell. Serv. Robot., 6–13*, 3.
16. Tao, W., Al-Amin, M., Chen, H., Leu, M. C., Yin, Z., & Qin, R., (2020). Real-time assembly operation recognition with fog computing and transfer learning for human-centered intelligent manufacturing. *Procedia Manufacturing, 48*, 926–931.
17. Angelo, E., Abhijit, G., & Susan, L. M., (2021). Total productive maintenance of make-to-stock production-inventory systems via artificial-intelligence-based iSMART. *International Journal of Systems Science: Operations & Logistics, 8*(2), 154–166.
18. Hassani, I. E., Mazgualdi, C. E., & Masrour, T., (2019). *Artificial Intelligence and Machine Learning to Predict and Improve Efficiency in Manufacturing Industry*. arXiv preprint arXiv:1901.02256.
19. Matthew, N. O. S., Sarhan, M. M., & Osama, M. M., (2019). Artificial intelligence in the manufacturing industry. *International Journal of Advances in Scientific Research and Engineering (IJASRE), 5*(6), 108–110.
20. Dash, R., Mcmurtrey, M., Rebman, C., & Kar, U. K., (2019). Application of artificial intelligence in automation of supply chain management. *Journal of Strategic Innovation and Sustainability, 14*(3), 43–53.
21. Virbahu, N. J., (2019). Artificial intelligence (AI) for supply chain industries and the future it holds. *International Journal of Engineering Research and Technology, 8*(3), 26–35.
22. Yubo, W., (2019). The application of artificial intelligence in mechanical manufacture industry *IOP Conf. Series: Materials Science and Engineering, 688*, 033058.
23. Li, B. H., Hou, B. C., Yu, W. T., Lu, X. B., & Yang, C. W., (2017). Applications of artificial intelligence in intelligent manufacturing: A review. *Frontiers of Information Technology & Electronic Engineering, 18*(1), 86–96.
24. Sustrova, T., (2016). A suitable artificial intelligence model for inventory level optimization. *Trends Economics and Management, 10*, 48–55.
25. Yu, M. C., (2011). A.B.C. multi-criteria, techniques. Analysis using artificial-intelligence-based classification. *Expert Systems with Applications, 38*(4), 3416–3421.
26. Min, H., (2010). Artificial intelligence in supply chain management: Theory and applications. *International Journal of Logistics: Research and Applications, 13*(1), 13–39.
27. Tan, Y., Wenhe, Y., Kohtaroh, Y., & Soemon, T., (2019). Application of IoT-aided simulation to manufacturing systems in cyber-physical system. *Machines, 7*(1), 2.
28. Yu-Sheng, K., Kazumitsu, N., & Chi-Yo, H., (2019). Evaluating the performance of systemic innovation problems of the IoT in manufacturing industries by novel MCDM methods. *Sustainability, 11*(18), 4970.

29. Caputo, A., Marzi, G., & Pellegrini, M. M., (2016). The internet of things in manufacturing innovation processes: Development and application of a conceptual framework. *Business Process Management Journal, 22*(2), 383–402.
30. Yang, C., Shen, W., & Wang, X., (2016). Applications of internet of things in manufacturing. In: *2016 IEEE 20th International Conference on Computer Supported Cooperative Work in Design (CSCWD)* (pp. 670–675).
31. Lee, J., Kao, H. A., & Yang, S., (2014). Service innovation and smart analytics for industry 4.0 and big data environment. *Procedia CIRP, 16*, 3–8.
32. Zhuming, B., & David, C., (2014). Big data analytics with applications. *Journal of Management Analytics, 1*(4), 249–265.
33. Lee, J., Lapira, E., Bagheri, B., & Kao, H. A., (2013). Recent advances and trends in predictive manufacturing systems in big data environment. *Manufacturing Letters, 1*, 38–41.
34. Buchmeister, B., Palcic, I., & Ojstersek, R., (2019). Artificial intelligence in manufacturing companies and broader: An overview. *DAAAM International Scientific Book*, 081–098.
35. Mao, S., Wang, B., Tang, Y., & Qian, F., (2019). Opportunities and challenges of artificial intelligence for green manufacturing in the process industry. *Engineering, 5*, 995–1002.
36. Dai, H. N., Wang, H., Xu, G., Wan, J., &. Imran, M., (2019). Big data analytics for manufacturing internet of things: Opportunities, challenges and enabling technologies. *Enterprise Information Systems, 14*(9, 10), 1279–1303.
37. Mourtzis, D., Vlachou, E., & Milas, N., (2016). Industrial big data as a result of IoT adoption in manufacturing. *Procedia CIRP, 55*, 290–295.
38. Chang, V. I., & Lin, W., (2019). How big data transforms manufacturing industry: A review paper. *International Journal of Strategic Engineering (IJoSE), 2*(1), 39–51. http://doi.org/10.4018/IJoSE.2019010104.
39. Singhal, D., Tripathy, S., Jena, S. K., Nayak, K. K., & Dash, A., (2018). Interpretive structural modelling (ISM) of obstacles hindering the remanufacturing practices in India. *Procedia Manufacturing, 20*, 452–457.
40. Aich, S., & Tripathy, S., (2014). An interpretive structural model of green supply chain management in Indian computer and its peripheral industries. *Int. J. Procure Management, 7*, 239.

CHAPTER 20

AN ISM APPROACH AMONG THE KEY FACTORS OF SCM FOR INDUSTRY 4.0 TOWARDS SUSTAINABILITY IN INDIAN MANUFACTURING SECTORS

DEOBRAT AKHOURI[1], NILABHRA GHOSH DASTIDAR[1], SUSHANTA TRIPATHY[1], DEEPAK SINGHAL[1], and DHYANADIPTA PANDA[2]

[1]*School of Mechanical Engineering, KIIT University, Bhubaneswar, Odisha, India*

[2]*School of Humanities, KIIT University, Bhubaneswar, Odisha, India*

ABSTRACT

Supply chain management (SCM) is a process by which the flow of goods and services can be managed between two locations or individuals. It essentially entails the initial storage of raw materials in inventory, followed by the delivery of goods in the form of finished items to the location. Many channels and nodes work together in this process to assist the customer's products and services flow as quickly as possible. This process of the supply is called SC. It is basically termed as a cross-functional approach of doing the work. The goal of SCM is to improve the collaboration and trust between different supply chain partners, to reduce the transportation cost and improve inventory management. This term was first used by Keith Oliver in 1982. After some years, SCM was also used as a service. It helps the organization to anticipate and analyze the problem and also to dynamically adjust the prices according to the situation. This chapter focuses on the relationship

Decision-Making Models and Applications in Manufacturing Environments. Pushpdant Jain, Kumar Abhishek, & Prasenjit Chatterjee (Eds.)

of key factors of SCM for industry 4.0 towards sustainability in Indian manufacturing industries.

20.1 INTRODUCTION

The management of the flow of Products and services between two locations or people is done through the supply chain management (SCM). In this process, many channels and nodes combine to help the products and services required by the customer to move it in an efficient manner. The supply chain is what we refer to as the supply process. It is generally referred to as a cross-functional method of carrying out the job. SCM aims to increase communication and trust between various supply chain partners, lower transportation costs, and enhance inventory control. Keith Oliver coined this phrase for the first time in 1982. SCM was eventually used as a service as well. It aids the company in anticipating and analysing the issue, as well as in dynamically adjusting rates in response to the circumstance. Apart from this, it improves the inventory management and fulfillment of the requirements. It reduces the cost of the material and also helps in increasing the revenue of a firm. It helps for the maximum utilization of the limited asset present in the organization. The five stages of SCM are: plan, source, make, deliver, and return. It is an old concept and it is a vital technology right from the 1st Industrial Revolution itself. When the mechanical production facility was introduced to the world during Industry 1.0, in the late 18th century. During the period, water, and steam power devices were designed to aid in the effective and efficient mass production of goods. Industry 1.0 is also regarded as the beginning of the global industry culture, which prioritized efficiency and quality. Then there was Industry 2.0, whose main contribution to the revolution was the creation of electric machinery. The primary source of power was electrical energy.

The first assembly line production also begins during this era. Mass production was carried out using an assembly line which became a standard practice in that era. Then came the Industry 3.0 which revolutionized the world with the help of the electronic devices which included the transistor. Integrated circuits significantly mechanized machinery, resulting in less work being expended on boosting speed, precision, and, in certain circumstances, completely replacing humans. Aside from that, the programmable logic controller (PLC), which was first built in the 1960s and signaled the use of electronics in automation, was a significant invention in the field of technology, and the most current industry 4.0, which altered the entire world, was introduced in 1990. CPSs have blurred the line, causing a slew of industry-wide technology changes. CPS enabled machines and organizations to interact with one another in the most efficient way possible, regardless of physical or geographical barriers, in order to increase efficiency. Industry 4.0 has provided transparent autonomy, collaboration, connectivity, and also the flexibility throughout the whole supply chain.

Also, many benefits of it include the advancement and enabling technologies that can be realized through enhanced integration as well as the digitization and automation, which led to the creation of the analytical capabilities and to improve the performance of the supply chain process level. The four-point advancement had also lead to the increase in the speed of work of an individual. Blockchain technology used in nowadays, is the use of storing the information in the form of blocks in several digital databases. It is a tool, and the sustainability consequences are mainly because of the strategies and the underlying visions that the business select to govern their daily operations in the country. The vital part of the today's SCM revolves around the technological advancement and other IT infrastructure in different sectors.

There are numerous barriers in the supply chains. Some of them are the barriers to the implementation of industry 4.0 techniques in the supply chains that is due to the lack of the skilled workforce, and short term goals in effective legislation. Lack of funds, resistance to the employees, growth in efficient performance, risk of investments and many others. To increase the efficiency of the SCM and also to survive in the market, many things are to be collaborated with SCM which includes the use of six Sigma technology, circular economy as well as reverse SCM. Discussion on the sustainable SCM and solutions to overcome the shortcomings can be taken with the help of the following.

In this post-pandemic world enhancing sustainability has become the sine qua non of the global supply chains. The principal path to attain that solution is through the use of Industry 4.0 technologies. Organizations around the world want to leverage Industry 4.0 technologies to make the supply chains more sustainable. In this globally competitive business atmosphere sustainability is of utmost importance to withstand customer pressure, government pressure, industry pressure and competition. In addition to that supply chains induce the economic growth of society. The primary objective of Industry 4.0 in the supply chain ecosystem is to improve resource utilization and information sharing, thus enhancing customer satisfaction. From a social and economic perspective, supply chain sustainability is vital to increase customer value in supply chains. There has never been more impetus than the sustainable development goals (2030) proposed by the United Nations to improve the sustainability of global supply chains.

20.2 LITERATURE REVIEW

A literature review is a systematic way of representing the data in an efficient manner for evaluating the data. Chalmeta et al. [1] identified the requirement of big data and i4.0 in the SSCM to identify the future requirements in different categories to make reasons whether to invest and improve the current methods [1]. It was only limited to the field of humanities or social sciences. Ralston et al. [2] focuses on the smart systems which improves the supply chain performance, It has also allowed the firms to be better prepared from the unexpected events and also source in improving performance as well as enhancing the resilience [2]. Another area which would be investigated here is the impact of i4.0 initiatives on the company's profits analyzing reports of smart systems in the companies.

Bag et al. [3] discussed about the training for skill development for the existing shortage in the mining sector for the South Africa mining sectors with the use of cross-sectional data are relatively small sample size a research team can be grounded to build a framework [3]. Luthra et al. [4] threw the limelight into the introduction of an IoT-based technology which can be helpful in the agriculture supply chain transparency at each step of the supply chain [4]. It will automatically increase with the help of IoT. It will guarantee to connect each and every link in the most efficient manner. However, in a developing country like India, It would be a costly measure to provide such facilities. The IoT-based technology has a significant scope in India in the near future in the industrial context.

Hopkins et al. [31] is mostly based on the Australian SCM and i4.0 technology. In India, most of the i4.0 technologies are in its initial phase. Yet the large of firms are at a rapid pace of grasping these technologies for a better opportunity in different fields [31]. The COVID-19 pandemic has impacted the global supply chain and has also restricted the movement of goods. Baryannis et al. [5] discussed about the possibility of utilizing the ML and IoT to provide the analytics of SCM to deliver the results with accuracy [5]. A data-driven AI has been proposed which could increase the efficiency level of the supply chain. Kumar et al. [6] identified the criteria of the sustainability in the circular economy and has also identified the barriers of the sustainable operations in i4.0 and has map this barriers for sustainable operations [6]. The lack of skilled labor, poor laws, short-term ineffectual goals, risk of mis-investing funds, and employees' unwillingness to change in the old production system owing to fear all contributed to the implementation of sustainability. Witkowski et al. [7] elaborated on the circumstances that led to the SCM's successful implementation [7].

They also said that the smart digital products like 3D printers have also revolutionized the Supply Chain process and are capable of producing technological and product innovation in the existing firms. Müller et al. [8] discussed about the challenges which will be faced by the Industry i4.0 is currently facing Engineer to Order industries which are providing a differential perspective for several stakeholders within the entire SCM [8]. The restrictions in resources and organizational capabilities were discussed. Sharma et al. [9] have highlighted about the use of circular and lean economy. SCM and the multi-tier manufacturing of the supply chain are also discussed here. Esmaeilian et al. [10] discussed about the capability of the i4.0 and IoT, smart business models, the smart transportation and logistics systems [10].

The capability of the blockchain where also discussed, i.e., increasing system efficiency, decreasing the development, enhancing the corporate performance and promoting the green behavior through designing specialized tokens. Though, there were shortcomings like the need of the analysis of the human machines interactions are there to be discussed. Fatorachian et al. [11] highlighted about the integration, digitization, and automation in i4.0 to achieve the maximum reserves and also the advancement of the SC performance with the help of i4.0 are highlighted [11]. It was found that the i4.0 attach with the IoT lead to the advanced level of transparency and high level of integration and accuracy in the supply chain. Although, the investigation on the culture and organizational factors which influence the i4.0 perspective are there to be checked.

Abdirad et al. [12] discussed about the improvements in the manufacturing of the supply chain logistics through the application of i4.0 by which the productivity can be increased and the labor cost will be reduced [12]. Though, there were shortcomings which could be adjusted in the future by adding the explanatory vs confirmatory as well as quantitative vs qualitative and lastly the technology level vs management level. Furstenau et al. [13] discussed about the challenges in the SCM and also told about the different perspective of the SCM and highlighted the use of i4.0 for sustainability [13]. Moreover the issues such as a Circular Economy, production resource scarcity, food production, waste reduction, social aspects and the raw materials used will arise.

We had discussed for the optimum utilization of them in the near future. Bag et al. [14] described about the integration of SCM with i4.0, also told about the development of performance enhancing systems as an identification of the enablers for i4.0, testing the framework using the data from the South African manufacturing industries can be considered for the future work [14]. Dev et al. [15] mainly highlighted about the connection of circular economy with the i4.0 and also told about system that enables computing products returns with the inbuilt Bass model algorithm which will help in the nearby future for sustainability [15].

Industry 4.0 is a concept that makes sure the transformation of traditional supply chain to a multi-level supply chain. It helps in creating a dynamic supply chain with the help of integration of different levels. In reality, managers with the knowledge of Industry 4.0 technologies are responsible for making efficient decisions, reduce risk and increase productivity. Abdirad & Krishna [32] have analyzed literature reviews on the application of Industry 4.0 in SCM [32]. Their research provides the current scenario of research on this topic and provides a baseline for implementation of Industry 4.0 and further research.

Now when we talk about integrating Industry 4.0 with Supply chain, there are some key concepts that need to be shared lights on like circular economy, reverse logistics. Ciliberto et al. [33] have clearly analyzed the Circular Economy principles to formulate a practical strategy to implement sustainable production ways [33].

Santana et al. [16] highlighted the study of capacity optimization and costing models as well as they also told about the costing model such as the ABC analysis [16]. Hereafter, the results obtained from the different models shows that the capacity issues demand restrictions cost, technical limitations, and profitability conditions are the main factors which are to be considered to optimize instead of just maximizing the capacity. It was further known that the manipulation of the external factors lead to the change in the investment and profitability of the system. Hahn [17] used the multiple in depth cases studies, and such an approach provided an insight into the future role in the digital transformation of the firm easily [17]. Kumar et al. [18] discussed the uses of the SCM by the small and medium enterprises [18]. Strandhagen et al. [20] described about the sustainability challenges in the building of a ship and the solutions to support sustainable operations in the shipbuilding [20]. It also included investigation on the environmental, economic, and social dimensions of the sustainability. Moreover, the investigation of the relationship between the dimensions and the ship building are to be discussed in the near future.

Also, the research methodologies that allow for the generalization of the building of the current study are topics for the future. Yadav et al. [21] described about the lean Six Sigma along with the sustainability and the solutions to overcome the SSCM were also discussed [21]. It was further found out that achieving the sustainability in the service-based Industry is much easier than achieving it in a Manufacturing Industry. With the use of Six Sigma, achievement of sustainability is discussed and the various approaches to tackle SSCM is also discussed here which are the points to ponder for the future.

Bär et al. [27] discussed a case study of SME where they developed and implemented a framework to analyze the status of Industry 4.0 implementation in a case company [27]. The research actually provides a five-step framework for eight Industry 4.0 realization factors to measure their present use and future in the company's supply chain. Thus it would give the SME an overview of external and internal requirements for Industry 4.0 realization for better alignment with customer satisfaction. However, further research is required on generalizing the framework for other industries along with a cost-benefit analysis (CBA) of the implementation on the organization structure in addition to branding and strategizing.

Ghadge, Er Kara, Moradlou, & Goswami [19] assessed the key drivers and barriers for Industry 4.0 implementation on supply chain performance with the help of a system dynamic model [19]. They have provided a base framework for future research, thereby bridging the existing gap in between research on Industry 4.0 and their implementation in Supply chains. The System Dynamics approach proposed by them has also helped in quantifying the inter-relationship between Industry 4.0 and Supply chain dynamics. The limitations to be worked upon in the future, as stated in the paper were – the use of more Industry 4.0 parameters and validation of the model with more empirical data. Borregan-Alvarado et al. [28] have analyzed the available literature to provide insights about Industry 4.0 and additive manufacturing (AM) [28]. Here the objective is to carry out a comprehensive analysis to identify previous academic research and corroborated scientific profiles of associated research fields framed along with Industry 4.0 and SCM.

20.3 METHODOLOGY

Multi-criteria decision making (MCDM) methods are techniques used in Operation research. These techniques are used for effective statistical interpretation when decision making is vague with little or unavailable data.

The reason why we have used such a technique in our research is that SCM is a multi-criteria decision-making problem. A supply chain is riddled with different criteria and sub-criteria are mostly conflicting in nature. So as we know that MCDM techniques are basically used for making decisions when multiple criteria are evaluated together and thus rank or choose between the different criteria being evaluated. In this case, the ISM technique is employed to determine the interrelationships between several inhibiting variables relating to long-term SCM. ISM is used to analyze the underlying inter-relationship among driving and dependent factors regarding the implementation of sustainability in the supply chain. It has not only helped us in solving the problem but has also added value through its simple structure. The ISM methodology starts with the identification of variables/factors relevant to the objective. In the second step a contextual relevant relation among the factors is selected. From the relation we create a SSIM based on pairwise comparison of the different factors. In the next step the SSIM is transformed into a Reachability Matrix and we check its transitivity. After completing the transitivity check we attain a new model. In the last step, we partition the levels and thus obtain the extraction of the structural model named ISM [30].

20.3.1 PROPOSED ISM METHODOLOGY: STRUCTURAL SELF-INTERACTION MATRIX (SSIM)

In ISM, we generally use the expert opinions which are based on different management techniques. For analyzing the factors, a contextual relationship of "leads to" is chosen for expressing the relationship between different critical factors for the performance of different factors. This reflects that one factor influences the other. Thus we obtain a contextual relationship among the chosen factors.

In the next phase, we need to determine the direction of the relationship between the factors. We have used some symbols here to establish a relationship between the parameters x and y (here X<Y).

- V: If X will influence Y;
- A: If Y will influence X;
- X: When parameters X and Y both influence each other;
- O: When parameters X and Y remains unchanged.

Thus, analyzing this relationship and the direction between any two related factors the SSIM is obtained.

20.4 FACTORS IDENTIFICATION (Table 20.1)

TABLE 20.1 Names of Different Factors Identified

SL. No.	Factors	References
1.	Industry 4.0 (I4.0)	[2, 6, 9, 20]
2.	Lack of management foresight (LMF)	[22, 25]
3.	Poor participation and initiation from management (PIM)	[22, 24]
4.	Supplier RISK (SR)	[23]
5.	Supply chain sustainability (SCS)	[6]
6.	Lack of skilled workforce (LSF)	[6]
7.	Fear of ROI (FROI) (risk of mis-investment)	[5]
8.	Supply chain resilience (SCR)	[2]
9.	Automation and digitization (AND)	[11]
10.	Logistics (LOG)	[7]
11.	Hazardous labor force (HLF)	[23]
12.	Production, planning, and control (PPC)	[22, 26]

20.4.1 REACHABILITY MATRIX

Here we have developed a Reachability matrix (Table 20.2). The previous table was converted into a simple binary table consisting of only 1 and 0 using the following conditions:

- If X influences Y and there is V in row, it will be 1.
- If Y influences X and there is A in row, it will lead to 0.
- If both X and Y influences each other leading to X, it will be 1.
- If both X and Y remain unaltered leading to O, it will be 0.

20.4.2 LEVEL PARTITIONS

The reachability Matrix aids us in identifying each factor's reachability and antecedent sets. The reachability set includes the element and any other elements it may affect, whereas the antecedent set includes the element and any other elements that may have an effect on it.

The intersection of these sets is calculated for all factors in the next stage. The top-level element in the ISM structure is the factors for which the reachability and intersection sets are the same. These factors do not lead to other factors that are higher in the hierarchy than they are. We can eliminate the top-level factor from subsequent iterations after deducing it. As a result, we can carry on with this approach to determine the amount of each element. Finally, we can obtain a digraph for factor alignment and build the model sequentially (Table 20.3).

TABLE 20.2 Self-Structural Interaction Matrix

	I4.0	LMF	PIM	SR	SCS	LSF	FROI	SCR	A&D	LOG	HLF	PPC
I4.0	–	A	A	V	V	A	A	V	V	V	A	V
LMF	–	–	X	V	V	V	V	V	V	V	V	V
PIM	–	–	–	V	V	V	V	V	V	V	V	V
SR	–	–	–	–	V	A	A	V	A	X	A	X
SCS	–	–	–	–	–	A	A	A	A	A	A	A
LSF	–	–	–	–	–	–	X	V	V	V	X	V
FROI	–	–	–	–	–	–	–	V	V	V	X	V
SCR	–	–	–	–	–	–	–	–	A	A	A	A
A&D	–	–	–	–	–	–	–	–	–	V	A	V
LOG	–	–	–	–	–	–	–	–	–	–	A	X
HLF	–	–	–	–	–	–	–	–	–	–	–	V
PPC	–	–	–	–	–	–	–	–	–	–	–	–

20.4.2.2 LEVEL PARTITION MATRIX

In the level partition matrix, by counting the no of ones and zeroes, we can determine the Driving Power and Dependence of each factor row-wise and column-wise respectively. This will help us in obtaining the Driver Dependence Diagram.

20.5 CLASSIFICATION OF FACTORS

The factors that we have obtained after analyzing our Literature Review are classified into four different clusters.

20.5.1 AUTONOMOUS FACTORS

Autonomous factors are those that have a low driving power and a low dependence on others. These elements are relatively detached from the system, with

TABLE 20.3 Binary Reachability Matrix

	I4.0	LMF	PIM	SR	SCS	LSF	FROI	SCR	A&D	LOG	HLF	PPC	Driving Power
I4.0	1	0	0	1	1	0	0	1	1	1	0	1	7
LMF	1	1	1	1	1	1	1	1	1	1	1	1	12
PIM	1	0	1	1	1	1	1	1	1	1	1	1	11
SR	0	0	0	1	1	0	0	1	0	1	0	1	5
SCS	0	0	0	0	1	0	0	0	0	0	0	0	1
LSF	1	0	0	1	1	1	1	1	1	1	1	1	10
FROI	1	0	0	1	1	0	1	1	1	1	1	1	9
SCR	0	0	0	0	1	0	0	1	0	0	0	0	2
A&D	0	0	0	1	1	0	0	1	1	1	0	1	6
LOG	0	0	0	0	1	0	0	1	0	1	0	1	4
HLF	1	0	0	1	1	0	0	1	1	1	1	1	8
PPC	0	0	0	0	1	0	0	1	0	0	0	1	3
Dependence	6	1	2	8	12	3	4	11	7	9	5	10	–

which they have only a few tenuous ties. They don't have much of an impact on other variables.

20.5.2 DEPENDENT VARIABLES

These factors have a low driving power but a high degree of reliance. They're largely visible near the top of the ISM model.

20.5.3 LINKAGE FACTORS

These Linkage factors provide a lot of driving strength as well as a lot of dependability. These elements have an impact on both the driving and dependent factors because any changes to them will have an impact on other components as well.

20.5.4 INDEPENDENT VARIABLES

Independent factors have a high driving force but low reliance. They occupy the bottom rung of the ISM hierarchy. They are mostly strategic in nature and have a direct bearing on the performance of other variables (Table 20.4).

TABLE 20.4 Level Partition Matrix

Factor	Notation	Reachability Set	Antecedent Set	Intersection Set	Level
1	I4.0	1,4,5,6,7,8,9,10,12	1,2,3,6,7,11	1,6,7	5
2	LMF	1,2,3,4,5,6,7,8,9,10,11,12	2	2	7
3	PIM	1,3,4,5,6,7,8,9,10,11,12	2,3	3	7
4	SR	4,5,8,10,12	1,2,3,4,6,7,9,11	4	3
5	SCS	5	1,2,3,4,5,6,7,8,9,10,11	5	1
6	LSF	1,4,5,6,7,8,9,10,11,12	2,3,6,	6	6
7	FROI	1,4,5,7,8,9,10,11,12	2,3,6,7	7	6
8	SCR	5,8	1,2,3,5,6,7,8,9,10,11,12	5,8	2
9	A&D	4,5,8,9,10,12	1,2,3,6,7,9,11	9	4
10	LOG	5,8,10,12	1,2,3,4,6,7,9,10,11	10	3
11	HLF	1,4,5,8,9,10,11,12	2,3,6,7,11	11	6
12	PPC	5,8,12	1,2,3,4,6,7,9,10,11,12	12	3

20.6 DRIVER DEPENDENCE DIAGRAM

After determining driving power and dependency for each factor we can now plot each factor on a graph to obtain the Driver-Dependence Diagram. Here we take the driving power on the Y axis and dependency on X axis. Thus plotting each factor on the diagram, we can easily determine which factor belongs to which cluster (Figure 20.1).

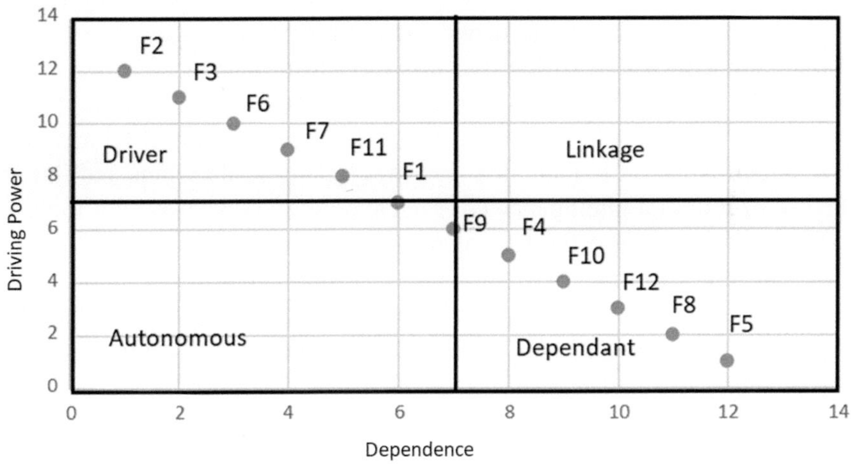

FIGURE 20.1 Driver dependence diagram.

20.7 ISM BASED MODEL

An ISM Model is developed by taking help from the digraph and level partition table. A digraph is a representation of factors and the inter-dependance in a graphical format in the form of edges and nodes with the help of level partition diagram. Figure shows an ISM Model.

20.8 RESULTS AND DISCUSSIONS

We have used the ISM methodology to identify the key relationship between different factors impacting the association of supply chain and sustainability in the Indian manufacturing sector [29]. From the model that we have developed, it is clear that supply chain sustainability is our long-term focus as it is present at the highest level of the ISM hierarchy. It is a dependent factor which is heavily influenced by Supply Chain Resilience present at the second level. Efficiency in logistics, low supplier risk, and proper production planning and control are the factors that can enable better resiliency in the supply chain and would thus help in performing better against supply chain disruption. Now we have identified Industry 4.0: Use of AI as an important driving factor which can be used to automate essential tasks and digitize activities which can directly affect supply chain resilience and thus impact better supply chain performance. Now if we come to the bottom of the ISM model we would find two factors lack of management foresight (LMF) and poor participation and initiation from management which are driving the entire supply chain and thus are our short-term focus. Improving these factors can drastically affect the hazardous condition of laborers and low skilled workforce. Thus they can make efficient decisions and initiate the use of artificial intelligence (AI) (technologies associated with Industry 4.0) which will reduce the risk of mis-investment and thus would reduce their fear of low Return On Invest while investing in new technologies. Thus analyzing these factors will surely help in mitigating risks associated with supply chain disruption and improve the association of supply chain and sustainability in the long term.

20.9 CONCLUSION

In the post-Covid, world sustainability is of utmost importance regarding Supply Chains. To withstand any kind of disruption the supply chains need to be more agile and resilient. In fact, when we are talking about Indian Manufacturing Sector, it has already been in doldrums for the past year. And a large part of the blame goes towards unorganized supply chains. In our research, both the driver-dependence diagram and ISM model proved that supply chain sustainability should be our long-term focus/ultimate goal. But to get there, we need to have a proper plan in place. So we can start by transforming the management and expect that they take more effective participation in investing for implementing tools relevant to Industry 4.0 tools and technologies to digitize logistics, production, planning, and control (PPC). This will help in reducing Supplier Risks. In the end, this can impact in making the supply chain more resilient and more sustainable. However, supply chain resilience is a matter of concern as more work can be done to analyze more factors affecting it as it is one of the immediate and significant factors which needs to be considered when we discuss about Sustainability in Supply Chain. Finally, it was found that the articles that were studied had mainly focused on the sustainability scenario in the pre-covid world. But very little research has been done on Supply Chain sustainability after the pandemic. Thus, it is imperative that more research should be done on the integration of sustainability and supply chain in Indian manufacturing sectors.

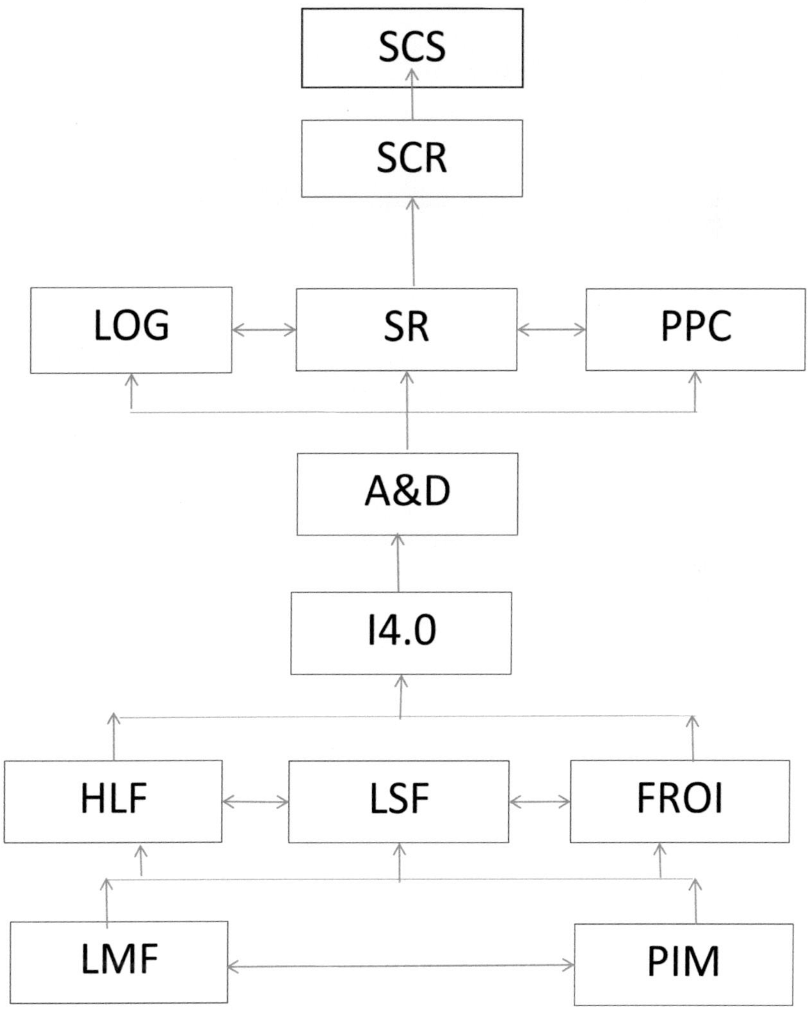

FIGURE 20.2 ISM model.

KEYWORDS

- **industry 4.0**
- **interpretive structural modeling**
- **multi-criteria decision making**
- **programmable logic controller**
- **structural self-interaction matrix**
- **supplier risk**
- **supply chain management**
- **supply chain resilience**
- **sustainability**

REFERENCES

1. Chalmeta, R., & Santos-De, L. N. J., (2020). Sustainable supply chain in the era of industry 4.0 and big data: A systematic analysis of literature and research. *Sustainability, 12*(10), 4108.
2. Ralston, P., & Jennifer, B., (2020). Industry 4.0 and resilience in the supply chain: A driver of capability enhancement or capability loss? *International Journal of Production Research, 58*(16), 5006–5019.
3. Bag, S., et al., (2020). Big data analytics as an operational excellence approach to enhance sustainable supply chain performance. *Resources, Conservation and Recycling, 153*, 104559.
4. Luthra, S., et al., (2018). Internet of things (IoT) in agriculture supply chain management: A developing country perspective. *Emerging Markets from a Multidisciplinary Perspective,* 209–220. http://dx.doi.org/10.1007/978-3-319-75013-2_16.
5. Baryannis, G., Samir, D., & Grigoris, A., (2019). Predicting supply chain risks using machine learning: The trade-off between performance and interpretability. *Future Generation Computer Systems*, *101*, 993–1004.
6. Kumar, P., Rajesh, Kr. S., & Vikas, K., (2021). Managing supply chains for sustainable operations in the era of industry 4.0 and circular economy: Analysis of barriers. *Resources, Conservation and Recycling, 164.* https://doi.org/10.1016/j.resconrec.2020.105215.
7. Witkowski, K., (2017). Internet of things, big data, industry 4.0–innovative solutions in logistics and supply chains management. *Procedia Engineering, 182*, 763–769 https://doi.org/10.1016/j.proeng.2017.03.197.
8. Müller, J. M., & Kai-Ingo, V., (2018). The impact of industry 4.0 on supply chains in engineer-to-order industries-an exploratory case study. *IFAC-PapersOnLine, 51*(11), 122–127. https://doi.org/10.1016/j.ifacol.2018.08.245.
9. Sharma, M., et al., (2021). Industry 4.0 adoption for sustainability in multi-tier manufacturing supply chain in emerging economies. *Journal of Cleaner Production, 281*, 125013.

10. Esmaeilian, B., et al., (2020). Blockchain for the future of sustainable supply chain management in industry 4.0. *Resources, Conservation and Recycling, 163.*
11. Fatorachian, H., & Hadi, K., (2021). Impact of industry 4.0 on supply chain performance. *Production Planning & Control, 32*(1), 63–81.
12. Abdirad, M., & Krishna, K., (2020). Industry 4.0 in logistics and supply chain management: A systematic literature review. *Engineering Management Journal*, 1–15. http://dx.doi.org/10.1080/10429247.2020.1783935.
13. Furstenau, L. B., et al., (2020). Link between sustainability and industry 4.0: Trends, challenges and new perspectives. *IEEE Access, 8*, 140079–140096 https://doi.org/10.1109/ACCESS.2020.3012812.
14. Bag, S., Telukdarie, A., Pretorius, J. H. C., & Gupta, S., (2021). Industry 4.0 and supply chain sustainability: Framework and future research directions. *Benchmarking: An International Journal, 28*(5), 1410–1450. https://doi.org/10.1108/BIJ-03-2018-0056.
15. Dev, N. K., Ravi, S., & Fahham, H. Q., (2020). Industry 4.0 and circular economy: Operational excellence for sustainable reverse supply chain performance. *Resources, Conservation and Recycling, 153*. https://doi.org/10.1016/j.resconrec.2019.104583.
16. Santana, A., et al., (2017). Costing models for capacity optimization in industry 4.0: Trade-off between used capacity and operational efficiency. *Procedia Manufacturing, 13,* 1183–1190.
17. Hahn, G. J., (2020). Industry 4.0: A supply chain innovation perspective. *International Journal of Production Research, 58*(5), 1425–1441.
18. Kumar, R., Rajesh, Kr. S., & Yogesh, Kr. D., (2020). Application of industry 4.0 technologies in SMEs for ethical and sustainable operations: Analysis of challenges. *Journal of Cleaner Production, 275*. https://doi.org/10.1016/j.jclepro.2020.124063.
19. Ghadge, A., Er Kara, M., Moradlou, H., & Goswami, M. (2020). The impact of Industry 4.0 implementation on supply chains. *Journal of Manufacturing Technology Management, 31*(4), 669-686.
20. Strandhagen, J. W., et al., (2020). Sustainability challenges and how industry 4.0 technologies can address them: A case study of a shipbuilding supply chain. *Production Planning & Control*, 1–16.
21. Yadav, G., et al., (2020). A framework to overcome sustainable supply chain challenges through solution measures of industry 4.0 and circular economy: An automotive case. *Journal of Cleaner Production, 254*, 120112. http://dx.doi.org/10.1016/j.jclepro.2020.120112.
22. Singhal, D., et al., (2018). Interpretive structural modelling (ISM) of obstacles hindering the re manufacturing practices in India. *Procedia Manufacturing, 20*, 452–457.
23. Martinez, V. H. V., & Dennis, A. G., (2020). *A More Sustainable Supply Chain Companies Tend to Focus on Their Top-Tier Suppliers, But the Real Risks Come Lower Down* (pp. 1–11). Harvard Business Review.
24. Mukherjee, K., & Mondal, S., (2009). Analysis of issues relating to remanufacturing technology–a case of an Indian company. *Technology Anal. Strateg. Manag., 21*(5), 639–652.
25. Balon, V., Sharma, A. K., & Barua, M. K., (2016). Assessment of barriers in green supply chain management using ISM: A case study of the automobile industry in India. *Global Business Review, 17*(1), 1–20.
26. Van der Laan, E., & Solomon, M., (1997). Production planning and inventory control with remanufacturing and disposal. *European Journal of Operational Research, 102*, 264–278.

27. Bär, K., Herbert-Hansen, Z. N. L., & Waqas, K., (2018). Considering industry 4.0 aspects in the supply chain for an SME. *Production Engineering, 12*(6), 747–758.
28. Borregan-Alvarado, J., et al., (2020). A bibliometric analysis in industry 4.0 and advanced manufacturing: What about the sustainable supply chain? *Sustainability, 12*(19), 7840.
29. ripathydrivers, S., Sahu, S., & Ray, P. K., (2013). Interpretive structural modelling for critical success factors of R&D performance in Indian manufacturing firms. *Journal of Modeling in Management, 8*(2), 212–240. https://doi.org/10.1108/JM2-11-2011-0061.
30. Mathiyazhagan, K., Govindan, K., Noorul, H. A., & Geng, Y., (2013). An ISM approach for the barrier analysis in implementing green supply chain management. *Journal of Cleaner Production, 47*, 283–297. http://dx.doi.org/10.1016/j.jclepro.2012.10.042.
31. Hopkins, J. L., (2021). An investigation into emerging industry 4.0 technologies as drivers of supply chain innovation in Australia. *Computers in Industry, 125.* doi: Https://doi.org/10.1016/j.compind.2020.103323.
32. Abdirad, M., & Krishna, K., (2021). Industry 4.0 in logistics and supply chain management: A systematic literature review. *Engineering Management Journal, 33*(3), 187–201.
33. Ciliberto, C., et al., (2021). Enabling the circular economy transition: A sustainable lean manufacturing recipe for industry 4.0. *Business Strategy and the Environment.*

CHAPTER 21

SELECTION OF GREEN SUPPLIER USING INTEGRATED MULTI-CRITERIA OPTIMIZATION METHOD: WITH SPECIAL REFERENCE TO PLASTIC EXTRUSION AND VACUUM FORMING COMPANIES IN INDIA

REEMA AGARWAL[1], VIPIN TYAGI[1], and ANKUR AGRAWAL[2]

[1]*Department of Applied Sciences, Shobhit University, Meerut, Uttar Pradesh, India*

[2]*School of Business Studies, Sharda University, Greater Noida, Uttar Pradesh, India*

ABSTRACT

The manufacturer ideally should have a fair idea about the rating of its suppliers. Today's demands include cost-effectiveness, premium-quality goods, timely delivery, and superior services after the sales and engage in environmental responsibility activity. In recent years, environmental consciousness has increased significantly. Hence, while assessing a supplier, its sensitivity towards the environment must be considered along with economic factors. That is why it can be termed as green supplier selection (GSS). The GSS is considered to be a multi-criteria optimization problem, which is more popularly known as Multi-Criterion Decision-Making (MCDM) problem.

Decision-Making Models and Applications in Manufacturing Environments. Pushpdant Jain, Kumar Abhishek, & Prasenjit Chatterjee (Eds.)

MCDM tools are required to solve such kinds of problems. A case study in India's industrial plastic component manufacturing company was carried out to address the issue. An integrated MCDM approach was deployed for green supplier assessment and selection. The research study, novel in nature, suggests how to rank suppliers as well as how low-rank suppliers incorporating environmental consciousness can improve their performance.

21.1 INTRODUCTION

Protection of the environment is the main issue and competitive factor in the manufacturing industry nowadays. Green supply chain strategies and actions are required for every facet of supply chain management (SCM), right from acquiring raw materials from suppliers to delivering finished goods to the customer [1]. Environmental regulatory bodies and the government are continuously trying and forcing manufacturing industries to restrain traditional SCM practices' ill effects and adopt green supply chain management (GSCM) practices [2]. Being in the 21st century, deploying GSCM practices is the need of the hour. GSCM methodology is nothing but a judicial blend of traditional SCM with green environmental protection practices for sustainable development (SD) [3]. Tyagi et al. [4] explained GSCM as "incorporating environmental thinking into SCM, including product design, material sourcing and selection, manufacturing practices, delivery of the final products to the consumers, and end-of-life management of the product after its intended life." Since the beginning of the 21st century, GSCM has redefined business activities to build a sound economic-environment-friendly ecosystem [5]. Nevertheless, developed countries are in a continuous engagement in adopting and successfully implementing GSCM in their organizations, whereas India is still struggling to adapt and adequately implement GSCM practices [6, 7]. In the current global scenario, there is tremendous pressure from international environmental bodies, raising the flag when GSCM practices are not implemented. They are encouraging businesses to recognize the factors that will allow them to properly and effectively follow GSCM guidelines [8].

The green supplier selection (GSS) system is critical to the GSCM value chain. Green supplier firm, having a high environmental sensitive image. That is why they have a competitive edge over non-environmentally performing suppliers. A green supplier focuses on using renewable and environment-friendly energy resources, recyclable, and reusable raw materials and consumables, green designing and green packaging and packaging material, etc., in their supply chain operations [9]. It is not easy to select a green supplier in compression to select a conventional supplier [10, 11] as it has to consider both qualitative as well as opposing selection standards [12, 13]. That is why GSS is acknowledged as a multi-criterion decision-making (MCDM) issue. This issue had drawn many researchers' attention in recent times [8, 9, 14]. Additionally, a remarkable increase in research on GSS [15–17] in recent years has highlighted the significance of the issue.

MCDM strategy is used in making decisions through setting up and fulfilling multiple and conflicting criteria. Its problems are common in daily routine; for instance, if anyone buys a mobile phone, it is qualified by cost, features, storage, looks, camera, etc. While in business, problems are much completed. In recent years this method has been used for better business modeling. Purchase departments of many companies use this method to select their suppliers based on a vast range of criteria such as cost on material, quality, after-sales service, financial stability, etc. [18, 19].

In the recent past, different techniques have been used for suppliers' ranking and selection. ANP, TOPSIS, AI, and integrated techniques are the most used methods for this purpose [20]. The TOPSIS-based model has been put forward to score, rank, and select suppliers [21]. This model lays out the criterion of cost, quality, delivery, relationship, and closeness for evaluating suppliers. The main advantage of TOPSIS is easy to apply. The steps of the TOPSIS remain the same, whereas many attributes can be changed. TOPSIS has been applied in manufacturing systems, engineering, SCM, trading, promotion-based marketing, and transportation and logistics. Further, it has also been deployed for evaluating human resources, ranking environmental factors, and assessing the suppliers [18, 19].

The suggested approach was applied to the industrial plastic component manufacturing company (Plastic Extrusion and Vacuum Forming Company) to select green suppliers. The products have been broadly used for home appliances, like refrigerators' inner plastic bodies and other necessities. This is a raw material usage-intensive industry and the finished product is used in the domestic market and export also. It is the reason why selecting the appropriate supplier plays a major role in evaluating how much the businesses accomplish. For this purpose, this chapter covers studies and methods in the literature review section and later opines a multi-phase MCDM model for an effective assessment and selection of a green supplier for a plastic sheet manufacturing company. The case design approach was carried out by selected managers and suppliers [22, 23]. This chapter meets the following objectives:

i. What are the important and applicable criteria for green suppliers' selection (GSS);
ii. Ranking the green suppliers using TOPSIS, an MCDM technique;
iii. How the performance of the lowest-ranked supplier can be improved.

21.2 LITERATURE REVIEW

Supplier selection is a crucial facet in the management of supplies and purchasing function [24]. Researchers have used a variety of MCDM methods for handling the problems of GSS. Various criteria for supplier selection were explored in the first section of the literature review. These criteria, including quality, timely delivery, and economic pricing, have been used by different companies to assess their suppliers [25]. The GSCM approach takes into consideration the environmental hazards, ecological balance, and climate change for all the phases of SCM, covering its entire cycle [26, 27]. GSCM increases buyers and suppliers' various opportunities because all corporate and industrial activities are environment protection-centric [5]. The process of GSCM starts with procurement and green purchasing. It exhibits a critical impact on the supply chain environmental effect [28, 29]. As far as GSS criteria are concerned, many researchers have taken different criteria based on different experts and different industry inputs across different economies [30]. Lee et al. [31], considered the net cost of the product through its lifecycle, quality as well as technology capabilities, along with green criteria, such as pollution control, green image, green product, environmental management, and green proficiencies. Grisi et al. [51] also took into consideration the green image and environmental management, along with ISO14001, environmental policies, ecological material, clean-technology, and environmental planning. Chen et al. [32], identified green design, ISO14001, clean production, R&D on green products, green purchasing, quality, flexibility, and delivery. Kannan et al. [33], included environment protection, green image, green product, green innovation, corporate social responsibility, hazard management and pollution control as criteria to implement GSS. A green supplier makes efforts primarily on using renewable energy sources and saving, reusing, and recycling the materials. Also, its focus remains on green designing as well as green packing while performing GSCM activities [9]. The process of selecting and verifying a green supplier is more complicated than that of a conventional supplier [10, 11, 34]. This is the reason why the problem of GSS is treated as a multiple-criteria decision-making

(MCDM) issue [12]. This issue has drawn many researchers' attention in recent times [9, 14, 30]. Additionally, a remarkable increase in the research on GSS [15–17] recently has highlighted the significance of the matter. Numerous economic and environmental facets have also been taken into consideration to verify and validate green suppliers [35].

The second part of the literature review deals with the decision-making issue of MCDM problems. As far as green supplies selection methods or approaches are concerned, rich literature is available. Zhang [36], classified supplier selection methods into five groups: mathematical programming models, statistical approach, linear weighting models, cost-driven models, and artificial intelligence (AI)-driven models. Kannan et al. [37], performed a dense review of the literature, exploring MCDM techniques for selecting a green supplier. He found that the linear programming (LP), analytic hierarchy process (AHP), as well as data envelopment analysis (DEA) were the most favored techniques. Similarly, in another resembling study performed by Govindan et al. [17], selecting a green supplier was deemed to be an MCDM problem. And to solve this issue, TOPSIS, LP, DEA, network analysis, and AHP can be used. The name of the model was Fuzzy PIvot Pairwise Relative Criteria Importance Assessment (Fuzzy PIPRECIA) model.

He identified seven criteria for the industry: environmental image, recycling, environmentally friendly products, environmental management system, resource consumption, pollution control, and green competencies. In a study in 2019, three MCDM methods were applied for GSS, which included fuzzy TOPSIS fuzzy ARAS methods and fuzzy WASPS, providing a new dimension to achieve the objective [40].

21.3 RESEARCH METHODOLOGY

The study is carried through a Case study by analyzing and observing the single company, Ambar enterprise ltd. The case study technique has a unique benefit in situations when the "how," "which," and "what" kind of inquiry is being inquired [41]. This study's questions are intended to show "what" are the important and applicable criteria for GSS, "how" should Ambar enterprise ltd. Rank the supplier and "how" should improve the performance of the suppliers. This study conducted a questionnaire survey to measure the important and applicable selection criteria and rank the suppliers. The research design of this study will be a combination of both the case study and survey. The questionnaire was designed and fabricated on the basis of the results and crux of the literature review. The first questionnaire survey has been sent to five managers of related departments, and the intended objective will be to get a list of important and applicable criteria according to their opinion. Similarly, the second survey will be sent to the purchase manager to rank the suppliers against the identified selection criteria list. Cronbach's alpha has been deployed to ensure the statistical reliability of the collected data set. Cronbach's alpha is coming to 0.78, which shows a higher internal reliability level. Mann-Whitney U-test, TOPSIS (technique for order performance by similarity to ideal solution) analysis, and Parametric Analysis have been used to analyze the data.

21.3.1 MANN–WHITNEY U-TEST

- **H_0:** Statistically, there should be no difference in the importance and applicability of the criteria for GSS.
- **H_1:** The importance and applicability of the criteria for GSS must be different, statistically.

To assess whether the mean scores of the collected information, concerning significance and relevance, Mann–Whitney U-test has been applied. It is a non-parametric evaluation method used to compare two data means from the same population. It also examines whether the mean of the two data sets is equal.

21.3.2 TOPSIS ANALYSIS

Hwang and Yoon incepted the idea to design TOPSIS to solve MCDM problems. It is used for ranking the decision-making alternatives. This technique's basic philosophy states that "the best alternative should have the minimum distance to the positive ideal solution (PIS) and the maximum distance from the negative ideal solution (NIS)." The initial step of this process is to establish a benchmark, i.e., the "ideal positive solution (IPS)" and "ideal negative solutions (INS)." The second step is to form a normalized decision matrix showing these numbers. The third step is to choose the largest normalized and weighted score to obtain each criterion's PIS. Similarly, choose the least normalized and weighted score to obtain the NIS for each attribute. In the final step, one needs to calculate and figure out how far or close each alternative is to the PIS and NIS.

21.3.3 PARAMETRIC ANALYSIS

The basis of this analysis is to change only one criterion while keeping the values of all other evaluation factors constant for entire data sets.

21.3.4 LIST OF CRITERIA

A survey was conducted to verify and certify the proposed green supplier evaluation criteria. To evaluate, the important and applicable green supplier evaluation criteria, a questionnaire was developed. It contains two aspects, namely economic and environmental aspects (Table 21.1).

21.4 ANALYSIS

To analyze the first objective, a survey was conducted. The purchase managers (respondents) had given their preferences on a scale of 1 to 5 for each criterion. The mean value of each criterion was taken to evaluate the importance level and applicability level. The results are depicted in Figures 21.1 and 21.2. The graph depicts important green supplier evaluation criteria. The manager gave importance to the economic aspect during supplier evaluation rather than the environmental aspect. It can be easily observed cost is on the top of the list with the highest score of 4.6; in contrast, the green product image was at the bottom of the list with a 3.9 score. Quality 4.6, flexibility 4.4, Delivery and Service 4.2, with a percentage of 92%, 88%, and 88%, followed respectively. Green transportation had more importance compared to other environmental criteria in the list. Green technology and eco-design had equal importance.

Figure 21.2 depicts the applicable supplier evaluation criteria. As it can be observed, cost is the most applicable criteria, with a score of 3.7 on the list. This was followed by Quality 3.2, flexibility; delivery service and green product image have same score 3.5. It can also be observed from Figure 21.2 that green image had the top score of 3.5, succeeded by green transport 3.1, Eco design 2.9, and green technology 2.4.

Figure 21.3 shows that although all nine identified GSS criteria are important and applicable cost, quality, environmental management, and green transportation are more important than the remaining criteria. So, the company should give more emphasis on these criteria for the evaluation of suppliers.

TABLE 21.1 List of Green Supplier Selection Criteria

	Main Criteria	Sub-Criteria (Alternatives)	References
Economic aspect	Cost	Cost of material	[33, 42]
		Ordering and holding cost	[43]
		Freight cost	[33]
	Quality	Rate of rejection	[44]
		Product performance	[31, 33, 38]
		Quality inspection methods	[43]
	Delivery service	On-time delivery	[45]
		After-sales service	[46]
		Delivery speed	[45]
		Responsiveness in solving complaints	[45]
	Flexibility	Flexibility in ordering	[33]
		Flexibility in delivery time	[33, 39, 44]
		Flexibility in giving discount	[33]
Environment aspect	Environmental management system	ISO14001 certification	[43]
		Ozone-depleting chemical used	[33]
		Eco-labeling	[45]
		Use of environment-friendly raw materials	[45]
	Green product image	Green Certification	[47]
		Reuse	[33]
		Green packaging	[46]
		Air emissions	[33]
		Wastewater	[48]
		Hazardous wastes	[47]
	Eco-design	Recycle of products	[48]
		Re-manufacturing of products when design	[49]
		Decrease the use of hazard materials in production	[50]
	Green technology	Capability of R&D	[43]
		Process alteration to save natural resources	[31]
		Use of green raw material	[31]
	Green Transportation	Using a modern eco-efficient transportation fleet	[33]
		Use of green fuels	[47]

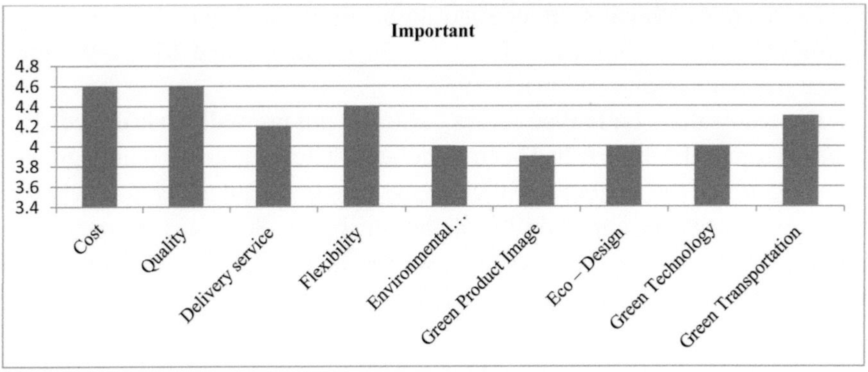

FIGURE 21.1 Important green supplier evaluation criteria.

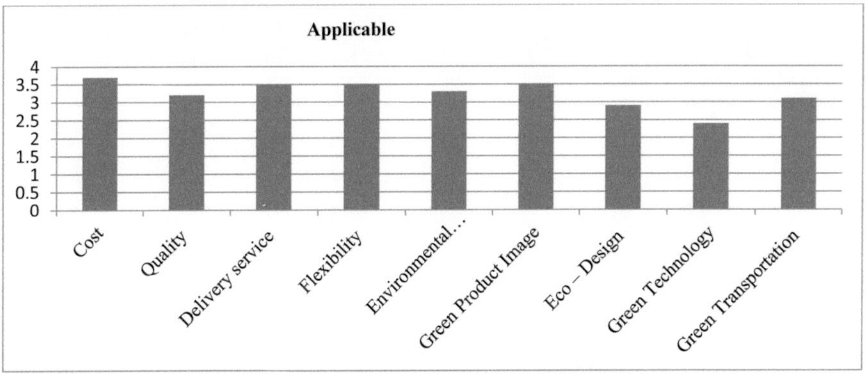

FIGURE 21.2 Applicable green supplier evaluation criteria (analysis).

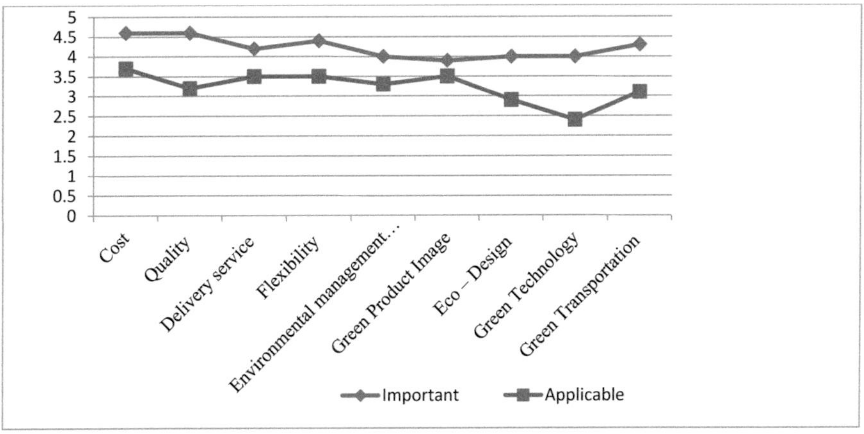

FIGURE 21.3 Important and applicable green supplier evaluation criteria.

To check, whether importance and applicability are different. Mean scores of importance and applicability have been collected, and Mann–Whitney U-test has been applied by using SPSS.

The test is done on ranked scores, which is not normally distributed. The p-Value for all the criteria has been calculated. The p-value is coming greater than 0.05 for all the attributes, which shows no significant difference between the importance and applicability of the attributes. So, we can conclude a good correlation between the criteria, we have taken for GSS with respect to importance and applicability (Table 21.2).

TABLE 21.2 The Results of the Mann-Whitney U-Test

	Mann–Whitney U-Test			P-Value
		Mean Value of Importance	Mean Value of Applicability	
Economic	Cost	4.6	3.7	0.76
	Quality	4.6	3.2	0.67
	Delivery service	4.2	3.5	0.59
	Flexibility	4.4	3.5	0.78
Environmental	Environmental management system	4	3.3	0.71
	Green product image	3.9	3.5	0.35
	Eco-design	4	2.9	0.21
	Green technology	4	2.4	0.47
	Green Transportation	4.3	3.1	0.88

To analyze the second object, ranking the suppliers, the multi-criteria decision-making method TOPSIS has been applied. Seven leading suppliers are known as S1, S2, S3, S4, S5, S6, and S7 have been taken. Selecting the best supplier among these seven alternatives based on nine criteria is very difficult and complicated. That is why; the company's purchase manager cannot decide the right strategic decision for making long-term collaboration with the suppliers. Therefore, this section aims to evaluate and select the best suppliers of the company using TOPSIS. It is a simple mathematical equation for determining the best alternative. It has a straightforward computation process. The philosophy of this method is that "the best alternative should have the minimum distance to the PIS and the maximum distance from the NIS." The closeness Coefficient (CCI) of each alternative is derived as closer to the PIS and farther from NIS and CCI approaches to 1.

The steps-wise results of TOPSIS are as follows:

- **Step 1:** The company's purchasing manager was asked to rate the performance of the suppliers against each attribute so that the decision matrix can be formed on a Likert scale of 5 (Table 21.3).
- **Step 2:** Obtaining a normalized decision matrix (Table 21.4).
- **Step 3:** Calculate the distance from PIS and NIS for each alternative. Calculate ranking and closeness coefficient (CC_i) (Table 21.5).

Based on nine criteria, the ranking of the suppliers has been obtained. S1 is found to be the best supplier with a CCI score of 0.8286, whereas supplier S7 was found to be the lowest rank supplier with a CCI score of 0.0598.

To analyze, the third objective, i.e., to improve the Performance lowest-ranked supplier's performance (S7). A parametric analysis technique has been applied to determine which criterion should be focused on so the performance of the supplier can be improved. This is done by changing the weight of one criterion while keeping other criteria constant and seeing how much suppliers' performance would change, which is denoted by the difference in Table 21.6. We applied parametric analysis to find out among nine main criteria that have more effect on the supplier's performance (Table 21.6). In the table, we can see some criteria such as the cost being not alien with the performance, which means increasing the weight of Cost, Performance is decreased. Similarly, other criteria like quality, delivery service, and flexibility are alien to performance, which meaning by increasing the weight, and performance is also increased. Table 21.6 shows the performance of the lowest rank supplier S7 by changing the weights of each criterion. First of all, weights of different criteria are given by the option of company purchase managers. In the second step, an average mean value was used to weight each criteria.

Figure 21.4 shows the performance of the lowest-ranked supplier (S7) at a different weight (1 to 5). It has been observed that cost has reverse relation with the performance of the supplier. Whereas other green supplier evaluation criteria such as quality, delivery service, flexibility, green product image, eco-design, green technology, and green transportation directly correlate with the supplier's performance.

Figure 21.5 shows the difference between the lowest and highest values of Performance of S7. As it is mentioned in Table 21.6, calculated by changing the weight (1 to 5) of one criterion and keeping other criteria's constant.

If the performance difference is high, then any GSS criteria are more effective. Figure 21.5 shows that flexibility has the highest performance difference, whereas cost is considered a negative criterion in terms of flexibility. Flexibility

TABLE 21.3 The Opinions of an Expert About the Suppliers

Criteria	Cost	Quality	Delivery Service	Flexibility	Environmental Management System	Green Product Image	Eco-Design	Green Technology	Green Transportation
S-1	4	4	3	4	5	4	4	2	5
S-2	3	5	4	2	4	3	3	2	4
S-3	2	4	3	3	3	4	3	3	4
S-4	4	3	2	3	4	5	2	2	5
S-5	5	3	5	4	2	5	2	1	3
S-6	2	2	2	5	3	4	1	3	4
S-7	5	1	3	1	2	4	3	2	4

TABLE 21.4 The Normalized Data Set

Criteria	Cost	Quality	Delivery Service	Flexibility	Environmental Management System	Green Product Image	Eco-Design	Green Technology	Green Transportation
S-1	0.406	0.547	0.344	0.486	0.547	0.385	0.652	0.530	0.456
S-2	0.205	0.547	0.468	0.245	0.695	0.480	0.374	0.530	0.342
S-3	0.309	0.328	0.229	0.245	0.289	0.480	0.496	0.397	0.456
S-4	0.503	0.219	0.344	0.370	0.168	0.480	0.378	0.268	0.342
S-5	0.409	0.110	0.574	0.122	0.148	0.193	0.289	0.268	0.342
S-6	0.213	0.437	0.229	0.608	0.277	0.290	0.124	0.268	0.342
S-7	0.509	0.110	0.344	0.364	0.134	0.195	0.124	0.268	0.342
The Best	0.213	0.547	0.574	0.608	0.695	0.480	0.652	0.530	0.456
The Worst	0.503	0.110	0.229	0.122	0.148	0.193	0.124	0.268	0.342

TABLE 21.5 Ranking the Suppliers

	Distance between the Worst Alternatives	Distance between the Best Alternatives	Distance between the Worst and the Best	Ranking the Suppliers
		Suppliers		
S-1	1.456739	0.352071	0.828671	1
S-2	1.442791	0.698345	0.678423	2
S-3	0.972341	0.652311	0.601284	3
S-4	0.295671	0.855076	0.307531	4
S-5	0.479612	1.286531	0.287643	5
S-6	0.619571	1.270653	0.195632	6
S-7	0.092316	1.478235	0.059832	7

TABLE 21.6 Performance of S7 by Changing the Weight (1–5)

Criteria / Weight	Cost	Quality	Delivery Service	Flexibility	Environmental Management System	Green Product Image	Eco-Design	Green Technology	Green Transportation
1	0.135	0.037	0.077	0.047	0.046	0.043	0.046	0.049	0.042
2	0.090	0.034	0.018	0.046	0.064	0.046	0.058	0.046	0.044
3	0.076	0.042	0.045	0.054	0.098	0.054	0.087	0.058	0.047
4	0.058	0.065	0.087	0.078	0.147	0.073	0.126	0.089	0.047
5	0.049	0.097	0.122	0.097	0.198	0.099	0.168	0.127	0.057
Difference	–0.091	0.681	0.119	0.148	0.072	0.120	0.076	0.029	0.056

is decreased by increasing the cost. Suppliers should try to decrease the flexibility in the delivery for improvement. Delivery services and green product image have almost equal importance for performance improvement. Moreover, the remaining criteria are also helpful in S7 for performance improvement.

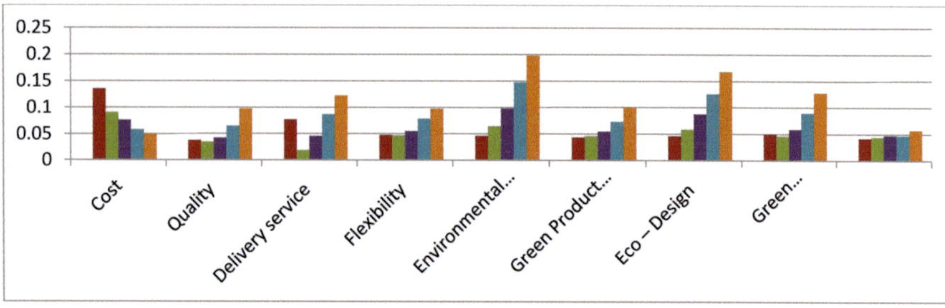

FIGURE 21.4 Performance of S7 by changing the weight (1–5).

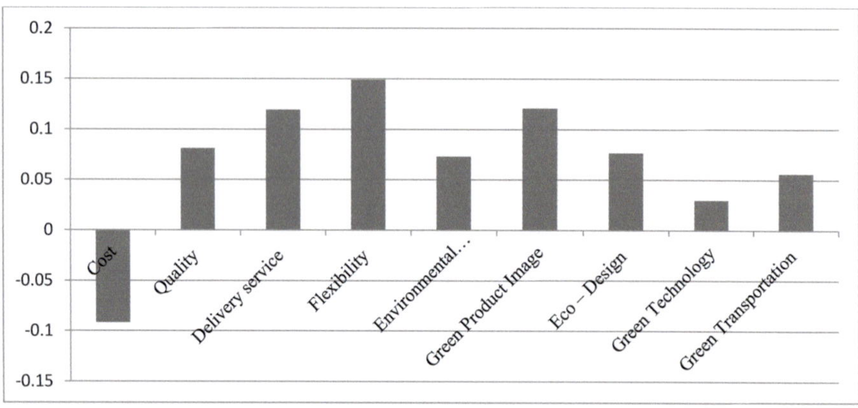

FIGURE 21.5 Difference of green performance of S7.

21.5 CONCLUSION

On the basis dense of literature reviewed, a comprehensive list consisting of nine main green supplier evaluation criteria and thirty-one sub-criteria were finalized. Based on the questionnaire-based survey, the inputs of the managers have been collected. The importance and applicability have been measured based on these inputs.

The study reveals that the most effective criteria for green supplier evaluation are economic criteria, followed by environmental criteria. Mann–Whitney

U-test demonstrates no significant difference between the mean scores of importance and applicability. Therefore, the developed list of the GSS criteria and their corresponding sub-criteria can be used to evaluate the green supplier's performance. It is always difficult for the decision to make to decide by considering thirty-one conflicting criteria. A multi-criteria decision making method, TOPSIS, had been applied to rank the supplier as it is simple to apply, mathematical structure modeling technique. The supplies have been ranked considering all the thirty-one sub-criteria. This ranking is beneficial for various strategic decisions and long-term collaboration with suitable suppliers. At the same time, the performance of the lower rank suppliers can also be measured. It has been suggested that the lowest performer supplier S7 take care of flexibility, delivery service, and green image by applying the parametric analysis. Similarly, the performance of the lower-ranked supplier can also be improved by concentrating upon specific suggested criteria.

The study has a vital future scope as the top 1,000 listed companies in India have to produce business responsibility and sustainability reporting (BRSR) from the financial year 2021–2022. Ranking the green supplier will definitively give an edge to the companies.

KEYWORDS

- **analytic hierarchy process**
- **green supplier selection**
- **green supplier selection criteria**
- **green supply chain management**
- **multi-criterion decision-making**
- **supply chain management**

REFERENCES

1. Qureshi, M. N., Kumar, D., & Kumar, P., (2008). An integrated model to identify and classify the key criteria and their role in the assessment of 3PL services providers. *Asia Pacific Journal of Marketing and Logistics*.
2. Zhu, Q., Sarkis, J., & Geng, Y., (2005). Green supply chain management in China: Pressures, practices and performance. *International Journal of Operations & Production Management.*

3. Madaan, J., & Mangla, S., (2005). Decision modeling approach for eco-driven flexible green supply chain. *Systemic Flexibility and Business Agility*, 343–364.
4. Tyagi, M., Kumar, P., & Kumar, D., (2015). Assessment of critical enablers for flexible supply chain performance measurement system using fuzzy DEMATEL approach. *Global Journal of Flexible Systems Management, 16*(2), 115–132.
5. Rao, P., & Holt, D., (2005). Do green supply chains lead to competitiveness and economic performance? *International Journal of Operations & Production Management.*
6. Chien, M. K., & Shih, L. H., (2007). An Empirical Study of the Implementation of Green Supply Chain Management Practices in the Electrical and Electronic Industry and Their Relation to Organizational Performances. *International Journal of Environmental Science and Technology, 4*(3), 383–394.
7. Dubey, R., & Ali, S. S., (2014). Identification of flexible manufacturing system dimensions and their interrelationship using total interpretive structural modeling and fuzzy MICMAC analysis. *Global Journal of Flexible Systems Management, 15*(2), 131–143.
8. Mathiyazhagan, K., & Haq, A. N., (2013). Analysis of the influential pressures for green supply chain management adoption—An Indian perspective using interpretive structural modeling. *The International Journal of Advanced Manufacturing Technology, 68*(1–4), 817–833.
9. Wu, Q., Zhou, L., Chen, Y., & Chen, H., (2019). An integrated approach to green supplier selection based on the interval type-2 fuzzy best-worst and extended VIKOR methods. *Information Sciences, 502*, 394–417.
10. Mendoza, A., & Ventura, J. A., (2013). Modeling actual transportation costs in supplier selection and order quantity allocation decisions. *Operational Research, 13*(1), 5–25.
11. Aretoulis, G. N., Kalfakakou, G. P., & Striagka, F. Z., (2010). Construction material supplier selection under multiple criteria. *Operational Research, 10*(2), 209–230.
12. Cao, Q., Wu, J., & Liang, C., (2013). An intuitionistic fuzzy judgement matrix and TOPSIS integrated multi-criteria decision-making method for green supplier selection. *Journal of Intelligent & Fuzzy Systems, 28*(1), 117–126.
13. Fahimnia, B., Sarkis, J., & Davarzani, H., (2015). Green supply chain management: A review and bibliometric analysis. *International Journal of Production Economics, 162*, 101–114.
14. Chatterjee, S., Maji, B., & Pham, H., (2019). A fuzzy rule-based generation algorithm in interval type-2 fuzzy logic system for fault prediction in the early phase of software development. *Journal of Experimental & Theoretical Artificial Intelligence, 31*(3), 369–391.
15. Tseng, M. L., Islam, M. S., Karia, N., Fauzi, F. A., & Afrin, S., (2019). A literature review on green supply chain management: Trends and future challenges. *Resources, Conservation and Recycling, 141*, 145–162.
16. Malviya, R. K., & Kant, R., (2015). Green supply chain management (GSCM): A structured literature review and research implications. *Benchmarking: An International Journal.*
17. Govindan, K., Rajendran, S., Sarkis, J., & Murugesan, P., (2015). Multi-criteria decision-making approaches for green supplier evaluation and selection: A literature review. *Journal of Cleaner Production, 98*, 66–83.
18. Govindan, K., Khodaverdi, R., & Jafarian, A., (2013). A fuzzy multi-criteria approach for measuring sustainability performance of a supplier based on a triple bottom line approach. *Journal of Cleaner Production, 47*, 345–354.

19. Gunasekaran, & Gallear, D., (2012). Special issue on sustainable development of manufacturing and services. *International Journal of Production Economics, 140*(1), 1–6.
20. Akcan, S., & Taş, M. A., (2019). Green supplier evaluation with SWARA-TOPSIS integrated method to reduce ecological risk factors. *Environmental Monitoring and Assessment, 191*(12), pp.1–22.
21. Boran, F. E., Genç, S., Kurt, M., & Akay, D., (2009). A multi-criteria intuitionistic fuzzy group decision making for supplier selection with TOPSIS method. *Expert Systems with Applications, 36*(8), 11363–11368.
22. Tian, C., Zhang, W. Y., Zhang, S., & Peng, J. J., (2019). An extended single-valued neutrosophic projection-based qualitative flexible multi-criteria decision-making method. *Mathematics, 7*(1), 39.
23. Liu, P., & Chen, S. M., (2018). Multiattribute group decision making based on intuitionistic 2-tuple linguistic information. *Information Sciences, 430*, 599–619.
24. Banaeian, N., Mobli, H., Fahimnia, B., Nielsen, I. E., & Omid, M., (2018). Green supplier selection using fuzzy group decision making methods: A case study from the agri-food industry. *Computers & Operations Research, 89*, 337–347.
25. Hlioui, R., Gharbi, A., & Hajji, A., (2017). Joint supplier selection, production and replenishment of an unreliable manufacturing-oriented supply chain. *International Journal of Production Economics, 187*, 53–67.
26. Min, H., & Galle, W. P., (1997). Green purchasing strategies: Trends and implications. *International Journal of Purchasing and Materials Management, 33*(2), 10–17.
27. Mangla, S., Madaan, J., Sarma, P. R. S., & Gupta, M. P., (2014). Multi-objective decision modelling using interpretive structural modelling for green supply chains. *International Journal of Logistics Systems and Management, 17*(2), 125–142.
28. Günther, E., & Scheibe, L., (2006). The hurdle analysis–A self-evaluation tool for municipalities to identify, analyze and overcome hurdles to green procurement. *Corporate Social Responsibility and Environmental Management, 13*(2), 61–77.
29. Min, H., & Kim, I., (2012). Green supply chain research: Past, present, and future. *Logistics Research, 4*(1), 39–47.
30. Mathiyazhagan, K., Sudhakar, S., & Bhalotia, A., (2018). Modeling the criteria for selection of suppliers towards green aspect: A case in Indian automobile industry. *Operation Research, 55*(1), 65–84.
31. Lee, A. H., Kang, H. Y., Hsu, C. F., & Hung, H. C., (2009). A green supplier selection model for high-tech industry. *Expert Systems with Applications, 36*(4), 7917–7927.
32. Chen, C. C., Tseng, M. L., Lin, Y. H., & Lin, Z. S., (2010). Implementation of green supply chain management in uncertainty. *IEEE International Conference on Industrial Engineering and Engineering Management* (pp. 260–264).
33. Kannan, D., Govindan, K., & Rajendran, S., (2015). Fuzzy axiomatic design approach based green supplier selection: A case study from Singapore. *Journal of Cleaner Production, 96*, 194–208.
34. Yousefi, S., Rezaee, M. J., & Solimanpur, M., (2019). Supplier selection and order allocation using a two-stage hybrid supply chain model and game-based order price. *Operational Research*, 1–36.
35. Yu, F., Yang, Y., & Chang, D., (2018). Carbon footprint based green supplier selection under dynamic environment. *Journal of Cleaner Production, 170*, 880–889.
36. Zhang, S. F., Liu, S. Y., & Zhai, R. H., (2011). An extended GRA method for MCDM with interval-valued triangular fuzzy assessments and unknown weights. *Computers & Industrial Engineering, 61*(4), 1336–1341.

37. Kannan, D., De Sousa, J. A. B. L., & Jabbour, C. J. C., (2014). Selecting green suppliers based on GSCM practices: Using fuzzy TOPSIS applied to a Brazilian electronics company. *European Journal of Operational Research, 233*(2), 432–447.
38. Durmić, E., (2019). Evaluation of criteria for sustainable supplier selection using FUCOM method. *Operational Research in Engineering Sciences: Theory and Applications, 2*(1), 91–107.
39. Dalić, I., Stević, Z., Karamasa, C., & Puška, A., (2020). A novel integrated fuzzy PIPRECIA–interval rough SAW model: Green supplier selection. *Decision Making: Applications in Management and Engineering, 3*(1), 126–145.
40. Paul, A., Nagesh, S., Sanjoy, K. P., & Andrea, T., (2021). Sustainable supply chain management and multi-criteria decision-making methods: A systematic review. *Sustainability 13*(13).
41. Yin, R. K., (2013). Validity and generalization in future case study evaluations. *Suppliers'-Evaluation, 19*(3), 321–332.
42. Fallahpour, A., Telugu, E. U., Musa, S. N., Wong, K. Y., & Noori, S., (2017). A decision support model for sustainable supplier selection in sustainable supply chain management. *Computers & Industrial Engineering, 105*, 391–410.
43. Humphreys, P., McCloskey, A., McIvor, R., Maguire, L., & Glackin, C., (2006). Employing dynamic fuzzy membership functions to assess environmental performance in the supplier selection process. *International Journal of Production Research, 44*(12), 2379–2419.
44. Feyzioglu, O., & Büyüközkan, G., (2010). Evaluation of green suppliers considering decision criteria dependencies. In: *Multiple Criteria Decision Making for Sustainable Energy and Transportation Systems* (pp. 145–154).
45. Yang, Y. Z., & Wu, L. Y., (2008). Extension method for green supplier selection. *International Conference on Wireless Communications, Networking and Mobile Computing*, 1–4.
46. Büyüközkan, G., & Çifçi, G., (2012). A novel hybrid MCDM approach based on fuzzy DEMATEL, fuzzy ANP and fuzzy TOPSIS to evaluate green suppliers. *Expert Systems with Applications, 39*(3), 3000–3011.
47. Rostamzadeh, R., Govindan, K., Esmaeili, A., & Sabaghi, M., (2015). Application of fuzzy VIKOR for evaluation of green supply chain management practices. *Ecological Indicators, 49*, 188–203.
48. Hongjun, L., & Bin, L., (2010). May notice of retraction: A research on supplier assessment indices system of green purchasing. *International Conference on E-Business and E-Government*, 3335–3338.
49. Handfield, R., Walton, S. V., Sroufe, R., & Melnyk, S. A., (2002). Applying environmental criteria to supplier assessment: A study in the application of the analytical hierarchy process. *European Journal of Operational Research, 141*(1), 70–87.
50. Bai, C., & Sarkis, J., (2010). Integrating sustainability into supplier selection with grey system and rough set methodologies. *International Journal of Production Economics, 124*(1), 252–264.
51. Gatti, A. B., Ferreira, A. G., Arduin, M., Perez, S. C. J. G. A. (2010). Allelopathic effects of aqueous extracts of *Artistolochia esperanzae* O. Kuntze on development of *Sesamum indicum* L. seedlings. *Acta Bot Bras. 24*(2), 454–461. Doi: http://dx.doi.org/10.1590/S0102- 33062010000200016.

INDEX

F

G

T

U

V

W